D0074947

ACS SYMPOSIUM SERIES **624**

Step-Growth Polymers for High-Performance Materials

New Synthetic Methods

James L. Hedrick, EDITOR
IBM Almaden Research Center

Jeff W. Labadie, EDITOR
Argonaut Technologies, Inc.

Developed from a symposium sponsored
by the Division of Polymer Chemistry, Inc.,
at the 209th National Meeting
of the American Chemical Society,
Anaheim, California,
April 2–7, 1995

American Chemical Society, Washington, DC 1996

Library of Congress Cataloging-in-Publication Data

Step-growth polymers for high-performance materials: new synthetic methods / James L. Hedrick, editor, Jeff W. Labadie, editor.

p. cm.—(ACS symposium series, ISSN 0097–6156; 624)

"Developed from a symposium sponsored by the Division of Polymer Chemistry, Inc., at the 209th National Meeting of the American Chemical Society, Anaheim, California, April 2–7, 1995."

Includes bibliographical references and indexes.

ISBN 0–8412–3394–2

1. Polymers—Congresses.

I. Hedrick, James L., 1959– . II. Labadie, Jeff W., 1955– . III. American Chemical Society. Division of Polymer Chemistry. IV. American Chemical Society. Meeting (209th: 1995: Anaheim, Calif.) V. Series.

TP1081.S84 1996
668.9—dc20

95–54170
CIP

This book is printed on acid-free, recycled paper.

PRINTED IN THE UNITED STATES OF AMERICA

Foreword

THE ACS SYMPOSIUM SERIES was first published in 1974 to provide a mechanism for publishing symposia quickly in book form. The purpose of this series is to publish comprehensive books developed from symposia, which are usually "snapshots in time" of the current research being done on a topic, plus some review material on the topic. For this reason, it is necessary that the papers be published as quickly as possible.

Before a symposium-based book is put under contract, the proposed table of contents is reviewed for appropriateness to the topic and for comprehensiveness of the collection. Some papers are excluded at this point, and others are added to round out the scope of the volume. In addition, a draft of each paper is peer-reviewed prior to final acceptance or rejection. This anonymous review process is supervised by the organizer(s) of the symposium, who become the editor(s) of the book. The authors then revise their papers according to the recommendations of both the reviewers and the editors, prepare camera-ready copy, and submit the final papers to the editors, who check that all necessary revisions have been made.

As a rule, only original research papers and original review papers are included in the volumes. Verbatim reproductions of previously published papers are not accepted.

ACS BOOKS DEPARTMENT

Contents

GENERAL STEP-GROWTH POLYMERIZATION TOPICS

POLYIMIDES AND HIGH-TEMPERATURE POLYMERS

INDEXES

Preface

THE USE OF POLYMERS IS WIDESPREAD in modern society, and their applications continue to grow. Many of the important advances in polymeric material involve imparting desirable properties through the control of the polymer structure. Polymer structure is determined and controlled by polymer synthesis. Step-growth polymerization represents one of the two major mechanisms by which polymers are synthesized. The other is chain-growth polymerization. Polymers prepared by step-growth polymerization, often referred to as step-growth polymers, represent some of the most important commercial polymer systems, including nylon, polyesters, and polycarbonates. Many high-technology applications in the areas of aviation, electronics, medicine, and apparel rely on step-growth polymers as key constituents. Hence, fundamental and applied research on step-growth polymers and their synthesis continues to be very active worldwide in both academic and industrial environments.

In recent years much of the focus in step-growth polymers has been in the area of high-performance materials, in which tailoring polymer structure to give a specific set of properties is paramount. This research has led to many advances in synthetic methodology, including both new polymer-forming reactions as well as enhancement of established methods of polymerization. Major areas of research activity include transition-metal-catalyzed coupling, dendritic and hyperbranched polymers, poly(aryl ether)s, and high-temperature polymers, including polyimides.

This book is organized by sections that include transition-metal-catalyzed polymerizations, dendritic and hyperbranched systems, poly(aryl ether)s, polyimides and high-temperature polymers, and general step-growth topics. With the exception of the section on high-temperature polymers, each section contains a chapter surveying important advances in that subject area during the past five years. In addition to providing a very useful review of the field as a whole, these overview chapters should provide the reader some context for understanding how the individual research papers impact the field.

Acknowledgments

We thank the authors and their support staffs for their efforts in preparing their manuscripts and the referees for reviewing each chapter. Our special thanks also go to Michelle Althuis and the production staff of ACS Books for their efforts in assembling this volume.

JAMES L. HEDRICK
IBM Almaden Research Center
650 Harry Road
San Jose, CA 95120-6099

JEFF W. LABADIE
Argonaut Technologies, Inc.
887 Industrial Road
San Carlos, CA 94070-3305

December 6, 1995

TRANSITION-METAL-CATALYZED POLYMERIZATION

Chapter 1

Step-Polymerization Reactions via Nickel- and Palladium-Catalyzed Carbon–Carbon Bond Formation

Virgil Percec and Dale H. Hill

W. M. Keck Laboratories for Organic Synthesis, Department of Macromolecular Science, Case Western Reserve University, Cleveland, OH 44106–7202

Recent developments in the application of Ni(0) and Pd(0) catalyzed carbon-carbon bond forming reactions to the preparation of well defined polymer structures by step polymerization reactions are discussed. Three types of coupling reactions using these catalysts are especially applicable to step polymerizations: (a) homocoupling reactions of organic electrophiles, (b) cross-coupling reactions of organic electrophiles with alkenes or alkynes, and (c) cross-coupling reactions of organic electrophiles with organometallic compounds. The organometallic compounds which have been most useful to polymer chemists include: organomagnesium, organozinc, organostannane, and organoboronic acids (and their esters). This survey of the scope of these reactions includes, for each type of reaction, a discussion of: the types of electrophiles and leaving groups, and their reactivity; reaction mechanism; catalysts; selectivities; and limitations.

The formation of carbon-carbon bonds between monomers is the key step in a number of polymerization reactions. In the case of chain polymerizations the formation of carbon-carbon bonds is the most frequently encountered when the reaction proceeds via radical, nucleophilic, electrophilic, and coordinative mechanisms. However, there are a relatively small number of basic reactions which are effective for carbon-carbon bond formation applicable to a step polymerization reaction. Transition metal catalysts have emerged as a versatile means of carbon-carbon bond formation in organic synthesis (*1-5*). The discovery and development of transition metal catalysts for the stereospecific chain polymerization of olefins by Ziegler (*6-8*) and Natta (*9-11*) was a milestone and marked the beginning of the wide spread use of transition metal catalysts in polymerization reactions. Since that time the field of organometallic chemistry has experienced explosive growth. This growth has occurred both in an understanding of the basic principles of transition metal mediated reactions as well as in the application of transition metal reagents and catalysts to organic synthesis (*1*). Among the synthetically most important catalysts are those which involve palladium or nickel. Many widely used homocoupling and cross-coupling reactions have been developed which utilize zerovalent palladium or nickel catalysts. As a

0097–6156/96/0624–0002$20.75/0

consequence of their efficiency, these catalysts have been applied to the development of step polymerization reactions.

This paper surveys the use of Ni(0)- and Pd(0)-catalysts in step polymerization reactions which involve carbon-carbon bond formation. The discussion centers on the basic reaction types which are most widely and effectively utilized for polymer synthesis. This chapter is organized in two parts. First, the basic features of these coupling reactions are discussed. Then their application to polymer synthesis is examined. The discussion of each type of reaction includes a brief description of the basic reaction, the scope of reaction, the nature of leaving groups involved and their reactivity, limitations to the reaction, and side reactions.

Ni(0) and Pd(0) Catalyzed Reactions of Organic Electrophiles

The Ni(0) and Pd(0) catalyzed reactions of organic electrophiles can be divided into three general groups. Organic electrophiles undergo catalytic homocoupling reactions. They also participate in coupling reactions with alkenes and alkynes. Finally, they participate in several cross-coupling reactions with organometallic compounds. The most useful organometallic compounds for polymer synthesis include organomagnesium, organozinc, organostannane, and organoboron compounds.

Homocoupling reactions. Ni(0)-catalysts have been applied to the synthesis of symmetrical biaryls (*12-18*). Prior to the development of Ni(0) catalysts symmetric biaryls were obtained by the Ullmann reaction (equation 1).

$$\mathrm{ArX} \xrightarrow{\text{"Cu"}} \mathrm{Ar{-}Ar} \qquad (1)$$

For examples of the Ullmann reaction see references 19-22. Development of the Ni-catalyzed reaction began with the homocoupling of aryl iodides and bromides with stoichiometric amounts of air sensitive Ni(0) complexes (equations 2 and 3) (*12-14*). The reaction is slowed by the presence of ortho substituents. Substrates

$$\mathrm{CH_3{-}C(=O){-}C_6H_4{-}Br} \xrightarrow[\text{DMF, 45 °C, 36 h}]{\mathrm{Ni(COD)_2}\text{ (55 mol \%)}} \mathrm{CH_3{-}C(=O){-}C_6H_4{-}C_6H_4{-}C(=O){-}CH_3} \quad 93\% \qquad (2)$$

$$\mathrm{C_6H_5{-}Br} \xrightarrow[\text{DMF, 52 °C, 25 h}]{\mathrm{Ni(COD)_2}\text{ (55 mol \%)}} \mathrm{C_6H_5{-}C_6H_5} \quad 82\% \qquad (3)$$

containing hydroxy groups (i.e. alcohol, phenol, carboxylic acid), their sodium salts (sodium phenoxides, sodium benzoates), or nitro groups do not give useful yields of coupled products (usually 0%) (*12,15*). The reaction was improved with the realization that the Ni(0) species could be generated in situ from air stable precursors (*15*). Thus a stoichiometric amount of $Ni(0)(PPh_3)_2$ was generated in situ by the reduction of $NiCl_2(PPh_3)_2$ with an equivalent amount of Zn. It was then discovered that the reaction could be performed with catalytic amounts of Ni(0) complexes formed in situ when a stoichiometric quantity of Zn was present (equation 4) (*16*).

$$\text{PhBr} + \text{Zn} \xrightarrow[\text{50 °C, 20 h}]{\text{NiCl}_2(\text{PPh}_3)_2\ (5\ \%),\ \text{PPh}_3\ (40\ \%),\ \text{DMF}} \text{Ph–Ph}\ (89\ \%) \qquad (4)$$

The use of a Ni(0)-catalyst generated by the reduction of a Ni(II) salt with excess Zn in the presence of PPh_3 in a dipolar aprotic solvent allowed the high yield coupling of aryl chlorides (equation 5) (*17*). The highest yields were obtained

$$\text{PhCl} \xrightarrow[\text{Zn (1.5 equiv.), DMAc, 80 °C, 2 h}]{\text{NiCl}_2\ (5\ \%),\ \text{PPh}_3\ (38\ \%)} \text{Ph–Ph}\ (99\ \%) + \text{PhH}\ (1\ \%) \qquad (5)$$

when the Ni salt was $NiCl_2$ or $NiBr_2$ (e.g. $NiCl_2(PPh_3)_2$ or $NiBr_2(PPh_3)_2$). The reduction potential of the reducing agent was important. The best results were obtained with Zn. Triaryl phosphines were the best ligands. Evidence of several side reactions was provided. Reduction of the aryl halide occurred in the presence of good proton sources such as water or acidic protons (equation 6) (*17*). Phenols could be coupled in high yields by a protection-deprotection sequence that involved the coupling of 4-chlorophenyl acetate (equation 7) (*17*).

$$\text{HO–C}_6\text{H}_4\text{–Br} \xrightarrow[\text{Zn (1.5 equiv.), DMAc, 80°C}]{\text{NiCl}_2\ (5\%),\ \text{PPh}_3\ (38\%)} \text{HO–C}_6\text{H}_4\text{–C}_6\text{H}_4\text{–OH}\ (25\ \%) + \text{HO–C}_6\text{H}_5\ (62\ \%) \qquad (6)$$

$$\text{CH}_3\text{–C(=O)–O–C}_6\text{H}_4\text{–Cl} \xrightarrow[\text{Zn (1.5 equiv.), DMAc}]{\text{NiCl}_2\ (5\ \%),\ \text{PPh}_3\ (38\ \%)} \text{CH}_3\text{–C(=O)–O–C}_6\text{H}_4\text{–C}_6\text{H}_4\text{–O–C(=O)–CH}_3\ (90\ \%) + \text{CH}_3\text{–C(=O)–O–C}_6\text{H}_5\ (5\ \%) \qquad (7)$$

A side product formed from the coupling of the aryl halide with one of the aryl groups from the phosphine ligand was identified (equation 8) (*17*).

$$\text{CH}_3\text{O–C}_6\text{H}_4\text{–Cl} \xrightarrow[\text{Zn (1.5 eq.), DMAc, 80°C}]{\text{NiCl}_2\ (5\%),\ \text{L}\ (38\ \%)} \text{CH}_3\text{O–C}_6\text{H}_4\text{–C}_6\text{H}_4\text{–OCH}_3 + \text{CH}_3\text{O–C}_6\text{H}_4\text{–C}_6\text{H}_5 + \text{CH}_3\text{O–C}_6\text{H}_5 \qquad (8)$$

L	$CH_3O–C_6H_4–C_6H_4–OCH_3$	$CH_3O–C_6H_4–C_6H_5$	$CH_3O–C_6H_5$
L = PPh_3	69 %	19 %	12 %
L = $P(p\text{-}C_6H_5OCH_3)_3$	100 %	–	–
L = bipyridine	96 %	–	3 %

Larger amounts of this side product formed with arylhalides containing strong electron-donating groups and as the temperature was increased. The formation of the phenyl substituted product is related to the concentration of coordinatively unsaturated Ni-species (equation 9) (*17*). The formation of this side product was suppressed by performing reactions at the lowest temperature possible, increasing the concentration of PPh_3, or by using other ligands.

$$CH_3O\text{-}C_6H_4\text{-}Ni(PPh_3)_2 \longrightarrow CH_3O\text{-}C_6H_4\text{-}Ni(Ph)(PPh_2)(PPh_3) \xrightarrow{PPh_3} CH_3O\text{-}C_6H_4\text{-}Ph + Ni(PPh_2)(PPh_3)(PPh_3) \quad (9)$$

The reaction can be performed in THF in the absence of added PPh_3 in the presence of Et_4NI (equation 10) (*18*). This has several advantageous features. Traces of water are more readily removed from THF than aprotic apolar solvents such as DMF and DMAc. The product may be purified more readily in the absence of a large excess of PPh_3. Also phosphonium salt formation is reduced. The soluble iodide ion provided by Et_4NI functions as a ligand for Ni complexes. Thus I^- stabilizes the catalyst and may facilitate the reduction of an ArNi(II)X species in a key mechanistic step by acting as a bridging ligand between Ni and Zn.

$$CH_3\text{-}C(=O)\text{-}C_6H_4\text{-}Cl \xrightarrow[\text{THF, 50 °C, 20 h}]{NiBr_2(PPh_3)_2\ (10\%)\ /\ Zn\ (1.5\ equiv.),\ Et_4NI\ (10\%)} CH_3\text{-}C(=O)\text{-}C_6H_4\text{-}C_6H_4\text{-}C(=O)\text{-}CH_3 \quad 73\% \quad (10)$$

Other leaving groups have also been demonstrated to be effective. A Pd-catalyst was effective in the desulfonative homocoupling of arylsulfonyl chlorides (*23*). Aryl triflates participate in Ni- and Pd-catalyzed cross-coupling (*24,25*) and homocoupling reactions (*26-33*), for example see equation 11 (*26*).

$$CH_3O\text{-}C_6H_4\text{-}OTf \xrightarrow[\text{Zn (1.5 equiv.), NaI (60\%), } PPh_3 \text{ (60\%), DMF, Ultrasound, 2 h}]{NCl_2(PPh_3)_2\ (8\%)} CH_3O\text{-}C_6H_4\text{-}C_6H_4\text{-}OCH_3\ (95\%) + CH_3O\text{-}C_6H_5\ (\text{trace}) \quad (11)$$

The participation of aryl triflates in this reaction is important from a synthetic standpoint as many substituted phenols are readily available from either commercial sources or well established synthetic procedures. The first report of the homocoupling of aryl triflates utilized ultrasonication to effectuate the reaction (*26,27*). Later, reaction conditions were developed which did not require ultrasonication. Alternatively, Pd-catalysts were effective in the electrochemically (*32*) and chemically (*33*) driven coupling of aryl triflates. Unfortunately, the high cost of the triflic anhydride reagent necessary for the formation of aryl triflates from phenols has hindered the widespread application of this methodology. Therefore, the utilization of less expensive sulfonate leaving groups was expected to facilitate the more extensive application of this methodology.

The first report of the homocoupling of aryl tosylates (73% yield for PhOTs) and aryl mesylates (21% yield for p-$CH_3C_6H_5OMs$) involved the use of ultrasonication (*27*). It was subsequently shown that ultrasonication was unnecessary and that a number of aryl sulfonates including one example of an aryl mesylate undergo Ni(0) catalyzed homocoupling reactions in polar aprotic solvents such as DMF with up to 83 % yield (*34*). Aryl mesylates and other aryl sulfonates were demonstrated to participate in high yield (greater than 99% in some cases) Ni(0) catalyzed homocoupling reactions in etheral solvents such as THF and dioxane (*35,36*). The Ni(0) species was generated from the reduction of $NiCl_2(PPh_3)_2$ with Zn in the presence of either PPh_3 and NaBr in DMF (*34*) or Et_4NI in THF and dioxane (*35,36*). PPh_3 and Br^- function to stabilize the catalyst and were required to prevent premature catalyst decomposition in DMF. The added phosphine ligand was not required in THF or dioxane when the Ni(0) species was generated in the presence of Et_4NI (equation 12) (*35,36*). In this case the Ni(0)

$$CH_3O-\overset{O}{\overset{\|}{C}}-C_6H_4-OSO_2CH_3 \xrightarrow[\text{67 °C, 10 h}]{NiCl_2(PPh_3)_2,\ Zn \quad Et_4NI,\ THF} CH_3O-\overset{O}{\overset{\|}{C}}-C_6H_4-C_6H_4-\overset{O}{\overset{\|}{C}}-OCH_3 \quad (> 99\ \%) \qquad (12)$$

species was highly coordinatively unsaturated. Aryl mesylates underwent facile oxidative addition to this $Ni(0)(PPh_3)_2$ species. The reaction of aryl mesylates tolerates a large number of common functional groups: esters, ethers, fluoro, ketone and nitrile. No coupled product was obtained in the presence of the nitro group. The highest yields were obtained when electron-withdrawing groups were in the para position. Yields were reduced slightly by steric hindrance when ortho substituents were present. However, high yields (greater than 99 % in some cases) could be obtained by adding PPh_3 and increasing the reaction time. Side reactions occurred to a greater degree in the reaction of less reactive aryl mesylates (i.e. those with electron-donating groups or large ortho substituents). Products formed by reduction and transarylation of the aryl mesylates were identified (equation 13). Surprisingly, it was discovered that "wet" THF (i.e. THF used directly out of a previously opened bottle with no drying step) could be used when extra PPh_3 was added (usually 20 % was sufficient) (*35*).

$$CH_3-\overset{O}{\overset{\|}{C}}-C_6H_4-OSO_2CH_3 \xrightarrow[\text{THF, 67°C, 10h}]{NiCl_2(PPh_3)_2,\ Zn,\ Et_4NI}$$

$$CH_3-\overset{O}{\overset{\|}{C}}-C_6H_4-C_6H_4-\overset{O}{\overset{\|}{C}}-CH_3\ (73\ \%) + CH_3-\overset{O}{\overset{\|}{C}}-C_6H_5\ (20\ \%) + CH_3-\overset{O}{\overset{\|}{C}}-C_6H_4-C_6H_5\ (7\ \%) \qquad (13)$$

Reaction Mechanism. Several different reaction mechanisms have been suggested for Ni(0) catalyzed homo-coupling reactions of aryl halides.(*17,37-39*) The mechanism outlined in Figure 1 has been proposed for the Ni(0) mediated coupling reaction of aryl chlorides in polar aprotic solvents in the presence of excess Zn (*17*). The mechanistic details have been supported by an electrochemical study (*39*). The mechanism outlined in Figure 2 has been proposed for the coupling reaction of ArX (X = halide) in nonpolar solvents in the absence of excess reducing metal (*37,38*). The primary mechanistic pathway followed is highly dependent on the reaction conditions (*14*). Under the conditions utilized for the coupling of aryl halides and sulfonates, the most plausible mechanism is shown in Figure 1 (*17,35,39*).

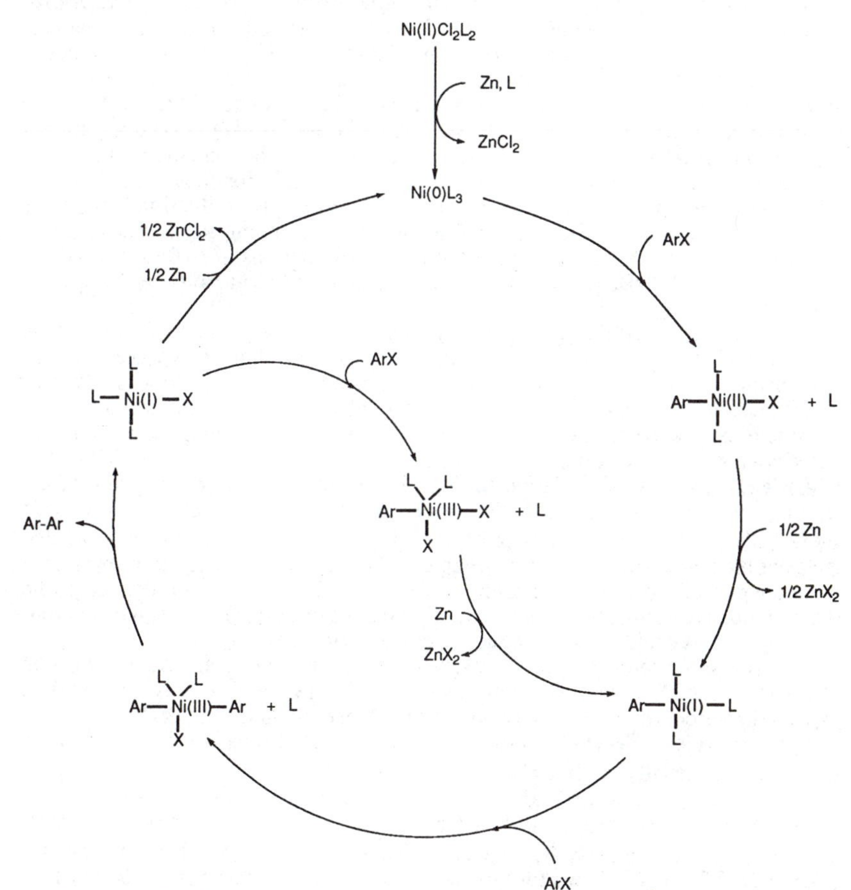

Figure 1. Mechanism of Homocoupling Reaction of Organic Electrophiles in the Presence of Excess Zinc.

The first step of the mechanism involves the reduction of Ni(II) to Ni(0) by Zn. This is followed by the oxidative addition of ArX (X= I, Br, Cl, OTf, OMs or other leaving group) to the Ni(0) species to form $ArNi(PPh_3)_2X$. When a sulfonate leaving group is involved this Ni(II) species may have an ionic structure, i.e., $[ArNi(PPh_3)_2]^+[OSO_2R]^-$. The oxidative addition of vinyl triflates to $Pt(PPh_3)_4$ results in the formation of the ionic Pt(II) complexes containing a s-vinyl ligand, three phosphine ligands and a noncoordinating triflate anion (*40*). A similar Pd(II) complex $[ArPdL_n]^+[OTf]^-$ has been proposed to result from the oxidative addition of ArOTf to Pd(0) in the presence of PPh_3 (*41*). In a reaction analogous to the reaction of $[RPtL_3]^+[OTf]^-$ with R_4NX (*40*), $ArNi(II)OSO_2R$ complexes may react quickly with Et_4NI to form $ArNi(II)I(PPh_3)_2$. The Ni(II) species then undergoes a one electron reduction to $ArNi(I)L_3$ (L = PPh_3 or solvent). ArX oxidatively adds to this species to give a diaryl Ni(III) complex which undergoes rapid reductive elimination, resulting in the formation of the biaryl product and the generation of $Ni(I)XL_3$. There are two productive reaction pathways available to this Ni species. $Ni(I)XL_3$ can be reduced by Zn to regenerate $Ni(0)L_3$, which can then repeat the catalytic cycle. Alternatively, ArX can undergo direct oxidative addition to $Ni(I)XL_3$ followed by reduction by Zn to form the $ArNi(I)L_3$ species once again.

The rate of oxidative addition of aryl halides to Ni(0) species is considered to be a fast reaction in other coupling reactions (*1,2,17*). The rate determining step in the homocoupling reaction of aryl chlorides in the presence of excess Zn, is the reduction of the aryl-Ni(II) species to the aryl Ni(I) species (*17*). However, at high conversions of ArX, the rate determining step becomes the oxidative addition of ArX to the Ni(I) species (*17*). Thus, the rate constants for these reactions are within an order of magnitude of each other. When the reduction occurs by electrochemical means, rather than via Zn, the rate determining step is the oxidative addition of ArBr to the ArNi(I) species at low concentrations of ArBr and at higher concentrations of ArBr is the reductive elimination of biaryl from the Ni species (*39*).

The rate determining step for the reaction of aryl mesylates is unknown (*35*). The highest yields in the coupling reaction of aryl mesylates were obtained with electron withdrawing groups in the para position. The reaction was inhibited by electron-donating groups as well as by sterically hindering ortho groups. None of these effects alone can be used to determine the rate determining step. Electron-withdrawing groups can increase the rate of oxidative addition of aryl halides to Ni(0) (*42*). The rate of oxidative addition is also influenced by the nature of the leaving group (*42*). The electronic properties of the aryl group have also been shown to influence the reactivity of $[ArPdL_n]^+[OTf]^-$ (*40*). Thus, it is possible that the electronic properties of the aryl group could affect the rate of electron-transfer in the reduction of the aryl-Ni(II) species to an aryl-Ni(I) species. Furthermore, the rate of reductive elimination is increased by positive charge (*5*). Thus, the electron-donating groups could slow the reductive elimination step.

The presence of sterically hindering ortho groups can also influence the rate of several steps. For example, the oxidative-addition of ArX to ArNi(I) would be expected to proceed more slowly with ortho substituents on the aryl groups. In addition, the rate of electron transfer to ArNi(II) could also be affected by an ortho substituent, especially if a bridging I^- ion is involved.

The coupling of aryl halides in nonpolar solvents in the absence of excess reducing metal has been established as following a double-chain mechanism involving the reaction of $Ar\text{-}Ni^{III}X_2$ and $ArNi^{II}X$ to form $Ar_2Ni^{III}X$ in the key step (Figure 2) (*37,38*). Evidence for this mechanism was obtained using solvents such as benzene, toluene and hexane. The bimolecular step involves Ni species expected to be present in trace quantities when excess Zn is present. Thus this mechanism is expected to be favored when large amounts of Ni catalyst and small amounts of Zn are present. This mechanism is expected to be operative in reactions involving stoichiometric amounts of preformed Ni(0) in nonpolar solvents.

Cross-coupling reactions involving alkenes and alkynes

Heck Reaction. The Pd-catalyzed arylation and vinylation of olefins in the presence of a weak base is known as the Heck reaction (equation 14) (*43*). Several excellent reviews are available on various aspects of this reaction (*44-47*).

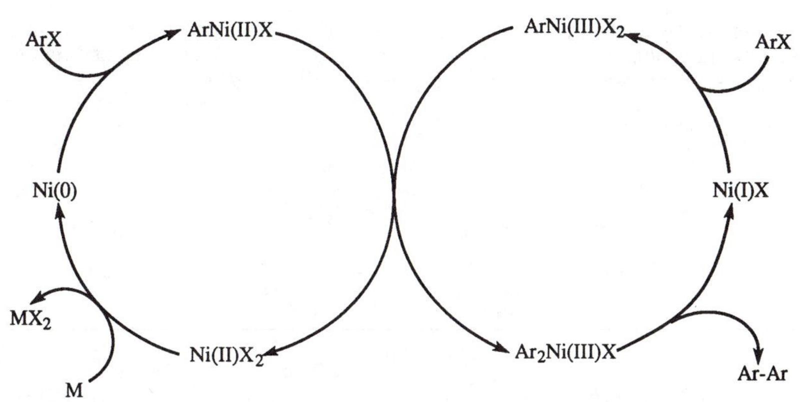

Figure 2. Mechanism of Homocoupling of Organic Halides in Nonpolar Solvents in the Presence of Large Amounts of Ni(0) Reagents.

$$\text{H(C=C)} + \text{R-X} \xrightarrow[\text{-HX}]{\text{Pd(0), Base}} \text{R(C=C)} \tag{14}$$

The R group can be an aryl, heteroaryl, benzyl or vinyl group. Alkyl groups with sp^3 hydrogens β to the leaving group undergo elimination reactions. The most common leaving groups are I (*45*), Br (*45*), and OSO_2CF_3 (OTf) (*24,25,49,50*). Other leaving groups such as COCl (*51,52*), SO_2Cl (*53*), N_2Cl (*54*), N_2BF_4 (*54,55*) and $OSO_2(CF_2)_2O(CF_2)_2H$ (*56*) can also be employed. Substituted olefins with electronic properties ranging from electron-deficient to electron-rich participate in the reaction. The base is typically a secondary or tertiary amine such as triethylamine. Weak bases such as sodium (or potassium) acetate, bicarbonate or carbonate are also used. In a typical Heck reaction, a solution of aryl iodide, olefin, $Pd(OAc)_2$, NEt_3 and solvent are reacted at an appropriate temperature (sometimes room temperature) until the reaction is complete. Triarylphosphines (i.e. PPh_3, P(*o*-tolyl)$_3$) are added when aryl bromides are reacted (*45*). The reaction proceeds stereospecifically with *syn* R group addition to the double bond (i.e. to the same face as is coordinated to Pd). The reaction in equation 15 proceeds with retention of configuration at the vinyl halide and substitution of the olefin occurs to give the E isomer (*57*).

$$\xrightarrow[\text{NEt}_3\text{ (3 equiv.), 100 °C, 4 h}]{\text{Pd(OAc)}_2\text{ (2\%), P(}o\text{-tolyl)}_3\text{ (4\%)}} \quad 86\% \tag{15}$$

The generally accepted mechanism for the reaction of aryl halides is shown is Figure 3 (*45-48*). The catalytic cycle begins with oxidative addition of aryl halide to Pd(0). The olefin coordinates to this Pd(II) species. The coordinated olefin then inserts into the Pd-R bond. When a hydrocarbon or electron-withdrawing group is attached to the double bond, the regioselectivity of the insertion of the olefin into the M-R bond is controlled by sterics (*45,58*). Consequently, substitution occurs predominately at the non-substituted carbon. In contrast, when electron-donating groups are attached to the double bond the regioselectivity varies depending on the exact alkene and sometimes depends upon the substrate (*45*).

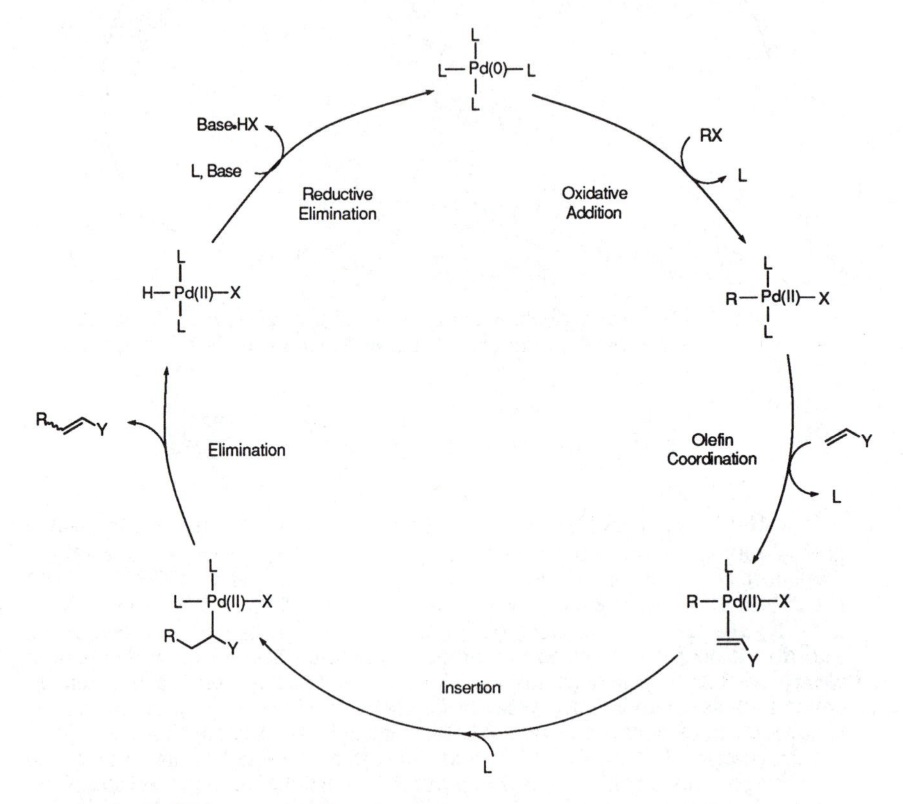

Figure 3. Mechanism of Heck Reaction.

There is evidence that an alternative mechanism for the coordination-insertion steps is operative when the leaving group is triflate (Figure 4) (*48,59-61*). As a result of the lability of this ligand, a cationic Pd(II) complex is formed after oxidative addition. Thus, in the absence of added halide salts, the olefin coordination and insertion steps involve cation Pd(II) species. This sequence of steps leads to product formation in which the regioselectivity is determined by electronic factors. These factors are more important because the polarization of the double bond is increased upon coordination to the cationic Pd species. The aryl group migrates to the carbon with the lowest charge density (*48,59*).

Aryl, heteroaryl (*62*) and vinyl (*49,50,63*) triflates readily participate in the Heck reaction. Examples are shown in equations 16 (*62*) and 17(*49*).

OTf + (ethyl acrylate) → $PdCl_2(PPh_3)_2$ (5%), LiCl (3 equiv.), DMF, 110 °C, 5.5 h → 85 %

(16)

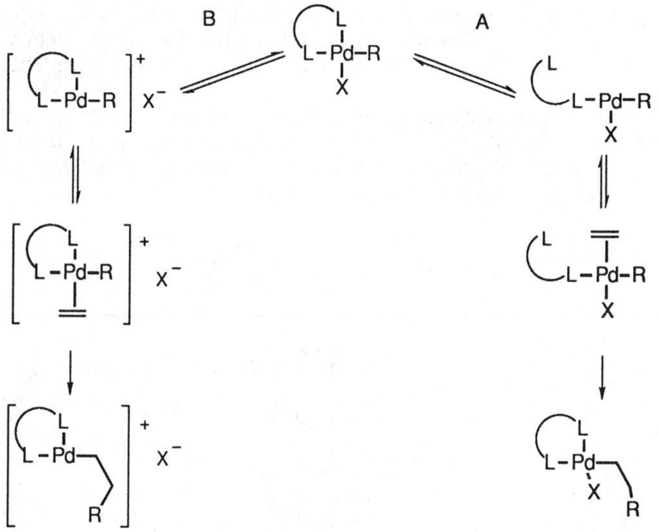
B
A
L
L-Pd-R
X
X⁻
+
R

Figure 4. Coordination-Insertion Steps of Heck Reaction for Aryl halides (Path A) or Aryl Triflates (Path B) in the Absence of Halide Ions.

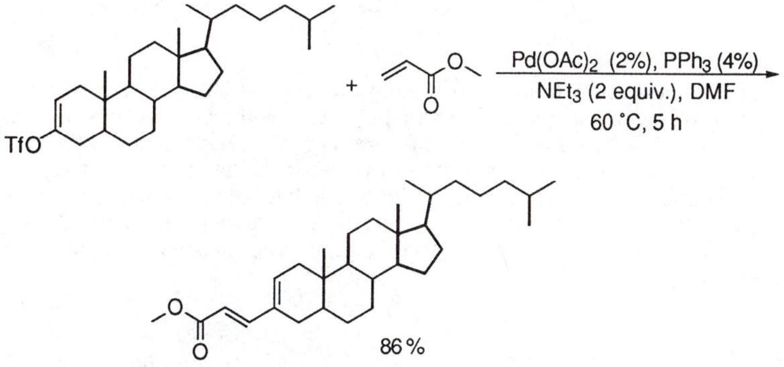
TfO
+
O
Pd(OAc)2 (2%), PPh3 (4%)
NEt3 (2 equiv.), DMF
60 °C, 5 h
86 %

(17)

Unless very strongly activated, aryl chlorides do not undergo oxidative addition to the Pd(0) catalysts typically utilized for the Heck reaction. Chlorobenzene participates in the Heck reaction when it is reacted with NaI in the presence of $NiBr_2$ prior to the addition of Pd-catalyst (equation 18) (*64*). In this example, a small equilibrium amount of iodobenzene was formed in a Ni-catalyzed

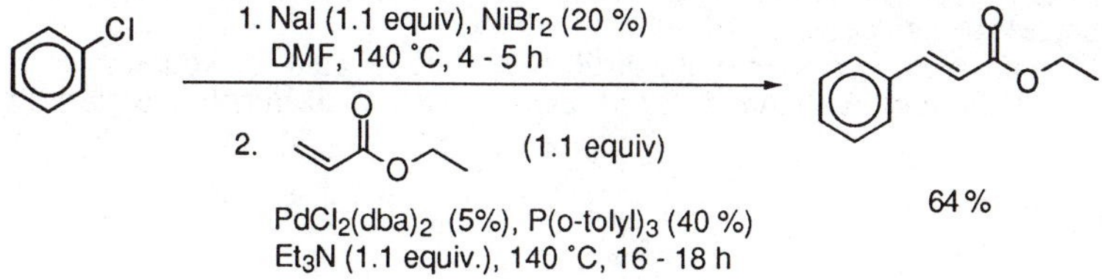
Cl
1. NaI (1.1 equiv), NiBr2 (20 %)
DMF, 140 °C, 4 - 5 h
2.
(1.1 equiv)
PdCl2(dba)2 (5%), P(o-tolyl)3 (40 %)
Et3N (1.1 equiv.), 140 °C, 16 - 18 h
O
64 %

(18)

reaction prior to the addition of Pd(0). Upon addition of Pd(0), the iodobenzene which is consumed is replenished in the Ni-catalyzed reaction. Thus a small equilibrium concentration of iodobenzene is present until the completion of the reaction.

Arylchlorides also participate in the Heck reaction if the appropriate ligand is coordinated to the Pd-catalyst. 4-Chlorobenzaldehyde participates in a high yield Heck reaction when sterically demanding 1,4-bis(diisopropylphosphino)butane (dippb) was used as the phosphine ligand (equation 19) (*65*).

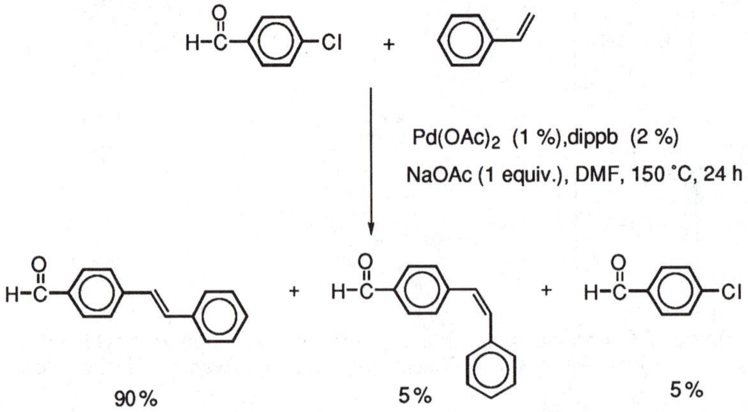

$Pd(OAc)_2$ (1 %), dippb (2 %)
NaOAc (1 equiv.), DMF, 150 °C, 24 h

90% 5% 5%

dippb = 1,4-bis(diisopropylphosphino)butane

(19)

Chloroarenes and bromoarenes undergo high yield Heck reactions in the presence of a palladacycle prepared from $Pd(OAc)_2$ and tri(*o*-tolyl)phosphine (equation 20) (*66*). When this catalyst was used at 140 °C none of the palladium deposits which are typical at this temperature with other Pd-catalysts were detected. The catalyst is also more active, allowing a lower catalyst concentration to be used. No added phosphine is necessary, so side reactions involving this ligand do not occur. 4-Chlorobenzaldehyde and *n*-butyl acrylate react to give 90 % yield of coupled product (*66*).

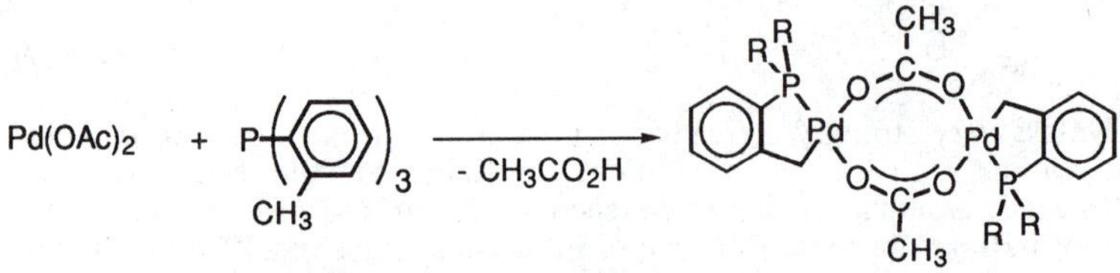

$Pd(OAc)_2$ + $P(C_6H_4CH_3)_3$ → ($- CH_3CO_2H$)

(20)

A few reports of the use of Ni-catalyst for the Heck reaction have appeared. For examples see references 67-70.

Side products derived from the disubstitution of the olefin sometimes occur. This type of product is obtained in the reaction of 2,2'-dibromobiphenyl with styrene. Evidently, the intermediate Pd(II) alkyl complex formed after insertion of the alkene into the Pd-aryl bond reacts faster with the adjacent aryl-bromide than the β–H elimination reaction occurs (equation 21). This problem was avoided by reversing the functional groups attached to each of the coupling partners. Thus, 2,2'-divinylbiphenyl was reacted with bromobenzene (equation 22) (*71*).

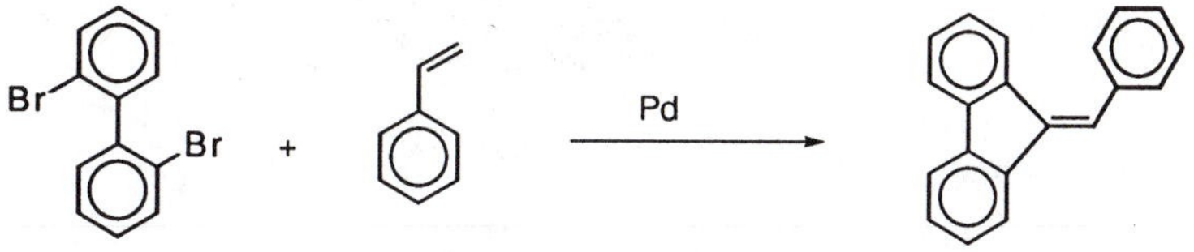

Br Br + Pd

(21)

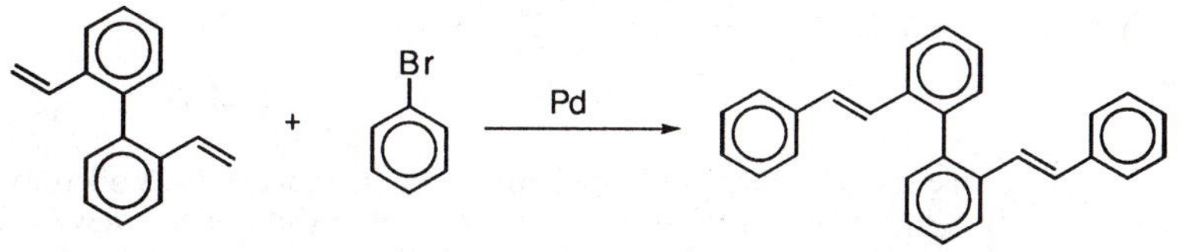

\+ Br Pd

(22)

Cross-coupling reactions of alkynes. Terminal alkynes undergo high yield Pd(0) catalyzed coupling reaction (*72-76*) with organic electrophiles. The electrophiles include: vinyl, aryl, and heteroaryl halides (*72-75*); vinyl triflates (*77,78*); aryl (*56,78*) and heteroaryl triflates (*75,78*); and polyfluorophenyl perfluoroalkane sulfonates (*79*). Several examples are shown in equation 23 - 27 (respectively references 78,72,73,78,79).

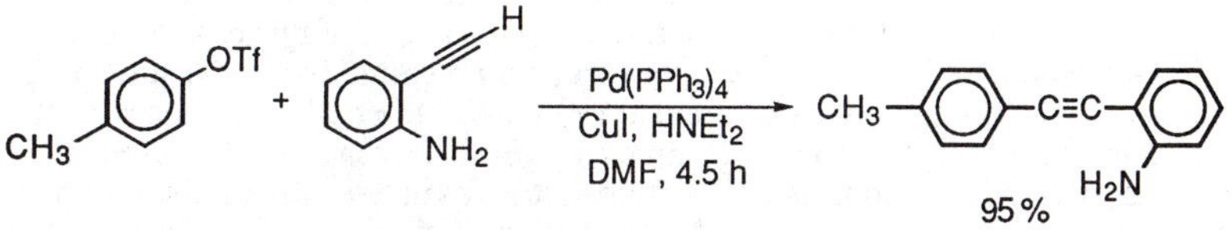

CH_3 OTf + H NH_2 $\xrightarrow[\text{CuI, HNEt}_2\text{; DMF, 4.5 h}]{\text{Pd(PPh}_3)_4}$ CH_3–C≡C– H_2N 95 %

(23)

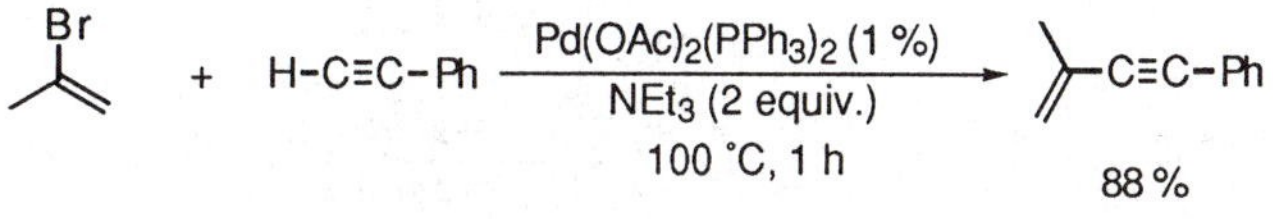

Br + H-C≡C-Ph $\xrightarrow[\text{NEt}_3\text{ (2 equiv.); 100 °C, 1 h}]{\text{Pd(OAc)}_2\text{(PPh}_3)_2\text{ (1 \%)}}$ –C≡C–Ph 88 %

(24)

I + H-C≡C– $\xrightarrow[\text{CH}_3\text{ONa (1.2 equiv.); 50 °C, 3 h}]{\text{Pd(PPh}_3)_4\text{ (4 \%)}}$ –C≡C– 97 %

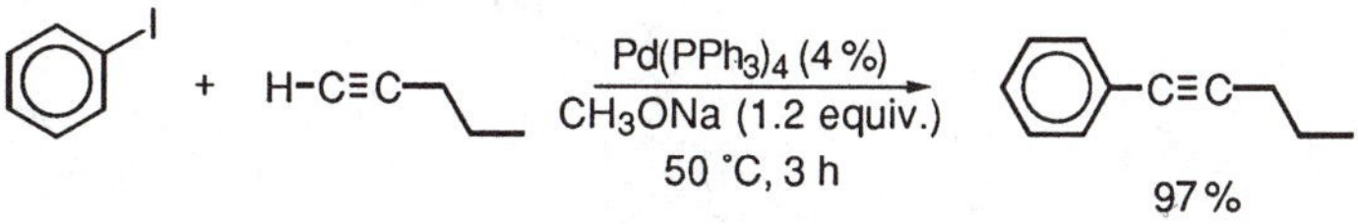

(25)

NH_2 + TfO– (OCH_3) → $Pd(PPh_3)_4$ (1 %), CuI (2 %), Et_2NH (20 equiv.), 5 h → NH_2, OCH_3 91 % (26)

$$C_6F_5OSO_2CF_3 + H{-}C{\equiv}C{-}Ph \xrightarrow[\text{NEt}_3\text{, DMF; 80 °C, 10 h}]{PdCl_2(PPh_3)_2} C_6F_5{-}C{\equiv}C{-}Ph \quad 92\% \tag{27}$$

One pot phase-transfer catalyzed synthetic procedures were developed for the synthesis of diheteroarylacetylenes (*80*), 1,2-(4,4'-dialkoxyaryl)acetylenes (*81*) and 1,4-bis[2-(4',4"-dialkoxyphenyl)ethynyl]benzenes (*81*).

The mechanistic features of Pd(0) catalyzed coupling reactions of organic electrophiles and terminal alkynes are controversial. In general, it is not regarded as involving an insertion mechanism similar to the Heck reaction (*73*). The terminal acetylene proton is readily deprotonated by base, therefore an oxidative addition, transmetallation, reductive elimination sequence (vide infra) is expected. A brief summary of the features of the most frequently proposed mechanism of Pd(0)/Cu(I) acetylene coupling reactions is contained in reference 81. However, an insertion mechanism was postulated for equation 27 (*79*).

The arylation and vinylation of internal alkynes has not been developed as much as the reactions of alkenes and 1-alkynes. An example of this type of reaction is depicted in equation 28 (*78*). The alkenylpalladium complexes formed from the reaction of Pd and internal alkynes are usually stable. Internal alkynes have been vinylated by intramolecular cascade reactions. In these reactions the product of insertion into an alkyne bond is able to subsequently insert into an alkene bond. Elimination from this complex gives the final reaction product. Equations 29 (*82*) and 30 (*82*) provide examples of this type of reaction.

HO, CH_3, CH_3 – C(=O)–N(CH_3)H + OCH_3 / I → $Pd(OAc)_2(PPh_3)_2$, piperidine, HCO_2H, DMF, 60 °C, 8 h → HO, CH_3, CH_3, C(=O)N(CH_3)H, OCH_3 90 % (28)

CH_3 ; I ; + ; O ; OCH_3 ; $Pd(PPh_3)_4$ (3 %) ; NEt_3 (2 equiv.) ; CH_3CN, reflux, 12 h ; CH_3 ; OCH_3 ; O

80 % (29)

EtO_2C ; EtO_2C ; I ; $SiMe_3$; $Pd(PPh_3)_4$; NEt_3, CH_3CN ; reflux, 6 h ; EtO_2C ; EtO_2C ; $SiMe_3$

80 % (30)

Cross-coupling of organic electrophiles with organometallics
A large number of reactions involving the Pd(0)- or Ni(0)-catalyzed cross-coupling of organic electrophiles with organometallic reagents. The general mechanism of the Pd-catalyzed reactions is shown in Figure 5 (1,2). These reactions involve a general sequence of (a) oxidative addition of the organic electrophile to the zero valent metal catalyst, (b) transfer of the organic group from the organometallic compound to the metal catalyst in a transmetallation step and (c) reductive elimination of the coupled product with the concurrent reformation of the original zero valent metal catalyst. The Pd catalyst is added to the reaction solution

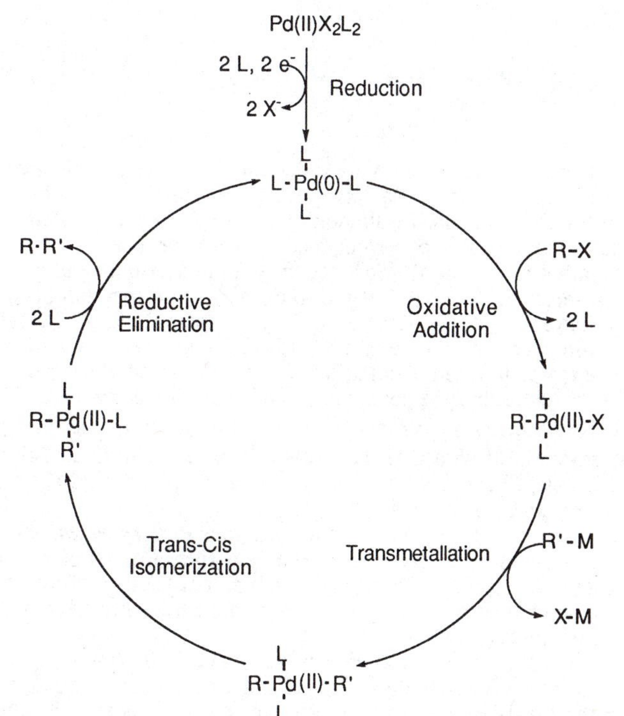

Figure 5. General Mechanism of Pd-Catalyzed Cross-Coupling of Organic Electrophiles and Organometallics

as a Pd(0) or Pd(II) complex. The reaction generally proceeds with an activation step in which the Pd(II) catalyst precursor is reduced to Pd(0). The oxidative addition of the organic electrophile readily occurs to this species to generate a Pd(II) species. The order of reactivity of the most common leaving groups in Pd(0)-catalyzed reactions is I > OTf > Br. The oxidative addition involves oxidation of the metal and reduction of the organic electrophile. Therefore, more electron deficient substrates undergo this reaction more readily. The transmetallation reaction is usually the slow step in the catalytic cycle. Therefore, oxidative addition product $RPd(II)XL_2$ may participate in side reactions prior to transmetallation.

Alkyl groups with β–hydrogens usually undergo rapid β-hydrogen elimination. As a result alkyl electrophiles with β–hydrogens are usually avoided. Another common side reaction involves exchange of the organic groups of the Pd(II) species and the organophosphine ligands (equation 31).

$$R\text{-}Pd(PPh_3)_2\text{-}X \underset{PPh_3}{\overset{-PPh_3}{\rightleftharpoons}} R\text{-}Pd(PPh_3)\text{-}X \rightleftharpoons R\text{-}Pd(PPh_2)(Ph)\text{-}X \rightleftharpoons Pd(PPh_2R)(Ph)\text{-}X \quad (31)$$

The R group can also transfer to the phosphorous atom of organophorous ligands to form phosphonium salts (i.e. $PR'_3R^+X^-$, equation 32). Phosphonium salts may also be formed by the reaction of organic electrophiles with non-coordinated phosphines.

$$R\text{-}Pd(PPh_3)_2\text{-}X + 2\ PPh_3 \longrightarrow [RPPh_3]^+[X]^- + Pd(PPh_3)_3 \quad (32)$$

The transmetallation step involves the transfer of the organic group from an organometallic species to a Pd(II) species. There are several important requirements for efficient transmetallation reactions (*2*). First, the $RPdL_2X$ species must have sufficient stability to prevent decomposition or other side reactions prior to transmetallation (vide supra). Second, the overall thermodynamics need to be favorable. In this regard, the ΔH of the transmetallation reaction becomes more negative by increasing the stabilities (i.e. lower enthalpy) of the products of transmetallation reaction. For example, lithium chloride is often an essential additive in the reaction of vinyl and aryl triflates with organostannane compounds (*83,84*). Lithium chloride apparently has several roles in this reaction. The product of the oxidative addition of ROTf to Pd(0) is $RPdL_2(OTf)$ or $[RPdL_2]^+[OTf]^-$. Ligand exchange of Cl^- with OTf^- converts this organo-palladium species to the more stable $RPdL_2Cl$ species. The product of transmetallation is thus R_3SnCl instead of the less stable R_3SnOTf.

The transmetallation reaction produces a trans Pd(II) species. Isomerization to a cis arrangement is necessary prior to the reductive elimination of the coupled product. The Pd(0) catalyst is regenerated upon reductive elimination. The reductive elimination reaction is very fast. Therefore competing reactions leading to side products are not usually a problem.

Recently the alternative mechanism shown in figure 6 was proposed based on elaborate electrochemical studies (*85*). This mechanism provides an explanation for many of the ligand effects and the effect of cocatalysts such as CuI. In this mechanism the catalytic cycle begins with the oxidative addition of aryl halide to a halide ligated zerovalent Pd species. This produces a pentacoordinate anionic aryl Pd(II) complex. This complex participates in the transmetallation reaction to give another pentacoordinate anionic Pd(II) complex. Reductive elimination of the coupled product regenerates the halide ligated Pd(0) species which began the catalytic cycle.

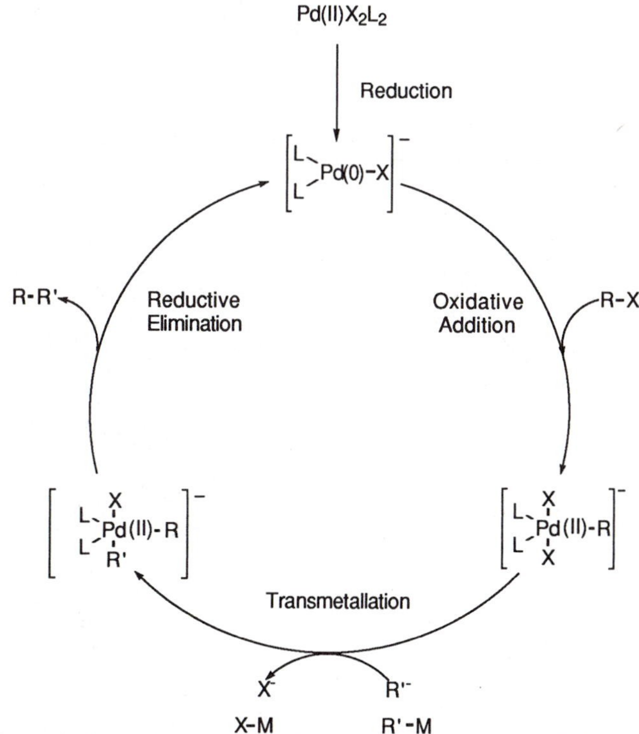

Figure 6. Cross-Coupling Mechanism Involving Anionic Pd-Species

Grignards. Aryl and vinyl halides can be coupled with Grignard reagents (equation 33) in a Ni- or Pd-catalyzed reaction (*86*) which proceeds via an oxidative addition - transmetallation sequence (*87*) (Figure 7).

$$R{-}Z \quad + \quad R'MgX \xrightarrow{Pd(0)} R{-}R' \quad + \quad MgXY \qquad (33)$$

Grignard reagents couple with vinyl halides (*87-91*), aryl halides (*87,88*), and heteroaryl halides (*92,93*), and vinyl halides (*86*). Alkyl iodides have been coupled with Grignards in the presence of $PdCl_2$(dppf) (*94*). The utility of the Grignard coupling reaction is limited by the reactivity of the reagent. The Pd catalysts are generated from PdX_2L_2 and PdL_4 where L = PPh_3 or other phosphine ligands. The halides in the NiX_2 catalyst may be Cl, Br, or I. Best results are usually obtained with Cl. The Pd-catalysts usually provide increased selectivity and yields. The Ni-catalyzed reaction proceeds with the retention of the geometry of the double bond in some cases. For example see equations 34 and 35 (*91*). The order of reactivity for aryl halides is ArI > ArBr > ArCl > ArF. The best results with Ni-catalysts were obtained with ArCl. Among the Grignard reagents used are allyl, alkenyl, aryl, alkyl (may have β-hydrogens) and heteroaryl.

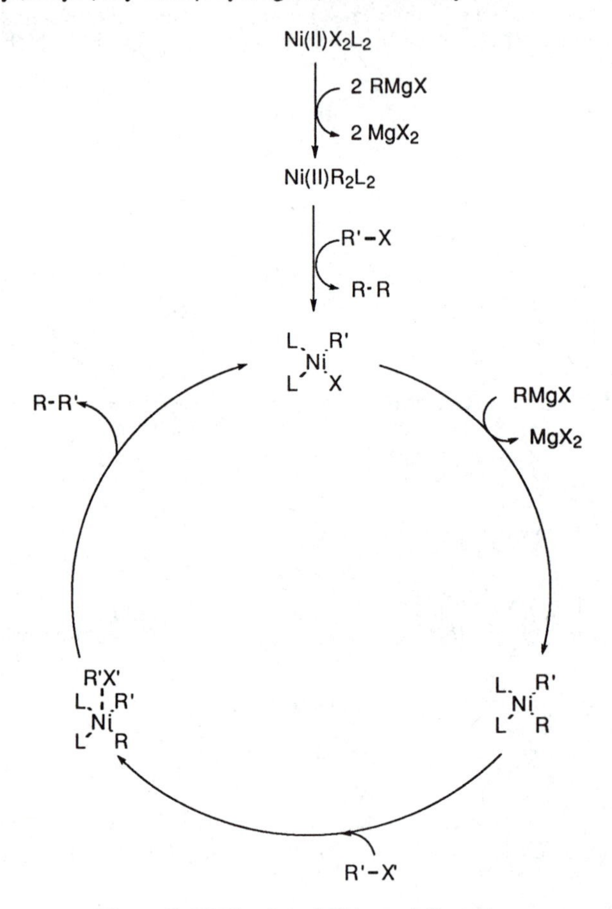

Figure 7. Ni-Catalyzed Grignard Coupling.

Ph–CH=CH–Br (cis) + PhMgBr $\xrightarrow[\text{Et}_2\text{O, 74\%}]{NiCl_2[Ph_2P(CH_2)_2PPh_2]}$ Ph–CH=CH–Ph (cis) + Ph–CH=CH–Ph (trans)

9 : 1 ratio (34)

Ph–CH=CH–Br (trans) + PhMgBr $\xrightarrow[\text{Et}_2\text{O, 100\%}]{NiCl_2[Ph_2P(CH_2)_2PPh_2]}$ Ph–CH=CH–Ph (cis) + Ph–CH=CH–Ph (trans)

7 : 93 ratio (35)

Organic electrophiles with other leaving groups in addition to halide leaving groups also participate in cross-coupling reactions with Grignard reagents. The organic electrophiles which participate in this reaction include aryl *O*-carbamates (equation 36) (*95*), aryl *t*-butyl sulfones (equation 37) (*96*), and aryl triflates (equation 38) (*95*).

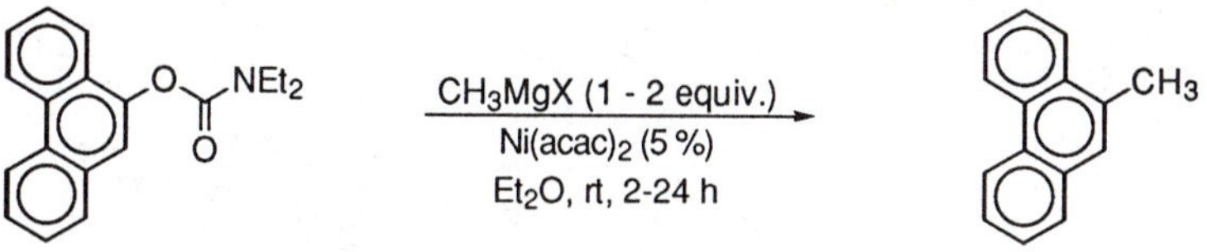

CH_3MgX (1 - 2 equiv.), $Ni(acac)_2$ (5 %), Et_2O, rt, 2-24 h; NEt_2, O; CH_3; 93 % (36)

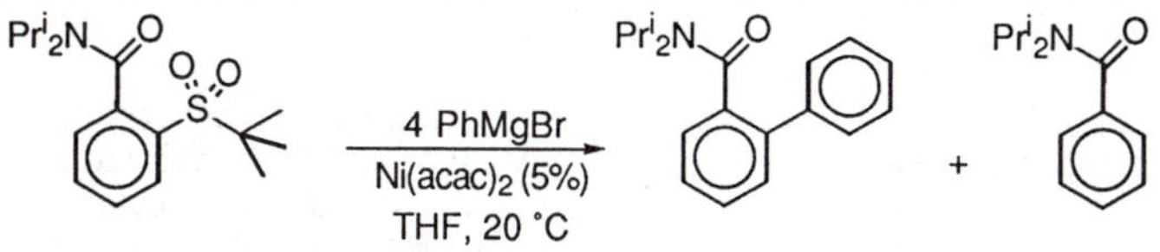

4 PhMgBr, $Ni(acac)_2$ (5%), THF, 20 °C; $Pr^i{}_2N$, O; 95 % + 2 % (37)

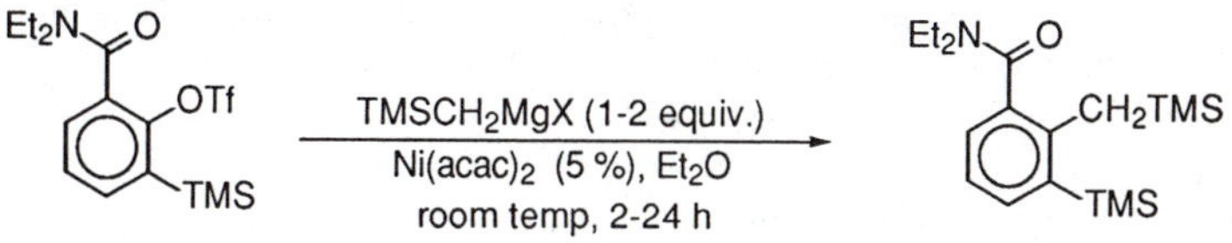

$TMSCH_2MgX$ (1-2 equiv.), $Ni(acac)_2$ (5 %), Et_2O, room temp, 2-24 h; Et_2N, OTf, TMS; CH_2TMS, TMS; 73 % (38)

Recently the mesylate group was developed as an inexpensive alternative to the triflate group in Ni(0)-catalyzed homocoupling and cross-coupling reactions. Aryl mesylates participated in coupling reactions with Grignard reagents (*97*).

Side reaction products include: homocoupled products derived from the organic electrophile or the Grignard reagent, reduction of the organic electrophile and Grignard reagent, and cleavage of the sulfur-oxygen bond of aryl sulfonates (*86-97*).

Organolithium compounds can also be coupled with organic electrophiles(*98,99*).

Organozinc. Organozinc reagents undergo Pd- or Ni-catalyzed cross-coupling reaction with organic electrophiles (equation 39). Several reviews are available (*100,101*). In this equation R can be aryl, alkenyl, alkynyl, allyl or acyl and R' can be alkyl, alkenyl, alkynyl or aryl. Unsatisfactory results are obtained when R = R' = alkynyl.

$$R{-}X \;+\; R'_2Zn \xrightarrow{Pd(0)} R{-}R' \;+\; ZnX_2 \qquad (39)$$

In an early example of this reaction aryl and benzyl zinc halides were coupled with aryl halides (equation 40) (*102*). Alkynyl zinc compounds cross-couple with aryl halides (iodides or bromides) in the presence of a Pd catalyst (equation 41) (*103*) or alkynyl iodides can be coupled with aryl and alkenyl zinc compounds (equation 42) (*100*).

$$RZnX + ArX' \xrightarrow[\text{cat } PdCl_2(PPh_3)_2 + (i\text{-}Bu)_2AlH]{\text{cat. } Ni(PPh_3)_4 \text{ or}} RAr$$

R = Ar, $ArCH_2$
X = Br, Cl
X' = Br, I (40)

$$R{-}C{\equiv}C{-}ZnCl + Ar{-}X \xrightarrow[25\ ^\circ C]{PdL_n} R{-}C{\equiv}C{-}Ar$$

R = H, Alkyl, aryl (41)

$$I{-}C{\equiv}C{-}C_4H_9 + R{-}ZnCl \xrightarrow[25\ ^\circ C]{PdL_n} R{-}C{\equiv}C{-}C_4H_9 \quad (42)$$

The catalysts are typically Pd or Ni phosphine complexes. Ni catalysts are more reactive and cost less. Pd catalysts usually give better results. Ni catalysts are more reactive with a number of functional groups. The use of Ni-catalysts with substrates with C-C triple bonds results in lower yields, presumably due to the formation of cyclic trimers, tetramers and oligomers (*104*). Lower yields of the desired cross-coupled product are obtained in the Ni catalyzed (RX- RZnX) coupling of alkenyl-alkynyl, alkynyl-aryl, and aryl-alkynyl, alkenyl-alkenyl. An examples of this cross-coupling reaction is shown in equation 43 (*105*).

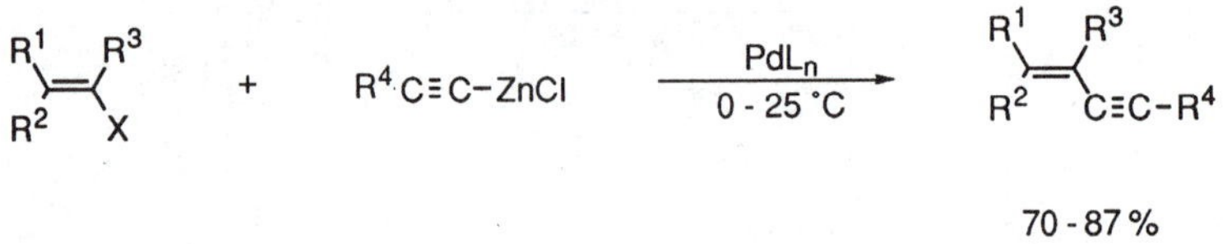

$$R^1R^2C{=}C(R^3)X + R^4{-}C{\equiv}C{-}ZnCl \xrightarrow[0-25\ ^\circ C]{PdL_n} R^1R^2C{=}C(R^3){-}C{\equiv}C{-}R^4$$

70 - 87 %

R^n = H, or organic group X = I or Br (43)

Allylic electrophiles also participate in coupling reaction (*106-109*). Examples are provided in equations 44 (*106,107*) and 45 (*108*). Acyl halides also participate in cross-coupling reactions (*110*).

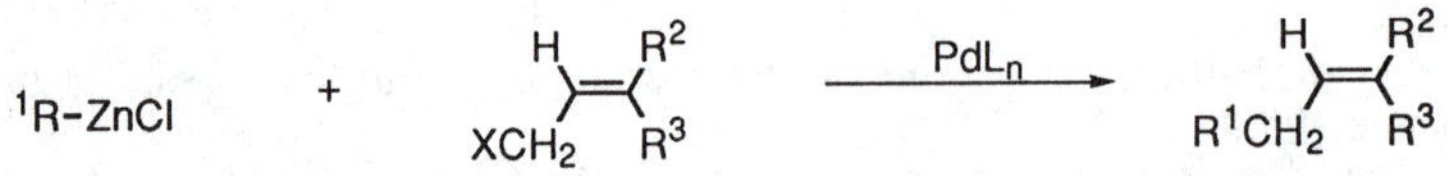

$$R^1{-}ZnCl + XCH_2(H)C{=}CR^2R^3 \xrightarrow{PdL_n} R^1CH_2(H)C{=}CR^2R^3$$

R^n = alkenyl or aryl
X = halogen, OAc, $OAlR_2$, $OP(O)(OR)_2$, $OSiR_3$ (44)

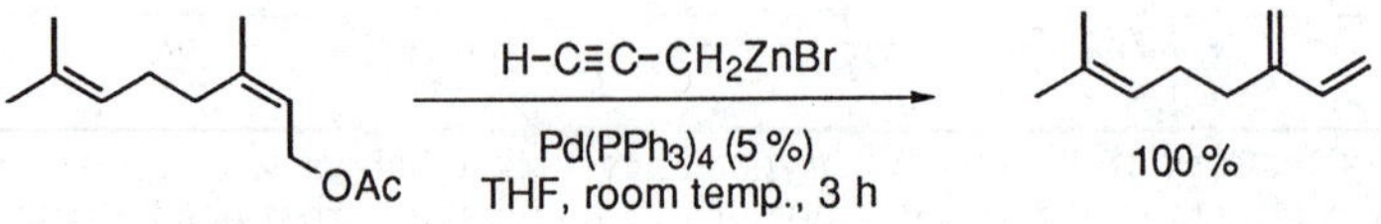

$$\text{geranyl-type allylic acetate (OAc)} \xrightarrow[\text{THF, room temp., 3 h}]{H{-}C{\equiv}C{-}CH_2ZnBr \quad Pd(PPh_3)_4\ (5\ \%)} \text{diene product} \quad 100\ \% \quad (45)$$

Zn reagents with β-sp^3 hydrogens are capable of undergoing the coupling reaction without significant amounts of β-elimination (equation 46) (*111*) and catalysts with the dppf ligand are particularly effective (*112-114*).

$$\text{(alkenyl iodide)} + \text{ClZn(CH}_2)_2\text{CH=CH}_2 \xrightarrow{Pd(PPh_3)_4} \text{(diene product)} \quad 81\,\% \tag{46}$$

Organoboron. The palladium-catalyzed cross-coupling reaction of organic electrophiles with organoboron compounds in the presence of a base (equation 47) is known as the "Suzuki" Reaction (for reviews see references 115-118). This reaction is an extremely important method for the formation of carbon-carbon bonds. High yields of cross-coupled product are obtained in a regiospecific and stereospecific manner. A number of different types of functional groups can be present on either coupling-partner. A wide range of organic halides [including alkyl (*119,120*), alkenyl (*121-123*), 1-alkynyl (*122*), allylic (*123,124*), aryl (*123,125*), benzylic (*124*), and heteroaryl (*126,127*) halides], and triflates (aryl and vinyl) (*128*) have been reported to participate in this reaction. The reaction has the advantage that synthetic procedures and/or methods have already been developed for the synthesis of a wide variety of organoboron compounds via hydroboration of alkynes and alkenes and the reaction of metallated arenes with alkyl borates. Organoboron compounds which couple include alkenyl (*121,122,129,130*), alkyl (*120,131,132*), and aryl boron (*125,126,133,134*) compounds. Sterically hindered boronic acids undergo cross-coupling reactions (see for example *135*). The generally proposed mechanism of the reaction is shown in Figure 8 (*115*). Recent studies have provided important mechanistic information (*136,37*).

$$R{-}X \; + \; R'{-}B(OH)_2 \xrightarrow[\text{Base}]{Pd(0)} R{-}R' \tag{47}$$

The most often used aryl halides in Suzuki reaction are aryl bromides and iodides (*115-118,*). Aryl chlorides do not participate in this cross-coupling reaction except when used in conjunction with electron deficient groups (*138,139*). However, recently promising results have been obtained with aryl chlorides in the Suzuki reaction when a palladacycle catalyst (see equation 20) is used (140).

The development of reaction conditions for the participation of aryl triflates in the Suzuki reaction (*128,141,142*) expands the synthetic utility of this reaction since triflates are readily obtained from phenols (*143*). Due to their inherent base sensitivity and thermally lability, mild reaction conditions have been developed for the cross-coupling reaction of arylboronic acids with triflates. High yields are obtained using weak bases, such as powdered K_3PO_4 or Na_2CO_3, suspended in non-aqueous polar solvents (THF, dioxane) (*128,142*). Alkali metal halides are added to promote the cross-coupling and/or to prevent the premature catalyst decomposition (*128,141*). The utilization of more efficient catalysts such as $PdCl_2$(dppf) also permit this reaction to proceed at lower temperatures (*128*).

Progress has been made in extending the Suzuki reaction from aryl triflates, to other less reactive aryl sulfonates, which show poor reactivity towards oxidative addition to Pd(0). A recent approach to this problem involves the activation of aryl triflates by complexation of electron-withdrawing $Cr(CO)_3$ to the arene moiety (*144*).

Alternative sulfonate leaving groups besides triflate have not been reported to be active in Pd catalyzed Suzuki-type reactions. Aryl mesylates, benzenesulfonates and tosylates are much less expensive than triflates and are usually unreactive towards palladium catalysts. However, in the first example of the use of a Ni-catalyst in the Suzuki reaction, aryl mesylates participated in cross-coupling reactions with arylboronic acids in good yields (equation 48) (*145*).

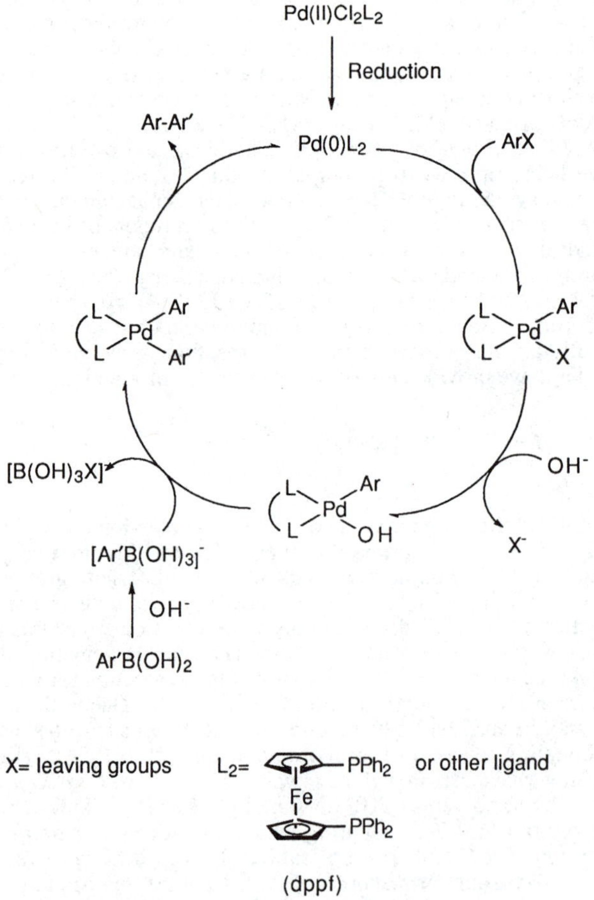

$Pd(II)Cl_2L_2$

Reduction

$Pd(0)L_2$

ArX

Ar-Ar'

$L_2Pd(Ar)(Ar')$

$L_2Pd(Ar)(X)$

OH^-

X^-

$L_2Pd(Ar)(OH)$

$[B(OH)_3X]^-$

$[Ar'B(OH)_3]^-$

OH^-

$Ar'B(OH)_2$

X= leaving groups L_2= PPh_2 Fe PPh_2 or other ligand

(dppf)

Figure 8. Suzuki Reaction Mechanism

OMs + B(OH)$_2$ / OCH$_3$ —[$NiCl_2(dppf)$, Zn; K_3PO_4, dioxane; 100°C, 24h]→ CH_3O–biphenyl 81% (48)

The Pd(0) catalyst precursor can be added to the reaction vessel in its zerovalent or 2+ oxidation states. Examples of typical catalyst precursors are $Pd(PPh_3)_4$, $Pd(dba)_2$, and $PdCl_2(dppf)$. Heterogeneous Pd-catalysts are also effective. For example palladium supported on carbon (Pd/C) is an effective catalyst for the coupling of aryl halides and triflates with phenyl boronic acid (*146,147*).

As in other Pd-mediated homocoupling and cross-coupling reactions, the phosphine ligands can participate in side reactions leading to the formation of phosphonium salts and organo phosphine ligands in which one (or more) of the original organic groups has exchanged with the organic group from the electrophile. These side reactions can be avoided by the use of a phosphine-free Pd-catalyst (*148,149*).

Areneboronic acids undergo a base-catalyzed protodeboronation reaction (*150*) which can lead to significant consumption of the aryl boronic acid under the basic reaction conditions of the Suzuki reaction (for example see reference 151 and references therein). In many reactions the addition of a slight excess of the organoboronic acid is sufficient for controlling this side reaction. In other reactions this problem is controlled by the use of the ester of the boronic acid and an anhydrous base (e.g. K_3PO_4 or Cs_2CO_3) in nonaqueous solvents (e.g. DMF) (*135*).

The ability of aryl and alkenyl boronic acids to oxidatively add to Pd(0) was demonstrated by their participation in cross-coupling reactions with alkenes (*152*). Homocoupled products derived from organoboronic acids have been identified in Suzuki reactions. This reaction was applied to the synthesis of coupled products derived from the homocoupling of organoboroxines (*153*).

Organostannane. Organic electrophiles usually unsaturated halides and triflates undergo a Pd catalyzed coupling reaction with organotin reagents in a reaction known as the Stille reaction (equation 49) (for reviews see references 154-156).

$$R-X + R''_3SnR' \xrightarrow{Pd(0)} R-R' + R''_3SnX \quad (49)$$

Although the tin is bonded to four organic groups, one group can be selectively transfered when the R' groups are alkyl (typically methyl or n-butyl) and R" is another type of group. This reaction is widely used because of many desirable features. High yields are obtained under mild reaction conditions. The reaction is not extremely air sensitive and tolerates water. The reaction can be performed in aqueous solvent mixtures (*157*). A variety of types of organic electrophiles can be utilized including the following: vinyl epoxides, halides and triflates; aryl and heteroaryl halides, triflates and arene sulfonates (*158,159*); acid chlorides; allyl halides and acetates; benzyl halides; a-haloketones and a-haloesters (unless noted otherwise see references *154* to *156*, for a review of vinyl triflates see *160*). Even less reactive substrates such as sterically hindered aryl triflates with electron-donating groups undergo high yield coupling reactions (161,162). The organotin reagents are relatively easy to prepare. A number of synthetic procedures are available for the preparation of highly functionalized organotin reagents. Many different types of groups are efficiently transfered. The organotin reagents are robust can be purified by conventional purification techniques such as distillation, column chromatography and recrystallization. A large number of different types of functional groups can be present on either coupling partner, including ester, nitrile, ketone, alcohol, nitro, aldehyde, epoxide and carboxylic acid groups. The coupling reaction is regioselective (allyl halides), stereospecific (sp^3 carbons), and proceeds with retention of geometry (vinyl halides, vinyl tin).

The most commonly proposed mechanism is shown in Figure 9. This mechanism involves the same basic steps (oxidative addition, transmetallation, isomerization, and reductive elimination) as discussed previously. It is important to note that the mechanism shown in Figure 6 is more consistent with the recent developments of the Stille reaction. The exact mechanism followed may vary depending on the exact substrates and reaction conditions.

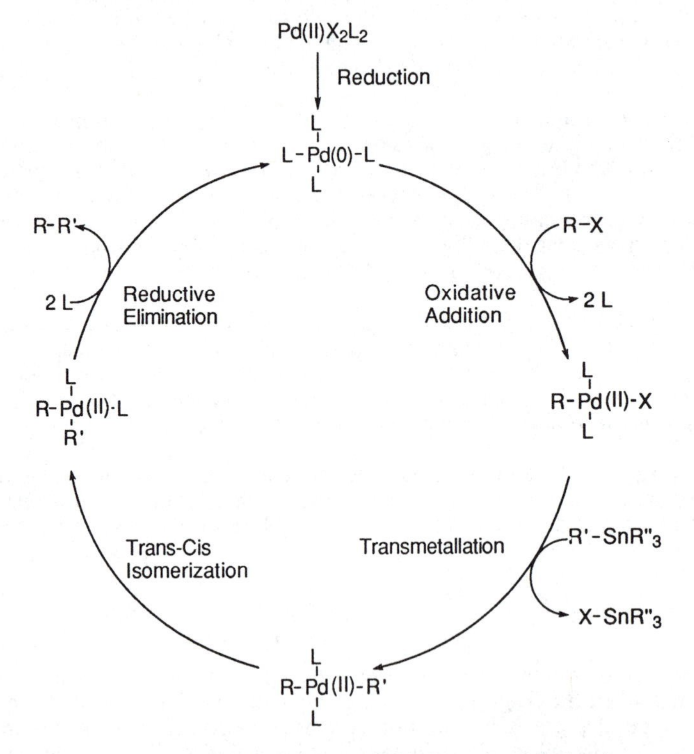

Figure 9. Commonly Proposed Stille Reaction Mechanism

A carbonylative Stille coupling reaction occurs in the presence of carbon monoxide (*155,163*). The CO insertion occurs after oxidative addition and before the transmetallation step (Figure 10).

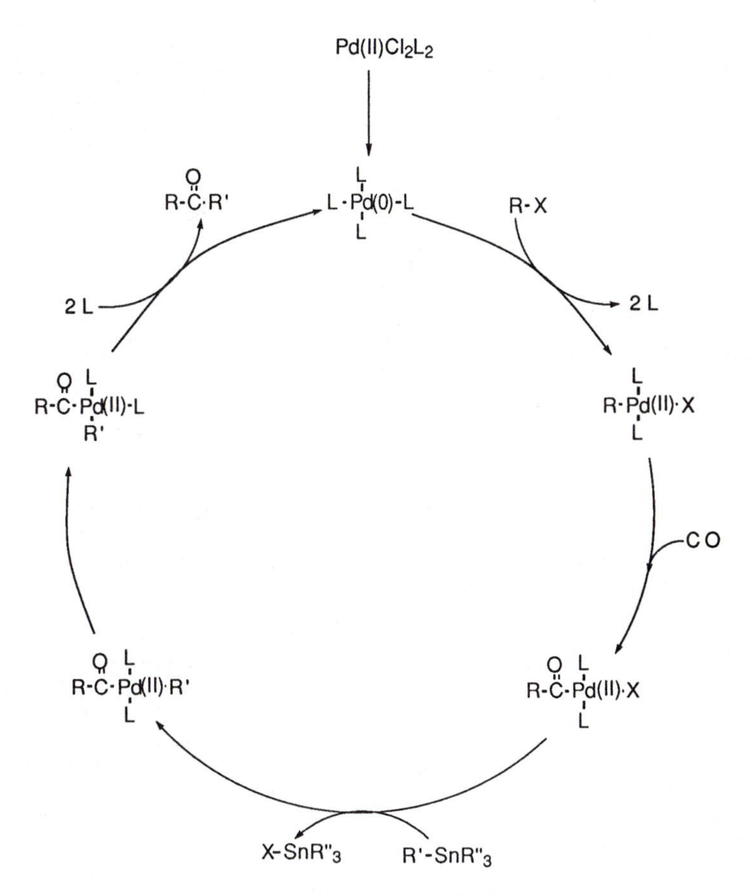

Figure 10. Carbonylative Stille Reaction

Both Pd(0) and Pd(II) catalyst precursors are effective. The Pd(II) is reduced to Pd(0) prior to entering the catalytic cycle. Triphenylphosphine (TPP) ligated catalysts such as $Pd(PPh_3)_4$,$(PhCH_2)PdCl(PPh_3)_2$, and $PdCl_2(PPh_3)_2$ are effective for many coupling reactions. In some cases substrates which are either unstable at the required temperatures or which have low reactivity with triphenylphosphine based catalysts have been coupled by modifying the ligands.

Rate accelerations of up to 10^3 were obtained with tri-2-furylphosphine or triphenylarsine ligands in cross-coupling reactions involving olefinic stannanes and organic electrophiles (*164*). The rate and yields of coupling products of arylstannanes and olefinic triflates were increased with triphenylarsine ligands (*165*). $AsPh_3$ is soft ligand and ligand dissociation occurs more easily. High yields of cross-coupled product have also been obtained under mild conditions when "ligandless" catalysts (*166*) were used in which solvent molecules functioned as weak donating ligands (*167*).

The addition of cocatalytic amounts of CuI (or other Cu(I) salts) facilitates the efficient cross-coupling of many organic electrophiles and organostannanes which are otherwise difficult to couple and in many other reactions accelerates the rate or increases the yield (*168-175*). One effect of Cu is to scavenge free ligand when strong ligands such as PPh_3 are present in ethereal solvents, thus increasing the rate of the transmetallation step. In highly polar solvent CuI reacts with organostannane to provide an organo-copper species which can participate in a transmetallation reaction to give the cross-coupled product (Figure 11) (*175*).

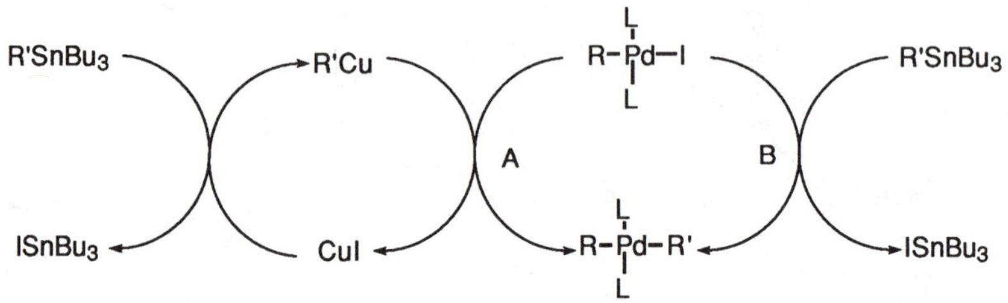

R'SnBu3 → R'Cu; ISnBu3 ← CuI; R–Pd(L)2–I; A; B; R–Pd(L)2–R'; R'SnBu3 → ISnBu3

Figure 11. The Effect of CuI in Highly Polar Solvents.

A side product can be formed from the cross-coupling of a phenyl group from triphenyl phosphine and the organotin reagent. In coupling reactions of aryl halides, the amount of side product formed is related to the electronic properties of the substituents. The lowest amount of side product is formed with strongly electron-withdrawing groups. Increased amounts are formed with electron-donating groups (*176*). In this case, a mechanism involving the reversible Pd-catalyzed formation of tetraaryl phosphonium salts was proposed. The oxidative addition of the phenyl-phosphorous bond of this phosphonium salt to Pd followed by transmetallation and reductive elimination would produce the side product. A different mechanism was proposed for the methyl-phenyl exchange which occurred in competition with the transmetallation reaction when *trans*-$CH_3Pd(PPh_3)_2I$ is treated with (*p*-anisyl)tributyltin. This exchange was found to occur in benzene without the participation of a free phosphonium ion and without phosphine dissociation (*177*).

Transition-Metal Mediated Polymerizations

Polymer Forming Homocoupling Reactions. The high yield homocoupling of aryl halides and sulfonates has been extensively applied to the synthesis of aromatic polymers. Two types of aromatic polymers are prepared by this method. The first type contains heteroatom linkages between aromatic groups and the second type contains only aromatic groups in the backbone.

In the first example of the application of the homocoupling reaction of aryl chlorides to polymer synthesis, high molecular weight polyarylether sulfones were prepared by the Ni(0) catalyzed homocoupling reaction of 4,4'-bis(p-chlorophenoxy)diphenylsulfone [equations 50 (*178*) and 51 (*179*)].

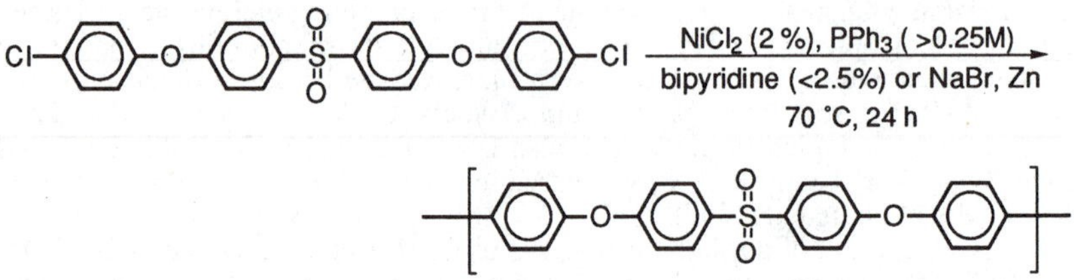

$NiCl_2$ (2 %), PPh_3 (>0.25M); bipyridine (<2.5%) or NaBr, Zn; 70 °C, 24 h

RV = 0.81 dL/g (NMP, 25°C) (50)

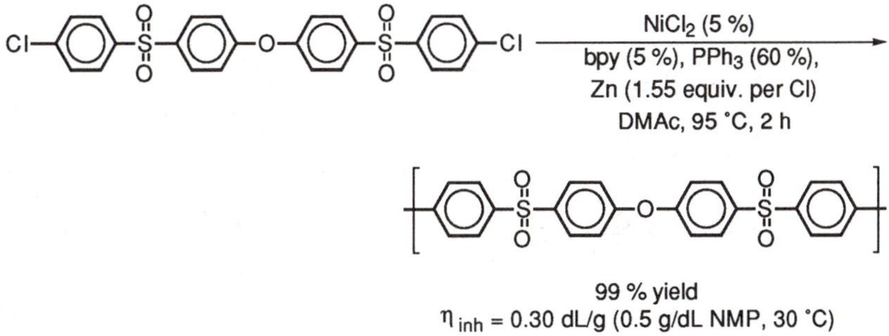

NiCl2 (5 %)
bpy (5 %), PPh3 (60 %),
Zn (1.55 equiv. per Cl)
DMAc, 95 °C, 2 h

99 % yield
η_{inh} = 0.30 dL/g (0.5 g/dL NMP, 30 °C)

(51)

The Ni(0) catalyst was generated by the reduction of $NiCl_2$ with Zn in the presence of PPh_3 and bipyridine in DMAc. Reaction conditions were chosen to minimize side reactions such as the reduction of ArCl to ArH and transarylation from triphenylphosphine. The most significant variable was the concentration of PPh_3. The reaction temperature also had a large effect on the molecular weight of the polymer. As the temperature increased the number of side reactions increased. The amount of Zn present was also very important.

Polymers with heteroaryl backbones could also be prepared by this method (*178,180*). 2,5-Dichlorothiophene was polymerized by this method (equation 52) (*178*). High molecular weights were not obtained due to premature precipitation of the polymer. The transarylation side reaction (equation 9) was inhibited by the presence of bipyridine and excess PPh_3 (vs. $NiCl_2$).

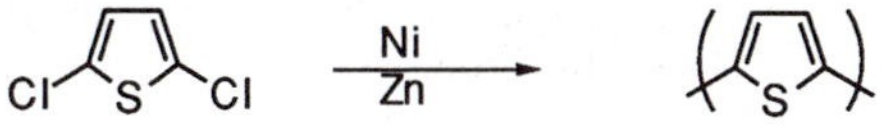

Ni / Zn

(52)

Almost simultaneous with the report of the preparation of polyarylether sulfones, aromatic polyether ketones were prepared via the Ni(0) catalyzed polymerization of aromatic ether ketones with chloro groups at each terminus (equation 53) (*181*). The reaction proceeds rapidly under mild conditions.

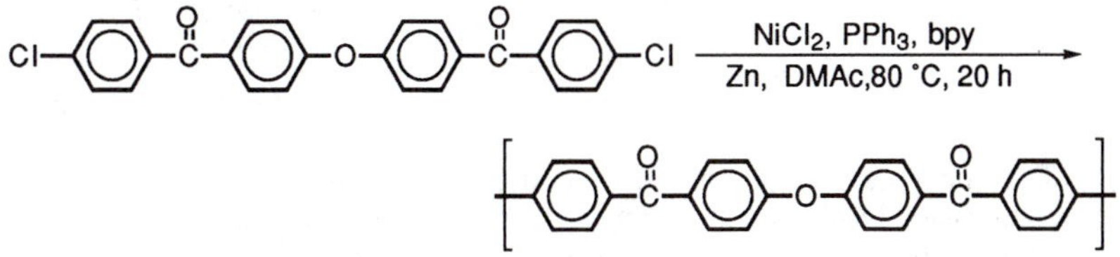

NiCl2, PPh3, bpy
Zn, DMAc,80 °C, 20 h

(53)

Molecular weights were limited by the solubility of the crystalline poly(ether ketone)s and by reduction and transarylation side reactions (*181*).

The Ni(0) catalyzed homocoupling reaction of aryl halides and sulfonates is also an excellent method for the preparation of high molecular weight soluble poly(*p*-phenylene)s and poly(arylene)s. The development of this methodology began with the synthesis of soluble functionalized PPP and polyarenes from substituted dichlorobenzene (*28,29,31*) and the triflates (*28-31,182*) of hydroquinones and bisphenols (equations 54 and 55).

R R Cl Cl Ni(0) (54)

CO_2CH_3 HO OH Tf_2O pyridine TfO OTf $NiCl_2(PPh_3)_2$, Zn Et_4NI, THF 67 °C, 1 h CO_2CH_3 n (55)

The polymerization of substituted dichlorobenzenes (equation 56) has been extensively utilized for the synthesis of PPP (*182-190*) and other related polymers (*183,187,189,191,192*).

CO_2CH_3 Cl Cl $NiCl_2(PPh_3)_2$, Zn Et_4NI, THF 67 °C, 1 h CO_2CH_3 n (56)

In one approach to the synthesis of soluble PPP, 2,7-dibromo-4,9-di-n-octyl-4,5,9,10-tetrahydropyrene was polymerized by a Ni(0) catalyst (equation 57) to give a polymer with an average DP of 40 (GPC vs. polystyrene standards) which corresponds to 80 phenyl units per chain (*193*). The ethandiyl bridges and octyl groups (R = octyl) increased the solubility of the polymer without reducing the conjugation between adjacent phenyl groups. Side reactions were decreased by having no substituent groups ortho to the bromo leaving groups of the monomer.

R Br Br R Ni(COD), 2,2'-bipyridyl COD, toluene, DMF R n R DP = 40 (57)

The polymerization of aryl bistriflate monomers derived from hydroquinones and bisphenols is important from a synthetic perspective. Synthetic procedures and/or methods for the preparation of a variety of substituted hydroquinones and bisphenols already exist. In most cases, these compounds can be easily converted in high yields to the corresponding bistriflates. However, the cost of the reagents necessary for this conversion has limited the application of this methodology. Reaction conditions for the polymerization of aryl bismesylates, which are much less expensive sulfonates, have recently been developed for the synthesis of high molecular weight soluble PPPs (equation 58) (*194*).

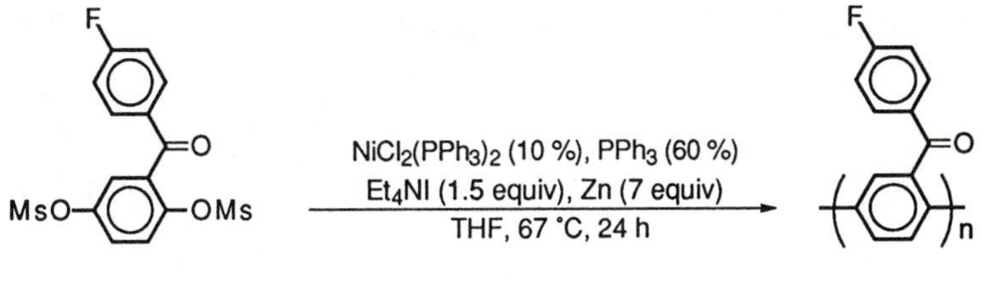

$NiCl_2(PPh_3)_2$ (10 %), PPh_3 (60 %)
Et_4NI (1.5 equiv), Zn (7 equiv)
THF, 67 °C, 24 h

DP = 101 (58)

Examples of the preparation of polythiophenes by the Ni(0) catalyzed polymerization of dihalothiophenes are shown in equations 59 (*159*) and 60 (*196*).

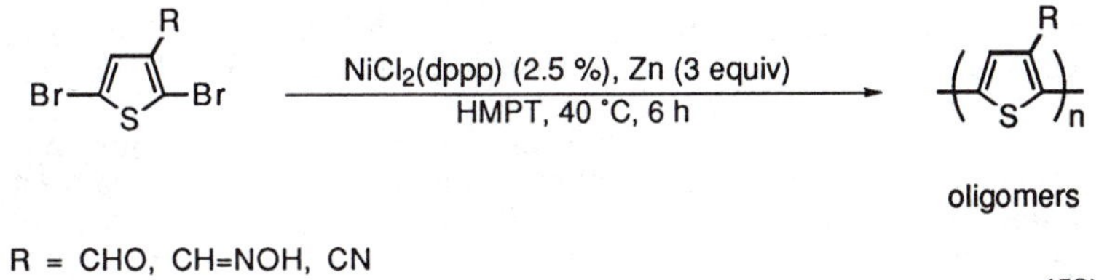

$NiCl_2$(dppp) (2.5 %), Zn (3 equiv)
HMPT, 40 °C, 6 h

oligomers

R = CHO, CH=NOH, CN

(59)

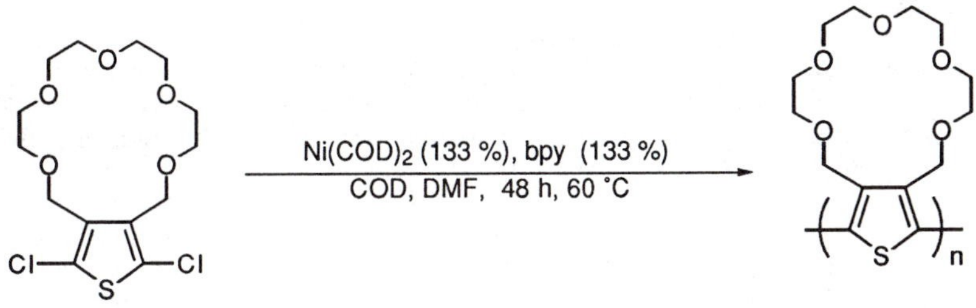

$Ni(COD)_2$ (133 %), bpy (133 %)
COD, DMF, 48 h, 60 °C

$M_w = 7.4 \times 10^5$
DP = 2460
(60)

Heck Reaction Polymerizations. The Heck reaction is readily applicable to the synthesis of poly(arylenevinylene)s and poly(heteroarylvinylene)s. Poly(1,4-phenylenevinylene)s have been prepared with both AA/BB-monomer pairs and AB-monomers. The AA/BB-monomer pairs are typically dihalogenated arenes and ethylene (*197*) or divinylarenes (*198*). Bromostyrene (equation 61) is an example of an AB-monomer (*199*). The molecular weight of the polymer obtained by the polymerization of 4-bromostyrene (equation 61) was limited by its poor solubility (*199*).

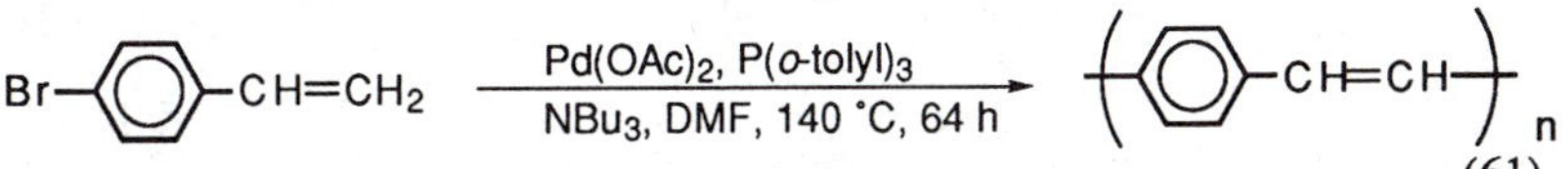

Br–C_6H_4–CH=CH_2 $\xrightarrow[\text{NBu}_3\text{, DMF, 140 °C, 64 h}]{\text{Pd(OAc)}_2\text{, P(}o\text{-tolyl)}_3}$ –(C_6H_4–CH=CH)$_n$– (61)

The attachment of substituent groups to the polymer backbone (*197,200,201*) resulted in more soluble polymers with higher molecular weights (equation 62)

(*197*). Similar polymers with alkyl or poly(oxyethylene) groups as spacers in the main chain have also been prepared by this method (*202*).

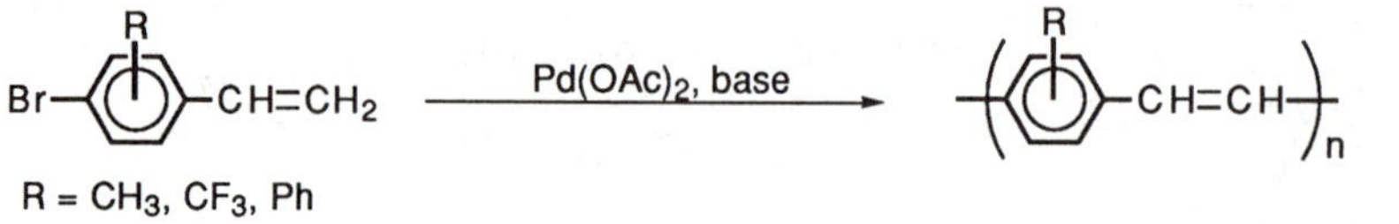

Br–C$_6$H$_3$(R)–CH=CH$_2$ —Pd(OAc)$_2$, base→ –[C$_6$H$_3$(R)–CH=CH]$_n$–

R = CH_3, CF_3, Ph

(62)

The preparation of soluble polymers with high molecular weights allowed the identification of structural irregularities produced by side reactions. The number of these defects was a function of several variables including reaction conditions and monomer structure. The most favorable reaction conditions in the reaction of dibromoarenes with ethylene or vinylarenes involved the use of low catalyst concentrations and low temperatures. Monomers containing COCl leaving groups were prone to side reactions (*203*).

Side reactions which limit the molecular weight of the polymer, by producing nonreactive end-groups, are reduction of the aryl halide and reactions involving the phosphine cocatalyst (*203*). These side reactions can also be minimized by keeping the reaction temperature and amount of catalyst as low as possible.

Heterogeneous catalyst are also effective for the preparation of polymers by the Heck reaction. Palladium-graphite (Pd-Gr) was used for the synthesis of polycinnamamide from N,N'-(3,4-oxydiphenylene)bis(acrylamide) and bis(4-iodophenyl) ether (equation 63). No phosphine or bipyridyl ligand was necessary. The presence of tri-*o*-tolylphosphine actually inhibited the reaction. The molecular weight M_w increased gradually to approximately 9×10^4 over 20 hours and then remained constant (*204*).

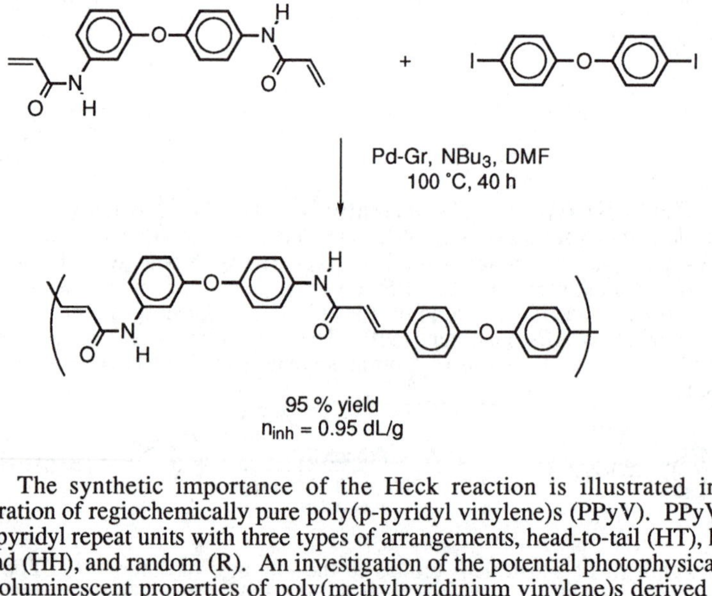

Pd-Gr, NBu_3, DMF
100 °C, 40 h

95 % yield
η_{inh} = 0.95 dL/g

(63)

The synthetic importance of the Heck reaction is illustrated in the preparation of regiochemically pure poly(p-pyridyl vinylene)s (PPyV). PPyV can have pyridyl repeat units with three types of arrangements, head-to-tail (HT), head-to-head (HH), and random (R). An investigation of the potential photophysical and electroluminescent properties of poly(methylpyridinium vinylene)s derived from PPyV necessitated the preparation of regiochemically pure polymers of these three types. In this case the Heck reaction was used to produce the HT isomer (equation 64) and the Stille reaction was used to produce the other two isomers (*205*).

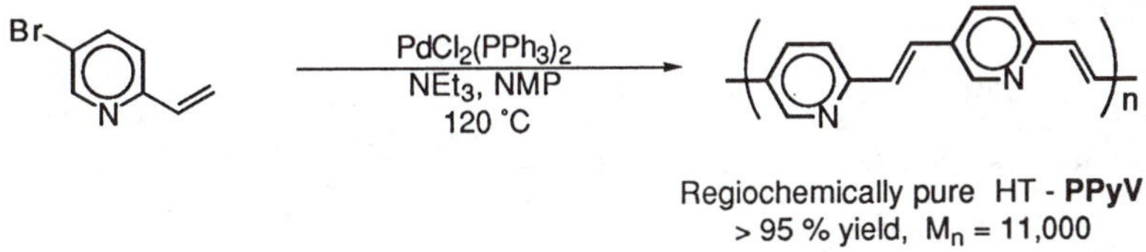

Br
N
PdCl2(PPh3)2
NEt3, NMP
120 °C
N
N
n
Regiochemically pure HT - PPyV
> 95 % yield, Mn = 11,000

(64)

The Heck reaction was also used to synthesize the polymers shown in equations 65 (*71*), 66 (*206*), and 67 (*207*).

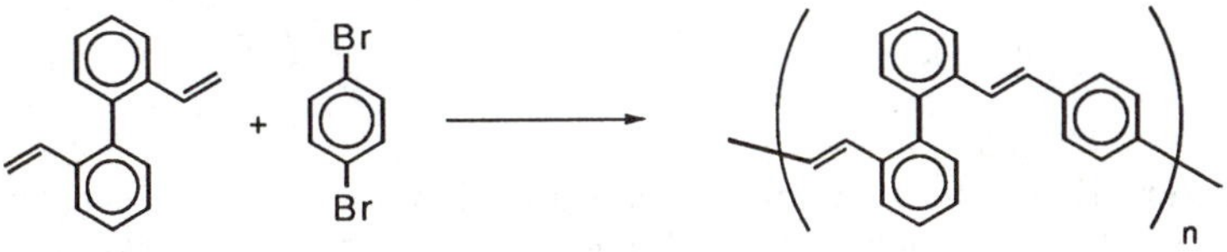

Br
Br
n

(65)

Br
Br
Pd(OAc)2
P(o-tolyl)3
NEt3
n
DP = 8

(66)

C6H13
C6H13
N
N
M
N
N
C6H13
C6H13
OC9H19
I
I
OC9H19
Pd(OAc)2, P(o-tolyl)3
NBu3, DMF, 100 °C, 3 - 4 h
n
M = H2, Zn, Cu, Ni

(67)

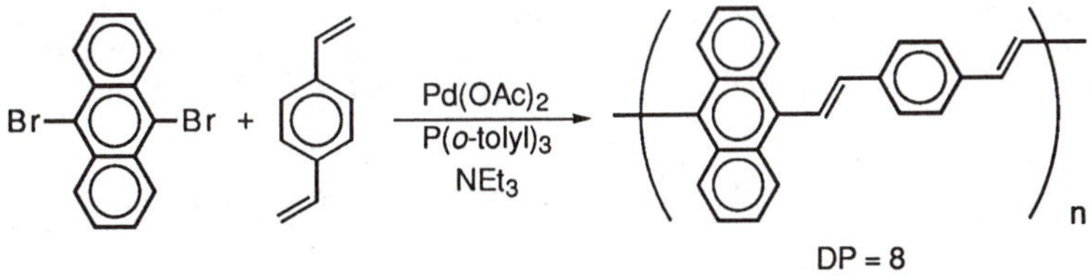

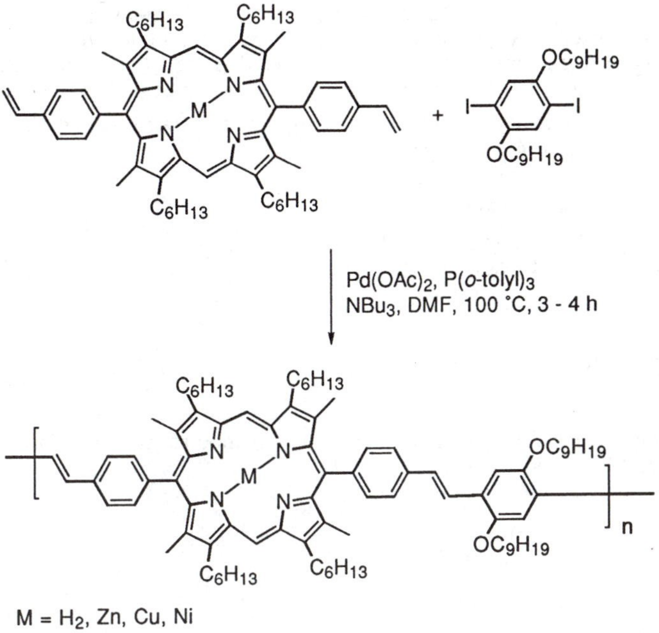

Polymers From Cross-Coupling Reactions of Alkynes. Examples of the preparation of polymers and oligomers from a cross-coupling reactions involving alkynes are shown in equation 68 (*208*) and Figure 12 (*209,210*).

$$\text{H-C}\equiv\text{C-C}_6\text{H}_2(\text{OR}^1)_2\text{-C}\equiv\text{C-H} + \text{Br-C}_6\text{H}_2(\text{OR}^2)_2\text{-Br} \xrightarrow[\text{or Pd(PPh}_3)_4\text{, CuI, NH}_3\text{, toluene}]{\text{PdCl}_2(\text{PPh}_3)_2\text{, CuI, PPh}_3} \left[\text{C}\equiv\text{C-C}_6\text{H}_2(\text{OR}^1)_2\text{-C}\equiv\text{C-C}_6\text{H}_2(\text{OR}^2)_2\right]_n \quad (68)$$

Polymerization of Organomagnesium Compounds. The coupling reaction of organic electrophiles with organomagnesium reagents is applicable to the synthesis of PPP. The first application of this reaction was the polymerization of 1,4-dibromobenzene (equation 69) (*211*). Stoichiometric amounts of Mg were added to form 4-bromomagnesium bromide, this substrate was polymerized by various transition metal catalysts including $NiCl_2(bpy)$ and $NiCl_2(PPh_3)_2$.

$$\text{Br-C}_6\text{H}_4\text{-Br} \xrightarrow[\text{2. NiCl}_2\text{(bpy), 67 °C, 5 h}]{\text{1. Mg, THF}} \left(\text{C}_6\text{H}_4\right)_n \quad (69)$$

This method was also used for the polymerization of disubstituted 1,4-dibromobenzene (equation 70) (*212*). This produced a polymer with a regioregular structure connected in a 1,4-manner. The maximum DP obtained was 13. Two types of end groups were identified (a) unreacted Br and (b) reduced Br. The proton for reduction may be from adventious water sources or from Ni-catalyzed cleavage of the α-CH bond of THF. The choice of reaction solvent is limited by the solubility of the Grignard reagent.

$$\text{Br-C}_6\text{H}_2\text{R}_2\text{-Br} \xrightarrow[\text{2. NiCl}_2\text{(bipy) (0.5 - 1.5\%) reflux, 48 h}]{\text{1. Mg (1.0 equiv) THF, reflux, 1h}} \text{X-C}_6\text{H}_2\text{R}_2\text{-}\left[\text{C}_6\text{H}_2\text{R}_2\right]_n\text{-C}_6\text{H}_2\text{R}_2\text{-X}$$

R = n-hexyl, n-octyl   bipy = 2,2'-bipyridyl   X = H or Br

(70)

Other examples of the synthesis of PPPs by this method are provided in equations 71 (*213*), 72 (*210*), and 73 (*210*).

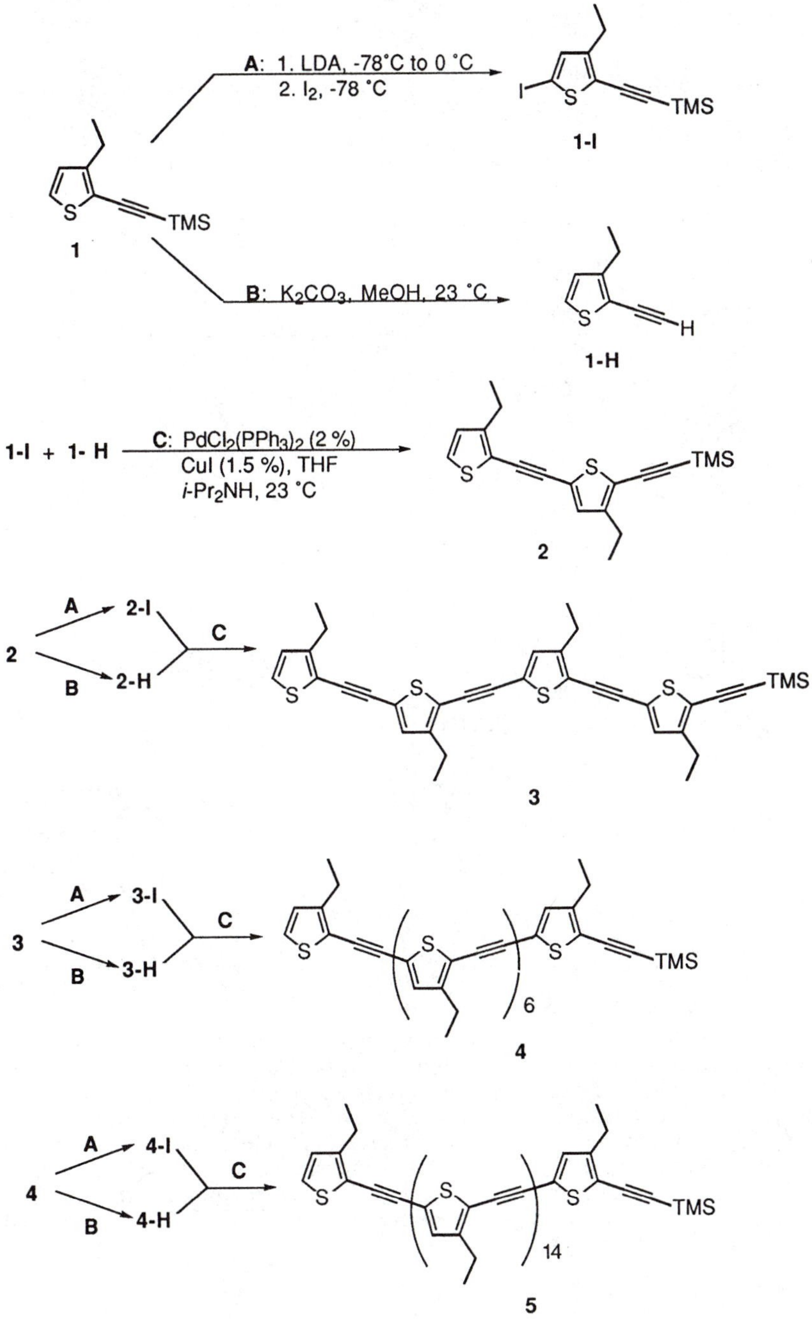

A: 1. LDA, -78 °C to 0 °C
2. I_2, -78 °C

TMS

1-I

TMS

1

B: K_2CO_3, MeOH, 23 °C

H

1-H

1-I + 1-H

C: $PdCl_2(PPh_3)_2$ (2 %)
CuI (1.5 %), THF
i-Pr_2NH, 23 °C

TMS

2

2 A → 2-I
2 B → 2-H
C

TMS

3

3 A → 3-I
3 B → 3-H
C

TMS

6

4

4 A → 4-I
4 B → 4-H
C

TMS

14

5

Figure 12. Step-wise Preparation of Oligo(Ethynyl Thiophene).

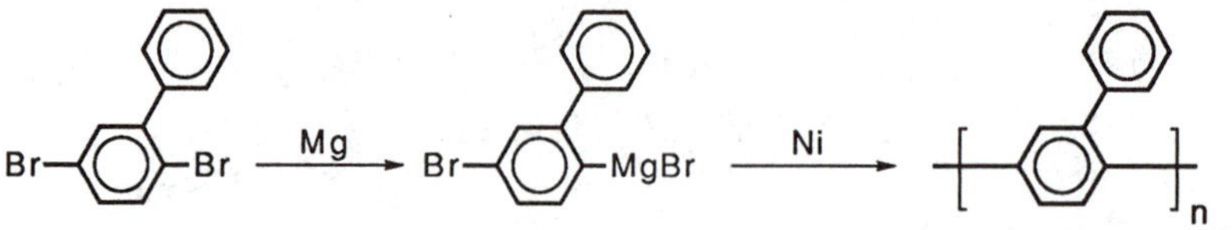

Mg → Ni → (71)

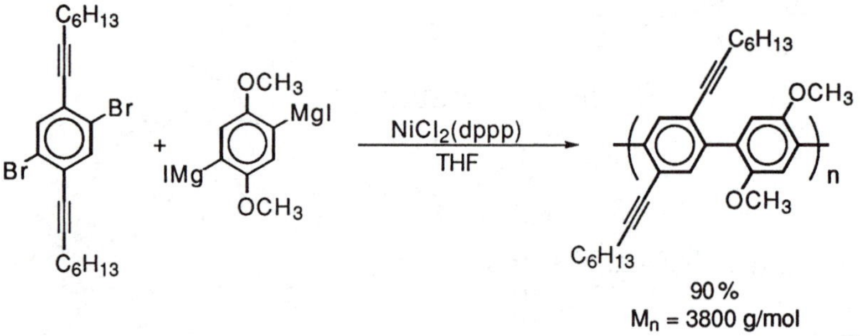

NiCl$_2$(dppp), THF → 90 %, M_n = 3800 g/mol (72)

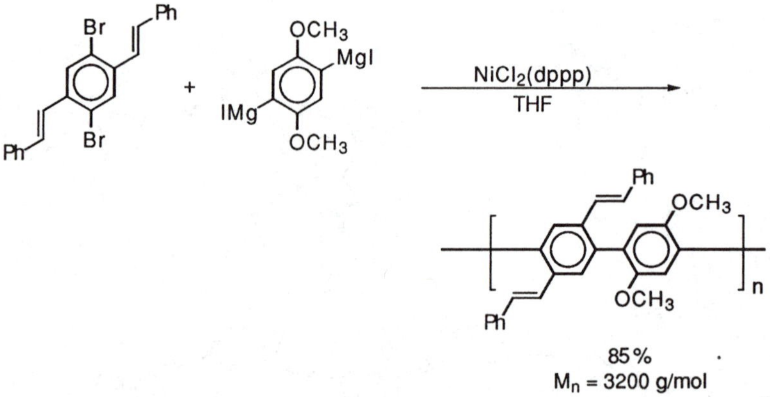

NiCl$_2$(dppp), THF → 85 %, M_n = 3200 g/mol (73)

Poly(thiophene)s have also been prepared from the coupling reaction of Grignard reagents with organic electrophiles (*214-220*). Poly(3-alkylthiophene)s can be prepared with a high degree of regioselectivity, to give highly ordered polymers with almost exclusive head-to-tail (HT) coupling (equation 74) (*218,220*). There are four possible structural arrangements for each triad of thiophene units (Figure 13). The polymer structure was determined to consist of 91% HT-HT coupling sequences when a dodecyl group was attached (R = $C_{12}H_{25}$). The molecular weight was determined by both GPC (M_w = 11,600, DP = 46) and NMR (DP = 64).

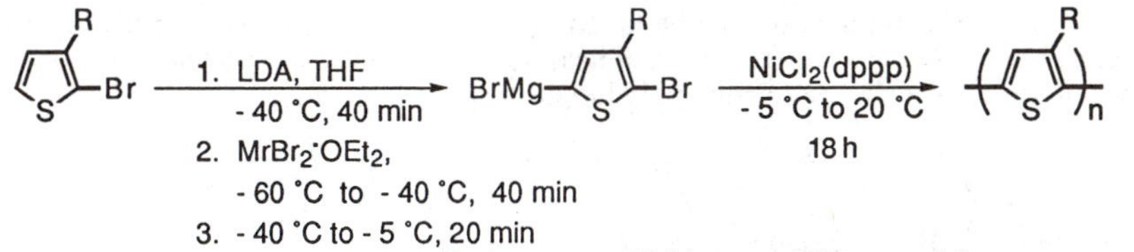

1. LDA, THF, - 40 °C, 40 min
2. $MrBr_2 \cdot OEt_2$, - 60 °C to - 40 °C, 40 min
3. - 40 °C to - 5 °C, 20 min

$NiCl_2(dppp)$, - 5 °C to 20 °C, 18 h

R	Yield	HT	M_w
n-butyl	69	93 %	
n-hexyl	36	98 %	10,000*
n-octyl	65	97 %	24,424
n-docecyl	33	95 %	

*of THF soluble fraction

(74)

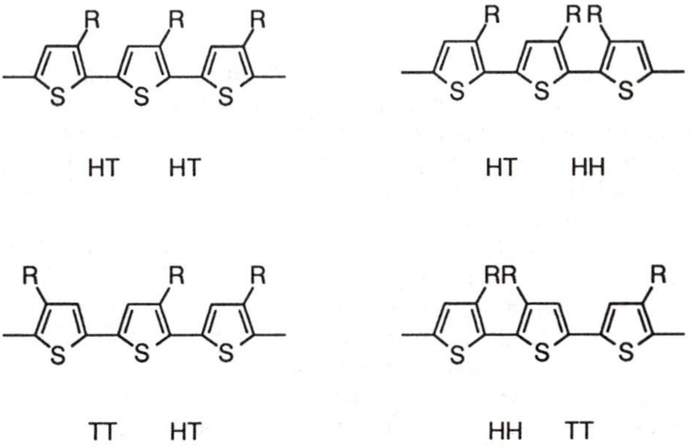

HT HT

HT HH

TT HT

HH TT

Figure 13. Four Possible Structural Arrangements for each Triad of Monosubstituted Thiophene Units.

The key to the regioselective synthesis is the regiospecific metallation of the 2-alkyl-1-bromothiophene monomer with LDA at the 5-position with no scrambling prior to the coupling reaction (*220,221*).

Polythiophenes with etheric side chains can also be synthesized (eq 75) (*222*).

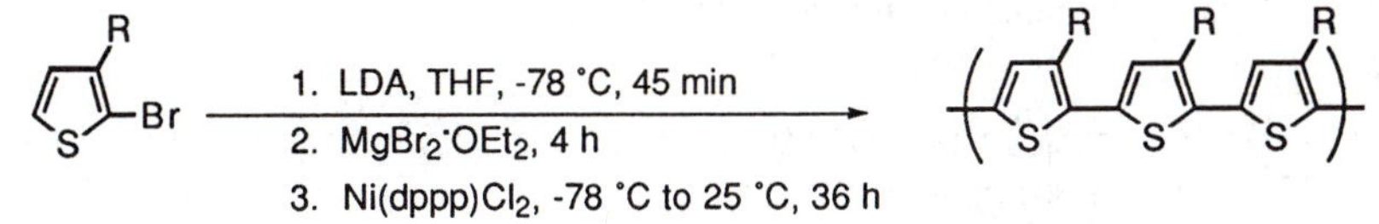

1. LDA, THF, -78 °C, 45 min
2. $MgBr_2 \cdot OEt_2$, 4 h
3. $Ni(dppp)Cl_2$, -78 °C to 25 °C, 36 h

R = $CH_2OCH_2CH_2OCH_2CH_2OCH_3$
$CH_2OCH_2CH_2OCH_3$
CH_2OCH_3
CH_2SCH_3

For R = $CH_2OCH_2CH_2OCH_2CH_2OCH_3$
65 % yield
M_w = 71,000
DP = 160
99 % HT

(75)

The polymer's conjugation can be tuned by mixing the copolymers appropriately (equation 76) (*222*). The formation of head-to-tail linkages is important in regard to obtaining electrically conducting polythiophenes.

$$\text{BrMg–(3-}R^1\text{-thienyl)–Br} + \text{BrMg–(3-}R^2\text{-thienyl)–Br} \xrightarrow[\text{2. } NiCl_2(dppp)\ (0.5\ \%),\ 25\ ^\circ C,\ 12\ h]{\text{1. } NiCl_2(dppp)\ (0.5\ \%),\ 25\ ^\circ C,\ 14\ h} \left[(\text{3-R-thiophene-2,5-diyl})_x (\text{3-R-thiophene-2,5-diyl})_y \right]_n$$

$R_1 = C_{12}H_{25}$;
$R_2 = C_6H_{13}$, CH_3, or H

(76)

Polymerization of Organozinc Monomers. The Ni(0)-catalyzed cross-coupling reaction of heteroarylhalides and organozinc compounds has been applied successfully to the synthesis of substituted poly(thiophene)s with very high degrees of regioregularity. Polymers with regioregularities of approximately 98.5% head-to-tail (HT) are obtained in four hours at 67 °C with a $NiCl_2$(dppe) catalyst [dppe = bis(diphenylphosphino)ethane] from a 9:1 mixture of 2-bromo-5-(bromozincio)-3-hexylthiophene and 2-(bromozincio)-5-bromo-3-hexylthiophene generated with Rieke Zinc (Zn*) (Figure 14) (*224*). The higher regioregularity than would be expected from a 9:1 mixture of monomers was rationalized by a rapid equilibrium reaction between the two monomers (equation 77). The regioselective control was rationalized on the basis of steric congestion at the reductive elimination step. Upon changing to the less sterically demanding triphenylphosphine ligand or upon changing the metal catalyst to the larger metal Pd, the regioselectivity decreased. A completely random structure was obtained using a $Pd(PPh_3)_4$ catalyst.

$$\text{BrZn–(3-}(CH_2)_6H\text{-thienyl)–Br} \underset{}{\overset{\text{Ni}}{\rightleftharpoons}} \text{Br–(3-}(CH_2)_6H\text{-thienyl)–ZnBr} \quad (77)$$

The Pd-catalyst was also effective for the polymerization of 1,4-diiodobenzene (equation 78), 2,5-dibromothiophene (equation 79) and several 3-alkyl-2,5-dibromothiophenes (equation 80) (*225*).

$$\text{I–}C_6H_4\text{–I} \xrightarrow[\text{THF, 30 min}]{\text{Zn*}} \text{IZn–}C_6H_4\text{–I} \xrightarrow[67\ ^\circ C,\ 4\ h]{Pd(PPh_3)_4\ (0.2\ \%)} \left(C_6H_4 \right)_n \quad (78)$$

$$\text{Br–(thienyl)–Br} \xrightarrow[\text{THF}]{\text{Zn*}} \text{BrZn–(thienyl)–Br} \xrightarrow[67\ ^\circ C,\ 4\ h]{Pd(PPh_3)_4\ (0.2\ \%)} \left(\text{thiophene-2,5-diyl} \right)_n \quad (79)$$

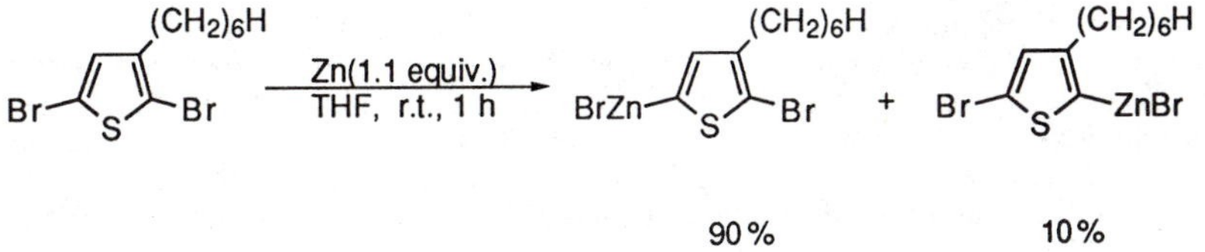

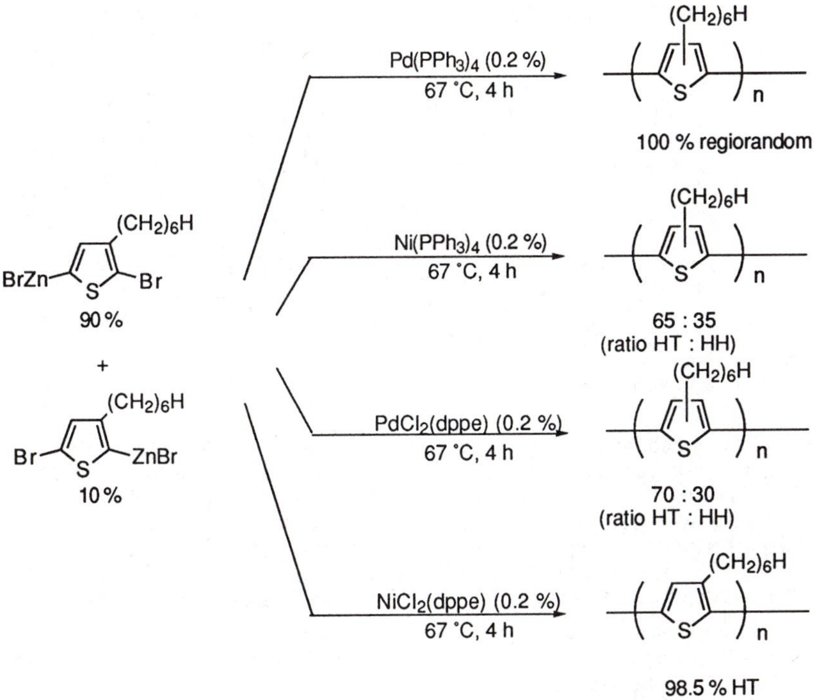

Figure 14. Effect of Metal and Ligands on Regioselectivity

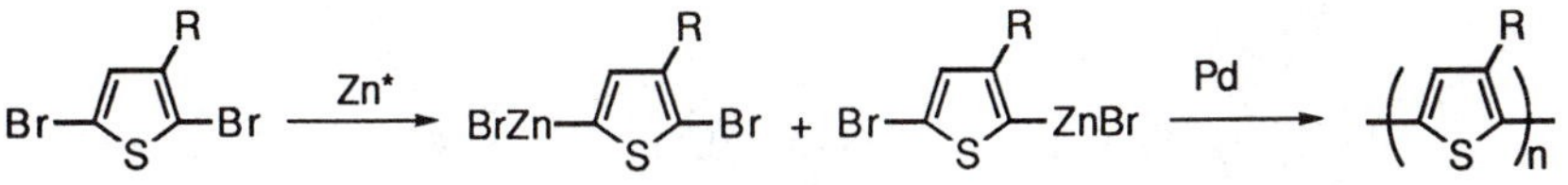

-R	Ratio			HT
$-CH_3$	80	:	20	98 %
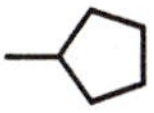 cyclopentyl	71	:	29	96 %
-Ph	66	:	34	98 %

(80)

The preparation of HT regioselective thiophene polymers results in an improvement of their electroconductivity, nonlinear optical and magnetic properties. Improved selectivity of the reaction of Zn* with 2,5-dibromo-3-alkyl thiophenes was achieved by performing the reaction at -78 °C and warming to 0 °C (equation 81) (*226*). The reaction of 3-alkyl-2-bromo-5-iodothiophene with Zn* at 0 °C resulted in the quantitative formation of 3-alkyl-2-bromo-5-(iodozincio)thiophene (equation 82).

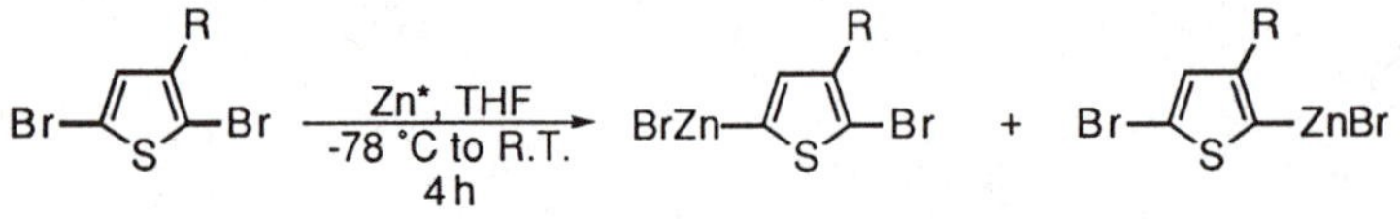

Zn*, THF; -78 °C to R.T.; 4 h

R	Ratio
n-butyl	94 : 6
n-hexyl	97 : 2
n-octyl	98 : 2

R	Ratio
n-decyl	98 : 2
n-dodecyl	98 : 2
n-tetradecyl	98 : 2

(81)

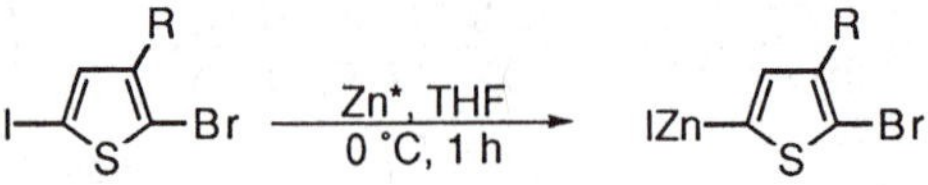

Zn*, THF; 0 °C, 1 h

> 99 %

(82)

The mechanism of polymer formation involves oxidative addition, transmetallation (or disproportionation) and reductive elimination. Reductive elimination was the rate determining step in the coupling reaction. The degree of stereoregularity was suggested as being controlled by sterics in the transmetallation (or disproportionation) step. The structure was determined by four different signals of the 4-proton of thiophene for the four triads possible (HT-HT, TT-HT, HT-HH, TT-HH). Alternatively two different diads (H-T and H-H) could be distinguished from the resonances of the α or β methylene protons. When the alkyl group was butyl, 97 % of the linkages were head-to-tail. For the longer alkyl groups greater than 98.5 % were head-to-tail.

The preparation of regioregular poly(3-alkylthio)thiophene has also been accomplished by this method (equation 83). The introduction of an alkylthio substituent which is directly bonded to the 3-position of the polythiophene provides a way to tune the electrical and physical properties of the polymer (*227*).

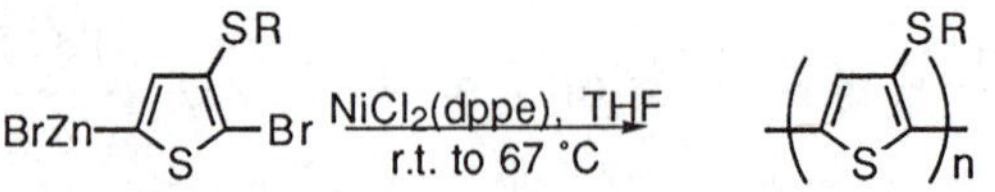

$NiCl_2$(dppe), THF; r.t. to 67 °C

> 90 % HT-HT (83)

The coupling reaction of organic-zinc compounds and organic bromides was also used to prepare a conducting polymer containing synthetic molecular receptors (Figure 15). This was the first example of the use of molecular recognition for producing reversible changes in the electroconductivity of a polymer (*228*). Polythiophene based pseudopolyrotaxanes have also been prepared (*229*).

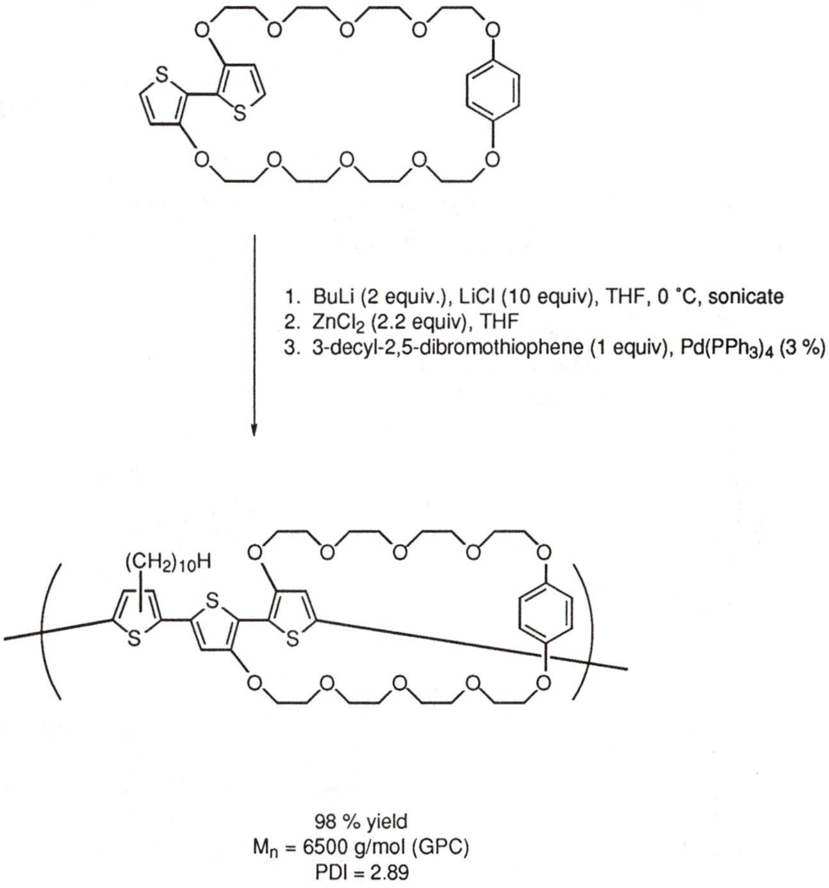

1. BuLi (2 equiv.), LiCl (10 equiv), THF, 0 °C, sonicate
2. $ZnCl_2$ (2.2 equiv), THF
3. 3-decyl-2,5-dibromothiophene (1 equiv), $Pd(PPh_3)_4$ (3 %)

$(CH_2)_{10}H$

98 % yield
M_n = 6500 g/mol (GPC)
PDI = 2.89

Figure 15. Preparation of Conducting Polymer Containing Synthetic Molecular Receptors

Suzuki Reaction Polymerizations. The Pd-catalyzed cross-coupling reaction of an organic electrophile and an organoboron compound in the presence of base has been extensively applied to the synthesis of polyphenylenes and related polymers. Well defined structures are obtained as a result of the regiospecificity of the Suzuki reaction (84) (*230*). The solubility of the polymers is commonly enhanced by the presence of alkyl or other types of substituents [for example see equations 85-87 (*231,232,233* respectively and *234*)].

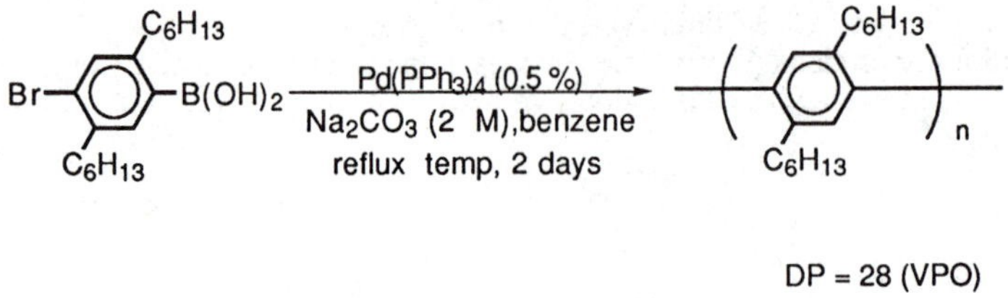

$$Br-C_6H_2(C_6H_{13})_2-B(OH)_2 \xrightarrow[\text{reflux temp, 2 days}]{Pd(PPh_3)_4\ (0.5\ \%),\ Na_2CO_3\ (2\ M),\ benzene} -[C_6H_2(C_6H_{13})_2]_n-$$

DP = 28 (VPO)

(84)

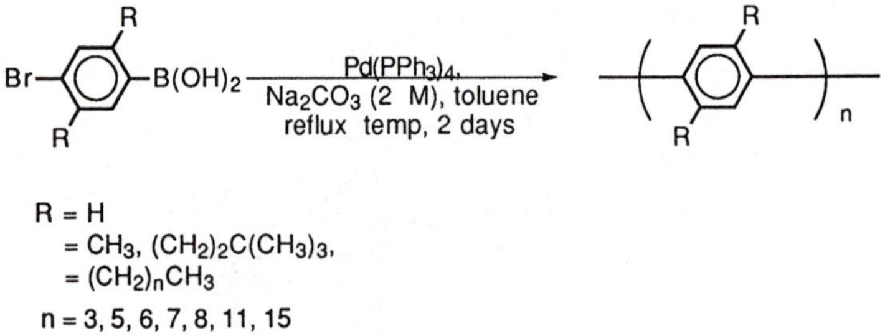

$$Br-C_6H_2R_2-B(OH)_2 \xrightarrow[\text{reflux temp, 2 days}]{Pd(PPh_3)_4,\ Na_2CO_3\ (2\ M),\ toluene} -[C_6H_2R_2]_n-$$

R = H
= CH_3, $(CH_2)_2C(CH_3)_3$,
= $(CH_2)_nCH_3$
n = 3, 5, 6, 7, 8, 11, 15

(85)

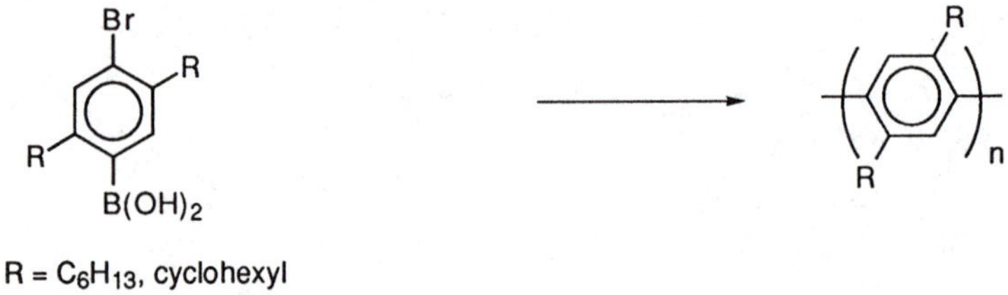

$$Br-C_6H_2R_2-B(OH)_2 \longrightarrow -[C_6H_2R_2]_n-$$

R = C_6H_{13}, cyclohexyl

(86)

$$Br-C_6H_2(CH_2O(CH_2)_4H)_2-B(OH)_2 \xrightarrow{Pd} -[C_6H_2(CH_2O(CH_2)_4H)_2]_n-$$

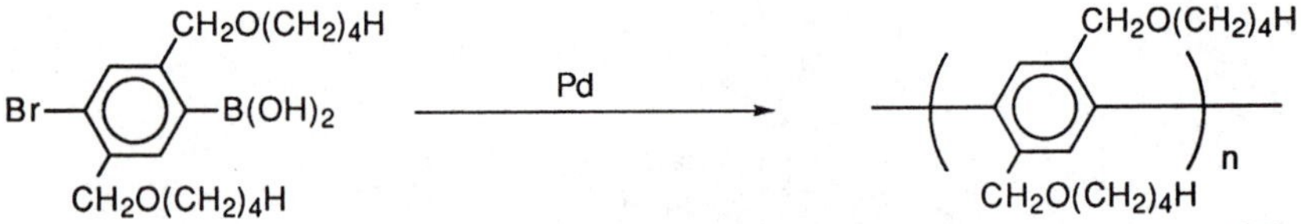

(87)

Statistical copolymers are obtained when two different AB monomers are copolymerized (equation 88) (*231*).

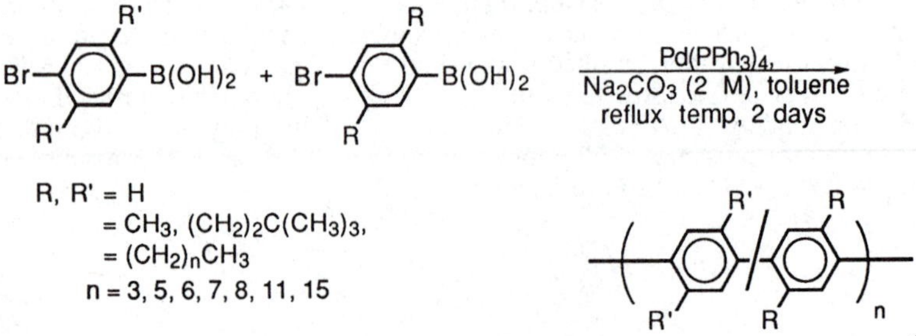

$$Br-C_6H_2R'_2-B(OH)_2 + Br-C_6H_2R_2-B(OH)_2 \xrightarrow[\text{reflux temp, 2 days}]{Pd(PPh_3)_4,\ Na_2CO_3\ (2\ M),\ toluene} -[C_6H_2R'_2 / C_6H_2R_2]_n-$$

R, R' = H
= CH_3, $(CH_2)_2C(CH_3)_3$,
= $(CH_2)_nCH_3$
n = 3, 5, 6, 7, 8, 11, 15

(88)

Alternating copolymers are obtained by the polymerization of AA/BB monomers with different substituent groups [equations 89 (*235*) and 90 (*232*)].

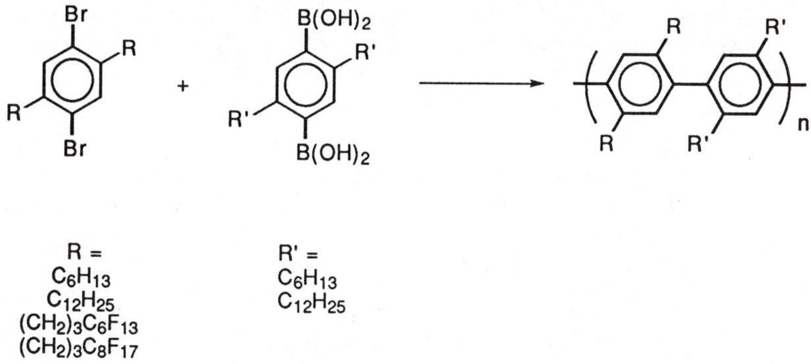

Br, R, R, Br + B(OH)$_2$, R', R', B(OH)$_2$ → R, R', R, R', n

R = C_6H_{13}, $C_{12}H_{25}$, $(CH_2)_3C_6F_{13}$, $(CH_2)_3C_8F_{17}$

R' = C_6H_{13}, $C_{12}H_{25}$

(89)

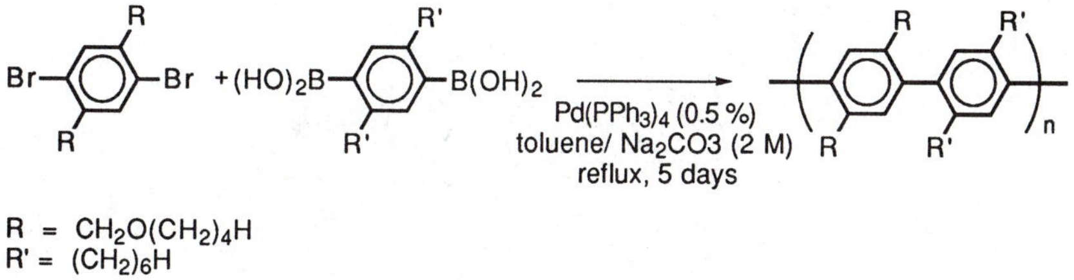

Br, R, R, Br + $(HO)_2B$, R', R', $B(OH)_2$ → R, R', R, R', n

$Pd(PPh_3)_4$ (0.5 %)
toluene/ Na_2CO3 (2 M)
reflux, 5 days

R = $CH_2O(CH_2)_4H$
R' = $(CH_2)_6H$

(90)

A polymer with random configurational isomerism is obtained in the polymerization shown in equation 91 (*233*).

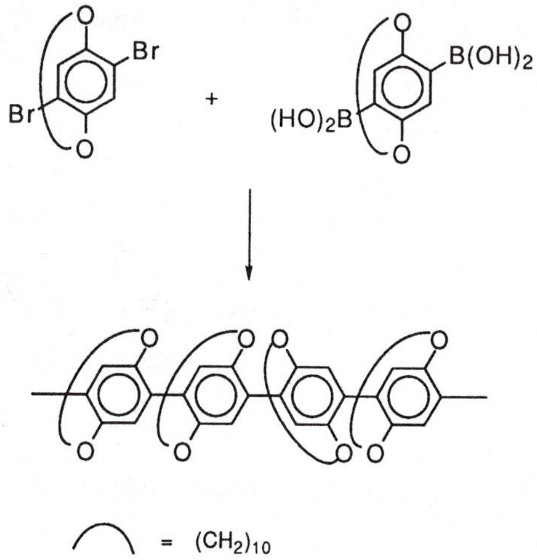

O, Br, Br, O + O, $B(OH)_2$, $(HO)_2B$, O

= $(CH_2)_{10}$

(91)

A variety of groups can be part of the main chain. The solubility of the polymer can also be enhanced by the presence of oxygen linkages (or other appropriate groups) in the main chain (equation 92) (*236*). A large number of functionalities can be inserted in the polymer since the Suzuki reaction tolerates them. The groups can be transformed in a subsequent step. For example, carbonyl groups are present in the reaction shown in equation 92.

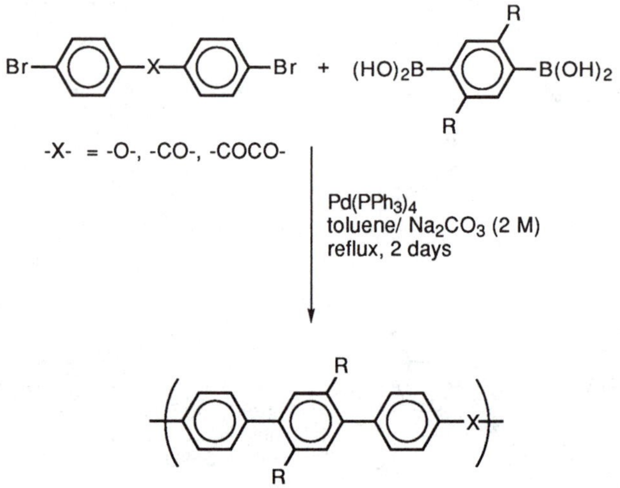
R
Br—X—Br + (HO)2B—B(OH)2
R
-X- = -O-, -CO-, -COCO-
Pd(PPh3)4
toluene/ Na2CO3 (2 M)
reflux, 2 days
R
X
R

(92)

Polymers incorporating polyimide groups can also be prepared by the Suzuki reaction (equation 93) (*237*).

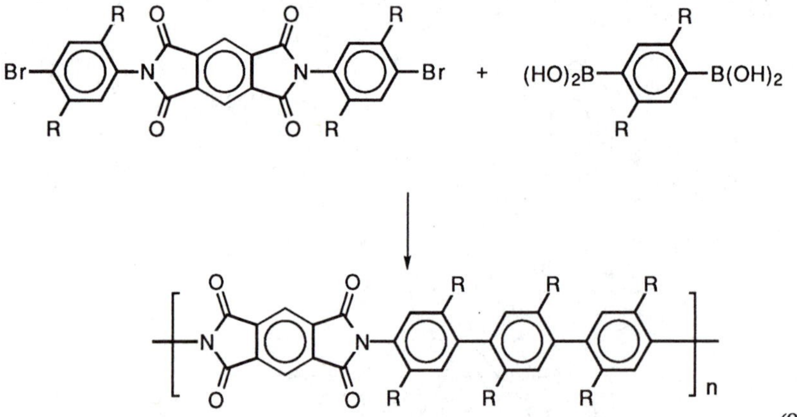
R O O R R
Br N N Br + (HO)2B B(OH)2
R O O R R
O O R R R
N N
O O R R R
n

(93)

Carboxylate groups can also be present. Water soluble (as the sodium salt) PPP was synthesized by the Suzuki reaction as shown in equation 94 (*238*).

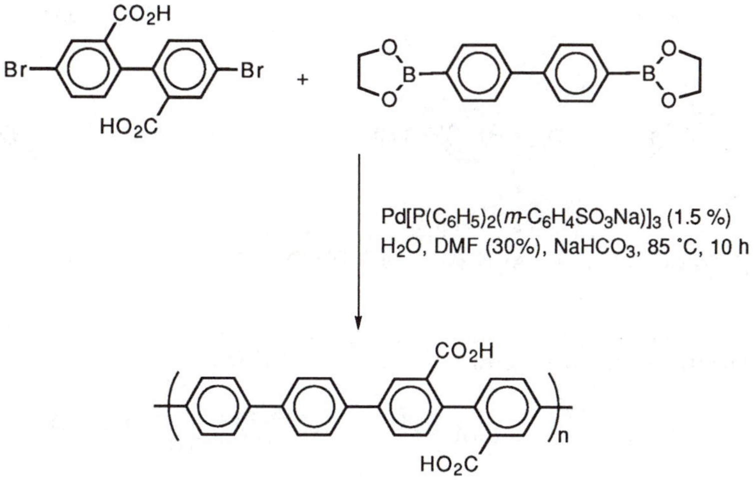

$Pd[P(C_6H_5)_2(m\text{-}C_6H_4SO_3Na)]_3$ (1.5 %)
H_2O, DMF (30%), $NaHCO_3$, 85 °C, 10 h

M_W = 50,000 g/mol (vs. DNA)
by poly(acylamide) gel electrophoresis

(94)

Polymers with sulfonate groups have also been prepared by the two methods shown in equations 95 (*239*) and 96 (*240*).

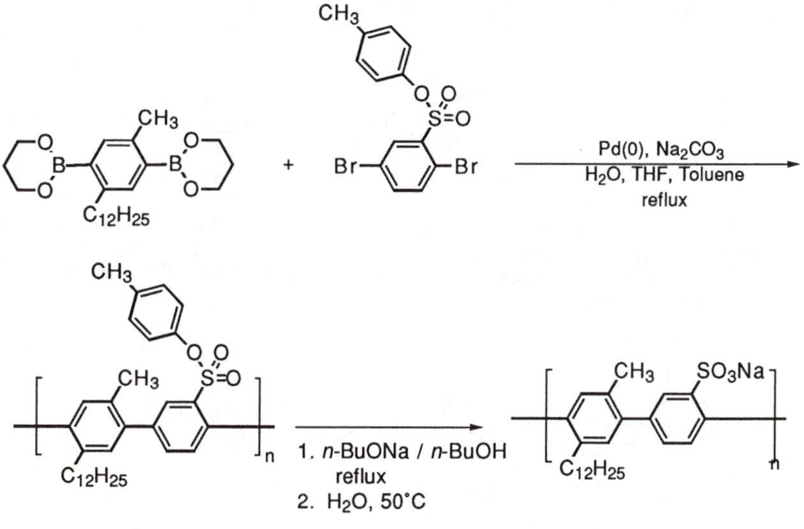

Pd(0), Na_2CO_3
H_2O, THF, Toluene
reflux

1. *n*-BuONa / *n*-BuOH
reflux
2. H_2O, 50°C

M_n = 36,200 (VPO)
M_n = 75,000 (GPC)

(95)

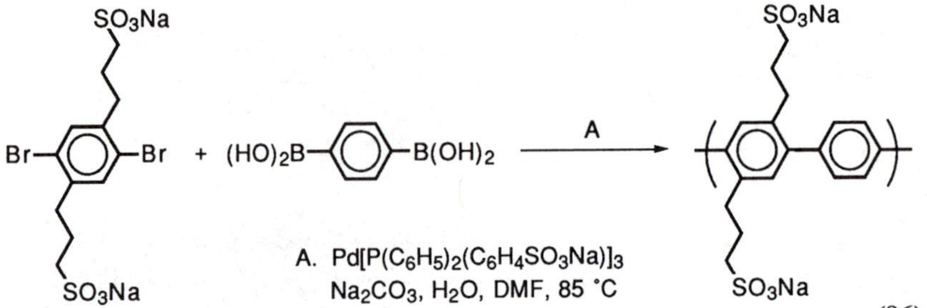

SO3Na
Br
Br
+ (HO)2B
B(OH)2
A
SO3Na
A. Pd[P(C6H5)2(C6H4SO3Na)]3
Na2CO3, H2O, DMF, 85 °C
SO3Na
SO3Na

(96)

Alkylboronates and aryl halides have been cross-coupled in the polymerization reactions shown in equations 97-99 (*241*).

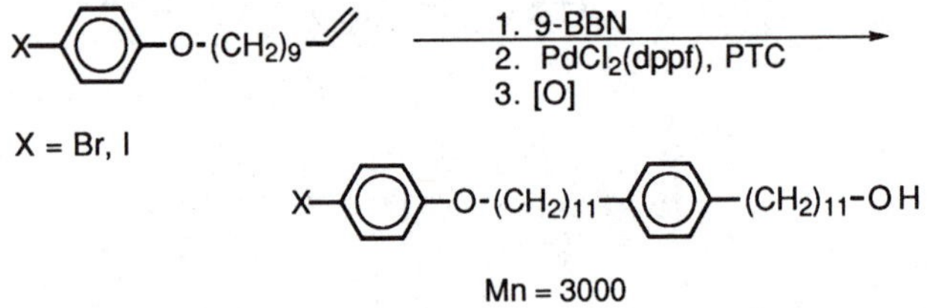

X
O-(CH2)9
1. 9-BBN
2. PdCl2(dppf), PTC
3. [O]
X = Br, I
X
O-(CH2)11
(CH2)11-OH
Mn = 3000

(97)

Br
1. 9-BBN
2. PdCl2(dppf), PTC
3. [O]
n

(98)

1. 9-BBN
2. 1,4-dibromobenzene
PdCl2(dppf), PTC
3. [O]
(CH2)8
n

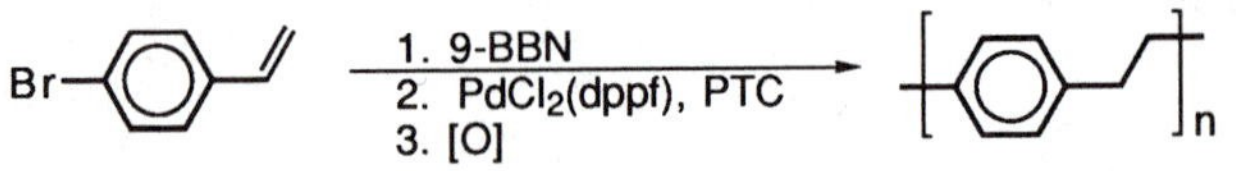

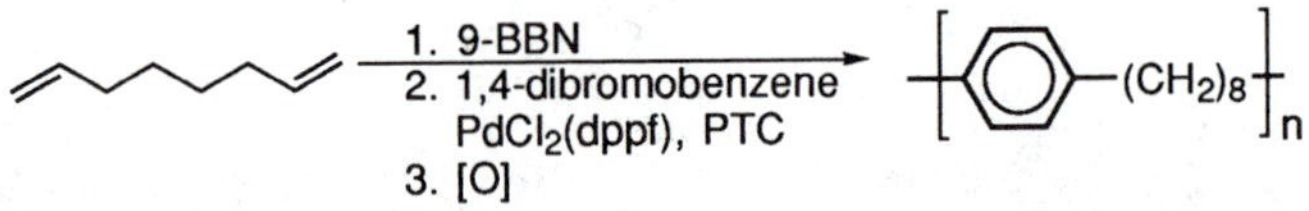

(99)

The variety of functional groups which can be present, and the regioselectivity of the reaction make this reaction the key step in the synthesis of well defined ladder-type polymers as shown in equations 100, 101 and 102 (*242,243,244* respectively).

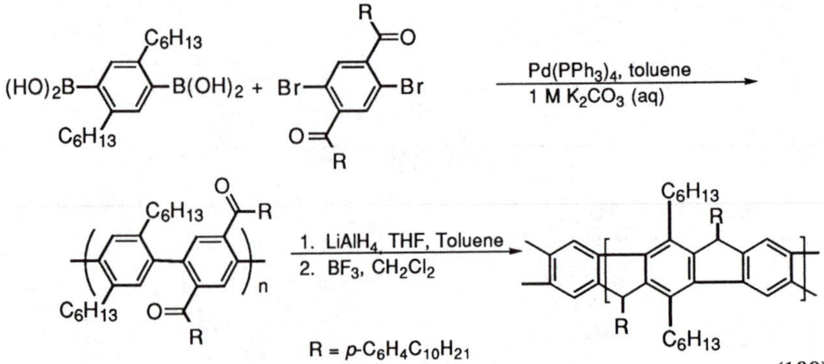

C6H13
(HO)2B
B(OH)2 + Br
Br
C6H13
R
O
O
R
Pd(PPh3)4, toluene
1 M K2CO3 (aq)
C6H13
O
R
n
C6H13
O
R
1. LiAlH4, THF, Toluene
2. BF3, CH2Cl2
R = p-C6H4C10H21
C6H13
R
R
C6H13

(100)

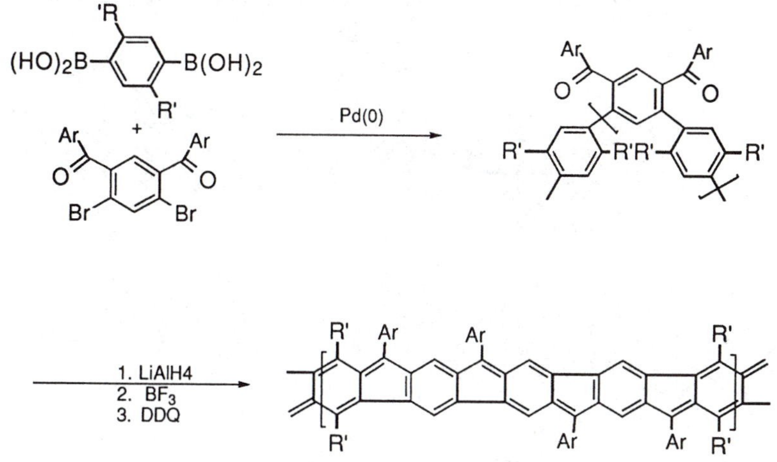

'R

$(HO)_2B$ — $B(OH)_2$

R'

\+

Ar Ar

O O

Br Br

Pd(0)

Ar Ar

O O

R' R'R' R'

1. $LiAlH_4$
2. BF_3
3. DDQ

R' Ar Ar R'

R' Ar Ar R'

R = $-C_6H_{13}$

Ar = p-$C_6H_4-C_{10}H_{21}$

(101)

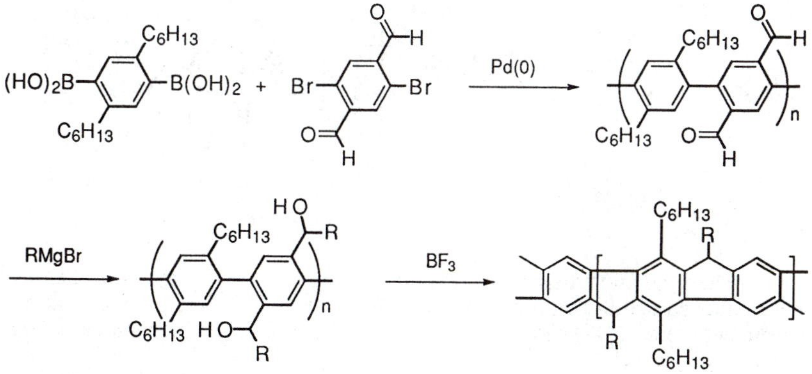

C_6H_{13}

$(HO)_2B$ — $B(OH)_2$ + Br — Br

C_6H_{13}

O H

O H

Pd(0)

C_6H_{13} O H

C_6H_{13} O H

n

RMgBr

H O

C_6H_{13} R

C_6H_{13} H O

R

n

BF_3

C_6H_{13}

R

R

C_6H_{13}

R = C_6H_5, $C_{10}H_{21}$, or p-$C_6H_4N(CH_3)_2$

(102)

Other examples are shown in equations 103 (*245*) and 104 (*246*).

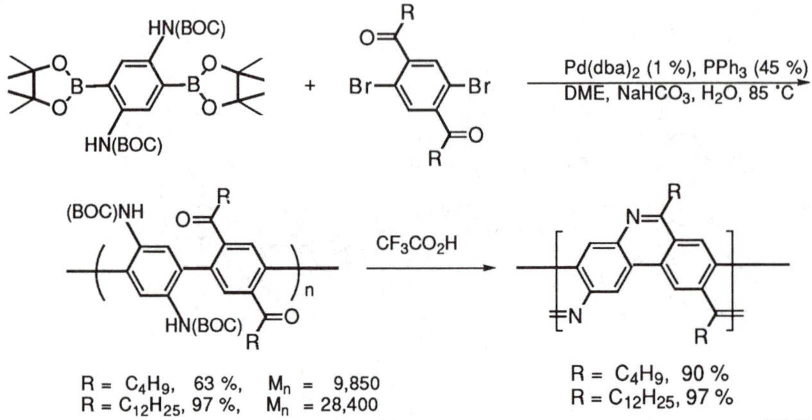

Pd(dba)$_2$ (1 %), PPh$_3$ (45 %)
DME, NaHCO$_3$, H$_2$O, 85 °C

CF$_3$CO$_2$H

R = C$_4$H$_9$, 63 %, M$_n$ = 9,850
R = C$_{12}$H$_{25}$, 97 %, M$_n$ = 28,400

R = C$_4$H$_9$, 90 %
R = C$_{12}$H$_{25}$, 97 %

(103)

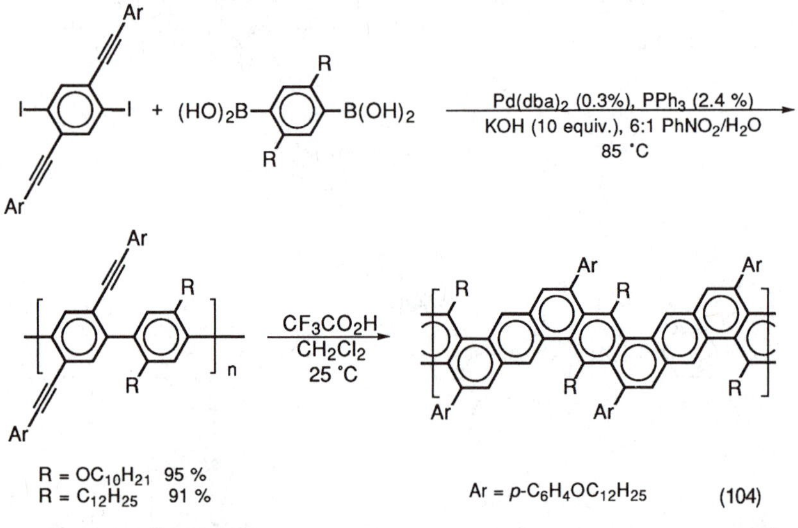

Pd(dba)$_2$ (0.3%), PPh$_3$ (2.4 %)
KOH (10 equiv.), 6:1 PhNO$_2$/H$_2$O
85 °C

CF$_3$CO$_2$H
CH$_2$Cl$_2$
25 °C

R = OC$_{10}$H$_{21}$ 95 %
R = C$_{12}$H$_{25}$ 91 %

Ar = p-C$_6$H$_4$OC$_{12}$H$_{25}$ (104)

The synthetic utility of the Suzuki reaction for polymer synthesis has also been demonstrated through the synthesis of polymers with Frechet-type dendritic substituent groups (*247*) and through progress made toward the preparation of large oligophenylene cycles (*248*).

The side reaction responsible for the coupling of phenyl groups from triphenylphosphine with the organic electrophile (see *148*) has also been identified as being capable of incorporating phosphorous into poly(p-phenylene)s synthesized by the Suzuki reaction (*249*).

Stille Reaction Polymerizations. Although the Stille reaction has not been as widely applied to polymer synthesis as the Suzuki reaction, a number of different monomer pairs can participate in the reaction to give a variety of polymer types (equation 105) (*250*).

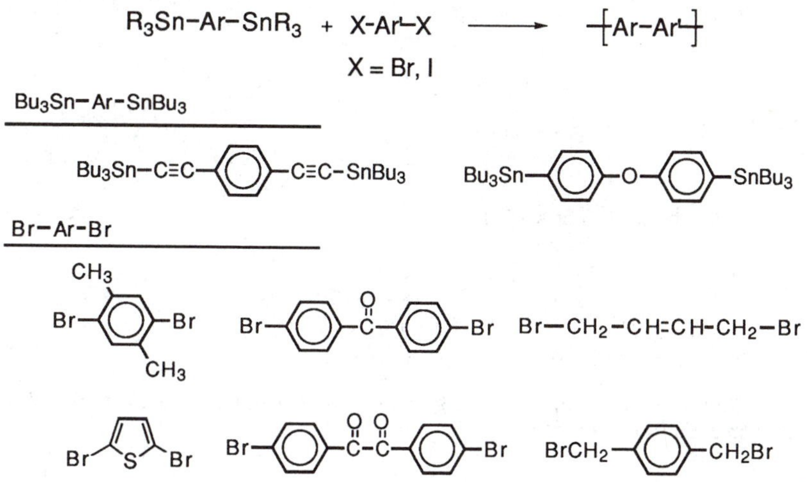

$$R_3Sn\text{-}Ar\text{-}SnR_3 + X\text{-}Ar'\text{-}X \longrightarrow \text{-}[Ar\text{-}Ar']\text{-} \quad X = Br, I$$

$Bu_3Sn\text{-}Ar\text{-}SnBu_3$

$Bu_3Sn\text{-}C{\equiv}C\text{-}C_6H_4\text{-}C{\equiv}C\text{-}SnBu_3$ $\quad Bu_3Sn\text{-}C_6H_4\text{-}O\text{-}C_6H_4\text{-}SnBu_3$

Br-Ar-Br

CH_3, Br, Br, CH_3 (dibromo-dimethylbenzene) $\quad Br\text{-}C_6H_4\text{-}C(=O)\text{-}C_6H_4\text{-}Br \quad Br\text{-}CH_2\text{-}CH{=}CH\text{-}CH_2\text{-}Br$

Br, S, Br (2,5-dibromothiophene) $\quad Br\text{-}C_6H_4\text{-}C(=O)\text{-}C(=O)\text{-}C_6H_4\text{-}Br \quad BrCH_2\text{-}C_6H_4\text{-}CH_2Br$

(105)

Monomers containing acylchloride groups have been polymerized (equation 106) (*251*).

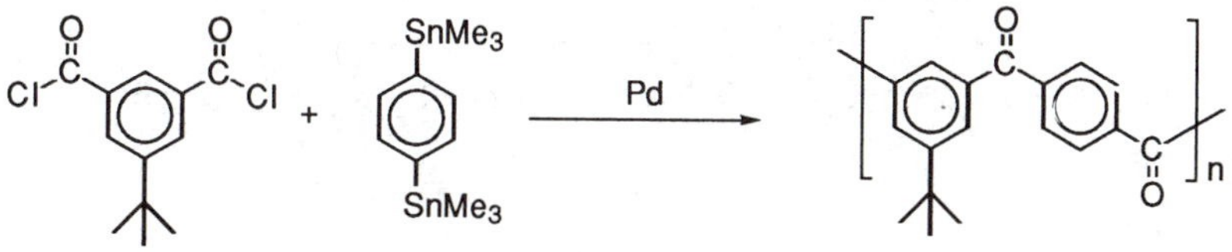

Cl-C(=O), C(=O)-Cl (5-tert-butylisophthaloyl chloride) + $SnMe_3$, $SnMe_3$ (1,4-bis(trimethylstannyl)benzene) $\xrightarrow{Pd}$ [polyketone]$_n$

(106)

The polymerization of aryl bistriflates with aryl bis(trimethylstannane) gave oligomers with an average DP of 11 (equation 107) (*252*). The low solubility of the polymer was cited as the cause of the low DP.

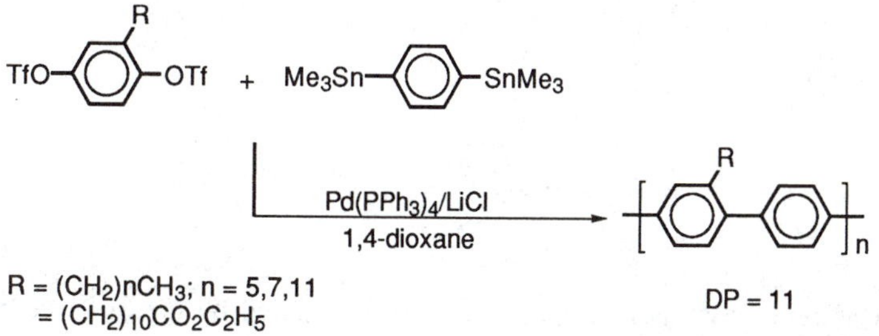

$$TfO\text{-}C_6H_3(R)\text{-}OTf + Me_3Sn\text{-}C_6H_4\text{-}SnMe_3 \xrightarrow[\text{1,4-dioxane}]{Pd(PPh_3)_4/LiCl} \text{-}[C_6H_3(R)\text{-}C_6H_4]_n\text{-}$$

$R = (CH_2)nCH_3$; n = 5,7,11
$= (CH_2)_{10}CO_2C_2H_5$

DP = 11

(107)

Copolymers with alternating benzene-thiophene units have been synthesized via the Stille reaction of both substituted aryl halide and triflate (equation 108) (*253,254*). The molecular weight was 14,000 by GPC when n was equal to 16. This corresponds to a DP of approximately 22.

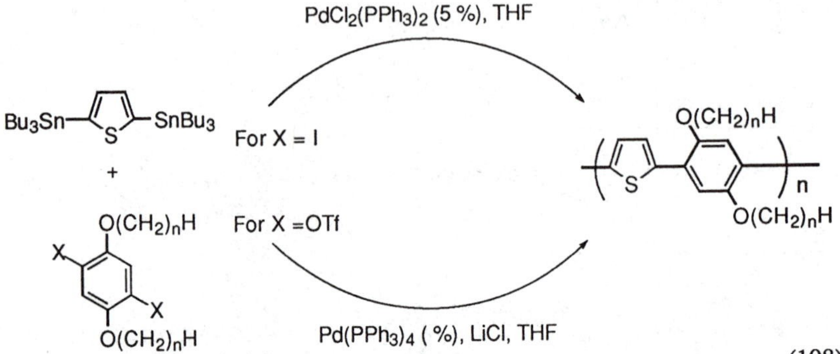

$PdCl_2(PPh_3)_2$ (5 %), THF

Bu_3Sn–(thiophene)–$SnBu_3$ For X = I

+

$O(CH_2)_nH$ For X =OTf

X, X

$O(CH_2)_nH$ $Pd(PPh_3)_4$ (%), LiCl, THF

$O(CH_2)_nH$, S, n, $O(CH_2)_nH$

n = 4, 5, 6, 7, 8, 9, 12, 16. (108)

Polythiophenes incorporating crown ether groups were synthesized as shown in equation 109 (*255*).

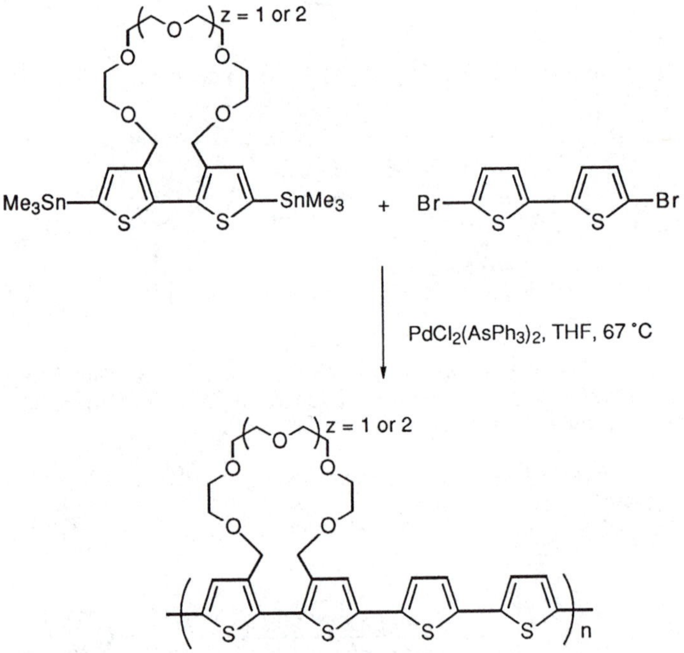

z = 1 or 2

Me_3Sn, S, S, $SnMe_3$ + Br, S, S, Br

$PdCl_2(AsPh_3)_2$, THF, 67 °C

z = 1 or 2

S, S, S, S, n

(109)

A calix[4]arene-based polythiophene was also prepared by the Stille coupling reaction (*256*).

Poly(*p*-pyridyl vinylene)s with random (equation 110) and with head-to-head (equation 111) couplings of the pyridyl units were synthesized by the Stille reaction (*205*). Soluble poly(1,4-phenyleneethynylene)s were prepared by the Pd(0) catalyzed cross-coupling of bis(tributylstannyl)acetylene with 2,5-dialkoxy-1,4-dibromobenzenes. (equation 112) (*208*)

Br, Br + Bu_3Sn–CH=CH–$SnBu_3$ $\xrightarrow[\text{LiCl, NMP, 115 °C}]{Pd(PPh_3)_4}$ polymer, n

Random
> 95 % yield
M_n = 7300

(110)

Br, Br $\xrightarrow[\text{NMP, 115 °C}]{Bu_3Sn\text{–CH=CH–}SnBu_3,\ Pd(PPh_3)_4,\ LiCl}$ polymer, n

Head-to-Head
> 95 % yield
M_n = 8300

(111)

OR, Br, Br, OR + Bu_3Sn–C≡C–$SnBu_3$ $\xrightarrow{Pd}$ OR, C≡C, OR, n

(112)

Synthesis of Dendritic Polymers by Pd(0) Catalysis.
Dendritic macromolecules constitute a new class of polymers which possess a branching point in each repeat unit. For a recent general review see reference *257*. Suzuki (*258-260*) and the cross-coupling of aryl halides with phenylacetylene (*261-262*) were applied to the synthesis of dendritic polymers. Several reviews are available (*257-262*).

Conclusions.
Ni(0) and Pd(0) catalyzed carbon-carbon bond forming reactions have unique characteristics that have enabled advances in the synthesis of aromatic and heteroaromatic polymers with well defined microstructures. Several recent reviews discuss the structure properties dependence in this class of polymers (*263-265*). The advantages of using these catalysts are primarily the following: (a) facile formation of carbon-carbon bonds under mild conditions in the presence of many different functionalities and (b) high regio- and chemo-selectivities leading to well defined structures. The major problem with these polymerizations is the control of the molecular weight and chain ends. Most frequently only low molecular weight polymers are obtained. The source of the low molecular weight is sometimes determined by the inherent insolubility of the polymer that is not related to the efficiency of the catalytic reaction. However, low molecular weights are obtained in many cases because of side reactions which terminate polymer chain growth prematurely. These side reactions can be classified as (a) reduction of monomer (either the organic electrophile or organometallic cross-coupling partner) or (b) coupling of monomer with a nonreactive chain end (examples: phenyl transfer from triphenylphosphine, phosphonium salt formation, etc.). A detailed elucidation of these reaction mechanisms and the development of new more stable and selective catalytic systems is required to overcome these problems.

Acknowledgments.
Financial support by the National Science Foundation (DMR-92-06781 and DMR-91-22227) is gratefully acknowledged.

Literature Cited

1. Collman, J. P.; Hegedus, L. S.; Norton, J. R.; Finke, R. G. *Principles and Applications of Organotransition Metal Chemistry*; University Science Books: Mill Valley, CA 1987.
2. Hegedus, L. S. In *Organometallics in Synthesis* M. Schlosser, Ed. John Wiley, Chichester, 1994, Chapter 5, p 383.
3. Heck, R. F. *Palladium Reagents in Organic Synthesis*; Academic Press: New York, 1985.
4. Heck, R. F. In *Comprehensive Organic Synthesis*, Trost, B. M. Ed.; Pergamon Press: Tarrytown, New York 1991, Vol. 4, p. 833.
5. Hegedus, L. S. *J. Organomet. Chem.* **1993**, *457*, 167.
6. Ziegler, K.; Holzkamp, E.; Breil, H.; Martin, H. *Angew. Chem.* **1955**, *67*, 426.
7. Ziegler, K.; Holzkamp, E.; Breil, H.; Martin, H. *Angew. Chem.* **1955**, *67*,541.
8. Ziegler, K. *Angew Chem.* **1964**, *76*, 545.
9. Natta, G.; Pino, P.; Corradini, P.; Danusso, F.; Mantica, E.; Mazzanti, G.; Moraglio, G. *J. Am. Chem. Soc.* **1955**, *77*, 1708.
10. Natta, G. *Science* **1965**, *147*, 261.
11. Natta, G. *J. Polym. Sci.* **1959**, *34*, 531.
12. Semmelhack, M. F.; Helquist, P. M.; Jones, L. D. *J. Am. Chem. Soc.* **1971**, *93*, 5908.
13. Semmelhack, M. F.; Ryono, L. S. *J. Am. Chem. Soc.* **1975**, *97*, 3873.
14. Semmelhack, M. F.; Helquist, P.; Jones, L. D.; Keller, L.; Mendelson, L.; Ryono, L. S.; Smith, J. G.; Stauffer, R. D. *J. Am. Chem. Soc.* **1981**, *103*, 6460.
15. Kende, A. S.; Liebeskind, L. S.; Braitsch, D. M. *Tetrahedron Lett.* **1975**, *39*, 3375.
16. Zembayashi, M.; Tamao, K.; Yoshida, J.; Kumada, M. *Tetrahedron Lett.* **1977**, *47*, 4089.
17. Colon, I; Kelsey, D. R. *J. Org. Chem.* **1986**, *51*, 2627.
18. Iyoda, M.; Otsuka, H.; Sato, K.; Nisato, N.; Oda, M. *Bull. Chem. Soc. Jpn.* **1990**, *63*, 80.
19. Fanta, P. E. *Synthesis* **1974**, 9.
20. Lindley, J. *Tetrahedron* **1984**, *40*, 1433.
21. Sainsbury, M. *Tetrahedron* **1980**, *36*, 3327.
22. Larock, R. C. *Comprehensive Organic Transformations*, VCH, New York, **1989**.
23. Miura, M.; Hashimoto, H.; Itoh, K.; Nomura, M. *Chem. Lett.* **1990**, 459.
24. Ritter, K. *Synthesis* **1993**, 735.
25. Cacchi, S. In *Seminars in Organic Synthesis: XVIII Summer School "A. Corbella"*, Polo Editoriale Chimico: Milan, Italy, **1993**, p.217.
26. Yamashita, J.; Inoue, Y.; Kondo, T.; Hashimoto, H. *Chem. Lett.* **1986**, 407.
27. Inoue, Y.; Yamashita, J.; Kondo, T.; Hashimoto, H. *Nippon Kagaku Kaishi*, **1987**, 197; *Chem. Abstr.* **1987**, *107*, 197686k.
28. Percec, V.; Pugh, C.; Cramer, E.; Weiss, R. *Polym. Prepr. (Am. Chem. Soc., Div. Polym. Chem.)* **1991**, *32(1)*, 329.
29. Percec, V.; Okita, S.; Weiss, R. *Macromolecules*, **1992**, *25*, 1816.
30. Percec, V.; Okita, S.; Bae, J. *Polym. Bull.* **1992**, *29*, 271.

31. Percec, V.; Pugh, C.; Cramer, E; Okita, S.; Weiss, R. *Makromol. Chem. Macromol. Symp.* **1992**, *54/55*, 113.
32. Jutand, A.; Négri, S.; Mosleh, A. *J. Chem. Soc. Chem. Commun.* **1992**, 1729.
33. Jutand, A.; Mosleh, A. *Synlett* **1993**, 568.
34. Eilingsfeld, H.; Patsch, M.; Siegel, B., Ger. Offen., DE, 3.941.494, **1990**; *Chem. Abstr.* **1991**, *114*, 23548w.
35. Percec, V.; Bae, J.-Y.; Zhao, M.; Hill, D. H. *J. Org. Chem.* **1995**, *60*, 176.
36. Percec, V.; Bae, J.-Y.; Zhao, M.; Hill, D. H. *J. Org. Chem.* **1995**, *60*, 1066.
37. Tsou, T. T.; Kochi, J. K. *J. Am. Chem. Soc.* **1979**, *101*, 7547.
38. Tsou, T. T.; Kochi, J. K. *J. Am. Chem. Soc.* **1979**, *101*, 6319.
39. Amatore, C.; Jutand, A. *Organometallics* **1988**, *7*, 2203.
40. Kowalski, M. H.; Stang, P. J. *Organometallics* **1986**, *5*, 2392.
41. Aoki, S.; Fujimura, T.; Nakamura, E.; Kuwajima, I. *J. Am. Chem. Soc.* **1988**, *110*, 3296.
42. Foà, M.; Cassar, L. *J. Chem. Soc., Dalton Trans.* **1975**, 2572.
43. Heck. R. F.; Nolly, J. P., Jr. *J. Org. Chem.* **1972**, *37*, 2320.
44. Dieck, H. A.; Heck, R. F. *J. Am. Chem. Soc.* **1974**, *96*, 1133.
45. Heck, R. F. *Acc. Chem. Res.* **1979**, *12*, 146.
46. Heck, R. F. *Org. React.* **1982**, *27*, 345.
47. Daves, G. D., Jr.; Hallberg, A. *Chem. Rev.* **1989**, *89*, 1433.
48. Cabri, W.; Candiani, I. *Acc. Chem. Res.* **1995**, *28*, 2.
49. Cacchi, S.; Morera, E.; Ortar, G. *Tetrahedron Lett.* **1984**, *25*, 2271.
50. Cacchi, S.; Ciattini, P. G.; Morera, E.; Ortar, G. *Tetrahedron Lett.* **1987**, *28*, 3039.
51. Spencer, A. *J. Organomet. Chem.* **1983**, *247*, 117.
52. Spencer, A. *J. Organomet. Chem.* **1984**, *265*, 323.
53. Kasahara, A.; Izumi, T.; Ogihara, T. *Chem. Ind.* **1988**, 1433.
54. Kikukawa, K.; Nagira, K.; Wada, F.; Matsuda, T. *Tetrahedron* **1981**, *37*, 31.
55. Sengupta, S.; Bhattacharya, S. *J. Chem. Soc. Perkin Trans. 1* **1993**, 1943.
56. Chen, Q.-Y.; Yang, Z.-Y. *Tetrahedron Lett.* **1986**, *27*, 1171.
57. Kim, J.-I. I.; Patel, B. A.; Heck, R. F. *J. Org. Chem.* **1981**, *46*, 1067.
58. Albeniz, A. C.; Espinet, P.; Lin, Y.-S. *Organometallics* **1995**, *14*, 2977.
59. Cabri, W.; Candiani, I.; Bedeschi, A. *J. Org. Chem.* **1993**, *58*, 7421.
60. Cabri, W.; Candiani, I.; DeBernardinis, S.; Francalanci, F.; Penco, S. *J. Org. Chem.* **1991**, *56*, 5796.
61. Ozawa, F.; Kubo, A.; Hayashi, T. *J. Am. Chem. Soc.*, **1991**, *113*, 1417.
62. Draper, T. L.; Bailey, T. R. *Synlett* **1995**, 157.
63. Scott, W. J.; Peña, M. R.; Swärd, K.; Stoessel, S. J.; Stille, J. K. *J. Org. Chem.* **1985**, *50*, 2302.
64. Bozell, J. J.; Vogt, C. E. *J. Am. Chem. Soc.* **1988**, *110*, 2655.
65. Ben-David, Y.; Portnoy, M.; Gozin, M.; Milstein, D. *Organometallics* **1992**, *11*, 1995.
66. Herrmann, W. A.; Brossmer, C.; Öfele, K.; Reisinger, C.-P.; Priermeier, T.; Beller, M.; Fischer, H. *Angew. Chem. Int. Ed. Eng.* **1995**, *34*, 1844.
67. Boldrini, G. P.; Savoia, D.; Tagliavini, E.; Trombini, C.; Ronchi, A. U. *J. Organomet. Chem* **1986**, *301*, C62.
68. Mori, M.; Kudo, S.; Ban, Y. *J. Chem. Soc. Perkin Trans. 1* **1979**, 771.
69. Colon, I. U.S. Patent 4 334 081, **1982**.
70. Lebedev, S. A.; Pedchenko, V. V.; Lopatina, V. S.; Berestova, S. S.; Petrov, E. S. *J. Org. Chem. USSR* **1990**, *26*, 1312.
71. Scherf, U.; Müllen, K. *Synthesis* **1992**, 23.
72. Dieck, H. A.; Heck, R. F. *J. Organomet. Chem.* **1975**, *93*, 259.

73. Cassar, L. *J. Organomet. Chem.* **1975**, *93*, 253.
74. Nguyen, B. V.; Yang, Z.-Y.; Burton, D. J. *J. Org. Chem.* **1993**, *58*, 7368.
75. Tilley, J. W.; Zawoiski, S. *J. Org. Chem.* **1988**, *53*, 386.
76. Sonogashira, K.; Tohda, Y.; Hagihara, N. *Tetrahedron Lett.* **1975**, 4467.
77. Cacchi, S.; Morera, E.; Ortar, G. *Synthesis* **1986**, 320.
78. Arcadi, A.; Cacchi, S.; Marinelli, F. *Tetrahedron Lett.* **1989**, *30*, 2581.
79. Chen, Q.-Y.; Li, Z.-T. *J. Chem. Soc. Perkin Trans. 1* **1992**, 2931.
80. Carpita, A.; Lessi, A.; Rossi, R. *Synthesis* **1984**, 571.
81. Pugh, C.; Percec, V. *J. Polym. Sci.: Part A: Polym. Chem. Ed.* **1990**, *28*, 1101.
82. Zhang, Y.; Negishi, E. *J. Am. Chem. Soc.* **1989**, *111*, 3454.
83. Scott, W. J.; Stille, J. K. *J. Am. Chem. Soc.* **1986**, *108*, 3033.
84. Echavarren, A. M.; Stille, J. K. *J. Am. Chem. Soc.* **1987**, *109*, 5478.
85. Amatore, C.; Jutand, A.; Suarez, A. *J. Am. Chem. Soc.* **1993**, *115*, 9531.
86. Kumada, M. *Pure Appl. Chem.* **1980**, *52*, 669.
87. Tamao, K.; Sumitani, K.; Kiso, Y.; Zembayashi, M.; Fujioka, A.; Kodama, S.; Nakajimi, I.; Minato, A.; Kumada, M. *Bull. Chem. Soc. Jpn.* **1976**, *49*, 1958.
88. Tamao, K.; Sumitani, K.; Kumada, M. *J. Am. Chem. Soc.* **1972**, *94*, 4374.
89. Corriu, R. J. P.; Masse, J. P. *J. Chem. Soc. Chem. Comm.* **1972**, 144.
90. Dang, H. P.; Linstrumelle, G. *Tetrahedron Lett.* **1978**, 191.
91. Tamao, K.; Zembayashi, M.; Kiso, Y.; Kumada, M. *J. Organometal. Chem.* **1973**, *55*, C91.
92. Tamao, K.; Kodama, S.; Nakajima, T.; Kiso, Y.; Kumada, M. *J. Am. Chem. Soc.* **1975**, *97*, 4405.
93. Tamao, K.; Kodama, S.; Nakayimam I.; Kumada, M.; Minato, A.; Suzuki, K *Tetrahedron* **1982**, *38*, 3347.
94. Castle, P. L.; Widdowson, D. A. *Tetrahedron Lett.* **1986**, *27*, 6013.
95. Sengupta, S.; Leite, M.; Raslan, D. S.; Quesnelle, C.; Snieckus, V. *J. Org. Chem.* **1992**, *57*, 4066.
96. Clayden, J.; Julia, M. *J. Chem. Soc., Chem. Comm.* **1993**, 1682.
97. Percec, V.; Bae, J.-Y.; Hill, D. H. *J. Org. Chem.* in press.
98. Murahashi, S.-I.; Yamamura, M.; Yanagisawa, K.; Mita, N.; Kondo, K. *J. Org. Chem.* **1979**, *44*, 2408.
99. Morell, D. G.; Kochi, J. K. *J. Am. Chem. Soc.* **1975**, *97*, 7262.
100. Negishi, E. *Acc. Chem. Res.* **1982**, *15*, 340.
101. Knochel, P.; Singer, R. D. *Chem. Rev.* **1993**, *93*, 2117.
102. Negishi, E.; King, A. O.; Okukado, N. *J. Org. Chem.* **1977**, *42*, 1821.
103. King, A. O.; Negishi, E.; Villani, F. J., Jr.; Silveira, A., Jr. *J. Org. Chem.* **1978**, *43*, 358.
104. Vollhardt, K. P. C. *Acc. Chem. Res.* **1977**, *10*, 1.
105. King, A. O.; Okukado, N.; Negishi, E. *J. Chem. Soc., Chem. Commun.* **1977**, 683.
106. Matsushita, H.; Negishi, E. *J. Am. Chem. Soc.* **1981**, *103*, 2882.
107. Negishi, E.; Chatterjee, S.; Matsushita, H. *Tetrahedron Lett.* **1981**, 3737.
108. Matsushita, H.; Negishi, E. *J. Org. Chem.* **1982**, *47*, 4161.
109. Kobayashi, M.; Negishi, E. *J. Org. Chem.* **1980**, *45*, 5223.
110. Negishi, E.; Bagheri, V.; Chatterjee, S.; Luo, F.-T.; Miller, J. A.; Stoll, A. T. *Tetrahedron Lett.* **1983**, *24*, 5181.
111. Negishi, E.; Valente, L. F.; Kobayashi, M. *J. Am. Chem. Soc.* **1980**, *102*, 3298.
112. Hayashi, T.; Konishi, M.; Kobori, Y.; Kumada, M.; Higuchi, T.; Hirotsu, K. *J. Am. Chem. Soc.* **1984**, *106*, 158.
113. Park, K.; Yuan, K.; Scott, W. J. *J. Org. Chem.* **1993**, *58*, 4866.
114. Yuan, K.; Scott, W. J. *Tetrahedron Lett.* **1991**, *32*, 189.

115. Suzuki, A. *Acc. Chem. Res.* **1982**, *15*, 178.
116. Suzuki, A. *Pure Appl. Chem.* **1985**, *57*, 1749.
117. Suzuki, A. *Pure Appl. Chem.* **1991**, *63*, 419.
118. Martin, A. R.; Yang, Y. *Acta Chem. Scand.***1993**, *47*, 221.
119. Ishiyama, T.; Miyaura, N.; Suzuki, A. *Tetrahedron Lett.* **1991**, *32*, 6923.
120. Ishiyama, T.; Abe, S.; Miyaura, N.; Suzuki, A. *Chem. Lett.* **1992**, 691.
121. Miyaura, N.; Suginome, H.; Suzuki, A. *Tetrahedron* **1983**, *39*, 3271.
122. Miyaura, N.; Yamada, K.; Suginome, H.; Suzuki, A. *J. Am. Chem. Soc.* **1985**, *107*, 972.
123. Satoh, M.; Miyaura, N.; Suzuki, A. *Chem. Lett.*, **1986**, 1329.
124. Miyaura, N.; Yano, T.; Suzuki, A. *Tetrahedron Lett.* **1980**, *21*, 2865.
125. Miyaura, N.; Yanagi, T.; Suzuki, A. *Synth. Commun.* **1981**, *11*, 513
126. Gronowitz, S.; Lawitz, K. *Chem. Scr.* **1983**, *22*, 265.
127. Thompson, W. J.; Gaudino, J. *J. Org. Chem.* **1984**, *49*, 5237.
128. Oh-e, T.; Miyaura, N.; Suzuki, A. *J. Org. Chem.* **1993**, *58*, 2201.
129. Miyaura, N.; Satoh, M.; Suzuki, A. *Tetrahedron Lett.* **1986**, *27*, 3745.
130. Ishiyama, T.; Miyaura, N.; Suzuki, A. *Chem. Lett.* **1987**, 25.
131. Miyaura, N.; Ishiyama, T.; Sasaki, H.; Ishikawa, M.; Satoh, M.; Suzuki , A. *J. Am. Chem. Soc.* **1989**, *111*, 314.
132. Miyaura, N.; Ishiyama, T.; Ishikawa, M.; Suzki, A. *Tetrahedron Lett.* **1986**, *27*, 6369.
133. Miller, R. B.; Dugar, S. *Organometallics* **1984**, *3*, 1261.
134. Sharp, M. J.; Snieckus, V. *Tetrahedron Lett.* **1985**, *26*, 5997.
135. Watanabe, T.; Miyaura, N.; Suzuki, A. *Synlett* **1992**, 207.
136. Aliprantis, A. O.; Canary, J. W. *J. Am. Chem. Soc.* **1994**, *116*, 6985.
137. Smith, G. B.; Dezeny, G. C.; Hughes, D. L.; King, A. O.; Verhoeven, T. R. *J. Org. Chem.* **1994**, *59*, 8151.
138. Grushin, V. V.; Alper, H. *Chem. Rev.* **1994**, *94*, 1047.
139. Mitchell, M. B.; Wallbank, P. J. *Tetrahedron Lett.* **1991**, *32*, 2273.
140. Beller, M.; Fischer, H.; Herrmann, W. A.; Öfele, K.; Brossmer, C. *Angew. Chem. Int. Ed. Engl.* **1995**, *34*, 1848.
141. Huth, A.; Beetz, I.; Schumann, I. *Tetrahedron* **1989**, *45*, 6679.
142. Fu, J.-m.; Snieckus, V. *Tetrahedron Lett.* **1990**, *31*, 1665.
143. Stang, P. J.; Hanack, M.; Subramanian, L. R. *Synthesis* **1982**, 85.
144. Gilbert, A. M.; Wulff, W. D. *J. Am. Chem. Soc.* **1994**, *116*, 7449.
145. Percec, V.; Bae, J.-Y.; Hill, D. H. *J. Org. Chem.* **1995**, *60*, 1060.
146. Marck, G.; Villiger, A.; Buchecker, R. *Tetrahedron Lett.*, **1994**, *35*, 3277.
147. Roth, G. D.; Farina, V. *Tetrahedron Lett.* **1995**, *36*, 2191.
148. Wallow, T. I.; Novak, B. M. *J. Org. Chem.* **1994**, *59*, 5034.
149. Moreno-Mañas, M.; Pajuelo, F.; Pleixats, R. *J. Org. Chem.* **1995**, *60*, 2396.
150. Kuivila, H. G.; Reuwer, J. F., Jr.; Mangravite, J. A. *Can. J. Chem.* **1963**, *41*, 3081.
151. Gronwitz, S.; Hornfeldt, A.-B.; Yang, Y. *Chem. Scr.* **1988**, *28*, 281.
152. Cho, C. S.; Uemura, S. *J. Organometal. Chem.* **1994**, *465*, 85.
153. Song, Z. Z.; Wong, H. N. C. *J. Org. Chem.* **1994**, *59*, 33.
154. Stille, J. K. *Pure Appl. Chem.* **1985**, *57*, 1771.
155. Stille, J. K. *Angew. Chem. Int. Ed. Engl.* **1986**, *25*, 508.
156. Mitchell, T. N. *Synthesis* **1992**, 803.
157. Zhang, H.-C.; Daves, G. D., Jr. *Organometallics* **1993**, *12*, 1499.
158. Badone, D.; Cecchi, R.; Guzzi, U. *J. Org. Chem.* **1992**, *57*, 6321.
159. Nagatsugi, F.; Uemura, K.; Nakashima, S.; Maeda, M.; Saski, S. *Tetrahedron Lett.* **1995**, *36*, 421.
160. Scott, W. J.; McMurry, J. E. *Acc. Chem. Res.* **1988**, *21*, 47.

161. Martorell, G.; García-Raso, A.; Saá, J. M. *Tetrahedron Lett.* **1990**, *31*, 2357.
162. Saá, J. M.; Martorell, G.; Garcia-Raso, A. *J. Org. Chem.* **1992**, *57*, 678.
163. Echavarren, A. M.; Stille, J. K. *J. Am. Chem. Soc.* **1988**, *110*, 1557.
164. Farina, V.; Krishnan, B. *J. Am. Chem. Soc.* **1991**, *113*, 9585.
165. Farina, V.; Krishnan, B.; Marshall, D. R.; Roth, G. P. *J. Org. Chem.* **1993**, *58*, 5434.
166. Beletskaya, I. P. *J. Organomet. Chem.* **1983**, *250*, 551.
167. Farina, V.; Roth, G. P. *Tetrahedron Lett.* **1991**, *32*, 4243.
168. Liebeskind, L. S.; Fengl, R. W. *J. Org. Chem.* **1990**, *55*, 5359.
169. Gómez-Bengoa, E.; Echavarren, A. M. *J. Org. Chem.* **1991**, *56*, 3497.
170. Tamayo, N.; Echavarren, A. M.; Paredes, M. C. *J. Org. Chem.* **1991**, *56*, 6488.
171. Johnson, C. R. Adams, J. P.; Braun, M. P.; Senanayake, C. B. W. *Tetrahedron Lett.* **1992**, *33*, 919.
172. Liebeskind, L. S.; Riesinger, S. W. *J. Org. Chem.* **1993**, *58*, 408.
173. Saá, J. M.; Martorell, G. *J. Org. Chem.* **1993**, *58*, 1963.
174. Ye, J.; Bhatt, R. K.; Falck, J. R. *J. Am. Chem. Soc.* **1994**, *116*, 1.
175. Farina, V.; Kapadia, S.; Krishnan, B.; Wang, C.; Liebeskind, L. S. *J. Org. Chem.* **1994**, *59*, 5905.
176. Segelstein, B. E.; Butler, T. W.; Chenard, B. L. *J. Org. Chem.* **1995**, *60*, 12.
177. Morita, D. K.; Stille, J. K.; Norton, J. R. *J. Am. Chem. Soc.* **1995**, *117*, 8576.
178. Colon, I.; Kwiakowski, G. T. *J. Polym. Sci., Part A : Polym. Chem. Ed.* **1990**, *28*, 367.
179. Udea, M.; Ito, T. *Polym. J.* **1991**, *23*, 297.
180. Yamamoto, T.; Kashiwazaki, A.; Kato, K. *Makromol. Chem.* **1989**, *190*, 1649.
181. Ueda, M.; Ichikawa, F. *Macromolecules* **1990**, *23*, 926.
182. Percec, V. U.S. Pat. 5 241 044, **1993**; *Chem. Abstr.*, **1994**, *120*, 108 108.
183. Yamamoto, T.; Morita, A.; Miyazaki, Y.; Maruyama, T.; Wakayama, H.; Zhou, Z.; Nakamura, Y.; Kanbara, T.; Sasaki, S.; Kubota, K. *Macromolecules* **1992**, *25*, 1214.
184. Chaturvedi, V.; Tanaka, S.; Kaeriyama, K. *J. Chem. Soc., Chem. Commun.* **1992**, 1658.
185. Chaturvedi, V.; Tanaka, S.; Kaeriyama, K. *Macromolecules* **1993**, *26*, 2607.
186. Ueda, M.; Seino, Y.; Sugiyama, J. *Polym. J. (Japan)* **1993**, *25*, 1319.
187. Williams, D. J.; Colquhoun, H. M.; O'Mahoney, C. A. *J. Chem. Soc., Chem. Comm.* **1994**, 1643.
188. Marrocco, M.; Gagne, R. R. *U. S. Pat.* 5 227 457, **1993**.
189. Phillips, R. W.; Sheares, V. V.; Samulski, E. T.; DeSimone, J. M. *Macromolecules* **1994**, *27*, 2354.
190. Wang, Y.; Quirk, R. P. *Macromolecules* **1995**, *28*, 3495.
191. Percec, V.; Okita, S. *J. Polym. Sci. Part A: Polym. Chem. Ed.* **1993**, *31*, 1087.
192. Percec, V.; Okita, S. *J. Polym. Sci. Part A: Polym. Chem. Ed.* **1993**, *31*, 877.
193. Kreyenschmidt, M.; Uckert, F.; Müllen, K. *Macromolecules* **1995**, *28*, 4577.
194. Percec, V.; Bae, J.-B.; Zhao, M.; Hill, D. H. *Macromolecules* in **1995**, *28*, 6726.
195. Yamamoto, T.; Kashiwazaki, A.; Kato, K. *Makromol. Chem.* **1989**, *190*, 1649.

196. Miyazaki, Y.; Kanbara, T.; Osakada, K.; Yamamoto, T.; Kubota, K. *Polym. J.* **1994**, *26*, 509.
197. Greiner, A.; Heitz, W. *Makromol. Chem. Rapid Commun.* **1988**, *9*, 581.
198. Jpn Pat 57 207 618, **1981**, Asahi Glass Co. Ltd.; *Chem Abstr.* **1983**, *99*, 23139k.
199. Heitz, W.; Brügging, W.; Freund, L.; Gailberger, M.; Greiner, A.; Jung, H.; Kampschulte, U.; Niessner, N.; Osan, F.; Schmidt, H.-W.; Wicker, M. *Makromol. Chem.* **1988**, *189*, 119.
200. Martelock, H.; Greiner, A.; Heitz, W. *Makromol. Chem.* **1991**, *192*, 967.
201. Bao, Z.; Chen, Y.; Cai, R.; Yu, L. *Macromolecules* **1993**, *26*, 5281.
202. Suzuki, M.; Lim, J. C.; Saegusa, T. *Macromolecules* **1990**, *23*, 1574.
203. Brenda, M.; Greiner, A.; Heitz, W. *Makromol. Chem.* **1990**, *191*, 1083.
204. Jikei, M.; Miyauchi, M.; Ishida, Y.; Kakimoto, M.; Imai, Y. *Macromol. Rapid Commun.* **1994**, *15*, 979.
205. Marsella, M. J.; Fu, D. K.; Swager, T. M. *Adv. Mater.* **1995**, *7*, 145.
206. Weitzel, H.-P.; Müllen, K. *Makromol. Chem.* **1990**, *191*, 2837.
207. Bao, Z.; Chen, Y.; Yu, L. *Macromolecules* **1994**, *27*, 4629.
208. Giesa, R.; Schulz, R. C. *Makromol. Chem.* **1990**, *191*, 857.
209. Pearson, D. L.; Schumm, J. S.; Tour, J. M. *Macromolecules* **1994**, *27*, 2348.
210. Goldfinger, M. B.; Swager, T. M. *Polym. Prepr. (Am. Chem. Soc., Div. Polym. Chem.)* **1993**, *34*(2), 755.
211. Yamamoto, T.; Hayashi, Y.; Yamamoto, A. *Bull. Chem. Soc. Jpn.* **1978**, *51*, 2091.
212. Rehahn, M.; Schlüter, A.-D.; Wegner, G.; Feast, W. J. *Polymer* **1989**, *30*, 1054.
213. Noll, A.; Siegfried, N.; Heitz, W. *Makromol. Chem., Rapid Commun.* **1990**, *11*, 485.
214. Lin, J. W.-P.; Dudek, P. *J. Polym. Sci., Polym. Chem. Ed.* **1980**, *18*, 2869.
215. Yamamoto, T.; Sanechika, K.; Yamamoto, A. *Chem. Lett.* **1981**, 1079.
216. Kobayashi, M.; Chen, J.; Moraes, T.-C.; Heeger, J.; Wudl, F. *Synth. Met.* **1984**, *9*, 77.
217. Elsenbaumer, R. L.; Jen, K. Y.; Oboad, R *Synth. Met.* **1986**, *15*, 169.
218. McCullough, R. D.; Lowe, R. D. *J. Chem. Soc., Chem. Commun.* **1992**, 70.
219. Mao, H.; Holdcroft, S. *Macromolecules* **1992**, *25*, 554.
220. McCullough, R. D.; Lowe, R. D.; Manikandan, J.; Anderson, D. L. *J. Org.Chem.* **1993**, *58*, 904.
221. McCullough, R. D.; Tristram-Nagle, S.; Williams, S. P.; Lowe, R. D.; Jayaraman, M. *J. Am. Chem. Soc.* **1993**, *115*, 4910.
222. McCullough, R. D.; Williams, S. P. *J. Am. Chem. Soc.* **1993**, *115*, 11608.
223. McCullough, R. D.; Jayaraman, M. *J. Chem. Soc., Chem. Commun.* **1995**, 135.
224. Chen, T. A.; Rieke, R. D. *J. Am. Chem. Soc.* **1992**, *114*, 10087.
225. Chen, T.-A.; O'Brien, R. A.; Rieke, R. D. *Macromolecules* **1993**, *26*, 3462.
226. Chen, T. A.; Wu, X.; Rieke, R. D. *J. Am. Chem. Soc.* **1995**, *117*, 233.
227. Wu, X.; Chen, T. A.; Rieke, R. D. *Macromolecules* **1995**, *28*, 2101.
228. Marsella, M. J.; Carrol, P. J.; Swager, T. M. *J. Am. Chem. Soc.* **1994**, *116*, 9347.
229. Marsella, M. J.; Carrol, P. J.; Swager, T. M. *J. Am. Chem. Soc.* **1995**, *117*, 9832.
230. Rehahn, M.; Schlüter, A.-D.; Wegner, G.; Feast, W. J. *Polymer*, **1989**, *30*, 1060.

231. Rehahn, M.; Schlüter, A.-D.; Wegner, G. *Makromol. Chem.* **1990**, *191*, 1991.
232. Rau, I. U.; Rehahn, M. *Makromol. Chem.* **1993**, *194*, 2225.
233. Huber, J.; Scherf, U. *Macromol. Rapid Commun.* **1994**, *15*, 897.
234. Vahlenkamp, T.; Wegner, G. *Macromol. Chem. Phys.* **1994**, *195*, 1933.
235. Witteler, H.; Lieser, G.; Wegner, G.; Schulze, M. *Makromol. Chem., Rapid Commun.* **1993**, *14*, 471.
236. Rehahn, M.; Schlüter, A.-D.; Wegner, G. *Makromol. Chem., Rapid Commun.* **1990**, *11*, 535.
237. Helmer-Metzmann, F.; Rehahn, M.; Schmitz, L.; Ballauff, M.; Wegner, G. *Makromol. Chem.* **1992**, 193, 1847.
238. Wallow, T. I.; Novak, B. M. *J. Am. Chem. Soc.* **1991**, 113, 7411.
239. Rulkens, R.; Schulze, M.; Wegner, G. *Macromol. Rapid Commun.* **1994**, *15*, 669.
240. Child, A. D.; Reynolds, J. R. *Macromolecules* **1994**, *27*, 1975.
241. Cramer, E.; Percec, V. *J. Polym. Sci. Part A: Polym. Chem. Ed.* **1990**, *28*, 3029.
242. Scherf, U.; Müllen, K. *Makromol. Chem. Rapid Commun.* **1991**, *12*, 489.
243. Scherf, U.; Müllen, K. *Polymer* **1992**, *33*, 2443.
244. Scherf, U.; Müllen, K. *Macromolecules* **1992**, *25*, 3546.
245. Lamba, J. J. S.; Tour, J. M. *J. Am. Chem. Soc.* **1994**, *116*, 11723.
246. Goldfinger, M. B.; Swager, T. M. *J. Am. Chem. Soc.* **1994**, *116*, 7895.
247. Claussen, W.; Schulte, N.; Schlüter, A.-D. *Macromol. Rapid Commun.* **1995**, *16*, 89.
248. Hensel, V.; Lützow, K.; Schlüter, A. D. *Polym. Prepr. (Am. Chem. Soc., Div. Polym. Chem.)* **1995**, *36*(1), 574.
249. Wallow, T. I.; Seery, T. A. P.; Goodson, F. E., III; Novak, B. M. *Polym. Prepr. (Am. Chem. Soc., Div. Polym. Chem.)* **1994**, *35*(1), 710.
250. Bochmann, M.; Kelly, K. *J. Chem. Soc., Chem. Commun.* **1989**, 532.
251. Moore, J. S.; Deeter, G. A. *Polym. Prepr. (Am. Chem. Soc., Div. Polym. Chem.)* **1991**, *32*(3), 213.
252. Quian, X.; Pena, M. *Macromolecules* **1995**, *28*, 4415.
253. Yu, L.; Bao, Z.; Cai, R. *Angew. Chem. Ed., Int. Ed. Engl.* **1993**, *32*, 1345.
254. Bao, Z.; Chan, W.; Yu, L. *Chem. Mater.* **1993**, *5*, 2.
255. Marsella, M. J.; Swager, T. M. *J. Am. Chem. Soc.* **1993**, *115*, 12214.
256. Marsella, M. J.; Newland, R. J.; Carrol. P. J.; Swager. T. M. *J. Am. Chem. Soc.* **1995**, *117*, 9842
257. Tomalia, D.; Durst, H. D. *Top. Curr. Chem.* **1993**, *165*, 193.
258. Kim, Y. H. *Macromol. Symp.* **1994**, *77*, 21.
259. Kim, Y. H. *Adv. Mater.* **1992**, *4*, 764.
260. Neenan, T. X.; Miller, T. M.; Kwock, E. W.; Bair, H. E. In *Advances in Dendritic Macromolecules*; Newkome, G. R., Ed.; JAI Press Inc.: Greenwsich, CT, 1994; Vol 1; 105.
261. Moore, J. S.; Zhang, J.; Wu. Z.; Dhandapani, V.; Lee, S. *Macromol. Symp.* **1994**, *77*, 295.
262. Xu, Z.; Kyan, B.; Moore, J. S. In *Advances in Dendritic Macromolecules*; Newkome, G. R., Ed.; JAI Press Inc.: Greenwich, CT, 1994; Vol. 1; 69.
263. Wegner, G.; Schluter, A.-D. *Acta Polym.* **1993**, *44*, 59.
264. Tour, J. M. *Adv. Mater.* **1994**, *6*, 190.
265. Percec, V.; Tomazos, D. In *Comprehensive Polymer Science, First Suppl.*; Allen, G., Ed.; Pergamon Press: Oxford, UK, 1992; p. 299.

RECEIVED December 12, 1995

Chapter 2

From ArylX to ArylH Activation in Metal-Catalyzed Polymerization Reactions

W. Heitz

Fachbereich Physikalische Chemie, Polymere, und Wissenschaftliches Zentrum für Materialwissenschaften, Philipps-Universität Marburg, Hans-Meerwein-Strasse, D–35032 Marburg, Germany

Metal catalyzed polymerizations are chain reactions necessitating a constant valence of the metal. In contrast to that metal catalyzed polycondensations are step growth reactions involving a change of the valence of the metal. Tuning of the reaction by structural variations of the metal catalyst are demonstrated with the Pd-catalyzed vinyl polymerization of norbornene, the alternating copolymerization of ethylene with carbon monoxide and the Heck reaction as examples. One and two electron processes can be involved in metal catalyzed polycondensations. The Heck reactions, the Ni-catalyzed synthesis of polyphenylenes, and Ru-catalyzed ArH-insertion reactions are discussed.

Metal catalyzed reactions are of indispensable importance in the value adding chain of chemical technology. In the area of macromolecular chemistry metal catalyzed polymerizations are of basic importance to produce commodity polymers. The elaboration of metalocene catalysts is causing a strong push in the development (*1*). Metal catalysis is also of importance in the synthesis of monomers and in polycondensation as well as in polyaddition.

Metal-catalyzed Polymerizations

Chain reaction with constant valence of the metal

Metal-catalyzed Polycondensations

Step growth reaction with change of valence of the metal

1-Electron processes $Cu^{I} \rightleftarrows Cu^{II}$

2-Electron processes $Pd^{0} \rightleftarrows Pd^{II}$

Metal catalyzed polymerization reactions are chain reactions and necessitate to keep the valence of the metal constant. According to this definition redox polymerizations are not included as metal catalyzed polymerizations; the metal is not participating in

0097–6156/96/0624–0057$12.00/0

the chain growth reaction. Polycondensations and polyadditions feasible by metal catalysis are chain reactions involving a change of the valence of the metal.

These reactions are predominantly two electron processes Pd(0) ⇌ Pd(II), Rh(I) ⇌ Rh(III), Ru(0) ⇌ Ru(II). One electron processes are prevailing in oxidation reactions (Cu(I) ⇌ Cu(II).

Polycondensation By Two Electron Processes

Metal catalyzed polycondensations are investigated with special intensity using palladium as example. With palladium the necessary differences of catalyst structure can be demonstrated very well between polymerization and polycondensation.

Heck reaction (1)

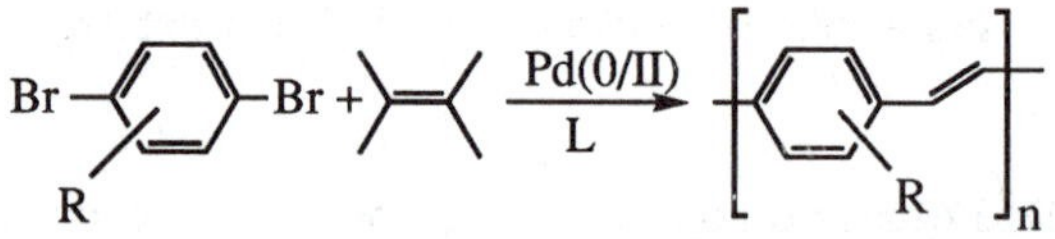

Pd(0/II), L

R, R, n

alternating copolymers (2)

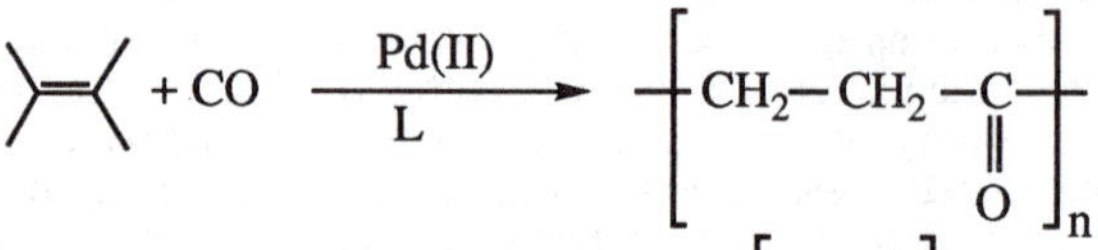

$$+ CO \xrightarrow[L]{Pd(II)} -[CH_2-CH_2-C(=O)]_n-$$

vinyl polymers (3)

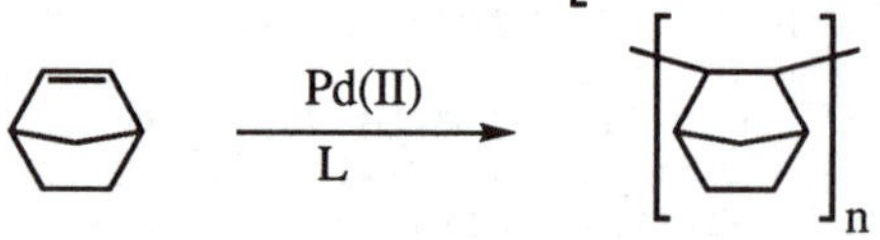

Pd(II), L, n

Derivatives of polyphenylene vinylene are formed in the Pd-catalyzed reaction of ethylene with substituted dihalogeno arenes. This requires a change of the valence of the metal. In contrast to this the alternating copolymerization of olefins with carbon monoxide and the vinylic polymerization of norbornene is successful only if the two valent state of Pd is maintained. The mechanism of the Heck reaction (*1*) (equation 4) shows the limiting requirements to tune the reaction in one or the other way.

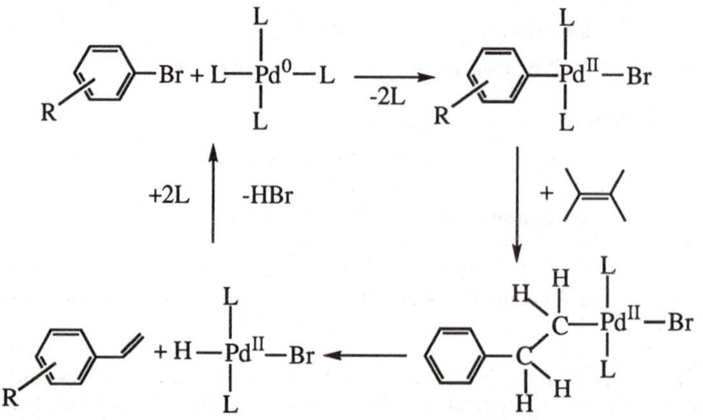

$$R-C_6H_4-Br + L-Pd^0L_3 \xrightarrow{-2L} R-C_6H_4-Pd^{II}L_2-Br$$

+2L, -HBr, + (olefin) (4)

$$R-C_6H_4-CH=CH_2 + H-Pd^{II}L_2-Br \longleftarrow C_6H_5-CH_2-CH_2-Pd^{II}L_2-Br$$

The halogen compound reacts with the Pd(O) species by oxidative addition to form a Pd(II) complex. After insertion of the olefin in the aryl-Pd-bond the target molecule is formed by β-H-elimination. At this stage palladium is still in the two valent state. Separation of HBr in presence of the base leads back to the Pd(O) complex. The application of the Pd chemistry in polymerization reactions requires that the β-H-elimination can be neglected.

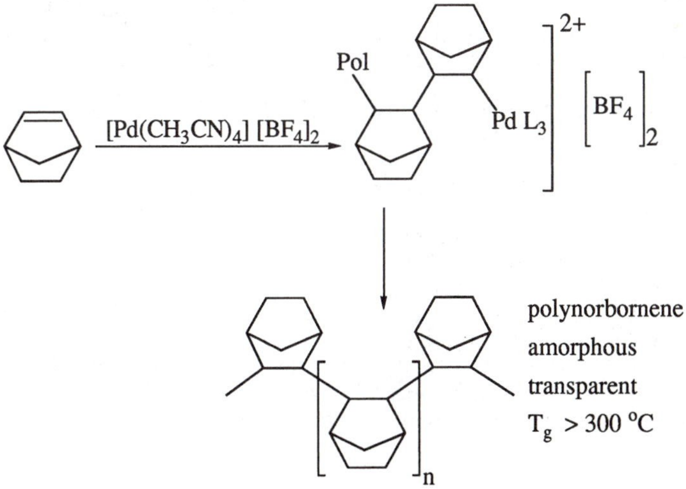

$[Pd(CH_3CN)_4][BF_4]_2$ Pol Pd L_3 $2+$ $[BF_4]_2$ (5)

polynorbornene
amorphous
transparent
$T_g > 300\ ^\circ C$

The β-H-elimination is impeded in the norbornene polymerization by the reformation of a highly strained ring system. The catalysts required for this reaction have a non-coordinating anion and weak nitril ligands as a characteristic. Such a palladium complex was described 1981 by Senn (*3*). The modification of this catalyst by Risse et al. resulted in a homogeneous reaction in the polymerization of norbornene (*4*). Vinyl polymerized norbornene is an example of conformationally restricted polymers having typically very high transition temperatures.

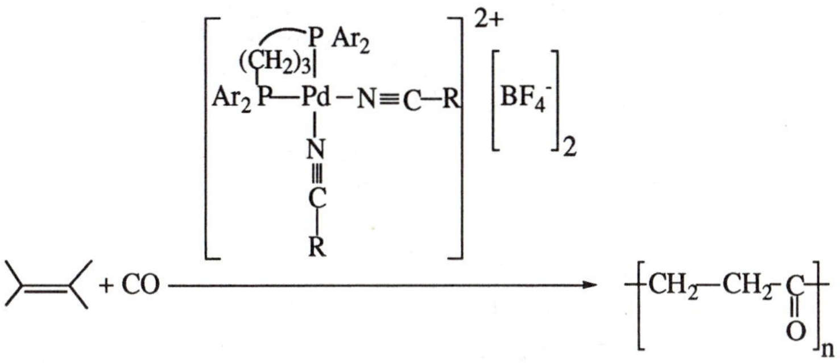

$[(CH_2)_3(PAr_2)Ar_2P{-}Pd(N{\equiv}C{-}R)_2]^{2+}[BF_4^-]_2$

+ CO → $[CH_2{-}CH_2{-}C(=O)]_n$

⟹ Pd-cation, non-coordinating anion
at least 2 ligand sites with weak ligands

Pd-complexes described for the alternating copolymerization of olefins with carbon monoxide shows similar characteristics: non- or low coordinating anions as well as two ligand sites occupied by nitril. The β-H-elimination is a relatively fast reaction but compared to the insertion of carbon monoxide it is slow. This reduces the β-H-elimination and high molecular weights can be obtained.

The Heck reaction opens the possibility to obtain substituted styrenes and stilbenes in a one step reaction.

Pd-catalyzed reactions tolerate a great variety of functional groups (*5*). Vinyl benzamide or vinyl anilin are examples obtained in good yields and at polymerization-grade purity. By substitution of ethylene with acrylic acid derivatives products with cinamic acid structures are obtained (*6*).

The warranted structure of a polymer obtained by the Heck reaction can be investigated by model reactions (*7*).

In the reaction of bromo benzene with ethylene trans-stilbene is formed as the target product. The formation of cis stilbene is not found within the detection limit of gas chromatography. The limitation of the molecular weight in the polyreaction is mainly caused by the dehalogenation of bromo benzene. By optimization of the reaction conditions (low reaction temperature, low catalyst concentration) the dehalogenation is pushed back below the detection limits of gas chromatography. The main defect structures in the polymer are caused by 1,1-disubstitution. This deficiency in the regioselectivity of the reaction can not be suppressed even by optimal reaction conditions. Therefore polyphenylene vinylene obtained by the Heck reaction has statistically a fold in the macromolecule after 30 - 100 monomer

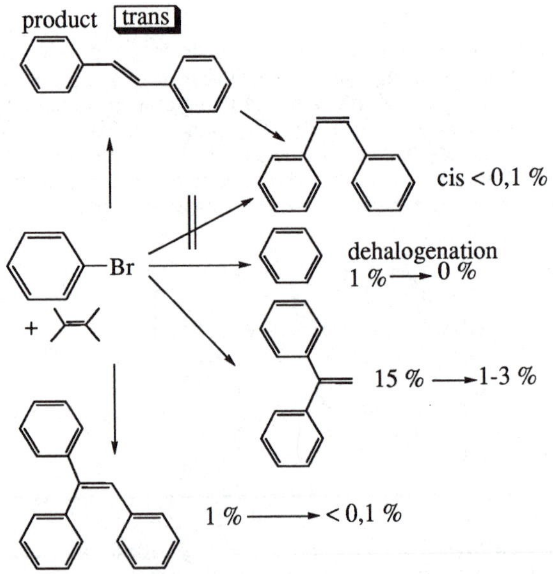

product trans

cis < 0,1 %

Br

+

dehalogenation
1 % → 0 %

15 % → 1-3 %

1 % → < 0,1 %

units. Metal catalyzed reactions can be tuned by the metal ligand or the educt besides the usual change of reaction conditions.

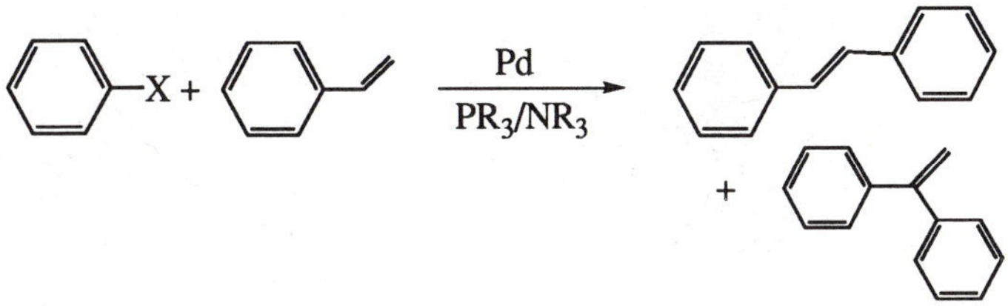

X + → Pd, PR_3/NR_3 → +

X =	J	Br	Cl	OTrif.
(1.2/1.1)	(94/6)	(95/5)	(94/6)	(58/42)
	↓ (99/1)			

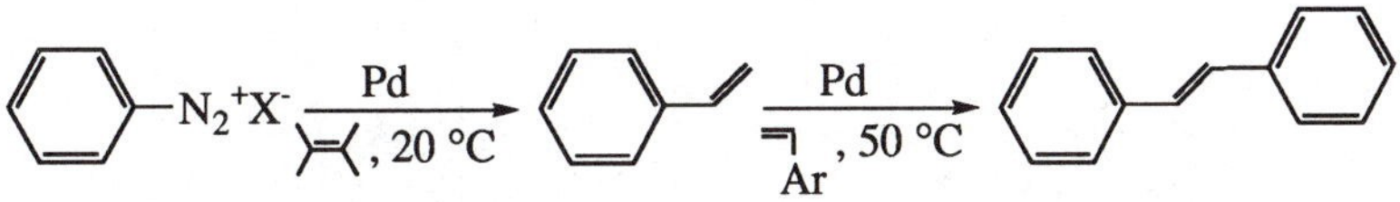

$N_2^+X^-$ → Pd, 20 °C → Pd, Ar, 50 °C →

X = OAc (98/2)

B(OH)(OH) + Br, Br → Pd, Ag_2O, r.temp →

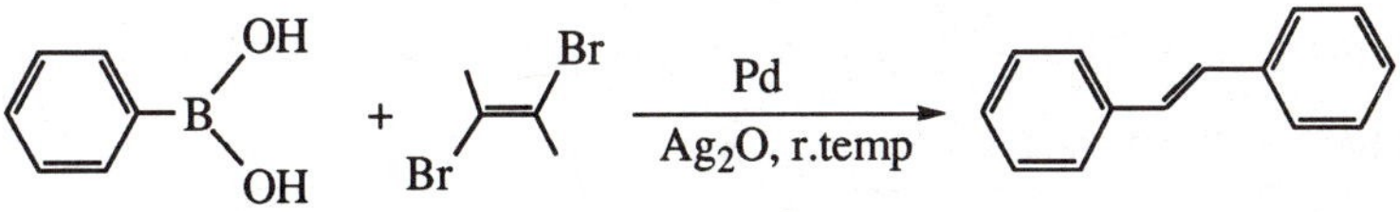

no cis
no 1,1-disubstitution

Aromatic iodo, bromo, fluoro and triflic compounds can be used for educt tuning in the Heck reaction. Using iodo arenes the reaction runs without problems and the use of phosphine ligands can be avoided. With bromo arenes the use of phosphine ligands is obligatory. Phosphines are oxidized in presence of catalyst metals to phosphine oxides very easily. This can happen by small contaminations with oxygen in the reaction mixture. In preparative scale work a higher amount of phosphine can be of advantage. With chloro arenes higher reaction temperatures are necessary (*8*). No quantitative conversions are obtained with chloro arenes. All the halogeno arenes result in similar regioselectivities. The higher tempertures in the sequence J<Br<Cl necessary to obtain oxidative addition may be counterproductive to the expected increase in regioselectivity. The highest regioselectivity was obtained with iodo arenes and Ag_2CO_3 as the base at 50 °C. Another approach is to lower the temperature of oxidative addition by use of diazonium salts. The reaction with ethylene results in high yields of styrene at room temperature but significant formation of stilbene is only observed at 50 °C. Aryl triflates result in a lower regio selectivity. Much higher proportions of 1,1-disubstitution products are formed (*9*). By reaction of phenyl boronic acid with trans-dichloro ethylene trans-stilbene is selectively obtained. This reaction shows no cis-addition and no 1,1-disubstitution. The limiting step with respect to molecular weight is the deboronation which is in the range of 0,1 %. The reaction temperature is reduced to room temperature using Ag_2O as the base (*10*). Substituted dibromo ethylenes can be reacted as well with aryl boronic acids. In this way polyphenylenevinylenes substituted at the double bond are accessable.

Br Br H

Pd | $PhB(OH)_2$ Pd | $PhB(OH)_2$

H

Br + $(HO)_2B$–C$_6$H$_4$–$B(OH)_2$ $\xrightarrow{Pd}$ [ ]$_n$
Br H H

Substituted polyphenylene vinylenes can be obtained by the Heck reaction (Table I).

Table I. Pd-catalyzed synthesis of poly(phenylene vinylene)s

polymer	M_w	therm. properties T_g	melt	solubility
–CH=CH–	—	—	—	—
–CH=CH– (CH_3)	12000	180	aniso.	NMP, 180 °C
–CH=CH– (CF_3)	3500	120	aniso.	NMP, 180 °C
–CH=CH– (CH_3, CH_3)	2600	—	—	NMP, 180 °C
–CH=CH– (C_6H_5)	8000	145	aniso.	$CHCl_3$, room temp.
–CH=CH– (CH_3, CH_3)	11000	130	aniso.	$CHCl_3$, room temp.

Unsubstituted dibromo benzene and centro symmetrically substituted 2,5-dimethyl-1,4-dibromo benzene result in materials with low molecular weight due to solubility problems. Amorphous substituted polyphenylene vinylenes show an anisotropic melt above the glass transition temperature. At about 200 °C this melt solidifies loosing the anisotropy. Insoluble products are formed in an irreversible reaction (*11*).

The optical properties of polymer conjugated hydrocarbons are of increasing importance due to the possibility to induce an electrically stimulated luminescence. Because the fluorescence is depended from the conjugation chain length the information of absorption dependence from the chain length is necessary.

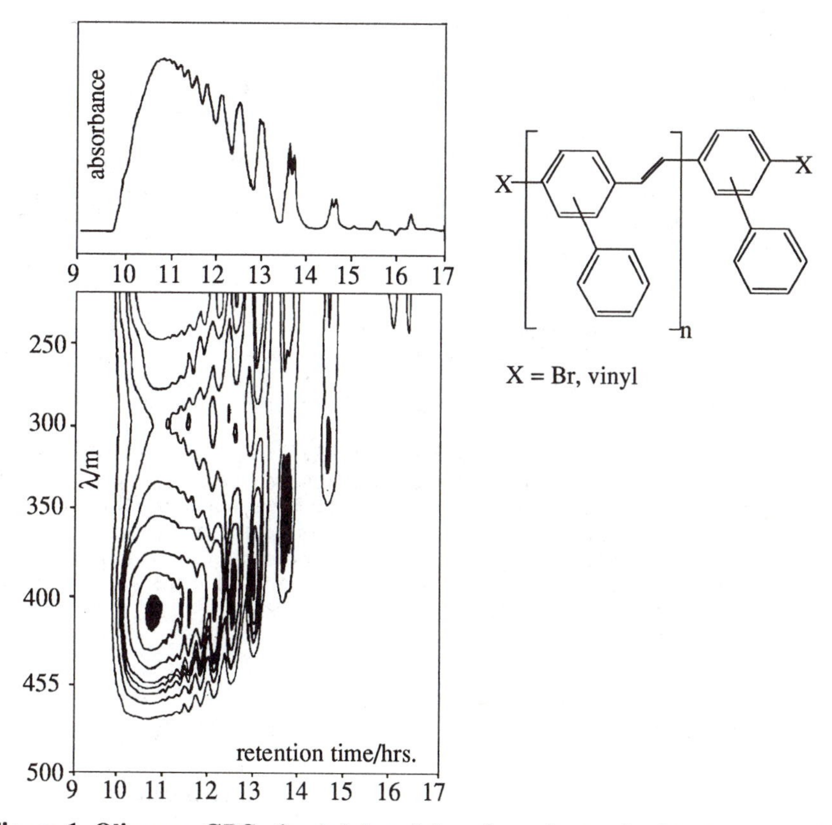

Figure 1. Oligomer GPC of poly(phenylphenylene vinylene) with chain length dependent UV/VIS spectra

This information is obtained from a single sample by high resolution oligomer GPC coupled with a diode array detector. As shown in figure 1 the absorption maximum of phenyl substituted polyphenylene vinylene is approached at a polymerization degree 7-8.

Electroluminescence is observed with very different polymer systems. Polyphenylenes, Polyphenylene ethinylenes or polyhetero cycles can be used besides polyphenylenevinylenes. The application of blends results in an increase of the relative quantum efficiency by a factor of 100 (12).

The synthesis of polyphenylene ethinylene by a Pd/Cu catalyzed reaction was described by Marvel et al. (*13*). The solubility problems of the para structure were reduced by the synthesis of the meta products. Soluble para connected polyphenylene ethinylene were synthesized by Schulz and Giesa in 1990 (*14*). Trimethylsilylacetylene have been used to avoid the preparative problems of stoichiometric use of acetylene.

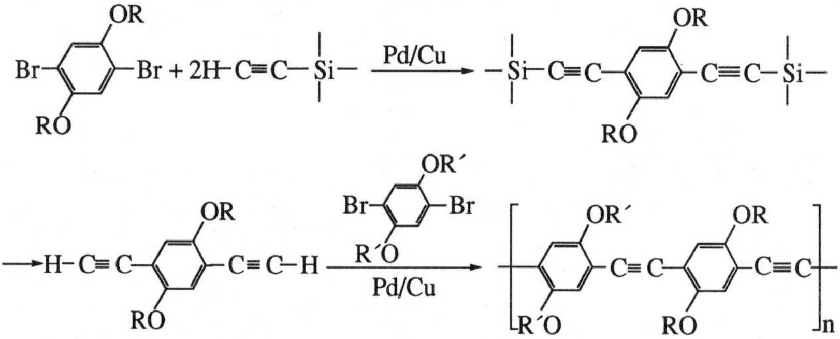

Pd/Cu

Pd/Cu

Using the same synthetic approach a wide variety of structures in this class of polymers were obtained. The soluble polymers were mainly investigated with respect to their NLO properties (*15*). The exponent of 1,92 obtained in the viscosity molecular weight relationship confirms the rod like structure of these polymers.

The model reaction between bromo benzene and phenylacetylene gives information about the chain limiting reactions and defect structures in the polymer (*16*). The dehalogenation can be neglected in contrast to the Heck reaction. Diin and enin formation as well as trimerization are demonstrated in the model reaction. The trimerization and the formation of enin can be suppressed quantitatively. The formation of diin occurs in the range < 0,5 %. It is not clear if the diin is formed during the reaction or on working up.

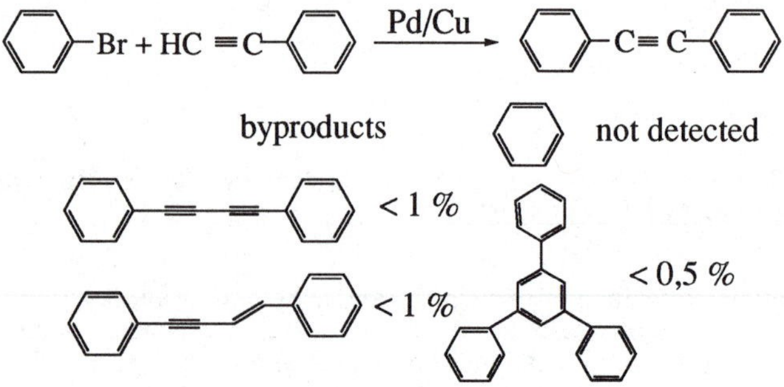

Pd/Cu

byproducts

not detected

< 1 %

< 1 %

< 0,5 %

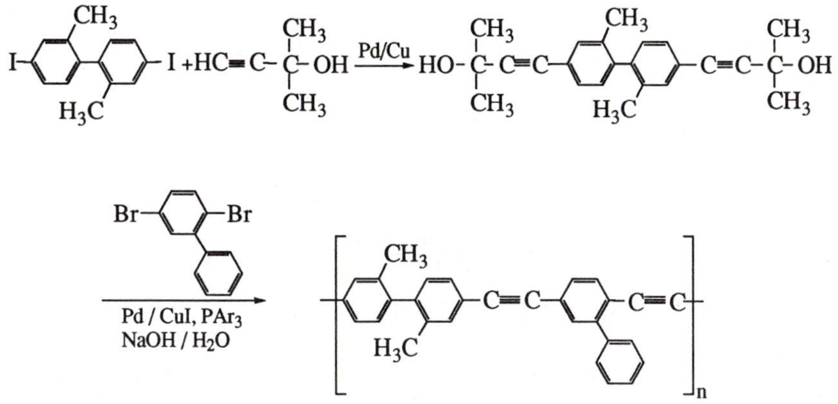

The educt of acetylene with acetone allows a simple synthesis of polyphenylene ethinylene (*17*). The absorption of these polymers is in the same range as polyphenylene vinylene. This absorption is reached at a polymerization degree 7 - 8 also in this case. Oligomers can be used as active components in LED films. The flexibility of catalytic reactions makes a wide variety of derivatives available.

Unsubstituted oligo- and polyphenylenes show the same problem as with the earlier described classes of polymers with stiff chain structure: they are insoluble and non-meltable. Even with methyl substitution the solubility can be drastically improved and the melting points are decreased. Oligo- and polyphenylenes form anisotropic melts (*18*).

The Ullmann reaction (*19*) and the metal mediated oxidative polymerization (*20*) were used for the synthesis. These methods are replaced by metal catalyzed reactions nowadays. Schlüter and Wegner (*21*) used in 1989 the coupling of boronic acids described by Susuki (*22*) to synthesize soluble polyphenylenes. The Pd catalyzed reaction tolerates a large number of functional groups. The diaryl coupling of Kumada can be used as well to synthesize polyphenylenes (*29,30*).

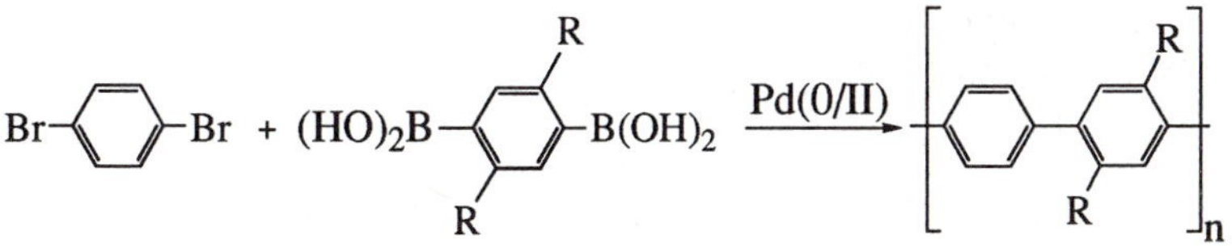

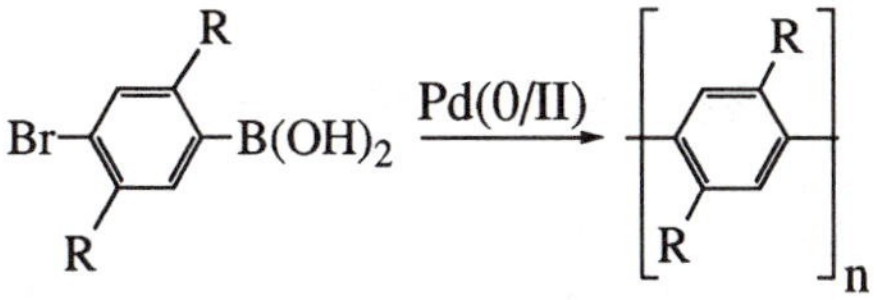

Br–C₆H₃(R)–Br $\xrightarrow[NiCl_2(PPh_3)_2]{Mg}$ [–C₆H₃(R)–]$_n$

catalysis cycle

$$Ni(I)X + ArX \xrightarrow{\text{oxidative addition}} ArNi(III)X_2$$

$$ArNi(III)X_2 + Ar'Ni(II)X \underset{}{\xrightleftharpoons{\text{aryl exchange}}} Ni(II)X_2 + Ar{-}Ni(III)X(Ar')$$

$$Ar{-}Ni(III)X(Ar') \xrightarrow{\text{reductive elimination}} Ar{-}Ar' + Ni(I)X$$

Ni(II) compounds are used usually in this Ni catalyzed reaction. Kochi disproved already in 1979 (25) the obvious assumption that the catalytic cycle is dominated by a change of Ni(0)/Ni(II). The shortened reaction scheme shows that the oxidative addition involves Ni(I)/Ni(III). In the reductive elimination the product fragments are involved. The halogen is eliminated from the reaction cycle as a Ni(II) compound. Methyl substituted polyphenylenes can be obtained according to this reaction scheme. The GPC of the THF soluble part is shown in figure 2. In this GPC the peaks of bi-, quarter-, sexi- and octiphenylene derivatives are included.

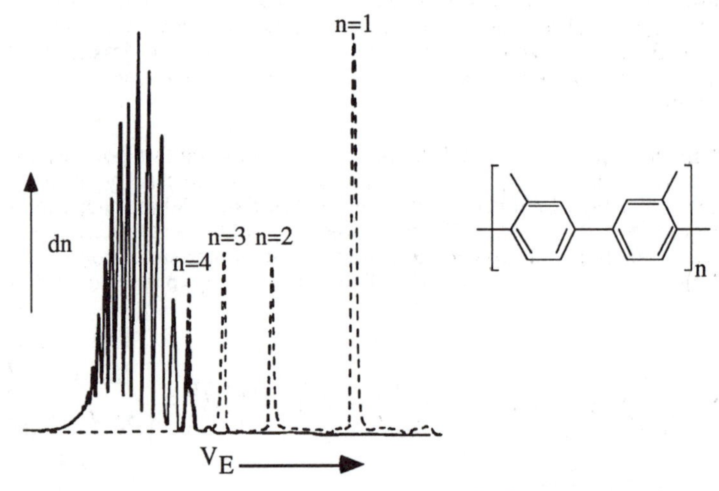

Figure 2. GPC of the THF soluble fraction of poly(1^2,2^3-dimethylbiphenyl) and of model oligomers

Oligophenylenes containing up to 30 p-linked benzene units are visible in the chromatogram.

Br–C₆H₃(C₆H₅)–Br $\xrightarrow[Ni]{Bog\text{-}Mg}$ [–C₆H₃(C₆H₅)–]$_n$

Methyl substituted polyphenylene has a limited thermal stability. The phenyl substituted polyphenylene is soluble in chloroform, anisol and toluene (*26*). DSC shows a step at 180 °C. The melting point of 280 °C is observable only during the first heating. The material solidifies to an anisotropic glass from anisotropic melt.

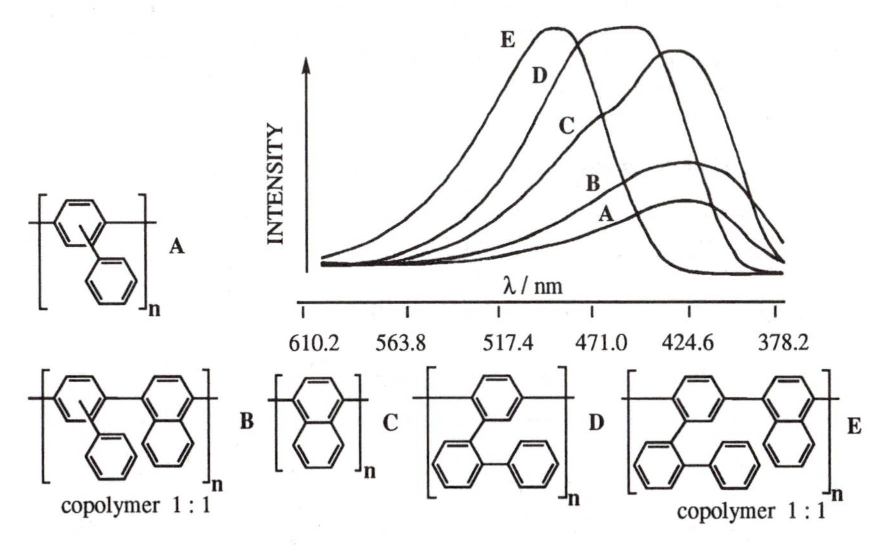

Figure 3. Fluorescence spectra of polyphenylene derivatives

By variation of the pattern of substitution the site of the fluorescence maximum of polyphenylenes can be influenced (*27*). Specially by use of copolymers the width of the range of emission can be changed (Fig. 3).

In the Heck reaction the educt participates in the oxidative addition. The reductive elimination occurs without the target molecule. In the Ni catalyzed polyphenylene synthesis the target molecule participates also in the reductive elimination and halogen is removed in another reaction step. Fundamentally different classes of products can be expected if oxidative addition and reductive elimination involves fragments of the product and no halogen participate in the reaction. This can be expected with ArH activated reactions. The regioselectivity of activation of

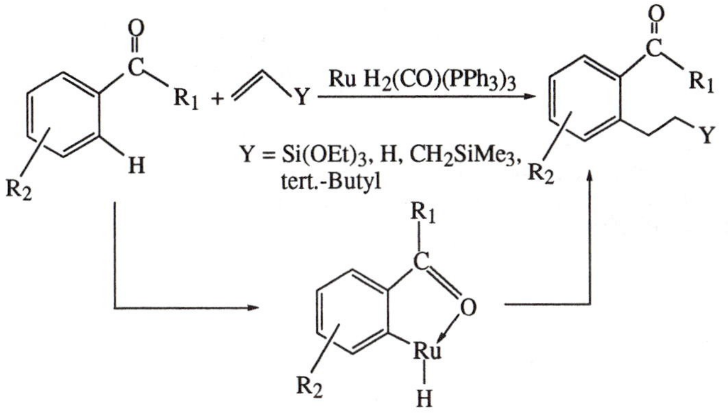

$Ru\,H_2(CO)(PPh_3)_3$

$Y = Si(OEt)_3$, H, CH_2SiMe_3, tert.-Butyl

hydrogen must be guaranteed by the structure of the aromatic units. The selective activation of hydrogen in o-position to carbonyl respectively azomethin groups is known for many years in stoichiometric reactions (*28*). By ruthenium complexes a catalytic reaction is possible (*29*).

Murai describes this reaction with acetophenone as an example. The insertion of the olefin in the o-ArH-bond is observed with high yields. In this reaction the change Ru(O)/Ru(II) is involved. The regioselectivity of the reaction is dependent from the olefin. With α-olefins a migration of the double bond is observed (Table II).

Table II. Model reaction in Ru catalyzed addition of acetophenone

Olefin	Product			
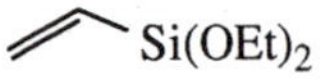 $Si(OEt)_2$	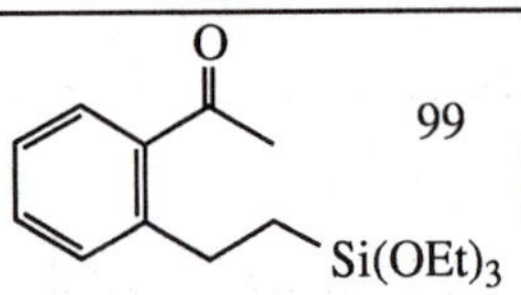 O, $Si(OEt)_3$	99	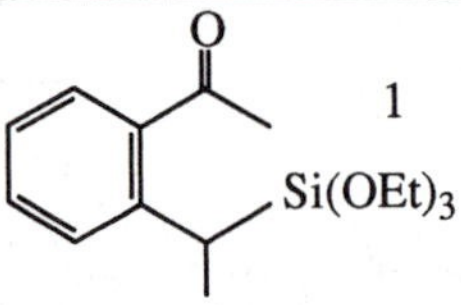 O, $Si(OEt)_3$	1
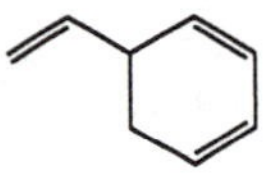	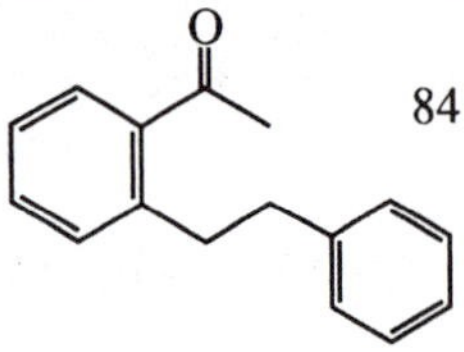 O	84	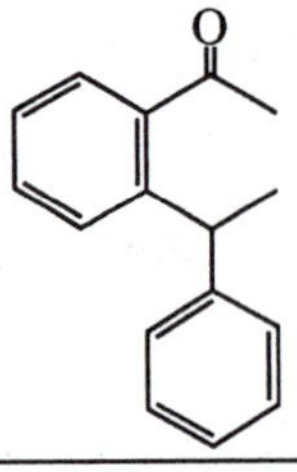 O	16
	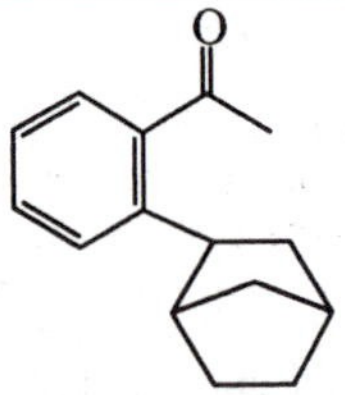 O	100 % exo		
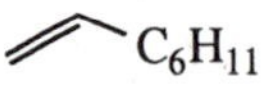 C_6H_{11}	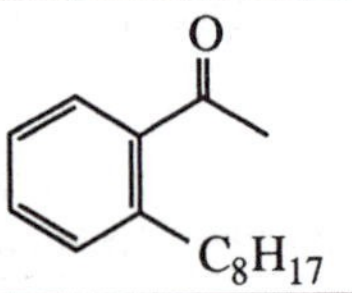 O, C_8H_{17}	+ Isomers of Octen		

Vinylsilylcomponents react with a high regioselectivity. Both ortho positions are accessible to the reaction. There is a strong difference in the reactivity between mono- and disubstitution.

Ar + R $\xrightarrow[90\ h]{Ru}$ $[ArR]_nAr$

At longer reaction times oligomers were formed in this reaction using divinyl compounds (figure 4)

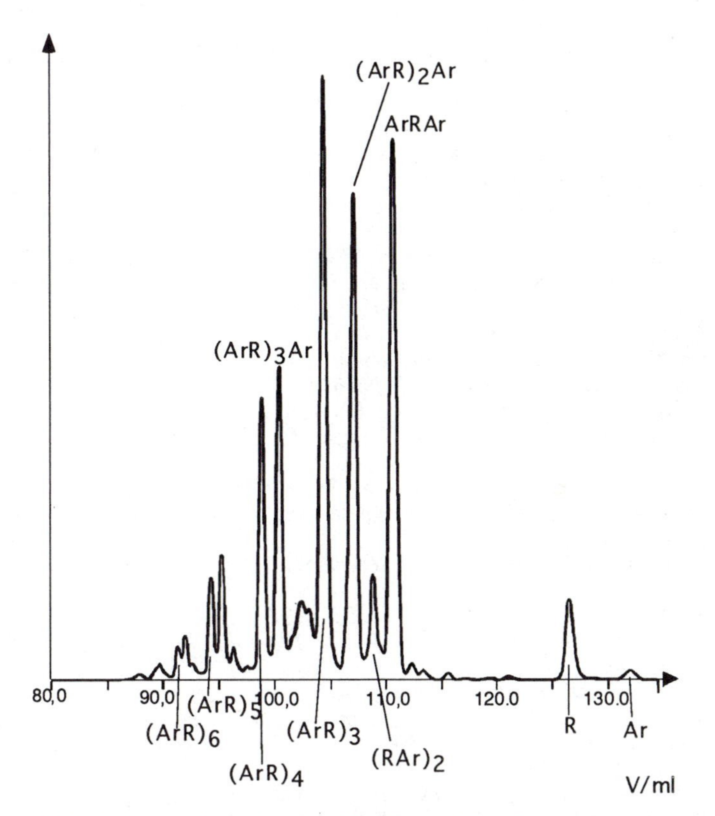

Figure 4. GPC of the product obtained by Ru-catalyzed reaction of acetophenone with 1.3-divinyl-1.1.3.3-tetramethyldisiloxan

Another approach is to distribute the reactive sites in two different aromatic units (*30*) or to use diacetyl compounds.

Literature Cited

1. Bochmann, M.; *Nachr. Chem. Techn. Lab.* **41**, 1220 (1993)
2. Heck, R. F.; *Palladium Reagents in Organic Synthesis*, Acad. Press, London 1985
3. Sen, A.; Lai, T.-W.; *J. Amer. Chem. Soc.* **103**, 4627 (1981)
4. Seehof, N.; Mehler, C.; Breunig, S.; Risse, W.; *J. Mol. Catal.* **76**, 219 (1992)
5. Heitz, W.; Brügging, W.; Freund, L.; Gailberger, M.; Greiner, A.; Jung, H.; Kampschulte, U.; Nießner, N.; Osan, F.; Schmidt, H.-W.; Wicker, M; *Makromol. Chem.* **189**, 119 (1988)
6. Reißig, H.-U.; *Nachr. Chem. Techn. Lab.* **34**, 1066 (1986)
7. Brenda, M.; Greiner, A.; Heitz, W.; *Makromol. Chem.* **191**, 1083 (1990)
8. Martelock, H.; *Dissertation Marburg 1992*
9. Cabri, W.; Candiani, I.; Bedeschi, A.; Santi, R.; *Tetrahedron Lett.* **32**, 1753 (1991)
10. Heitz, W.; Koch, F.; in press
11. Martelock, H.; Greiner, A.; Heitz, W.; *Makromol. Chem.* **192**, 967 (1991)
12. Vestweber, H.; Sander, R.; Greiner, A.; Heitz, W.; Mahrt, R. F.; Bässler, H; *Synthetic Metals* **64**, 141 (1994)
13. Trumbo, D. L.; Marvel, C. S.; *J. Polym. Sci., Polym. Chem. Ed.* **24**, 2311 (1986)
14. Giesa, R.; Schulz, R. C.; *Makromol. Chem.* **191**, 857 (1990)
15. Moroni, M.; Le Moigne, J.; Luzzati, S.; *Macromolecules* **27**, 562 (1994)
16. Varlemann, U.; *Dissertation Marburg 1992*
17. Solomin, V. A.; Heitz, W.; *Macromol. Chem. Phys.* **195**, 303 (1994)
18. Heitz, W.; *Chemiker Ztg.* **110**, 385 (1986)
19. Kern, W.; Gehm, R.; *Angew. Chem.* **62**, 337 (1950)
20. Heitz, W.; Ullrich, R.; *Makromol. Chem.* **98**, 29 (1966)
21. Rehahn, M.; Schlüter, A. D.; Wegner, G.; Feast, W. J.; *Polymer* **30**, 1060 (1989)
22. Miyaura, N.; Yanagi, T.; Susuki, A.; *Synth. Commun.* **11**, 513 (1981)
23. Tamao, K.; Sumitani, K.; Kumada, M.; *J. Amer. Chem. Soc.* **94**, 4374 (1972)
24. Yamamoto, T.; Hayashi, Y.; Yamamoto, A.; *Bull. Chem. Soc. Jpn.* **51**, 2091 (1978)
25. Kochi, J. K.; *Pure Appl. Chem.* **52**, 571 (1980)
26. Noll, A.; Siegfried, N.; Heitz, W.; *Makromol. Chem., Rapid Commun.* **11**, 485 (1990)
27. Fiebig, W.; *Dissertation Marburg 1993*
28. Collman, J. P.; Hegedus, L. S.; Norton, J. R.; Finke, R. G.; *Principles and Applications of Organotransition Metal Chemistry*, University Sci. Books, Mill Valley, Cal. 1987
29. Murai, S.; Kahiuchi, F.; Sekemi, S.; Tanaka, Y.; Kamatani, A.; Sonoda, M.; Chatani, N.; *Nature* **366**, 529 (1993)
30. Guo, H.; Tapsah, M.; Weber, W.; *Polymer Bulletin* **34**, 49 (1995)

RECEIVED December 11, 1995

Chapter 3

Palladium-Catalyzed Carbonylation Polymerizations

Robert J. Perry

Imaging Research and Advanced Development, Eastman Kodak Company, Rochester, NY 14650–1705

A variety of polymeric systems can be made from the Pd-catalyzed carbonylation and coupling reactions of aromatic iodides and amines or phenols. The relatively mild reaction conditions are tolerant of numerous functional groups and allow for the production of high molecular weight aramids, poly(imide-amides) and polybenzoxazoles. Polyimides and polyesters are also formed, although not as efficiently. Certain chloroaromatics will also undergo efficient polymerization reactions when accelerated by iodide ion. Aromatic triflates can also be used in a manner similar to the aromatic halides to give aramids, polyesters and poly(imide-amides). This metal-mediated method offers an alternate, and complementary, route to polymers that may be made through conventional means.

A large number of high-performance polymer systems are formed by the condensation reactions of diamines, or diols with aromatic diacids, or their derivatives. These include commercially successful products such as aramids, polyimides, poly(imide-amides), and polyesters. While some of the monomers used in the production of these polymers are readily available and made in large quantities, it is becoming more common to use smaller quantities of speciality monomers to tailor the properties of the resultant polymers to suit a specific application. To address this need, chemistries using unconventional monomers are being explored to offer alternative routes to a variety of polymeric systems.

In 1974, Heck reported on the formation of amides and esters via the palladium-mediated carbonylation and coupling reactions of aromatic halides and amines, or alcohols (*1*). The mechanism of the reaction, as outlined in Scheme I for amide formation, involves a coordinatively unsaturated Pd(0) species oxidatively adding to the aromatic halide producing a Pd(II) complex, **1**. Carbon monoxide then inserts in the aryl-palladium bond giving aroyl complex **2**. This is followed by attack of the amine leading to the regeneration of the active Pd(0) catalyst and liberation of the free amide **3** (*1,2*). (An alternate

0097–6156/96/0624–0071$12.00/0

possibility involves nucleophilic attack of the amine on a coordinated CO bound to the arylpalladium complex **1** to give an arylcarbamoylpalladium intermediate. This route has been established as a pathway when secondary aliphatic amines are used (*3*).) In this scheme L represents unspecified ligands, usually phosphines, amines or CO.

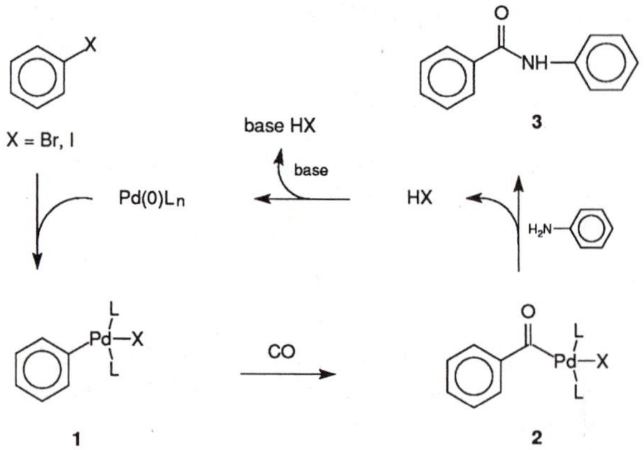

X = Br, I

Pd(0)Ln

base HX

base

HX

H_2N

CO

1

2

3

Scheme I

In the last several years a variety of polymeric systems have been prepared using variations on this chemistry. Herein are described several approaches to high-performance polymer systems using palladium-catalyzed polymerization reactions.

Aramids

Aromatic polyamides (aramids) were the first class of polymers synthesized by palladium-catalyzed carbonylation and condensation reactions. It was first shown that aromatic diamines and aryldibromides would react under 1 atm carbon monoxide (CO) in the presence of a tertiary amine base, a palladium catalyst and phosphine ligands in a dipolar aprotic solvent to form aramids of modest molecular weight (equation 1, X = Br) (*4*).

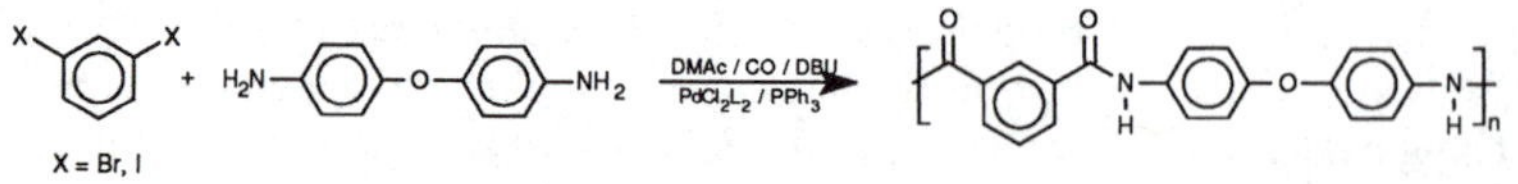

X = Br, I; + H_2N–O–NH_2; DMAc / CO / DBU; $PdCl_2L_2$ / PPh_3 (1)

At the time this work was published, we were engaged in exploratory research aimed at using iodinated aromatic compounds as monomers for high-performance polymers. Several reports had disclosed the preparation of a variety of regiospecifically bisiodinated aromatic systems (*5*). Model studies conducted in our laboratories between iodobenzene, aniline and CO indicated that, under appropriate conditions, amide formation was fast, clean, and quantitative (*6*).

We found that if aromatic diiodides were used, rather than bromides, and elevated CO pressures were employed, aramids of much higher molecular weight were obtained in much shorter reaction times (Figure 1) (*7*). The difference in rate of molecular weight build-up and the sensitivity difference to CO pressure between the diiodo and dibromobenzene is due to a change in the rate-determining step. It is well documented that zero-valent palladium compounds undergo oxidative addition to aryl iodides faster than the corresponding aryl bromides (*8*). It is also known that for the esterification of aryl bromides with CO, palladium and an alcohol, oxidative addition is the rate-limiting step (*9*). Higher CO pressures have been shown to promote a greater degree of CO coordination to Pd(0) (*10*) rendering the palladium less nucleophilic and thus slowing the oxidative addition reaction (*11*). In the dibromoaromatic polymerization, the rate-determining step is the oxidative addition of palladium, and increasing the CO pressure suppresses this reaction. In the diiodo case, the rate-determining step is CO insertion. Increasing the CO pressure increases the rate of reaction and the rate of molecular weight gain. This behavior mimics that seen in model compound studies. That is, higher CO pressures accelerate the overall amidation rate for iodoaromatics and suppresses amidation of the bromo analogs (*7*).

Use of diiodinated aromatics allowed the formation of a variety of aramids (Table I). The optimum stoichiometry for achieving maximum molecular weight in the reaction of m-dihalobenzene and 4,4'-diaminodiphenylether was to use a slight excess of the diamine. Model studies indicated that this decreased the likelihood of reduction of the aryl-halide bond that would effectively terminate the polymer chain (*6*). The other aramid reactions were not optimized but used conditions found optimal for equation 1 (X = I).

Poly(imide-amides)(PIAs)

The same technology used for aramid synthesis was applied towards PIAs (equation 2). In all these cases, the imide moiety was preformed. This eliminated any postreaction curing commonly seen in commercial PIAs prepared by the condensation of trimellitic acid chloride anhydride and diamines (*12*). High molecular weight PIAs were formed by the carbonylation and condensation of diamines and aryldiiodides, containing one or two preformed imide moieties. This provided the option of having a variable ratio of imide to amide groups, which might influence mechanical properties such as modulus and elongation. Table II indicates that this methodology was successful for homopolymers, although copolymers were also made (*13*).

$$\text{I-C}_6\text{H}_3(\text{CO})_2\text{N-C}_6\text{H}_4\text{-I} + H_2N\text{-Ar-}NH_2 \xrightarrow[\text{base / CO}]{\text{Pd catalyst}} \left[\text{-OC-C}_6\text{H}_3(\text{CO})_2\text{N-C}_6\text{H}_4\text{-CONH-Ar-NH-} \right]_n \quad (2)$$

Table I. Aramids from Diiodoaromatics[a]

Diiodo Monomer	*Diamine Monomer*	*Ratio (%)[b]*	*Mw[c]*	*Mn[c]*
1,3-$I_2C_6H_4$	H_2N–C_6H_4–O–C_6H_4–NH_2	95	202 000[d]	100 000[d]
I–C_6H_4–O–C_6H_4–I	H_2N–C_6H_4–O–C_6H_4–NH_2	95	86 000	44 500
I–C_6H_4–SO_2–C_6H_4–I	H_2N–C_6H_4–O–C_6H_4–NH_2	95[e]	30 800	19 100
2,6-$I_2C_{10}H_6$	H_2N–C_6H_4–O–C_6H_4–NH_2	95[f]	33 500	22 300
1,4-$I_2C_6H_4$	H_2N–C_6H_4–O–C_6H_4–NH_2	96[g]	26 500	17 200
I–C_6H_4–C_6H_4–I	H_2N–C_6H_4–O–C_6H_4–NH_2	97[h]	44 800	27 100
$I_2C_6H_2(CH_3)_2$ (H_3C, CH_3)	H_2N–C_6H_4–O–C_6H_4–NH_2	97[i]	30 700	8 200

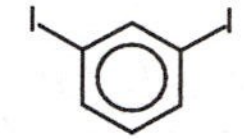	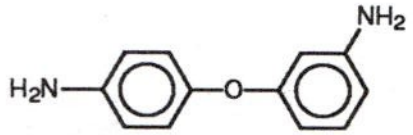	97[i]	68 300	26 700
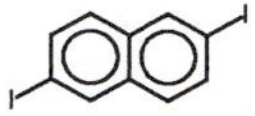	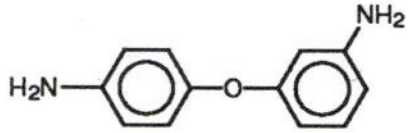	97[i]	88 400	36 600
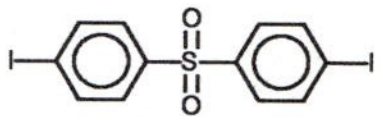	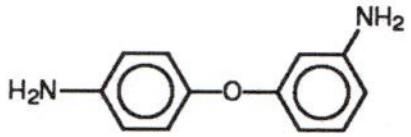	97[e]	54 100	31 700
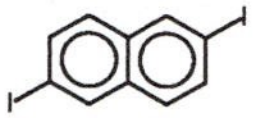	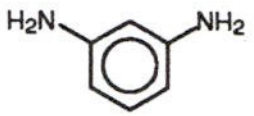	95	37 600	21 000
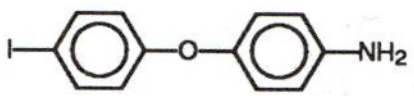		–	23 800[j]	14 200[k]

[a]Reaction conditions: DMAc (0.18–0.26 M), 115°C, 90 psig CO, 2.4 equiv DBU, 6% $PdCl_2(PPh_3)_2$, 4-6 h, reprecipitated 2x from MeOH. [b]% Of the stoichiometric amount of diiodo compound used relative to 100% diamine. [c]Absolute molecular weights. [d]PMMA equiv molecular weights. [e]2 Equiv LiCl added to reaction. [f]3 Equiv LiCl added to reaction. [g]Reaction run in NMP. [h]Reaction run in NMP with 4 equiv LiCl. [i]PMMA equiv molecular weights, single precipitation of polymer. [j]PMMA equiv Mw = 45 300. [k]PMMA equiv Mn = 25 700. Reproduced with permission from ref. 7a.

Table II. Poly(imide-amides) from Diiodoarylphthalimides[a]

Diiodoimide	*Diamine*	*Mn*[b]	*Mw*[b]
	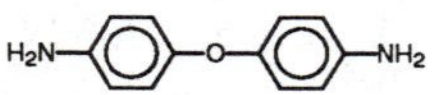	97 000[c]	40 400
	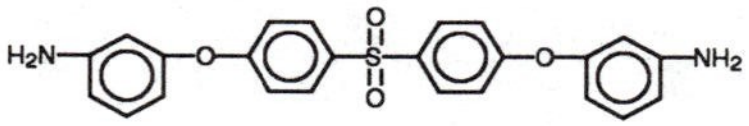	153 000[d]	59 000
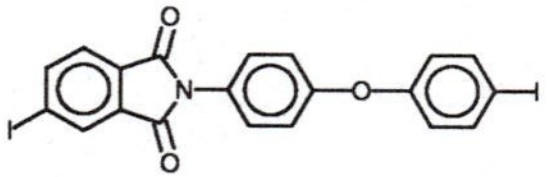	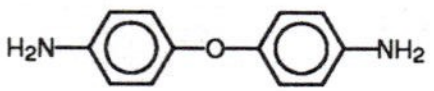	151 000[e]	45 000
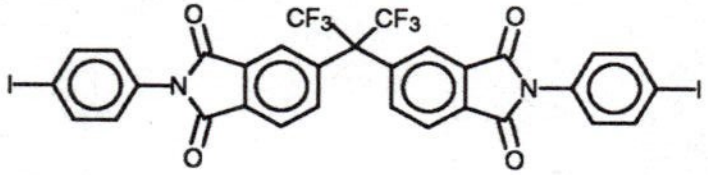	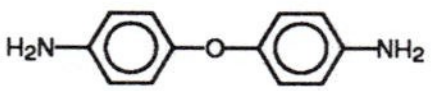	147 000[f]	71 200
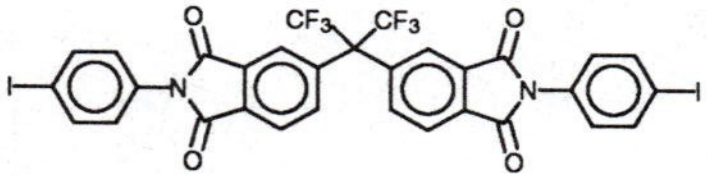	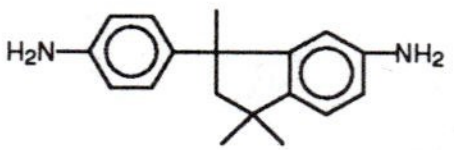	86 000[c]	44 000

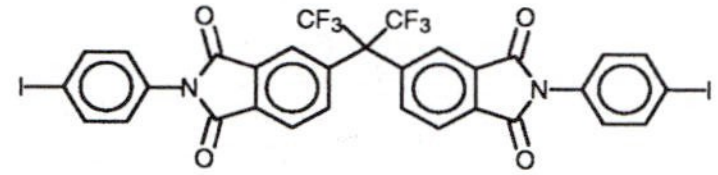	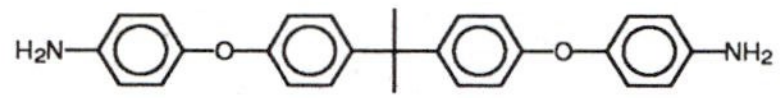	102 000[g]	43 600
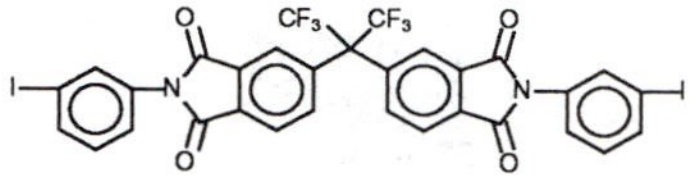	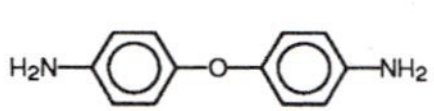	163 000[e]	72 300
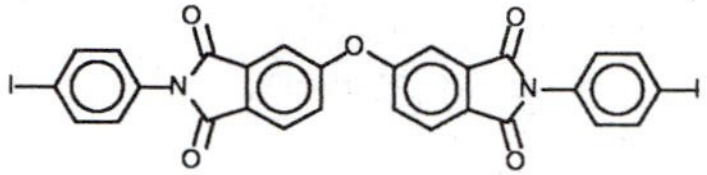	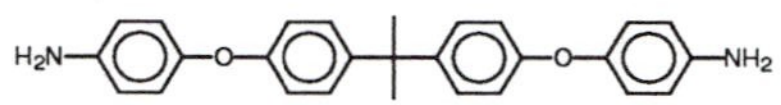	125 000[e]	48 700
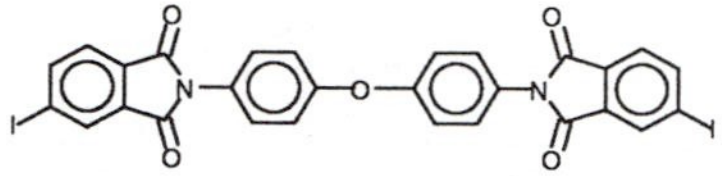	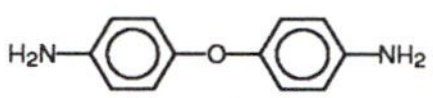	121 000[g]	49 900

[a]Reaction conditions: DMAc (0.18–0.26 M), 100°C, 95 psig CO, 2.4 equiv DBU, 3–6% $PdCl_2(PPh_3)_2$, 2–4 h. [b]PMMA equiv molecular weights. [c]5 mol% Excess diamine was used. [d]1.4 mol% Excess diiodide was used. [e]Equiv molar amounts of diamine and diiodiode were used. [f]3 mol% Excess diamine was used. [g]1 mol% Excess diamine was used.

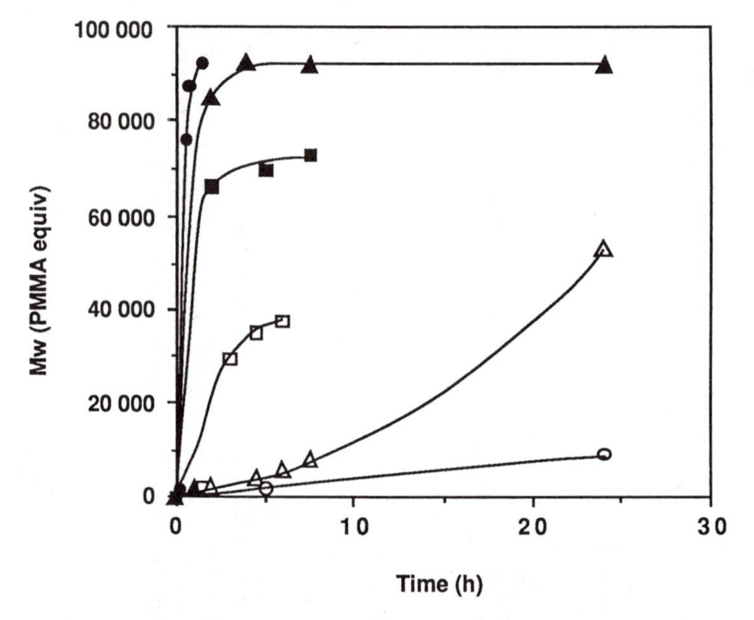

Figure 1. Effect of CO pressure on aromatic polyamide formation between diiodoaromatics and dibromoaromatics. Reaction in DMAc (0.33 M), 90°C, 3% $PdCl_2L_2$, 6% PPh_3 and 1.2 equiv DBU. □ bromo, 1 atm; ■ iodo, 1 atm; △ bromo, 20 psig CO; ▲ iodo, 20 psig CO; ○ bromo, 90 psig CO; ● iodo, 90 psig CO. Reproduced with permission from ref. 7a.

During the course of these reactions, we also found that the imide moiety was activating enough to allow chloro-substituted phthalimides to react in a manner similar to the diiodo compounds above. Enhanced rates were seen when iodide ion was present in the reaction mixture (*14*). Thus, a variety of PIAs derived from 4-chlorophthalic anhydride were made, (Table III).

The best yields and highest molecular weight materials were formed when low CO pressures were employed and iodide was present in the reaction mixture, either as an added salt or generated in-situ from one of the chloroiodo monomers. The reaction of iodide with a Pd(0) species was thought to form an anionic palladium(0) complex, which was more reactive in initial oxidative addition reactions than the neutral complex (*14*).

Polyimides

Several approaches were taken to extend the carbonylation chemistry to the formation of polyimides. Although variously substituted phthalimides could be easily made (*15*), preparation of high molecular weight linear polyimides was more difficult. Reaction of bis(o-diiodoaromatics) with diamines and CO (equation 3) gave materials with broad polydispersities, branching and residual iodide, and amide groups (*16*).

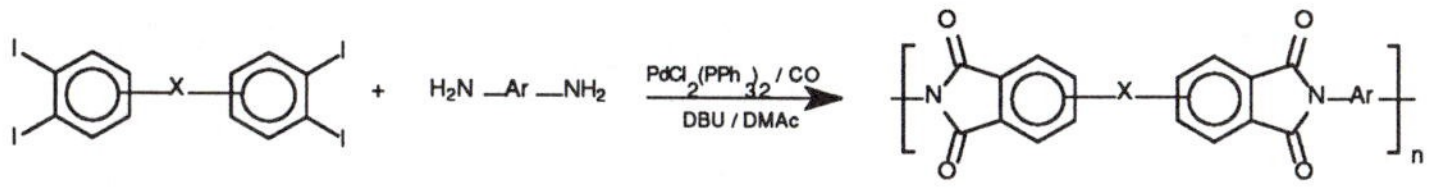

$$+ \; H_2N{-}Ar{-}NH_2 \xrightarrow[\text{DBU / DMAc}]{PdCl_2(PPh_3)_2 \,/\, CO} \qquad (3)$$

Attempts to circumvent these problems by using mixed halo-aromatic systems and varying CO pressures was not successful (equation 4).

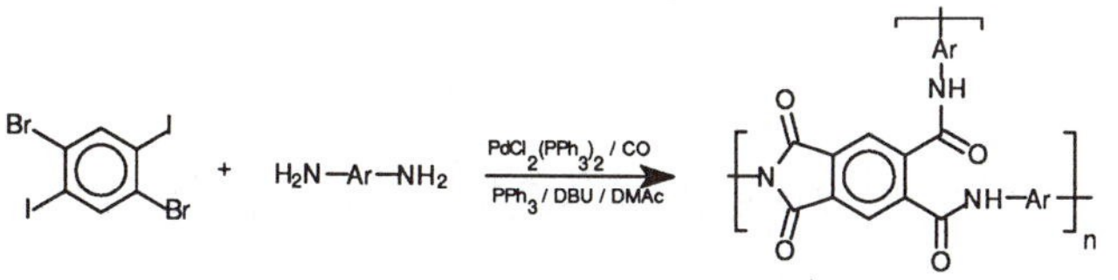

$$+ \; H_2N{-}Ar{-}NH_2 \xrightarrow[PPh_3 \,/\, \text{DBU / DMAc}]{PdCl_2(PPh_3)_2 \,/\, CO} \qquad (4)$$

A different approach to the imide structure incorporated the carbonylation chemistry with conventional poly(amide-ester) ring-closing procedures. Bis(*o*-iodoesters) were allowed to react with diamines and CO in the presence of a Pd catalyst. The intermediate poly(amide-esters) were unstable under the reaction conditions and spontaneously cyclized in the presence of base giving the fully imidized polymer (equation 5). It was found that *t*-butylesters gave the best results (Table IV) (*17*).

Table III. Poly(imide-amides) from chloroiodoimides[a]

Chloroiodoimide	*Diamine*	*Mn[b]*	*Mw[b]*
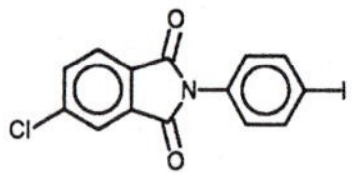 Cl, N, O, O, I	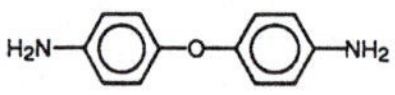 H_2N, O, NH_2	89 000[c]	47 800
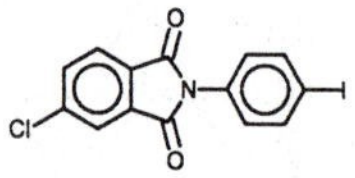 Cl, N, O, O, I	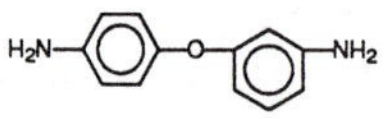 H_2N, O, NH_2	49 300[d]	27 000
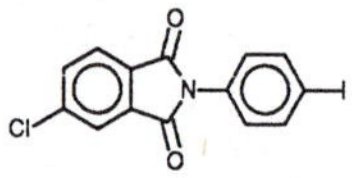 Cl, N, O, O, I	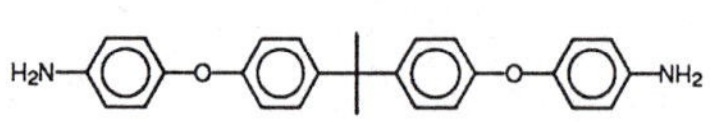 H_2N, O, O, NH_2	90 600[e]	42 100
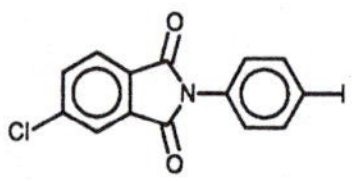 Cl, N, O, O, I	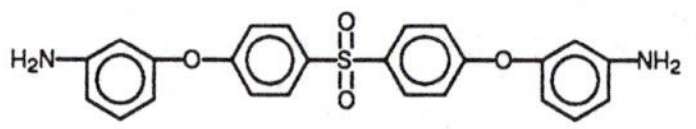 H_2N, O, O, S, O, O, NH_2	64 400[f]	32 000
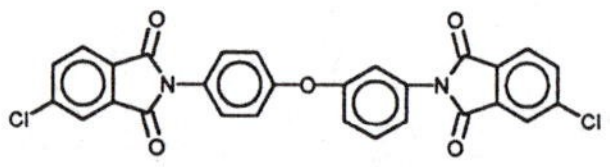 Cl, N, O, O, O, N, O, O, Cl	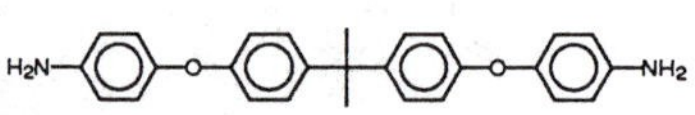 H_2N, O, O, NH_2	72 100[g]	38 600

[a]Reaction conditions: NMP or DMAc (0.18–0.26 M), 100°C, 95 psig then 20 psig CO after 2.6 h, 2.4 equiv DBU, 0.3–1.4% $PdCl_2(PPh_3)_2$, 24–48 h. [b]PMMA equiv molecular weights. [c]1 mol% Excess diamine was used. [d]1 mol% Excess iodochloro compound was used. [e]4 mol% Excess iodochloro compound was used. [f]6 mol% Excess diamine was used. [g]Equiv molar amounts of reactants were used.

Table IV. Polyimides from Bis(o-iodoesters)[a]

Bis(o-iodoester	*Diamine*	Mn^b	Mw^b
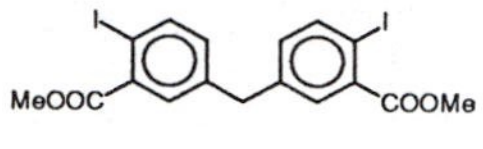 I, MeOOC, I, COOMe	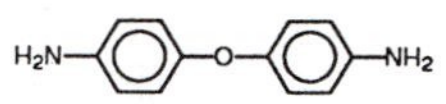 H_2N, O, NH_2	10 400	7 200
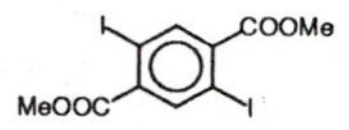 I, COOMe, MeOOC, I	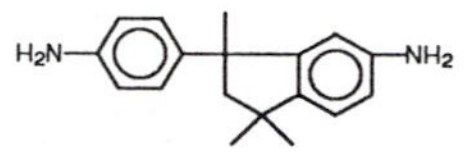 H_2N, NH_2	8 200	6 200
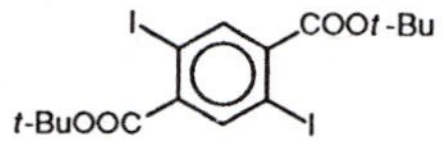 I, COO*t*-Bu, *t*-BuOOC, I	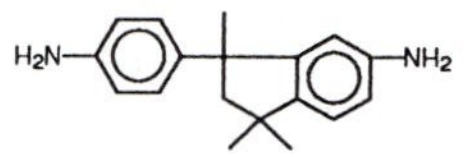 H_2N, NH_2	46 500	21 000
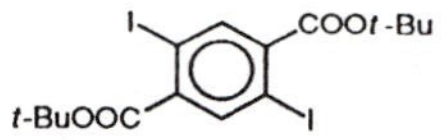 I, COO*t*-Bu, *t*-BuOOC, I	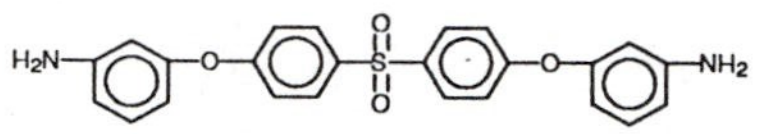 H_2N, O, O, S, O, O, NH_2	27 2000	13 900

[a]Reaction conditions: DMAc (0.33 M), 115–120°C, 95 psig CO, 2.4 equiv DBU, 3% $PdCl_2(PPh_3)_2$, 7–24 h. [b]PMMA equiv molecular weights.

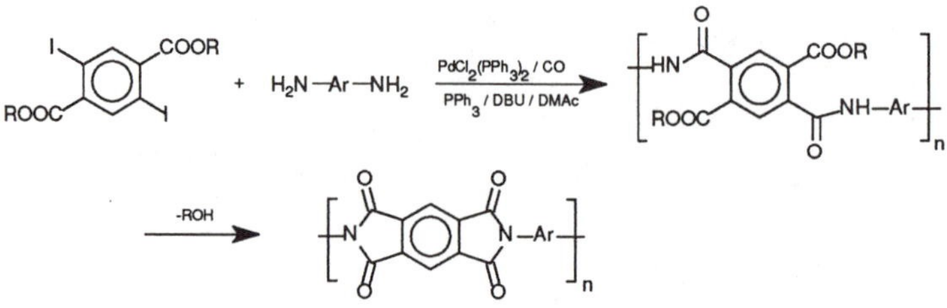

(5)

Poly(benzoxazoles) (PBOs)

Another class of fused-ring heterocyclic polymers produced by the carbonylation technology were the PBOs. Diiodoaromatics were again used but in conjunction with bis(*o*-aminophenols) (equation 6). Earlier work on the preparation of 2-arylbenoxazoles from amines and *o*-aminophenols showed that the intermediate 2-hydroxybenzamide did not spontaneously undergo cyclization to the benzoxazole (*18*).

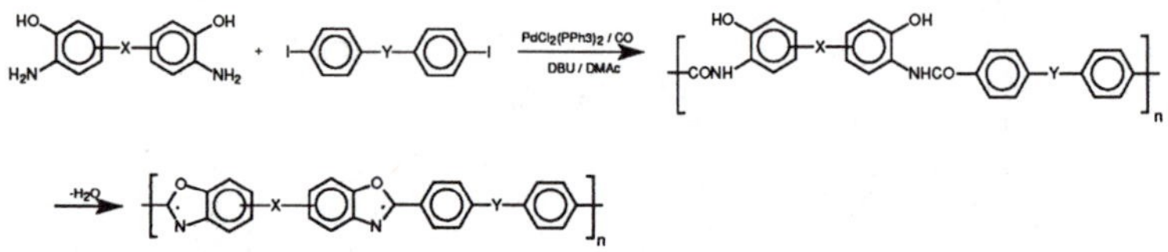

(6)

The phenol groups in the polymeric system were also sufficiently unreactive under the reaction conditions to allow isolation of the intermediate poly(amide-ol) (Table V). These PBO prepolymers were soluble in dipolar aprotic solvents and could be precipitated, purified, and stored until needed. Thermal gravimetric analysis (TGA) of these polymers showed a weight loss at 250-290°C, corresponding to the loss of two water molecules per repeat unit (Figure 2). Simple thermal or chemical cyclization could then be used to produce the thermally stable PBO (*19*). Figure 3 shows the same PBO prepolymers after heating at 325°C for 3 h. The resultant PBO polymers were stable in nitrogen to 460-540°C.

Polyesters

Much less work has been devoted to optimizing polyester formation. However, diols also react with aromatic diiodides and CO to form polyesters of modest molecular weight (equation 7). Longer reaction times are needed for the much less nucleophilic phenol (compared to the aniline derivatives), which allows more time for chain-limiting side reactions to occur (*20*).

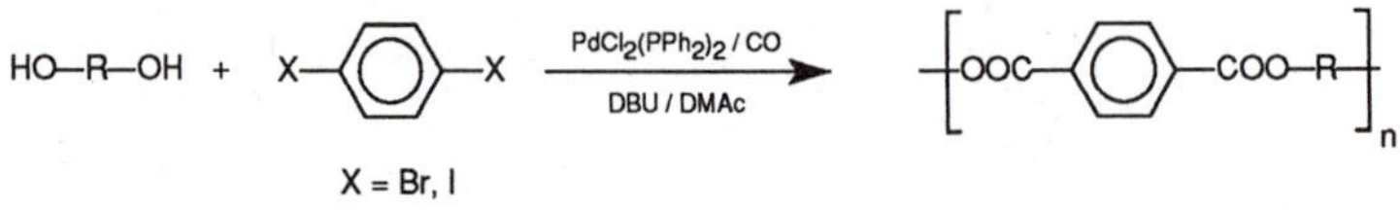

$$HO{-}R{-}OH + X{-}C_6H_4{-}X \xrightarrow[DBU\,/\,DMAc]{PdCl_2(PPh_2)_2\,/\,CO} {-}[OOC{-}C_6H_4{-}COO{-}R]_n{-} \quad (7)$$

X = Br, I

Table V. Poly(amide-ols)[a]

Entry	*Bis (o-aminophenol)*	*Diiodoaromatic*	Mn^b	Mw^b
a	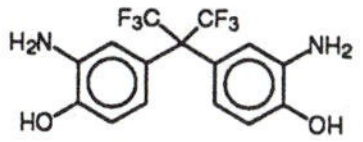	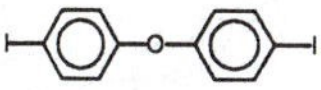	45 7000	24 4000
b	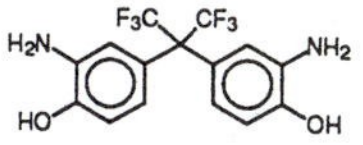		41 700	30 700
c	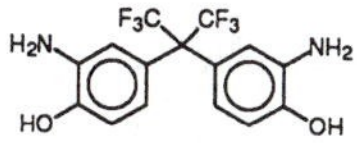	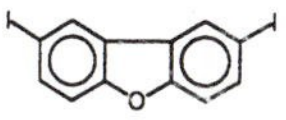	33 900	22 700
d	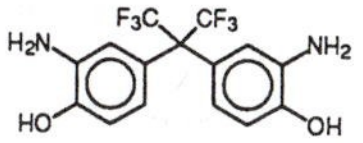	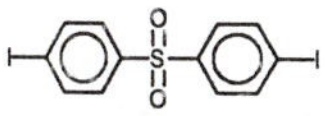	42 000	25 500
e	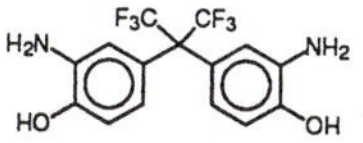	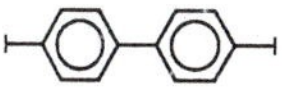	39 100	19 500
f	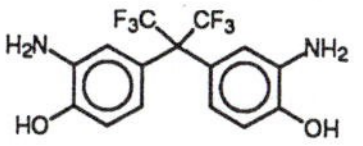		49 400	28 700
g	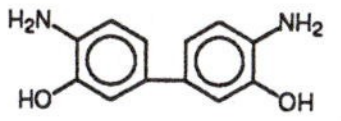	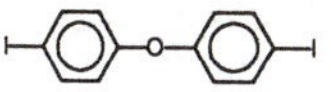	34 100	12 500

[a]Reaction conditions: DMAc (0.16 m), 115°C, 95 psig CO, 2.4 equiv DBU, 0.2% $PdCl_2(PPh_3)_2$, 4–6 h. [b]PMMA equiv molecular weights. Reproduced with permission from ref. 19.

Triflates

Aromatic bis(trifluoromethanesulfonates) (triflates) have also been used instead of aromatic dihalides for polymer syntheses. Easily prepared from the corresponding bisphenols, triflates are known to undergo oxidative addition and CO insertion reactions *(21)*, and were used in preparing aramids *(22)*, PIAs *(23)* and polyesters *(24)*. Reactions were performed in a manner similar to that reported for the iodo analogs but only modest molecular weights were achieved (Table VI).

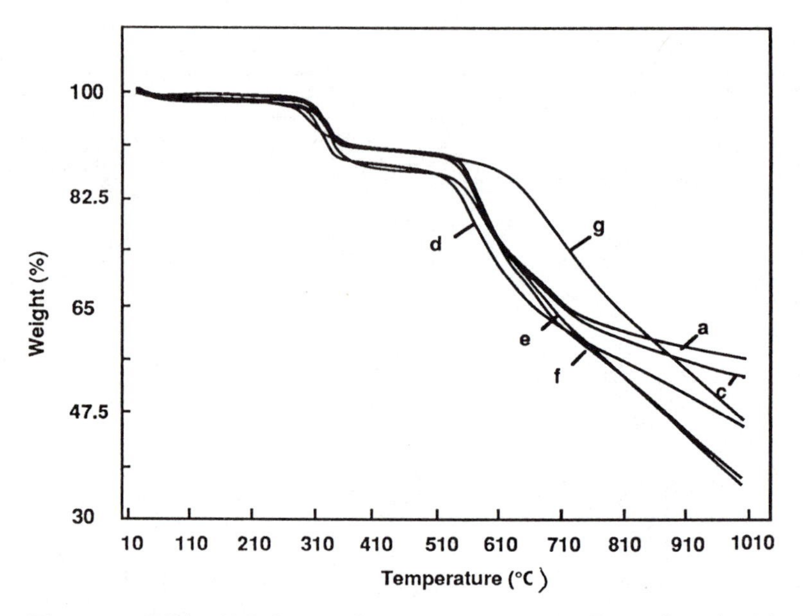

Figure 2. TGA of PBO prepolymers in nitrogen. Reproduced with permission from ref. 19.

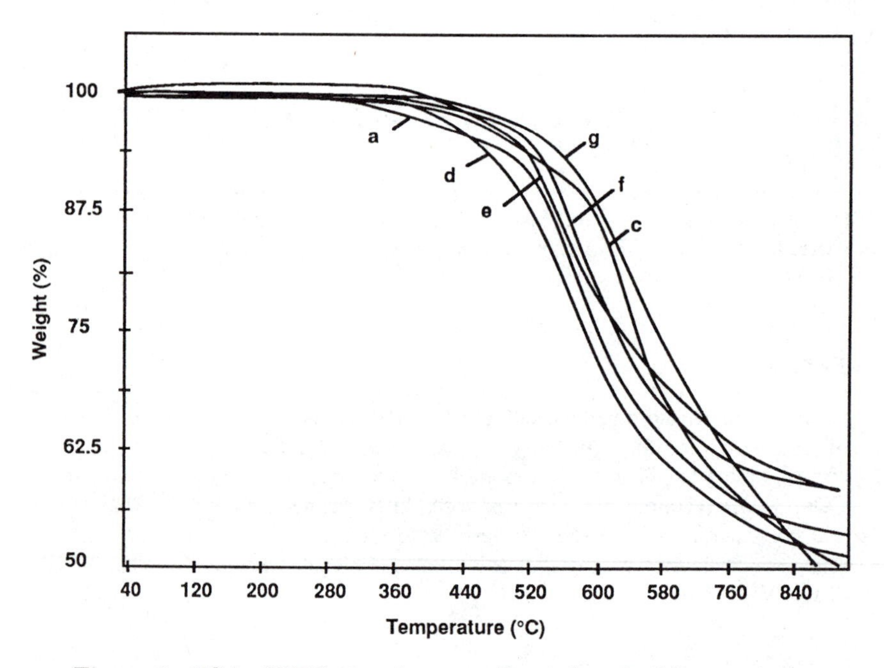

Figure 3. TGA of PBOs in nitrogen. Reproduced with permission from ref. 19.

Table VI. Polymers from Bistriflates[a]

Dinucleophile	*Bis(triflate)*	*Mn[b]*	*Mw[b]*
H_2N–C6H4–O–C6H4–NH_2	TfO–C6H4–C(Me)(Me)–C6H4–OTf	13 100	8 500
H_2N–C6H4–O–C6H4–C(Me)₂–C6H4–O–C6H4–NH_2	TfO–C6H4–C(Me)(Me)–C6H4–OTf	15 300	8 900
H_2N–C6H4–O–C6H4–SO_2–C6H4–O–C6H4–NH_2	TfO–C6H4–C(Me)(Me)–C6H4–OTf	12 300	8 300
H_2N–C6H4–O–C6H4–SO_2–C6H4–O–C6H4–NH_2	TfO–C6H4–C6H4–OTf	10 100	7 000
H_2N–C6H4–O–C6H4–C(Me)₂–C6H4–O–C6H4–NH_2	TfO–C6H4–N(phthalimide)–SO_2–(phthalimide)N–C6H4–OTf	11 300	6 900
H_2N–C6H4–O–C6H4–NH_2	TfO–C6H4–N(phthalimide)–C(CF_3)(CF_3)–(phthalimide)N–C6H4–OTf	6 900	4 600
HO–C6H4–C(Me)(Me)–C6H4–OH	TfO–C6H4–C(Me)(Me)–C6H4–OTf	7 500	4 200
HO–C6H4–C(CF_3)(CF_3)–C6H4–OH	TfO–C6H4–C(Me)(Me)–C6H4–OTf	3 400	2 300

[a]Reaction conditions: 95 psig CO, 115°C in DMAc, 2.4 equiv DBU, 3 mol% $Pd(OAc)_2$ with 6 mol% DPPP (1,3-diphenylphosphinopropane) ligand for amide reactions; 3 mol% $PdCl_2$(DPPE) with 3 mol% DPPE (1,2-diphenylphosphinoethane) ligand for PIA and polyester reactions. [b]PMMA equiv molecular weights for aramids, absolute molecular weights for PIAs, and polystyrene equiv molecular weights for polyesters.

Summary

This brief overview has shown that a variety of polymeric systems can be made from the Pd-catalyzed carbonylation and coupling of aromatic iodides and amines or phenols. The relatively mild reaction conditions are tolerant of numerous functional groups and allow for the production of high molecular weight aramids, poly(imide-amides) and polybenzoxazoles. Polyimides and polyesters were also formed, although not as efficiently. Certain chloroaromatics were also found to undergo efficient polymerization reactions when accelerated by iodide ion. Finally, aromatic triflates could also be used in a manner similar to the aromatic halides to give aramids, polyesters, and poly(imide-amides). Although there are many well-established processes for the syntheses of these classes of polymers, this metal-mediated method offers an alternate, and complementary, route to polymers that may otherwise be difficult to prepare.

Literature Cited

1. (a) Schoenberg, A.; Heck, R. F. *J. Org. Chem.* **1974**, *39*, 3327. (b) Schoenberg, A.; Bartoletti, I.; Heck, R. F. *J. Org. Chem.* **1974**, *39*, 3318.
2. Collman, J. P.; Hegedus, L. S.; Norton, J. R.; Finke, R. G. *Principles and Applications of Organotransition Metal Chemistry*; University Science Books: Mill Valley, CA, 1987: p. 721.
3. Ozawa, F.; Soyama, H.; Yanagihara, H.; Aoyama, I.; Takino, H.; Izawa, K.; Yamamoto, T.; Yamamoto, A. *J. Am. Chem. Soc.* **1985**, *107*, 3235.
4. Yoneyama, M.; Kakimoto, M.; Imai, Y. *Macromolecules* **1988**, *21*, 1908.
5. (a) Rule, M.; Lane, D. W.; Larkins, T. H.; Tustin, G. C. US Patent 4,746,758 (May 24, 1988) to Eastman Kodak Company. (b) Rule, M.; Lane, D. W.; Larkins, T. H.; Tustin, G. C. US Patent 4,792,641 (Dec. 20, 1988) to Eastman Kodak Company. (c) Rule, M.; Tustin, G. C.; Carver, D. L.; Fauver, J. S. US Patent 4,792,642 (Dec. 20, 1988) to Eastman Kodak Company.
6. Perry, R. J.; Wilson, B. D. *Macromolecules* **1993**, *26*, 1503.
7. (a) Perry, R. J.; Turner, S. R.; Blevins, R. W. *Macromolecules* **1993**, *26*, 1509. (b) Turner, S. R.; Perry, R. J.; Blevins, R. W. *Macromolecules* **1992**, *25*, 4819.
8. Fitton, P.; Rick, E. A. *J. Organomet. Chem.* **1971**, *28*, 287.
9. Moser, W. R.; Wang, A. W.; Kildahl, N. K. *J. Am. Chem. Soc.* **1988**, *110*, 2816.
10. Inglis, T.; Kilner, M. *Nature (London) Phys. Sci.* **1972**, *239*, 13.
11. Hidai, M.; Hikita, T.; Wada, Y.; Fujikura, Y.; Uchida, Y. *Bull. Chem. Soc. Jpn.* **1975**, *48*, 2075.
12. Walker, R. H. *Soc. Plast. Eng. Tech. Pap.* **1974**, *20*, 92.
13. Perry, R. J.; Turner, S. R.; Blevins, R. W. *Macromolecules* **1994**, *27*, 4058.
14. (a) Perry, R. J.; Turner, S. R.; Blevins, R. W. US Patents 5,266,678 and 5,266,679 (Nov. 30, 1993) to Eastman Kodak Company. (b) *Macromolecules* **1995**, *28*, 2607.
15. Perry, R. J.; Turner, S. R. *J. Org. Chem.* **1991**, *56*, 6573.

16. Perry, R. J.; Turner, S. R. *Makromol. Chem. Macromol. Symp.* **1992**, *54/55*, 159.
17. (a) Perry, R. J.; Turner, S. R.; Blevins, R. W. US Patent 5,216,118 (June 1, 1993) to Eastman Kodak Company. (b) *Macromolecules* **1995**, *28*, 3509.
18. Perry, R. J.; Wilson, B. D.; Miller, R. J. *J. Org. Chem.* **1992**, *57*, 2883.
19. Perry, R. J.; Wilson, B. D. *Macromolecules* **1994**, *27*, 40.
20. Perry, R. J.; Turner, S. R. US Patent 4,933,419 (June 12, 1990) to Eastman Kodak Company.
21. For ester and amide formation see (a) Dolle, R. E.; Schmidt, S. J.; Kruse, L. I. *J. Chem. Soc. Chem. Commun.* **1987**, 904. (b) Cacchi, S.; Moreva, E.; Ortar, G. *Tetrahedron Lett.* **1985**, *26*, 1109. (c) Cacchi, S.; Ciattini, P. G.; Moreva, E.; Ortar, G. *Tetrahedron Lett.* **1986**, *27*, 3931.
22. Perry, R. J. US Patent 5,159,057 (Oct. 27, 1992) to Eastman Kodak Company.
23. Perry, R. J. US Patent 5,210,175 (May 11, 1993) to Eastman Kodak Company.
24. Perry, R. J.; Turner, S. R.; Blevins, R. W. US Patent 5,214,123 (May 21, 1993) to Eastman Kodak Company.

RECEIVED December 6, 1995

Chapter 4

Synthesis of Polycinnamamide Catalyzed by Palladium-Graphite

Mitsutoshi Jikei, Masanori Miyauchi, Yuichi Ishida, Masa-aki Kakimoto, and Yoshio Imai

Department of Organic and Polymeric Materials, Tokyo Institute of Technology, Meguro-ku, Tokyo 152, Japan

Palladium-graphite (Pd-Gr) is useful heterogeneous catalyst for many organic reactions and polycondensations. The catalytic activities of Pd-Gr for the Heck reaction, oxidation, hydrogenation, and nucleophilic substitution of olefin through π-allyl complex were investigated. The polycondensation of *N,N'*-(3,4'-oxydiphenylene)bis(acrylamide) and bis(4-iodophenyl) ether catalyzed by Pd-Gr proceeded efficiently in the presence of tributylamine in DMF at 100 ^{0}C to form polycinnamamide. The polycondensation required longer reaction time compared to the polycondensation catalyzed by palladium acetate. The structure of the resulting polymer was confirmed as *trans*-polycinnamamide. The resulting polymer was almost white color, which means that it was less contaminated by palladium metal. The removal and recycling of Pd-Gr were much easier than the case of homogeneous catalysts.

Recently, polycondensations catalyzed by transition metal compounds deserve much attention to give a new methodology for polymer synthesis.(*1-12*) We have also reported the palladium-catalyzed polycondensations through C-C coupling reaction and insertion reactions of carbon monoxide.(*2,6,7*) However, there have been no reports about the use of heterogeneous catalysts in transition metal catalyzed polymerizations. The heterogeneous catalysts can give great advantages, such as removal from the reaction mixture, and recycling of the catalysts.

Metal-graphite combinations are one of the useful heterogeneous catalyst because of their high reactivity, easy preparation, and manipulation. Umani-Ronchi et al. have reported that transition metal-graphites prepared from potassium intercalated graphite are useful as heterogeneous catalysts in many organic reactions.(*13*) It is reported that palladium-graphite (Pd-Gr) showed high reactivity for the hydrogenation of olefins, (*14*) the Heck reaction, (*15*) and the nucleophilic substitution reactions through π-allyl complex.(*16*)

This paper describes organic reactions and polycondensation catalyzed by Pd-Gr used as a heterogeneous catalyst. The catalytic activity of Pd-Gr for organic reactions

0097–6156/96/0624–0088$12.00/0

especially for the Heck reaction was investigated in comparison with palladium-coated active carbon (Pd-C). Polycinnamamide **3** was synthesized by the polycondensation of bisacrylamide **1** and diiodide **2** catalyzed by Pd-Gr.

n **1** + n **2**

Pd-Gr, Bu_3N, - 2n HI → **3**

Results and Discussion

Characterization of Pd-Gr. Palladium-graphite (Pd-Gr) was prepared by the reaction of potassium intercalated graphite and palladium chloride (II) as described in the literature.(*16*) The palladium content by weight in palladium-graphite (Pd-Gr) catalyst was determined to be 17 %, which is about the half of the reported value (33 wt%).(*14*) The low content of palladium may be caused by the low conversion of the reaction of potassium and graphite to form potassium-graphite. The XPS spectrum of Pd-Gr showed typical peaks for zero valence palladium, which is the same as palladium-coated active carbon (Pd-C). It was also confirmed that any potassium atom could not be detected in Pd-Gr.

As it has been reported that palladium is highly dispersed on the graphite surface rather than intercalated between the graphite layers in Pd-Gr,(*17*) there was no diffraction peak corresponding to the intercalated layer in X-ray diffraction spectrum of Pd-Gr. The SEM backscattered images of Pd-Gr and Pd-C are shown in Figure 1. The palladium particles on Pd-Gr which were detected as white spots were

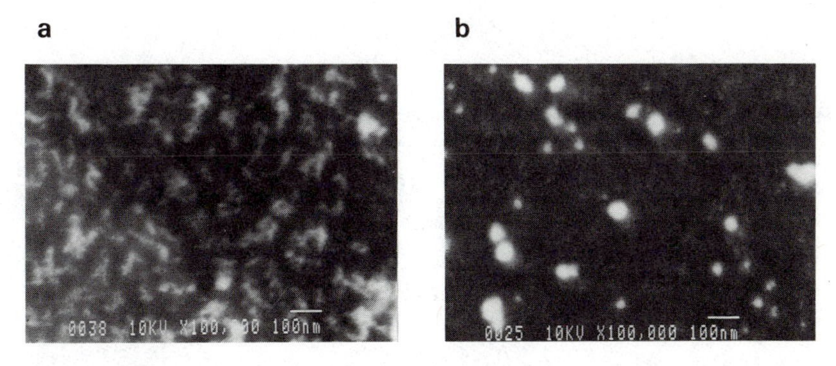

Figure 1 SEM backscattered images of palladium-graphite, a); palladium coated active carbon, b).

smaller than those on Pd-C. This difference must affect the catalytic activity for many reactions.

The Heck Reaction. In the beginning, the Heck reaction of ethyl acrylate and iodobenzene was examined in the presence of a catalytic amount of Pd-Gr. The yield of ethyl cinnamate was determined by HPLC measurements and plotted against the reaction time (Figure 2). Pd-C was also used in order to make a comparison. The activity of palladium-graphite was much higher than that of Pd-C for the Heck reaction. Both Pd-Gr and Pd-C were easily removed by filtration after the reaction because they were insoluble in the reaction mixture.

The effect of solvent and base on the Heck reaction catalyzed by Pd-Gr was investigated as shown in Table I. Aprotic polar solvents, such as DMF, DMAc, NMP, and DMSO, were available for this condensation. The reaction proceeded fast in high dielectric constant solvents such as DMSO. It is noteworthy that DMF containing 1% (volume) of water gave the high yield (86 %) of ethyl cinnamate. Small amount of water seems to increase in the polarity of DMF, which accelerates the reaction. A base was essential for the Heck reaction. Trialkylamines, such as tripropylamine and tributylamine were highly effective for the reaction. Triethylamine gave a lower yield of ethyl cinnamate because the reaction temperature at 80 ^{o}C was very close to the boiling point of triethylamine.

Other Organic Reactions. Table II shows the results of oxidation and hydrogenation in the presence of Pd-Gr, Pd-C and palladium chloride. The yield of 2-decanone by the oxidation of 1-decene catalyzed by Pd-Gr was higher than the case of Pd-C, though the highest yield was obtained in the case of palladium chloride. The hydrogenation of nitrobenzene to aniline proceeded quantitatively in the

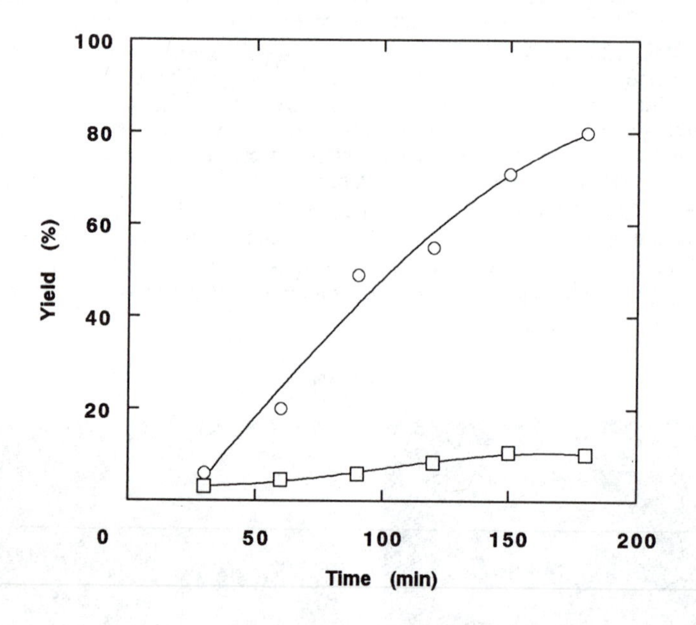

Figure 2 Formation of ethyl cinnamate with Pd-Gr (○), with Pd-C(□)

Table I. Condensation[a] of Ethyl Acrylate and Iodobenzene Catalyzed by Pd-Gr

solvent	base	yield (%)[b]
DMF	tributylamine	63
DMAc	tributylamine	76
NMP	tributylamine	69
DMSO	tributylamine	85
DMAc	triethylamine	37
DMAc	tripropylamine	89

[a] Reaction was carried out with 2.4 mmol of ethyl acrylate, 2.0 mmol of iodobenzene, 2.6 mmol of base, and 0.06 g of Pd-Gr at 80 °C. [b] Determined by HPLC measurement without isolation of the product.

Table II. Organic Reactions Catalyzed by Pd-Gr, Pd-C, and Palladium Chloride

catalyst	reaction	yield (%)[a]
Pd-Gr	1-octene $\xrightarrow{O_2}$ 2-octanone	93
Pd-C		70
$PdCl_2$		100
Pd-Gr	$C_6H_5-NO_2 \xrightarrow{H_2} C_6H_5-NH_2$	100[b]
Pd-C		100[c]

[a] The yield was calculated by gas chromatography without isolation. [b] The consumption of hydrogen was observed for 55 min. [c] The consumption of hydrogen was observed for 90 min.

presence of Pd-Gr and Pd-C. However, the hydrogenation catalyzed by Pd-Gr proceeded faster because the period of the hydrogen consumption was shorter than that in the case of Pd-C. The higher catalytic activities of Pd-Gr are caused by the wide dispersion of the palladium metal on the Pd-Gr catalyst .

Polymerization Catalyzed by Pd-Gr. The results of the polycondensation of bisacrylamide **1** and diiodide **2** catalyzed by Pd-Gr carried out in DMF at 100°C are shown in Table III. The polycondensation proceeded efficiently in the presence of tributylamine and without any ligand compounds. Polymer **3** was obtained with an inherent viscosity of 0.9 dL g^{-1} after 20 hours of polymerization. Among the bases, tributylamine was the most effective for this polymerization, while BDMAN afforded the polymer with the highest molecular weight in the case of previously reported homogeneous catalysts.(*2*) The aprotic polar solvents were effective for the polymerization, similar to the model reaction. The polymerization in DMF and DMAc gave the polymers with high inherent viscosities. DMSO which afforded the high yield in the model reaction was not suitable for the polymerization because of the poor solubility of the resulting polymer in DMSO. Since Pd-Gr is insoluble in the reaction mixture, it can be easily removed by filtration. In addition, the polymerization with the recovered Pd-Gr catalyst gave the polycinnamamide with an inherent viscosity of 0.7 dL g^{-1}.

Table III. Synthesis of Polycinnamamide Catalyzed by Palladium-Graphite[a]

no.	base	solvent	time (h)	yield(%)	η_{inh}(dL g^{-1})[b]
1	tributylamine	DMF	4	90	0.25
2	tributylamine	DMF	20	95	0.90
3	tributylamine	DMF	40	95	0.95
4	triethylamine	DMF	20	85	0.30
5	BDMAN[c]	DMF	20	96	0.45
6	DBU[d]	DMF	20	0	-
7	CH_3COONa	DMF	20	0	-
8	Na_2CO_3	DMF	20	0	-
9	tributylamine	DMAc	20	94	0.95
10	tributylamine	NMP	20	97	0.68
11	tributylamine	DMSO	20	86	0.25

[a] Polymerization was carried out with 1.25 mmol of the bisacrylamide and 1.25 mmol of the diiodobenzene, 3.25 mmol of base, and 6.87 x 10^{-3} mmol/Pd of Pd-Gr at 100 ^{o}C in 20 mL of the solvent. [b] Measured at a concentration of 0.5 g dL^{-1} in DMF at 30 ^{o}C. [c] BDMAN: 1,8-bis(dimethylamino)naphthalene. [d] DBU: 1,8-diazabicyclo[5.4.0]-7-undecene.

The weight average molecular weight of the resulting polymer was plotted against the reaction time (Figure 3). Polymers having a molecular weight about a hundred thousand were formed after the polymerization for 20 hours. As shown in Figure 4, the molecular weights of the resulting polymer increased with increasing the feed amount of Pd-Gr and the high molecular weight polymer was obtained when 2.5 mol% of Pd-Gr was used for the polymerization.

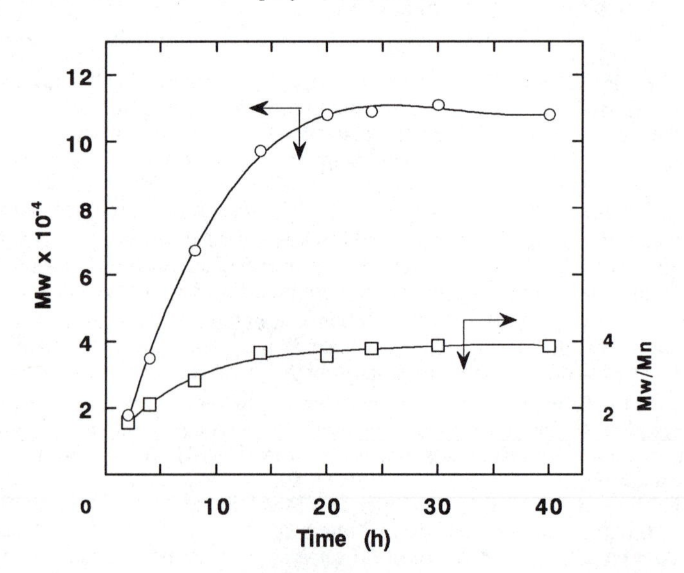

Figure 3 Time dependence of Mw (○) and Mw / Mn (□) of polycinnamamide formed with Pd-Gr.

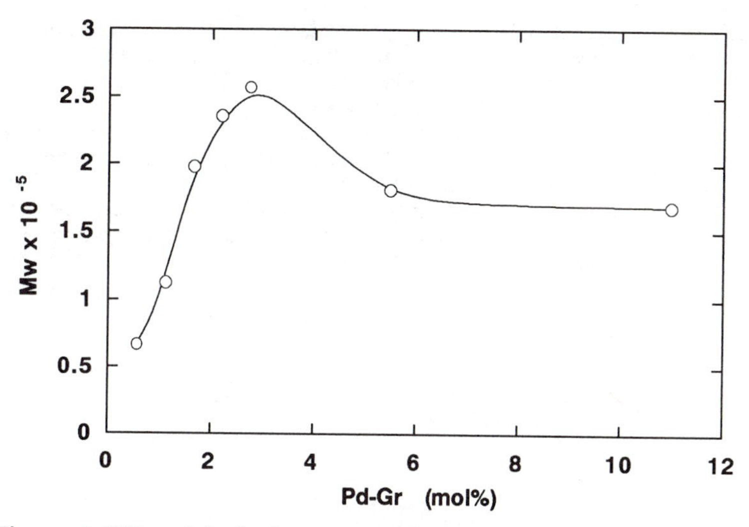

Figure 4 Effect of the feed amount of Pd-Gr on the molecular weight of polycinnamamide. Reaction conditions: 1.25 mmol of monomers and 3.25 mmol of Bu_3N in DMF (20 mL) at 100 °C for 8 h.

Table IV. Effect of the Molar Ratio of Diiodide on the Polymerization Catalyzed by Pd-Gr[a]

no.	diidodide/ bisacrylamide	η_{inh}(dL g^{-1})[b]	Mw x 10^{-5} [c]	Mn x 10^{-4} [c]	Mw/Mn [c]
12	0.99	0.43	0.24	0.11	2.13
2	1.00	0.90	1.13	3.43	3.28
13	1.01	0.90	1.84	4.17	4.41
14	1.05	d	d	d	d

[a] Polymerization was carried out with 1.25 mmol of the bisacrylamide, the prescribed amount of the diiodide, 3.25 mmol of tributylamine, and $6.87x10^{-3}$ mmol/Pd of Pd-Gr at 100 ^{o}C in DMF. [b] Inherent viscosity was measured at a concentration of 0.5 g dL^{-1} in DMF at 30 ^{o}C. [c] Absolute molecular weight was determined by GPC with a laser light scattering measurement. The dn/dc at 690 nm was determined to be 0.268 mL g^{-1}. [d] Gelation of the reaction mixture was occured.

The effect of the molar ratio of diiodide/bisacrylamide on the polycondensation was investigated as shown in Table IV. The polydispersity was evaluated by the polydispersity index (Mw/Mn) determined by GPC through a laser light scattering measurement. When the diiodide/bisacrylamide ratio was 0.99, the inherent viscosity and the molecular weight of the resulting polymer decreased compared with those prepared by the ratio of 1.00. In the case of the ratio of 1.01, the polydispersity index increased, while the inherent viscosity and the molecular weight were not affected. However, at the ratio of 1.05, the reaction mixture afforded gelation. It is obvious that the equimolar amount of two monomers is essential for the formation of the polymer with a high molecular weight, though the olefin is often used in small molar excess relative to the organic halide in order to achieve exclusive monosubstitution of olefin in the case of many organic syntheses using the Heck reaction. These results suggest that the second substitution of the residual vinyl proton groups by intermediate complex of the aromatic halide and palladium occurred during the polymerization, which enlarged the polydispersity indices. The second substitution was accelerated when excess amount of aromatic halide was used for the polycondensation, which caused to gelation at the ratio of 1.05.

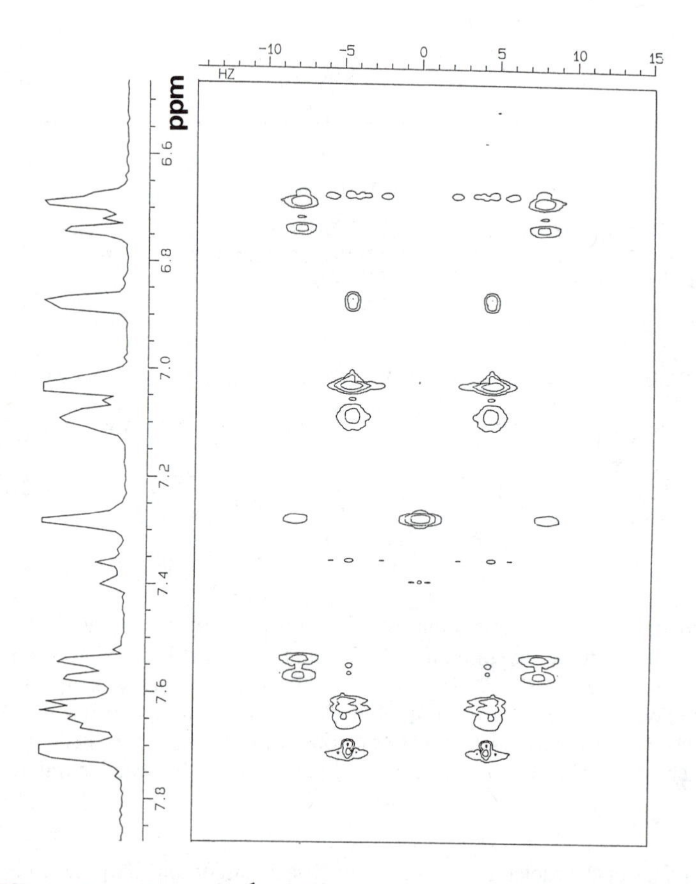

Figure 5 J-Resolved ^{1}H NMR spectrum of the polycinnamamide formed with Pd-Gr

The structure of the resulting polymer was determined by IR and NMR spectra and it was found that the polymer **3** had the identical structure to that of the previously prepared polycinnamamide using the homogeneous palladium catalysts.

The J-resolved two dimensional ^1H NMR spectrum was measured to determine coupling constants of protons around the double bond in the main chain. There were two different pairs of peaks with regard to the double bond as shown in Figure 5. The coupling constant for the peaks at 6.73 ppm and 7.54 ppm was 15.2 Hz and the constant for the peaks at 6.69 ppm and 7.57 ppm was 15.6 Hz. The both are typical coupling constants for olefinic protons connected by trans linkage. The two different pairs of peaks for olefinic protons was not changed to single peaks by the measurement at 100 ^oC, which denies that the two pairs arise from restriction of bond rotations. It is suggested that the pairs resulted from the unsymmetrical structure by the introduction of the meta-linked phenylene ring.

The resulting polymer prepared using Pd-Gr was less colored than the polymer prepared using the homogeneous palladium catalysts. The amount of residual palladium in the polymer prepared by Pd-Gr was determined to be *ca.* 50 ppm by the ICP plasma emission spectroscopy. However, the residual palladium of *ca.* 500 ppm was detected in the polymer prepared by palladium acetate. The fact that the amount of palladium in the polymer prepared by Pd-Gr was 10 times less than that in the polymer prepared by palladium acetate may affect the color of the polymer.

Polymerization Catalyzed by Various Palladium Compounds. Table V contains the results of the polycondensation catalyzed by Pd-C and palladium acetate. Pd-C showed less reactivity for the polymerization than the Pd-Gr, as similar to the model reaction. It is reported that ligand compounds such as tritolylphosphine accelerate the Heck reaction.(*18*) In the case of Pd-Gr, tritolylphosphine was not so effective on the inherent viscosity of the polymer. However, the polymerization using Pd-Gr required longer reaction time than the polymerization catalyzed by palladium acetate. The high molecular weight polymer was obtained within 5 h reaction time in the latter case.

Table VI shows the GPC results of the polymerization catalyzed by Pd-Gr and palladium acetate. When Pd-Gr was used as a catalyst, the polydispersity indices were lower than those of the polymer prepared by using palladium acetate. The control experiments using an equimolar amount of ethyl acrylate and iodobenzene with Pd-Gr or palladium acetate were carried out in the same condition as the polycondensation. In the case of Pd-Gr, the amount of the product by the second substitution was about three times less than the case of palladium acetate.(*19*) This fact supports the lower polydispersity indices for the polycondensation catalyzed by Pd-Gr.

Table V. Synthesis of Polycinnamamide with Various Palladium Catalysts[a]

no.	catalyst	ligand	time(h)	yield(%)	$\eta_{inh}(dL\ g^{-1})$[b]
1	Pd-Gr	-	4	90	0.25
2	Pd-Gr	-	20	95	0.90
15	Pd-Gr	$PTol_3$[c]	20	96	0.81
16	Pd-C	-	20	94	0.69
17	$Pd(OAc)_2$	$PTol_3$[c]	4	95	1.11

[a] Polymerization was carried out with 1.25 mmol of the bisacrylamide and 1.25 mmol of the diiodobenzene, 3.25 mmol of tributylamine, and 6.87 x 10^{-3} mmol of Pd (No. 1, 2, 12) or 1.25x10^{-2} mmol of Pd (No. 13, 14) at 100 ^oC in DMF. [b] Measured at a concentration of 0.5 g dL^{-1} in DMF at 30 ^oC. [c] $PTol_3$: tritolylphosphine (5.0x10^{-2} mmol).

Table VI. Polydispersity Indices[a] of the Polymers Prepared by Using Palladium Catalysts

no.	catalyst	η_{inh}(dL g^{-1})[b]	Mw x 10^{-5}	Mn x 10^{-4}	Mw/Mn
2	Pd-Gr	0.90	1.13	3.43	3.28
18	Pd-Gr	1.19	2.89	7.51	3.85
17	$Pd(OAc)_2$	1.11	2.20	5.10	4.32
19	$Pd(OAc)_2$	-	1.12	2.87	3.91

[a] Absolute molecular weight was determined by GPC with a laser light scattering measurement. The dn/dc at 690 nm was determined to be 0.268 mL g^{-1}.
[b] Measured at a concentration of 0.5 g dL^{-1} in DMF at 30 ^{o}C.

Conclusion

Pd-Gr can be used for many organic reactions as a heterogeneous catalyst alternative to Pd-C. The catalytic activity of Pd-Gr was higher than that of Pd-C from the present investigations because of the wide dispersion of palladium atom on the Pd-Gr catalyst. Especially, polycinnamamide was successfully prepared through the Heck reaction in the presence of the heterogeneous Pd-Gr catalyst. Although the polymerization reaction was slower compared with that of the homogeneous palladium catalysts, Pd-Gr possesses some benefits as follows: 1) The Pd-Gr catalyst could be easily separated from the reaction mixture. This fact caused lower content of the palladium atom in the resulting polymer, and gave a less colored polymer. 2) The polycondensation selectively proceeded to afford the narrower molecular weight distribution since the second substitution of vinyl proton occurred less than the case of the homogeneous palladium catalyst.

Experimental Part

Materials. Palladium-graphite (Pd-Gr) was prepared by the reaction of potassium intercalated graphite and palladium chloride (II) as described in the literature.(*16*) The content of palladium in Pd-Gr was determined to be 17 wt% by ICP plasma emission spectroscopy. *N,N'*-(3,4'-Oxydiphenylene)bis(acrylamide) **1** was prepared by the condensation of 3,4'-oxydianiline and acrylic acid chloride. The yield was 56 % and the structure was confirmed as described in the literature.(*2*) Bis(4-iodophenyl) ether **2** was prepared by the reaction of diphenyl ether, iodine, and bis[(bistrifluoroacetoxy)iodobenzene in carbon tetrachloride at room temperature. The precipitate was filtered, washed with methanol, and purified by recrystallization from n-hexane. The yield was 59 % and the structure was confirmed as described in the literature.[Yoneyama, 1989 #8] Ethyl acrylate, iodobenzene, trialkylamines, 1-decene, nitrobenzene, and all solvents used for the reaction were purified by distillation. Other materials were used as received.

The Heck Reaction. In a three-necked flask, 0.24 g (2.4 mmol) of ethyl acrylate, 0.41 g (2.0 mmol) of iodobenzene, 0.48 g (2.6 mmol) of tributylamine as a base, and 0.06 g (9.6 x 10^{-5} mol) of Pd-Gr (4.8 mol % of Pd with respect to iodobenzene) were added under nitrogen atmosphere. Biphenyl (40 mg) was also added to the mixture as an internal standard for HPLC measurements. The mixture was stirred and heated to 80 ^{o}C. The yield of ethyl cinnamate formed was determined by HPLC measurements without isolation of the product.

Oxidation of 1-Decene. In a three-necked flask, 0.137 g (0.5 mmol/Pd) of Pd-Gr, 0.495 g (5 mmol) of CuCl, 3.5 mL of DMF, and 0.5 mL of H_2O were added under oxygen atmosphere. After the mixture was stirred at room temperature for 1 hour, 1-decene was added to the mixture and stirred for 20 hours. The yield of 2-decanone formed was determined by GC measurements without isolation.

Hydrogenation of Nitrobenzene. In a three-necked flask, 0.21 mL (2 mmol) of nitrobenzene, 5 mL of methanol, and 7.1 mg (0.02 mmol/Pd) of Pd-Gr were added under hydrogen atmosphere. The mixture was stirred at room temperature and the hydrogen consumption was monitored by using a gas buret. The yield of aniline formed was determined by GC measurements without isolation.

Polymerization catalyzed by Pd-Gr. A typical procedure of the polymerization is as follows. In a three-necked flask, 0.385 g (1.25 mmol) of *N,N'*-(3,4'-oxydiphenylene)bis(acrylamide) **1** and 0.528 g (1.25 mmol) of bis(4-iodophenyl) ether **2** were dissolved in 20 mL of DMF. Pd-Gr (4.3 mg, 6.87×10^{-3} mmol/Pd), and 0.77 mL (3.25 mmol) of tributylamine were added to the solution. The mixture was stirred at 100 ^{o}C under nitrogen atmosphere. After 40 hours, the mixture was filtered to remove the catalyst, and then poured into 300 mL of methanol to precipitate polymer **3**. The polymer was filtered, washed with hot methanol, and dried *in vacuo*. The yield was 95 %. The inherent viscosity of the polymer was 0.95 dL g^{-1}, measured in DMF at a concentration of 0.5 g dL^{-1} at 30 ^{o}C; IR (KBr) 1665 (C=O), 976 cm^{-1} (C=C). Anal. Calcd for $(C_{30}H_{22}N_2O_4)_n$: C, 75.94; H, 4.67; N,5.90. Found: C, 75.83; H, 4.59; N, 5.70.

Polymerization catalyzed by Palladium acetate. In a three-necked flask, 0.385 g (1.25 mmol) of **1**, 0.528 g (1.25 mmol) of **2**, 2.8 mg (1.25×10^{-2} mmol) of palladium acetate, 15.2 mg (5.0×10^{-2} mmol) of tritolylphosphine, and 0.77 mL (3.25 mmol) of tributylamine were dissolved in 5 mL of DMF. The mixture was stirred at 100 ^{o}C under nitrogen atmosphere. After 4 hours, the mixture was poured into 300 mL of methanol to precipitate polymer **3**. The polymer was filtered, washed with hot methanol, and dried *in vacuo*. The yield was 95 %. The inherent viscosity of the polymer was 1.11 dL g^{-1}, measured in DMF at a concentration of 0.5 g dL^{-1} at 30 ^{o}C.

Measurements. High performance liquid chromatography (HPLC) measurements were performed by using C18 column with a Shimadzu LC-9A and SPD-6A (UV spectrophotometric detector). Gas chromatography (GC) measurements were performed by using a OV 101 column with Simadzu GC-7A with a flame ionization detector. Inductively coupled plasma (ICP) spectrometric measurements were carried out by using a SII SPS1500VR plasma Spectrometer. Infrared (IR) spectra were recorded on a JASCO FTIR-8100 Fourier transform infrared spectrophotometer. ^{1}H NMR spectra were recorded on a JEOL FX-90Q NMR spectrometer and homo J-resolved ^{1}H-^{1}H NMR spectra and homonuclear correlated (HH COSY) NMR spectra were recorded on a JEOL GSX-500 NMR spectrometer. Gel permeation chromatography (GPC) measurements were carried out by using a JASCO HPLC 880PU, polystyrene-divinylbenzene columns (two Shodex KD806M and KD802), and DMF containing 0.01 mol L^{-1} of lithium bromide as an eluent. Absolute molecular weight was determined by laser light scattering measurement using a miniDAWN apparatus (Wyatt Technology Co.) and a

Shimadzu RID-6A refractive index detector. A specific refractive index increment *(dn/dc)* of the polymer in DMF at 690 nm was determined to be 0.268 mL g^{-1} by using an Optilab 903 apparatus (Wyatt Technology Co.). XPS measurements were performed by using a ULVAC-PHI-5500MT system. The spectra were acquired using monochromated Al-Kα (1486.7 ev) radiation at 14 kV and 200 W. Scanning electron microscope (SEM) backscattered images were taken by using a JEOL T-220 microscope.

References and Notes

(1) Heitz, W.; Brugging, W.; Freund, L.; Gailberger, M.; Greiner, A.; Jung, H.; Kampschulte, U.; Niebner, M.; Osan, F.; Schmidt, H. W.; Wicker, M. *Makromol. Chem.* **1988**, *189*, 119.
(2) Yoneyama, M.; Kakimoto, M.; Imai, Y. *Macromolecules* **1989**, *22*, 4148.
(3) Bochmann, M.; Kelly, K. *J. Polym. Sci. Part A* **1992**, *30*, 2503.
(4) Bochmann, M.; Lu, J. *J. Polym. Sci., Part A: Polym. Chem.* **1994**, *32*, 2493.
(5) Cramer, E.; Percec, V. *J. Polym. Sci. Part A* **1990**, *28*, 3029.
(6) Yoneyama, M.; Kakimoto, M.; Imai, Y. *Macromolecules* **1988**, *21*, 1908.
(7) Yoneyama, M.; Kakimoto, M.; Imai, Y. *Macromolecules* **1989**, *22*, 4152.
(8) Perry, R. J.; Turner, S. R.; Blevins, R. W. *Macromolecules* **1993**, *26*, 1509.
(9) Perry, R. J.; Wilson, B. D. *Macromolecules* **1994**, *27*, 40.
(10) Suzuki, M.; Sawada, S.; Saegusa, T. *Macromolecules* **1989**, *22*, 1505.
(11) Suzuki, M.; Lim, J.-C.; Saegusa, T. *Macromolecules* **1990**, *23*, 1574.
(12) Suzuki, M.; Sawada, S.; Yoshida, S.; Eberhardt, A.; Saegusa, T. *Macromolecules* **1993**, *26*, 4748.
(13) Sovoia, D.; Trombini, C.; Umani-Ronchi, A. *Pure & Appl. Chem.* **1985**, *57*, 1887.
(14) Savoia, D.; Trombini, C.; Umani-Ronchi, A.; Verardo, G. *J. Chem. Soc., Chem. Comm.* **1981**, 540.
(15) Savoia, D.; Trombini, C.; Umani-Ronchi, A.; Verardo, G. *J. Chem. Soc., Chem. Comm.* **1981**, 541.
(16) Boldrini, G. P.; Savoia, D.; Tagliavini, E.; Trombini, C.; Umani-Ronchi, A. *J. Organometal. Chem.* **1984**, *268*, 97.
(17) Schaefer-Stahl, H. *J. Chem. Soc. Dalton* **1981**, 328.
(18) Dieck, H. A.; Heck, R. F. *J. Am. Chem. Soc.* **1974**, *96*, 1133.
(19) The amount of the product by the second substitution was evaluated by the ratio of the peak areas in HPLC measurements.

RECEIVED December 4, 1995

Chapter 5

Synthesis of High-Molecular-Weight Polymers by Ruthenium-Catalyzed Step-Growth Copolymerization of Acetophenones with α,ω-Dienes

Hongjie Guo, Mark A. Tapsak, Guohong Wang, and William P. Weber[1]

Donald P. and Katherine B. Locker Hydrocarbon Research Institute, Department of Chemistry, University of Southern California, Los Angeles, CA 90089–1661

Two approaches are successful for the preparation of high molecular weight polymers by ruthenium catalyzed step-growth copolymerization of acetophenones and α, ω-dienes. Acetophenones substituted with electron donating groups such as 4'-methoxy, 4'-phenoxy or 4'-dialkylamino groups lead to copolymers with respectable molecular weights. Alternatively, high molecular weight copolymers can be prepared by activation of the ruthenium catalyst with a stoichiometric amount of an alkene prior to addition of mixture of acetophenone and α,ω-diene monomers. The synthesis and characterization of these copolymers is reported. Higher molecular weight copolymers are found to be more thermally stable than those with lower molecular weight.

Murai et al. have reported that dihydridocarbonyl*tris*(triphenylphosphine)ruthenium (Ru) catalyzes the addition of the *ortho* C-H bonds of acetophenone across the C-C double bonds of olefins such as trimethylvinylsilane to yield *ortho* alkyl substituted acetophenones (*1-3*). We have shown that this reaction can be applied to achieve step-growth copolymerization (cooligomerization) of aromatic ketones and α,ω-dienes. For example, reaction of divinyldimethylsilane and acetophenone catalyzed by Ru at 150°C yields copoly(3,3-dimethyl-3-sila-1,5-pentanylene/2-aceto-1,3-phenylene), M_w/M_n = 3,500/2,430 as in equation 1 (*4*). We have carried out similar copolymerization

$$\text{acetophenone} + \text{divinyldimethylsilane} \xrightarrow{Ru(H)_2(CO)(Ph_3P)_3} \text{copolymer}_n \quad (1)$$

[1]Corresponding author

0097–6156/96/0624–0099$12.00/0

reactions between anthrone, fluorenone or xanthone and α,ω-dienes such as 3,3,6,6-tetra- methyl-3,6-disila-1,7-octadiene or 1,3-divinyltetramethyldisiloxane as in equation 2 (*5*).

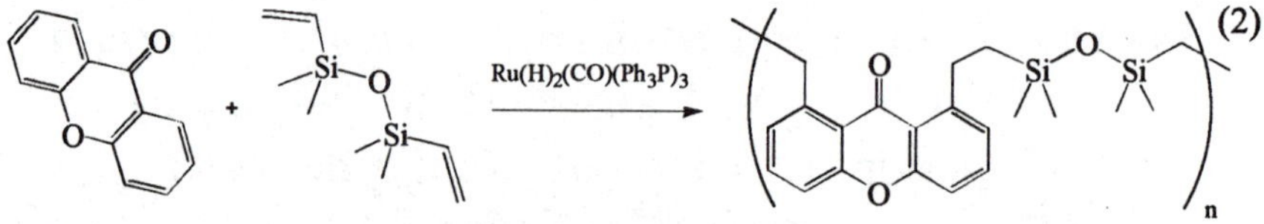

(2)

Unfortunately, the molecular weights of these copolymers (cooligomers) are also generally low ($M_w/M_n \sim 3,000/1,600$). This is not unexpected since exact stoichiometry is essential to achieve high molecular weights in step-growth copolymerization reactions (*6*). The occurrence of minor unknown side reactions will destroy the required balance of stoichiometry. Successful attainment of higher molecular weights is important since many polymer properties change rapidly until a minimum threshold value is reached. Frequently, constant polymer properties are observed when the molecular weight greater than 10,000.

In this chapter, we report two solutions to this problem. Ru catalyzed step-growth copolymerizations of 4'-dialkylacetophenones, 4'-methoxyacetophenone or 4'-phenoxyacetophenone with α,ω-dienes proceed more readily and yield significantly higher molecular weight copolymers than those previously reported examples as in equation 3. Alternatively, treatment of the Ru catalyst with one equivalent of an alkene

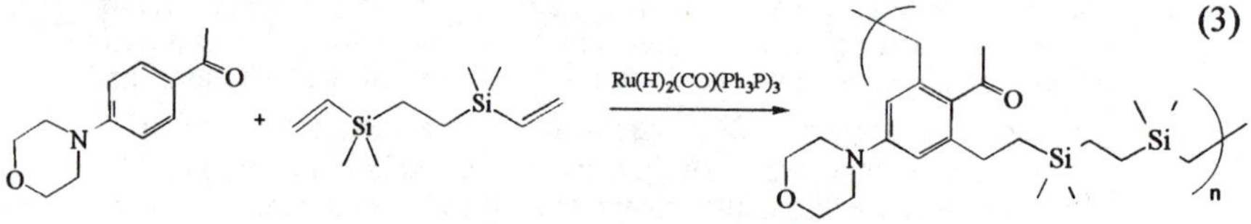

(3)

prior to addition of the acetophenone and α,ω-diene monomers leads to higher molecular weight copolymers as in equation 4.

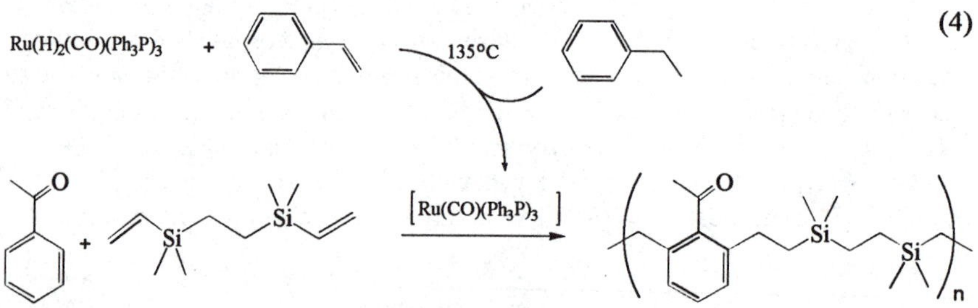

(4)

Experimental

Spectroscopic Measurements. ^1H and ^{13}C NMR spectra were obtained on either a Bruker AC-250 or an AM-360 spectrometer operating in the FT mode. ^{29}Si NMR spectra were recorded on an IBM Bruker WP-270-SY spectrometer. Five percent w/v solutions of copolymer in chloroform-*d* were used to obtain NMR spectra. ^{13}C NMR spectra were run with broad band proton decoupling. A heteronuclear gated decoupling pulse sequence (NONOE) with a 20 sec delay was used to acquire ^{29}Si NMR spectra (*7*). These were externally referenced to TMS. Chloroform was used as an internal standard for ^{1}H and ^{13}C NMR spectra. IR spectra of neat films on NaCl plates were recorded on an IBM FT-IR spectrometer. UV spectra of cyclohexane solutions were acquired on a Shimadzu UV-260 ultraviolet visible spectrometer.

Molecular Weight Distributions. Gel permeation chromatographic (GPC) analysis of the molecular weight distribution of these polymers was performed on a Waters system comprised of a U6K injector, a 510 HPLC solvent delivery system, a R401 refractive index detector and a model 820 Maxima control system. A series of three 7.8 mm x 30 cm columns packed with < 10 μm particles of monodisperse crosslinked styrenedivinylbenzene copolymer. These contain pore sizes of 1 x 10^4 Å (Waters Ultrastyragel), 1 x 10^3 Å (Waters Ultrastyragel) and finally 500 Å (Polymer Laboratories PLgel). The eluting solvent was HPLC grade THF at a flow rate of 0.6 mL/min. The retention times were calibrated against known monodisperse polystyrene standards: M_w 114,200; 47,500; 18,700; 5,120; and 2,200 whose M_w/M_n are less than 1.09.

Thermogravimetric Analysis (TGA) of the polymers was carried out on a Perkin-Elmer TGS-2 instrument with a nitrogen flow rate of 40 cc/min. The temperature program for the analysis was 50°C for 10 min followed by an increase of 4°C/min to 750°C.

Differential Scanning Calorimetry (DSC) was utilized to determine the glass transition temperatures (T_gs) of the copolymers. These measurements were carried out on a Perkin-Elmer DSC-7 instrument. The melting points of indium (156°C) and ice (0°C) were used to calibrate the DSC. The program for the analysis was -50°C for 10 min followed by an increase in temperature of 20°C/min to 150°C.

Reagents. All reactions were conducted in flame dried glassware under an atmosphere of purified argon. 4'-Piperidinoacetophenone, 4'-morpholinoacetophenone, 4'-piperazinoacetophenone, 4'-methoxyacetophenone, and 4'-phenoxyacetophenone were purchased from Aldrich. 1,3-Divinyltetramethyldisiloxane and 3,3,6,6-tetramethyl-3,6-disila-1,7-octadiene were obtained from United Chemical Technologies. Dihydridocarbonyl*tris*(triphenylphosphine)ruthenium was prepared from ruthenium trichloride (*8*).

Elemental Analysis was performed by Oneida Research Services Inc., Whitesboro, NY.

4'-(N'-Benzyl)piperazinoacetophenone. 4'-Piperazinoacetophenone (0.61 g, 3 mmol), benzyl chloride (0.38 g, 3 mmol), potassium carbonate (0.21 g, 1.5 mmol) and a Teflon-covered magnetic stirring bar were placed in a 100 mL flame dried roundbottomed flask. The flask was sealed with a rubber stopper. The reaction mixture was stirred for 24 h and was then poured into 100 mL of water. The precipitate was filtered and recrystallised from ethanol/water. In this way, 0.7 g, 68% yield of product mp 97.5-99°C was obtained. ^{1}H NMR δ: 2.49(s, 3H), 2.57(m, 4H), 3.34(m, 4H), 3.54 (s, 2H), 6.81-6.84(m, 2H), 7.27-7.33(m, 5H), 7.83-7.86(m, 2H). ^{13}C NMR δ: 26.05, 47.31, 52.68, 62.95, 113.31, 127.24, 127.53, 128.31, 129.13, 130.33, 137.69, 154.17,

196.40. IR ν: 2820, 2777, 1664, 1598, 1555, 1518, 1455, 1427, 1388, 1361, 1303, 1285, 1245, 1230, 1195, 1148, 1008, 926, 906, 824, 744, 737, 733 cm^{-1}. UV λ_{max} nm(ε): 311(25,870), 261(4,930), 255(5,110).

Copoly(3,3,5,5-tetramethyl-4-oxa-3,5-disila-1,7-heptanylene/2-aceto-5-piperidino -1,3-phenylene)(Copoly-I). 4'-Piperidinoacetophenone (0.51 g, 2.5 mmol), 1,3-divinyltetramethyldisiloxane (0.47 g, 2.5 mmol), xylene (2 mL), Ru catalyst (0.07 g, 0.076 mmol) and a Teflon-covered magnetic stirring bar were placed in an Ace pressure tube (15 mL, 10.2 cm long). A stream of argon was bubbled through the solution for 5 min. The tube and its contents were sealed with a Teflon bushing and FETFE "O"-ring. The reaction mixture was stirred for 48 h at 150°C. The color of the reaction mixture changed from colorless to black. The tube and its contents were cooled to rt and pentane (5 mL) was added. The reaction mixture was stirred for several min. This caused the catalyst to precipitate. After filtration, pentane was removed from the crude polymer, by evaporation under reduced pressure. The copolymer was purified three times by precipitation from tetrahydrofuran and methanol. In this way, 0.90 g, 91% yield of copolymer M_w/M_n = 20,880/9,540, T_g = 2°C was obtained. ^{1}H NMR δ: 0.06(s, 12H), 0.82(m, 4H), 1.55(s, 2H), 1.66(s, 4H), 2.42(s, 3H), 2.47(m, 4H), 3.13(s, 4H), 6.58(s, 2H). ^{13}C NMR δ: 0.22, 21.03, 24.29, 25.78, 27.27, 33.27, 50.25, 113.77, 132.21, 141.31, 152.44, 208.17. ^{29}Si NMR δ: 7.22. IR ν: 2940, 2859, 2809, 1690, 1598, 1563, 1557, 1467, 1453, 1444, 1413, 1385, 1352, 1256, 1224, 1183, 1121, 1062, 982, 911, 842, 791, 738, 649 cm^{-1}. UV λ_{max} nm(ε): 216(20,200), 243(8,950), 287(6,320). Elemental Anal. Calcd for $C_{21}H_{35}NO_2Si_2$: C, 64.78; H, 9.00; N, 3.60. Found: C, 64.02; H, 8.90; N, 3.49.

Copoly(3,3,5,5-tetramethyl-4-oxa-3,5-disila-1,7-heptanylene/2-aceto-5-morpholino-1,3-phenylene)(Copoly-II). 4'-Morpholinoacetophenone (0.52 g, 2.5 mmol), 1,3-divinyltetramethyldisiloxane (0.47 g, 2.5 mmol), xylene (2 mL) and Ru catalyst (0.07 g, 0.076 mmol) and a Teflon-covered magnetic stirring bar were placed in an Ace pressure tube as above. In this way, 0.85 g, 86% yield of copolymer M_w/M_n = 16,470/8,050, T_g = 14°C was obtained. ^{1}H NMR δ: 0.07(s, 12H), 0.81(m, 4H), 2.42(s, 3H), 2.47(m, 4H), 3.14(s, 4H), 3.81(s, 4H), 6.56(s, 2H). ^{13}C NMR δ: 0.25, 21.06, 27.27, 33.27, 48.98, 66.85, 113.00, 133.09, 141.50, 151.53, 208.00. ^{29}Si NMR δ: 7.22. IR ν: 2956, 2893, 2858, 1953, 1778, 1721, 1692, 1651, 1600, 1564, 1538, 1511, 1451, 1414, 1379, 1352, 1305, 1252, 1184, 1125, 1058, 906, 840, 788, 733, 707, 686, 647 cm^{-1}. UV λ_{max} nm(ε): 216(21,270), 241(3,220), 287(6,390). Elemental Anal. Calcd for $C_{20}H_{33}NO_3Si_2$: C, 61.38; H, 8.44; N, 3.60. Found: C, 60.90; H, 8.29; N, 3.47.

Copoly[3,3,5,5-tetramethyl-4-oxa-3,5-disila-1,7-heptanylene/2-aceto-4-(N'-benzyl)piperazino-1,3-phenylene](Copoly-III). 4-(N'-Benzyl)piperazinoacetophenone (0.41 g, 1.4 mmol), 1,3-divinyltetramethyldisiloxane (0.26 g, 1.4 mmol), xylene (2mL), Ru catalyst (0.04 g, 0.038 mmol) and a Teflon-covered magnetic stirring bar were placed in an Ace pressure tube as above. In this way, 0.57 g, 85% yield of copolymer M_w/M_n = 51,250/16,540, T_g = 25°C was obtained. ^{1}H NMR δ: 0.07(s, 12H), 0.81(m, 4H), 2.42(s, 3H), 2.45(m, 4H), 2.60(s, 4H), 3.21(s, 4H), 3.58(s, 2H), 6.56(s, 2H), 7.34(s, 5H). ^{13}C NMR δ: 0.18, 20.94, 27.17, 33.20, 48.52, 52.79, 62.78, 113.31,

127.22, 128.23, 129.22, 132.68, 137.22, 141.32, 151.42, 208.01. ^{29}Si NMR δ: 7.16. IR ν: 2977, 2869, 1693, 1600, 1456, 1385, 1352, 1255, 1180, 1147, 1069, 913, 843, 790, 734, 700 cm^{-1}. UV λ_{max} nm(ε): 215 (22,096), 243(7,213), 288(5,270). Elemental Anal. Calcd for $C_{27}H_{40}N_2O_2Si_2$: C, 67.54; H, 8.33; N, 5.83. Found: C, 65.66; H, 7.97; N, 5.61.

Copoly[3,3,6,6-tetramethyl-3,6-disila-1,8-octanylene/2-aceto-5-(N'-benzyl)piperazino-1,3-phenylene](Copoly-IV). 4'-(N'-Benzyl)piperazinoacetophenone (0.88 g, 3.0 mmol), 3,3,6,6-tetramethyl-3,6-disila-1,7-octadiene (0.59 g, 3.0 mmol), xylene (2mL), Ru catalyst (0.07 g, 0.076 mmol) and a Teflon-covered magnetic stirring bar were placed in an Ace pressure tube as above. In this way, 1.35 g, 92% yield of copolymer M_w/M_n = 25,790/7,490, T_g = 23°C was obtained. ^{1}H NMR δ: -0.02(s, 12H), 0.39(s, 4H), 0.80(m, 4H), 2.42(s, 3H), 2.45(m, 4H), 2.60(s, 4H), 3.21 (s, 4H), 3.56 (s, 2H), 6.58(s, 2H), 7.34(s, 5H). ^{13}C NMR δ: -4.16, 6.93, 17.57, 27.86, 33.24, 48.75, 52.94, 62.93, 113.32, 127.08, 128.19 , 129.11, 132.58, 137.75, 141.61, 151.54, 208.08. ^{29}Si NMR δ: 4.17. IR ν: 2954, 2873, 1692, 1599, 1495, 1455, 1385, 1352, 1302, 1248, 1188, 1134, 1103, 1067, 997, 911, 832, 782, 735, 699, 647 cm^{-1}. UV λ_{max} nm(ε): 214 (25,740), 243 (8,950), 290(6,330). Elemental Anal. Calcd for $C_{29}H_{44}N_2OSi_2$: C, 70.73; H, 8.92; N, 5.69. Found: 68.98; H, 8.84; N, 5.48.

Copoly(3,3,6,6-tetramethyl-3,6-disila-1,8-octanylene/2-aceto-5-piperidino-1,3-phenylene) (Copoly-V). 4'-Piperidinoacetophenone (1.00 g, 5 mmol), 3,3,6,6-tetramethyl-3,6-disila-1,7-octadiene (0.98 g, 5 mmol), Ru catalyst (0.1 g, 0.11 mmol) and a Teflon covered magnetic stirring bar were placed in an Ace #15 high pressure reaction tube as above. In this way, 1.59 g, 80% yield of copolymer M_w/M_n = 26,000/15,000, T_g = -5°C was isolated. ^{1}H NMR δ: -0.04(br.s, 12H), 0.37-0.41 (br.s, 4H), 0.76-0.82 (m, 4H), 1.56(s, 2H), 1.65(s, 4H), 2.42(s, 3H), 2.45 (m, 4H), 3.14(s, 4H), 3.56(s,2H), 6.58(s, 2H). ^{3}C NMR δ: -4.15, 6.97, 17.62, 24.26, 25.79, 27.90, 33.27, 50.30, 113.79, 133,19, 141.61, 152.43, 208.1. ^{29}Si NMR δ: 4.14. IR ν: 2942, 2811, 1688, 1682,1597, 1552, 1479, 1467, 1452, 1445, 1422, 1414, 1385, 1353, 1274, 1256, 1237, 1186, 1132, 1055, 967, 928, 888, 834, 781, 705, 655, 646 cm^{-1}. UV λ_{max} nm (ε): 215 (24,300), sh243(9,100), 291 (7,400). Elemental Anal. Calcd. for $C_{23}H_{39}ONSi_2$: C, 68.82; H, 9.73; N, 3.49. Found: C, 68.04; H, 9.41; N, 3.43.

Copoly(3,3,6,6-tetramethyl-3,6-disila-1,8-octanylene/2-aceto-5-morpholino-1,3-phenylene) (Copoly-VI). 4'-Morpholinoacetophenone (0.50 g, 2.4 mmol), 3,3,6,6-tetramethyl-3,6-disila-1,7-octadiene (0.48 g, 2.4 mmol), Ru catalyst (0.1 g, 0.11 mmol) and a Teflon-covered magnetic stirring bar were placed in an Ace pressure tube as above. In this way, 0.79 g, 80% yield of copolymer M_w/M_n = 38,000/25,000, T_g = 2.7 °C was isolated. ^{1}H NMR δ: -0.04(br.s, 12H), 0.37(br.s, 4H), 0.75-0.82(m, 4H), 2.42(m, 4H), 2.43(s, 3H) 3.15(s, 4H), 3.82(s, 4H), 6.56(s, 2H). ^{13}C NMR δ: -4.19, 6.91, 17.61, 27.85, 33.21, 48.95, 66.79, 112.93, 132.96, 141.72, 151.45, 207.94. ^{29}Si NMR δ: 4.16. IR ν: 2955, 2901, 1651, 1599, 1452, 1380, 1353, 1257, 1187, 1133, 1112, 910, 834, 782, 737, 647 cm^{-1}. UV λ_{max} nm (ε): 215(24,600), sh243(10,900), 283(6,500). Elemental Anal. Calcd for $C_{22}H_{37}O_2NSi_2$: C, 65.51; H, 9.18; N, 3.47. Found: C, 64.98; H, 8.88; N, 3.37.

Copoly(3,3,5,5-tetramethyl-4-oxa-3,5-disila-1,7-heptanylene/2-aceto-5-phenoxy-1,3-phenylene)(Copoly-VII). 4'-Phenoxyacetophenone (0.65 g, 3.0 mmol), 1,3-divinyltetramethyldisiloxane (0.56 g, 3.0 mmol), xylene (2 mL), Ru catalyst (0.07 g, 0.076 mmol) and a Teflon-covered magnetic stirring bar were placed in an Ace pressure tube as above. In this manner, 0.98 g, 87% yield of pure copolymer M_w/M_n = 19,550/10,490, T_g = -12°C was obtained. ^{1}H NMR δ: 0.04(s, 12H), 0.78(m, 4H), 2.44(m, 4 H), 2.48(s, 3H), 6.70(s, 2H), 6.97(d, 2H, J = 7.5 Hz), 7.07(t, 1H, J = 7.5 Hz), 7.31(t, 2H, J = 7.5 Hz). ^{13}C NMR δ: 0.18, 20.51, 26.70, 33.16, 116.19, 118.74, 123.19, 129.71, 136.22, 142.12, 156.93, 157.26, 207.77. ^{29}Si NMR δ: 7.29. IR ν: 2955, 1699, 1588, 1492, 1255, 1058, 840 cm^{-1}. UV λ_{max} nm(ε): 246 (8,070), 255(7,340), 261(6,600), 276 (3,910). Elemental Anal. Calcd for $C_{22}H_{30}O_3Si_2$: C, 66.33; H, 7.54. Found: C, 66.24; H, 7.24.

Copoly(3,3,6,6-tetramethyl-3,6-disila-1,8-octanylene/2-aceto-5-phenoxy-1,3-phen ylene)(Copoly-VIII). 4'-Phenoxyacetophenone (0.65 g, 3.0 mmol), 3,3,6,6-tetramethyl-3,6-disila-1,7-octadiene (0.59 g, 3.0 mmol), xylene (2 mL), Ru catalyst (0.07 g, 0.076 mmol) and a Teflon-covered magnetic stirring bar were placed in an Ace pressure tube as above. In this way, 0.87 g, 70% yield of pure copolymer M_w/M_n = 17,500/10,550, T_g = -1.5°C was obtained. ^{1}H NMR δ: -0.04(s, 12H), 0.36(s, 4H), 0.77(m, 4H), 2.39(m, 4H), 2.47 (s, 3H), 6.71(s, 2H), 6.98(d, 2H, J = 7.5 Hz) 7.08(t, 1H, J = 7.5 Hz), 7.32 (t, 2H, J = 7.5 Hz). ^{13}C NMR δ: -4.20, 6.88, 17.15, 27.37, 33.18, 116.27, 118.63, 123.11, 129.69, 136.22, 142.44, 157.03, 157.15, 207.90. ^{29}Si NMR δ: 4.23. IR ν: 2953, 2903, 1696, 1587, 1491, 1460, 1417, 1353, 1288, 1248, 1217, 1177, 1165, 1134, 1055, 832, 783, 695 cm^{-1}. UV λ_{max} nm(ε): 248(10,640), 255 (10,160), 261 (9,360), 276(5,770). Elemental Anal. Calcd for $C_{24}H_{34}O_2Si_2$: C, 70.24; H, 8.29. Found: C, 69.94; H, 7.98.

Copoly(3,3,5,5-tetramethyl-4-oxa-3,5-disila-1,7-heptanylene/2-aceto-5-methoxy-1,3-phenylene)(Copoly-IX). 4'-Methoxyacetophenone (0.50 g, 3.3 mmol), 1,3-divinyltetramethyldisiloxane (0.62 g, 3.3 mmol), toluene (1 mL), Ru catalyst (0.06 g, 0.066 mmol) and a Teflon covered magnetic stirring bar were placed in an Ace pressure tube as above. In this way, 0.95 g, 86% yield M_w/M_n = 13,860/6,700, T_g = -24°C was obtained. ^{1}H NMR δ: 0.07(s, 12H), 0.82(m, 4H), 2.43(s, 3H), 2.47(m, 4H), 3.77 (s, 3H), 6.58(s, 2H). ^{13}C NMR δ: 0.25, 20.73, 26.92, 33.26, 55.10, 111.33, 133.93, 141.86, 160.78, 207.95. ^{29}Si NMR δ: 7.21. IR ν: 3000, 2955, 2838, 1940, 1737, 1694, 1650, 1638, 1602, 1538, 1511, 1469, 1442, 1414, 1255, 1179, 1153, 1125, 1045, 954, 901, 788, 704, 623 cm $^{-1}$. UV λ_{max} nm(ε): 250(7,360), 256(7,850), 261(7,930). Elemental Anal. Calcd for $C_{17}H_{28}O_3Si_2$: C, 60.71; H, 8.33. Found: C, 59.82; H, 8.00.

Copoly(3,3,6,6-tetramethyl-3,6-disila-1,8-octanylene/2-aceto-5-methoxy-1,3-phenylene)(Copoly-X). 4'-Methoxyacetophenone (0.50 g, 3.3 mmol), 3,3,6,6-tetramethyl-3,6-disila-1,7-octadiene (0.65 g, 3.3 mmol), toluene (1 mL), Ru catalyst (0.06 g, 0.066 mmol) and a Teflon covered magnetic stirring bar were placed in an Ace pressure tube as above. In this manner, 0.89 g, 77% yield of pure copolymer M_w/M_n = 29,700/11,900, T_g = -6°C was obtained. ^{1}H NMR δ: -0.04(s, 12H), 0.37(s, 4H), 0.79 (m, 4H), 2.40(m, 4H), 2.43(s, 3H), 3.77(s, 3H), 6.57(s, 2H). ^{13}C NMR δ: -4.20, 6.92, 17.31, 27.54, 33.25, 55.09, 111.26, 133.63, 142.14, 159.61, 207.96. ^{29}Si NMR δ:

4.20. IR ν: 2951, 2900, 2838, 1702, 1698, 1694, 1682, 1674, 1619, 1600, 1574, 1468, 1455, 1441, 1422, 1414, 1350, 1327, 1293, 1246, 1192, 1176, 1152, 1133, 1075, 1054, 1037, 996, 953, 900, 828, 781 cm^{-1} . UV λ_{max} nm(ε): 250(5,710), 255 (6,010), 261(5,990). Elemental Anal. Calcd for $C_{19}H_{32}O_2Si_2$: C, 65.52; H, 9.20. Found: C, 64.94; H, 9.27.

High Molecular Weight Copoly(3,3,5,5-tetramethyl-4-oxa-3,5-disila-1,7-heptanylene/2-aceto-1,3-phenylene)(Copoly-XI). Ru catalyst (0.11g, 0.12 mmol) and toluene (1 mL) were placed in a 50 mL three-neck round-bottomed flask equippped with a reflux condenser and a Teflon-covered magnetic stirring bar. The other two necks of the flask were sealed with rubber septa. Vinyltrimethoxysilane (0.16 mL, 0.018 g, 0.12 mmol) was added via a syringe. The flask and its contents were heated to 135°C for 1 min. At this time, a mixture of acetophenone (0.48 g, 4 mmol) and 1,3-divinyltetramethyldisiloxane (0.74 g, 4 mmol) were added. The reaction was heated at 135°C for 12 h. After work-up, 0.9 g, 74% yield of copoly-XII, M_w/M_n = 40,600/14,800 and T_g = -33°C was obtained. It had spectral properties in complete agreement with low molecular weight copoly-XI previously reported (*4*).

Low Molecular Weight Copoly(3,3,6,6-tetramethyl-3,6-disila-1,8-octanylene/2-aceto-1,3-phenylene) (Copoly-XII). Acetophenone (0.3 g, 2.5 mmol), 3,3,6,6-tetramethyl-3,6-disila-1,7-octadiene (0.5 g, 2.5 mmol) and Ru catalyst (0.15 g, 0.16 mmol) were reacted as above. In this way, copoly-XIII, 0.72 g, 90% yield, M_w/M_n = 7,890/4,800, T_g = -20.5°C was obtained. ^{1}H NMR δ: -0.015(s, 12H), 0.40(s, 4H), 0.79-0.86(m, 4H), 2.42-2.56(m, 7H), 7.07(d, 2H, J = 7.5 Hz), 7.21(t, 1H, J = 7.5 Hz). ^{13}C NMR δ: -4.19, 6.90, 17.42, 27.29, 33.02, 125.89, 128.72, 139.87, 140.79, 208.11. ^{29}Si NMR δ: 4.17. IR ν: 3060, 2951, 2900, 1929, 1702, 1699, 1682, 1593, 1461, 1455, 1439, 1422, 1351, 1247, 1177, 1133, 1089, 1053, 995, 961, 902, 832, 780, 757, 695 cm^{-1}. UV λ_{max} nm(ε): 217 (18,600), 264(2,020). Elemental Anal. Calcd for $C_{18}H_{30}OSi_2$: C, 67.92; H, 9.43. Found: C, 67.19; H, 8.87.

High Molecular Weight Copoly-XII. Ru catalyst (0.073 g, 0.08 mmol) and styrene (0.009 mL, 0.008 g, 0.08 mmol) were placed in a 50 mL three-neck round-bottomed flask as above. The mixture was heated at 135°C for two min. At this time, a mixture of acetophenone (0.48 g, 4 mmol) and 3,3,6,6,-tetramethyl-3,6-disila-1,7-octadiene (0.79 g, 4 mmol) was added. The reaction mixture was heated at 135°C for 12 h. After work-up, 1.12 g, 88% yield of copoly-XIII, M_w/M_n = 22,900/21,300, T_g = -17°C was obtained. It had spectral properties idential to those reported above. Ethylbenzene was detected by GC/MS analysis of the solution by both its retention time and base peak which was found at m/e = 91.

Results and Discussion

Murai has suggested that the ruthenium catalyzed *ortho* alkylation reaction of acetophenone with alkenes proceeds by insertion of a carbonyl complexed coordinately unsaturated Ru species into an adjacent *ortho* C-H bond to yield an aryl-Ru-H intermediate (*1-3*). The well known insertion of palladium species into aromatic C-H bonds which are *ortho* to activating groups such as N,N-dimethylaminomethyl may be similar

(9,10). The ruthenium catalyzed reactions of aromatic ketones and alkenes or α,ω-dienes maybe related mechanistically to the palladium catalyzed Heck reaction of aryl halides with alkenes (*11*). Insertion of palladium into the C-X bond of the aryl halide leads to a reactive aryl palladium species which is the key intermediate in this reaction. The Heck reaction has also been applied to the synthesis of polymers (*12-16*).

Attempts to carry out ruthenium catalyzed copolymerization reaction between 4-acetylpyridine and 1,3-divinyltetramethyldisiloxane were unsuccessful (*17*). This may be rationalized if the insertion of the complexed coordinately unsaturated Ru species into the *ortho* C-H bond has some of the characteristics of an aromatic electrophilic substitution reaction. Thus pyridine undergoes aromatic electrophilic substitution reactions with difficulty (*18*). Based on this hypothesis, acetophenones substituted with electron donating groups, which facilitate electrophilic substitution, such as 4'-dialkylamino, 4'-methoxy or 4'-phenoxy should undergo this reaction with greater facility and might lead to higher molecular weight copolymers. To test this concept, we have carried out Ru catalyzed copolymerization reactions between 4'-piperidinoacetophenone, 4'-morpholinoacetophenone, 4'-(N'-benzyl)piperazinoacetophenone, 4'-methoxyacetophenone, or 4'-phenoxyacetophenone and 1,3-divinyltetramethyldisiloxane or 3,3,6,6- tetramethyl-3,6-disila-1,7-octadiene. In fact, the reactions proceed significantly faster and the **molecular weights of the copolymers obtained are significantly higher.**

Alternatively, treatment of the Ru catalyst with one equivalent of an alkene prior to additon of the acetophenone and α,ω-diene monomers leads to high molecular weight copolymers. This latter method was developed based on end group analysis of low molecular weight copoly[1,8-xanthonylene/3,3,5,5-tetramethyl-4-oxa-3,5-disila-1,7-heptanylene](M_w/M_n = 3,500/1,700)which had been previously prepared by the Ru catalyzed step-growth copolymerization of xanthone and 1,3-divinyltetramethyldisiloxane (*5*). Resonances consistent with an ethyl group attached to silicon were detected by NMR spectroscopy. Our hypothesis was that these might be formed by hydrogenation of one of the C-C double bonds of the α,ω-diene and that the hydrogen needed for this reduction must come from the catalyst itself. This hydrogenation may in fact be involved in the formation of the coordinately unsaturated catalytically active ruthenium species needed for copolymerization.

Based on this hypothesis, we have treated the Ru catalyst with a stoichiometric amount of an alkene such as styrene or vinyltrimethoxysilane. After heating this mixture at 135°C for a few minutes, a solution of acetophenone and α,ω-diene were added. In this way, significantly higher molecular weight copolymers were obtained. For example, copolymerization of acetophenone and 1,3-divinyltetramethyldisiloxane with Ru catalyst which had been previously activated with a stoichiometric amount of vinyltrimethoxysilane gave copoly-XII, M_w/M_n = 40,600/14,800 with a T_g = -33°C. Whereas, copoly-XII synthesized without prior activation of the catalyst had M_w/M_n = 8,300/6,700 and T_g = -44°C. Similarly, copolymerization of acetophenone and 3,3,6,6-tetramethyl-3,6-disila-1,7-octadiene catalyzed by Ru which had been previously activated by treatment with a stoichiometric amount of styrene gave copoly-XIII M_w/M_n = 22,900/21,300 with a T_g = -17°C. Ethylbenzene was detected by GC/MS in this reaction. Copoly-XIII prepared without prior activation of the catalyst had M_w/M_n = 7,800 /4,800 with T_g = -20.5°C. Catalyst activation with a stoichiometric amount of alkene provides a significant imporvement in copolymer molecular weight, T_g and thermal stability.

The structure of the copolymers reported herein as determined by ^{1}H, ^{13}C, and ^{29}Si NMR spectroscopy is consistent with predominant regioselective addition of the C-H bonds, which are *ortho* to the carbonyl group of the substituted acetophenones across the C-C double bonds of 1,3-divinyltetramethyldisiloxane or 3,3,6,6-tetramethyl-3,6-disila-1,7-octadiene such that the hydrogen becomes attached to the more substituted end of the double bond (see experimental) as in Figure 1.

The choice of α,ω-dienes is not random. Based on our previous work, it is apparent that most α,ω-dienes are NOT suitable substrates for this reaction (*19*). The ruthenium catalyst not only catalyzes the insertion of C-C double bonds of α,ω-dienes into the *ortho* C-H bonds of aromatic ketones - but also the isomerization of terminal C-C double bonds to internal double bonds which are much less reactive. 1,3-Divinyltetramethyldisiloxane and 3,3,6,6-tetramethyl-3,6-disila-1,7-octadiene have been utilized because isomerization of their C-C double bonds is blocked.

The TGAs of copoly-I, II, IV, V, and VI are quite similar. They are thermally stable to 200-210°C. Between 200-210°C and 385-400°C they lose about six percent of their initial weight. Above 385-400°C, rapid weight loss occurs.

Copoly-VII, VIII, IX, and X are thermally stable to 270°C. Between 270 and 380°C they each experience approximately a five percent weight loss. Above 400°C rapid weight loss occurs. By 550°C residues which amount to between eighteen and ten percent remain as in Figure 2. The copolymers which incorporate 4'-methoxy and 4'-phenoxyacetophenones are more thermally stable that those based on 4'-dialkylaminoacetophenones.

The achievement of higher molecular weight polymers is important since many polymer properties change rapidly until a minimum threshold molecular weight is achieved (*19*). Frequently this minimum polymer molecular weight for constant polymer properties occurs at about a molecular weight of 10,000. All of the copolymers reported herein have molecular weights M_w greater than 15,000.

The effect of polymer molecular weight on thermal stability of these polymers is particularly dramatic. The TGAs of low and high molecular weight samples of copoly-V and copoly-XII provide striking graphical comparisons. Low molecular weight samples of copoly-V have been prepared by carrying out the copolymerization with a non-stoichiometric balance of reactants. In one case, a three percent excess of 3,3,6,6-tetramethyl-3,6-disila-1,7-octadiene was utilized, while in the other a three percent excess of 4'-piperidinoacetophenone was employed. These reactions lead to lower molecular weight copolymers. In the case of the former, terminal vinyl groups could be detected by ^{1}H and ^{13}C NMR. In the latter, terminal 4-piperidino acetophenone units were observed ^{1}H and ^{13}C NMR spectroscopy. The thermal stability of these lower molecular weight copolymers was significantly lower than higher molecular weight copolymer as in Figure 3. High molecular weight copoly-XII is also found to be significantly more thermally stable than low molecular weight copoly-XII as in Figure 4.

Oxidative photodegradation is a major problem for polymers which are utilized as coatings in exterior applications. Perfectly alternating copolymers formed by ruthenium catalyzed step-growth copolymerization of acetophenones and α,ω-dienes have considerable potential in this area. Thus, these polymers contain *ortho* alkyl acetophenone units as an integral part of the copolymer backbone. It is well known that monomeric *ortho* alkyl acetophenones undergo light induced photoenolization and that this is a process which is reversible in the dark as in equation 5. Experiments to test the photoenolization and photostability of these new copolymers are in progress.

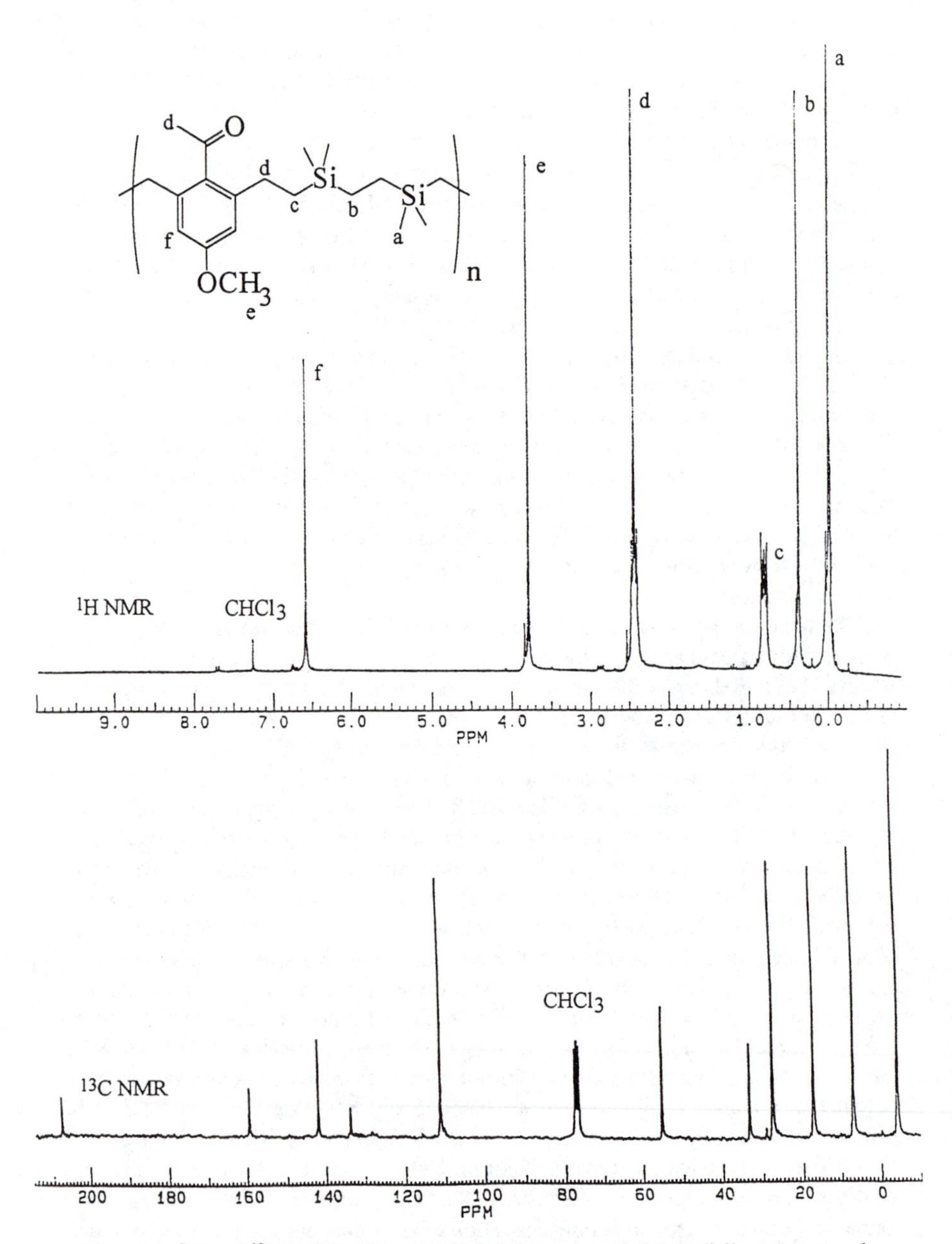

Figure 1. ^{1}H and ^{13}C NMR of Copoly(3,3,6,6,-tetramethyl-3,6-disila-1,8-octanyl-ene/2-aceto-5-methoxy-1,3-phenylene).

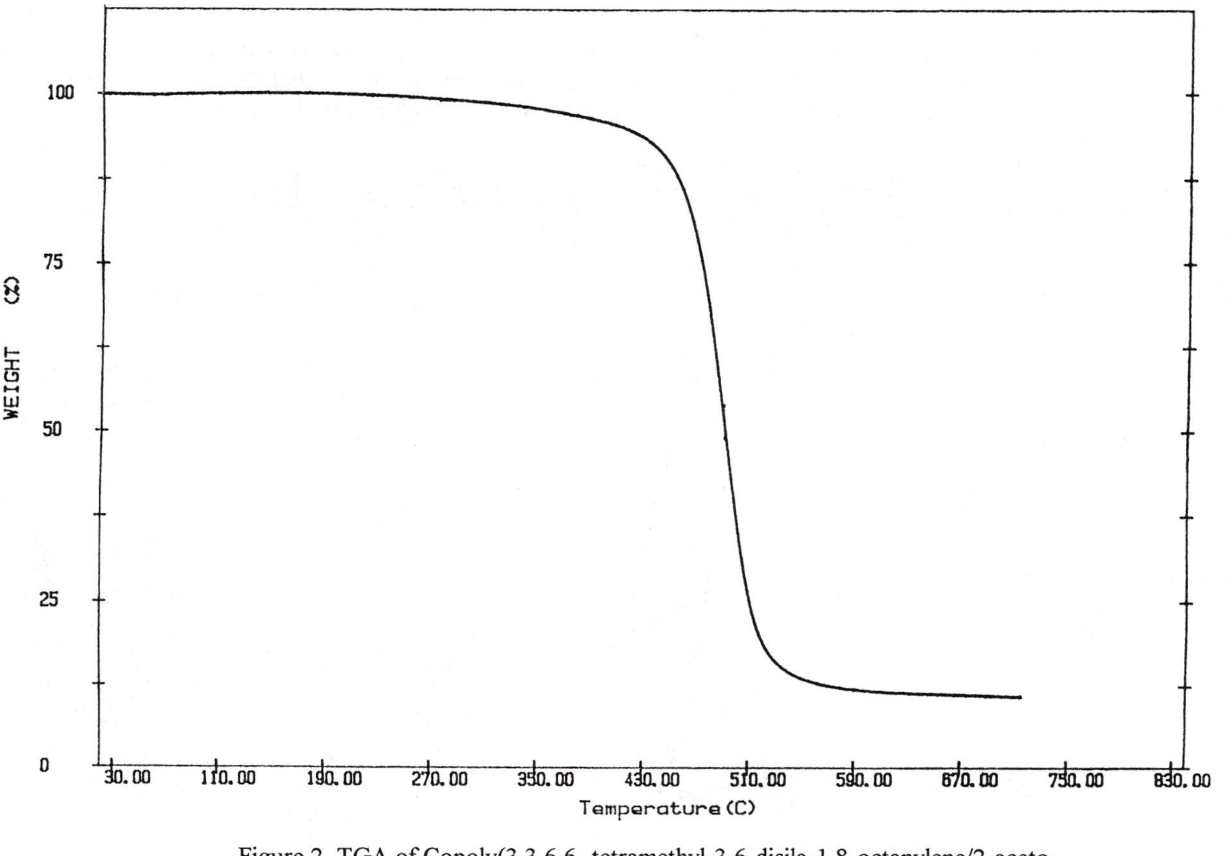

Figure 2. TGA of Copoly(3,3,6,6,-tetramethyl-3,6-disila-1,8-octanylene/2-aceto-5-methoxy-1,3-phenylene).

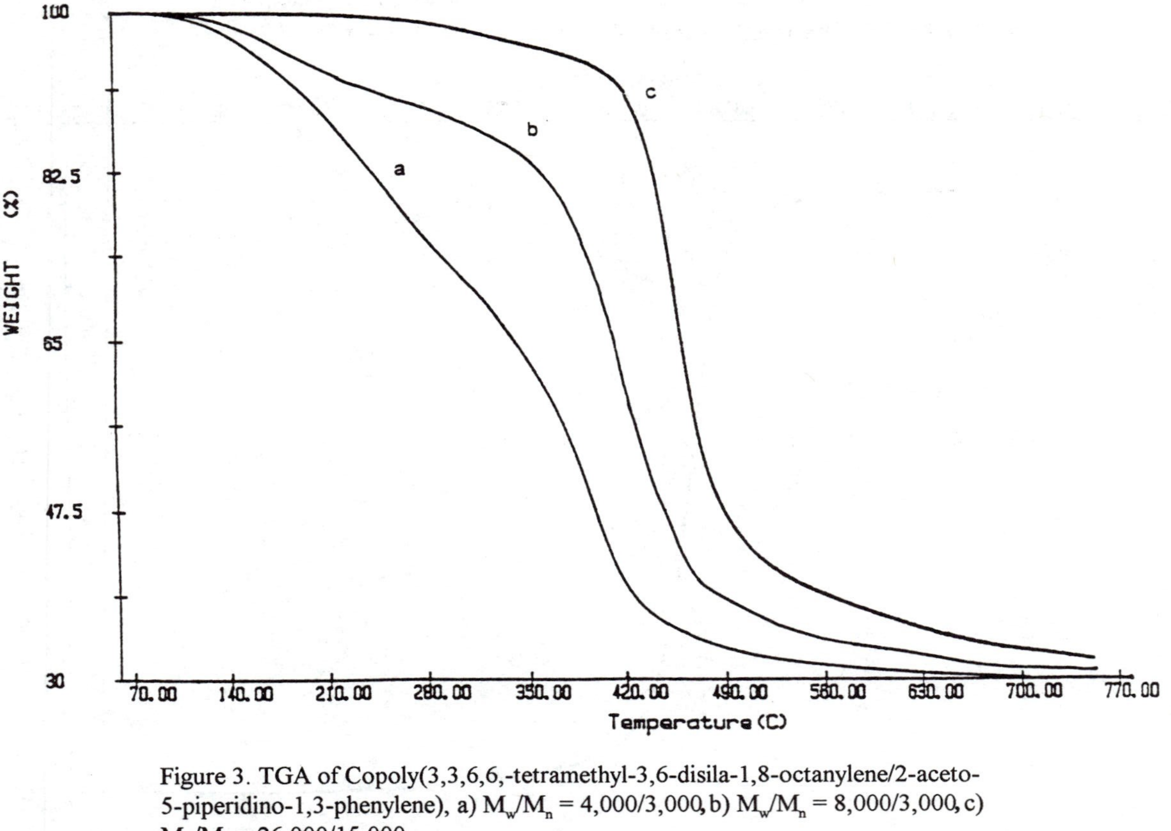

Figure 3. TGA of Copoly(3,3,6,6,-tetramethyl-3,6-disila-1,8-octanylene/2-aceto-5-piperidino-1,3-phenylene), a) M_w/M_n = 4,000/3,000, b) M_w/M_n = 8,000/3,000, c) M_w/M_n = 26,000/15,000.

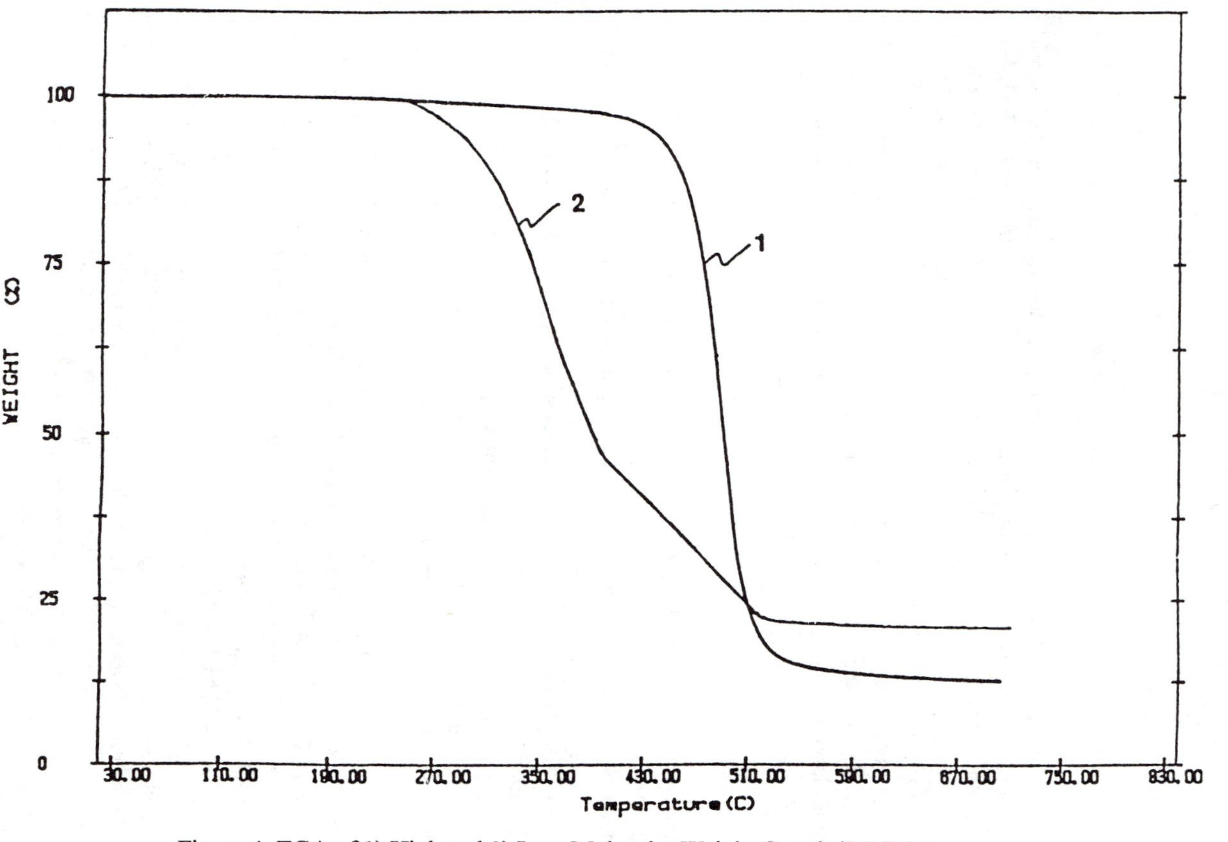

Figure 4. TGA of 1) High and 2) Low Molecular Weight Copoly(3,3,5,5-tetramethyl-4-oxa-3,5-disila-1,7-heptanylene/2-aceto-1,3-phenylene).

(5)

hv

dark

Conclusion

The ruthenium catalyzed copolymerization of acetophenones with α,ω-dienes is sensitive to electronic effects in the acetophenone. Specifically, electron donating dialkylamino, methoxy and phenoxy groups facilitate the copolymerization and result in high molecular weight polymers. In these systems, the molecular weights are sufficiently high to warrant legitimate use of the word copolymers rather than cooligomers. Alternatively, prior activation of the Ru catalyst with a stoichiometric amount of an alkene before addition of a mixture of the acetophenone and the α,ω-diene also leads to higher molecular weight polymers. Further studies on this step growth polymerization based on the Murai reaction are in progress.

Literature Cited

1. Murai, S.; Kakiuchi, S.; Sekine, S.; Tanaka, Y.; Sonoda, M.; Chatani, N. *Nature* **1993**, *366*, 529.
2. Murai, S.; Kakiuchi, F.; Sekine, S.; Tanaka, Y.; Kamatani, A.; Sonoda, M.; Chatani, N. *Pure & Applied Chem.* **1994**, *66*, 1527.
3. Kakiuchi, F.; Sekine, S.; Tanaka, Y.; Kamatani, A.; Sonoda, M.; Chatani, M.; Murai, S. *Bull. Chem. Soc. Jpn.* **1995**, *68*, 62.
4. Guo H,; Weber, W. P. *Polymer Bull.* **1994**, *32*, 525.
5. Guo, H.; Tapsak, M. A.; Weber,W. P. *Polymer Bull.* **1995**, *34*, 49.
6. Odian, G. *Principles of Polymerization*, J. Wiley & Sons: New York, NY, 1981, p 82-96
7. Freeman, F.; Hill, H. D. W.; Kaptein, R. *J. Magn. Reson.* **1972**, *7*, 327.
8. Levison, J. J.; Robinson, S. D. *J. Chem. Soc.* **1970**, *A*, 2947.
9. Cope, A. C.; Siekman, R. W. *J. Am. Chem. Soc.* **1965**, *87*, 3272.
10. Cope, A. C.; Friedrich, E. C. *J. Am. Chem. Soc.*, **1968**, *90*, 909.
11. Heck, R. F. *Org. React.* **1982**, *27*, 345.
12. Weitzel, H. P.; Mullen, K. *Makromol. Chem.*, **1990**, *191*, 2837.
13. Bao, Z. N.; Chen, Y. M.; Cai, R. B.; Yu, L. P. *Macromolecules* **1993**, *26*, 5281.
14. Suzuki, M.; Lim, J. C.; Saegusa, T. *Macromolecules* **1990**, *23*, 1574.
15. Martelock, H.; Geiner, A.; Heitz, W. *Makromol. Chem.* **1991**, *192*, 967.
16. Heitz, W.; Brugging, W.; Freund, L.; Gailberger, M.; Greiner, A.; Jung, H.; Kampschulte, U.; Nieber, N.; Osan, F. *Makromol. Chem.* **1990**, *189*, 119.
17. Lu, Q. J. unpublished results 1994.
18. Acheson, R. M. *An Introduction to the Chemistry of Heterocyclic Compounds,* J. Wiley & Sons: New York, NY, 1976, p 236-238.
19. Guo, H.; Tapsak, M. A.; Weber, W. P. *Polymer Bull.* **1994**, *33,* 417.

RECEIVED December 4, 1995

Chapter 6

Access to Silicon-Derived Polymers via Acyclic Diene Metathesis Chemistry

S. Cummings[1], E. Ginsburg[2], R. Miller[3], J. Portmess[1], Dennis W. Smith, Jr.[4], and K. Wagener[1]

[1]Department of Chemistry and Center for Macromolecular Science and Engineering, University of Florida, Gainesville, FL 32611–7200
[2]Research Laboratories, Eastman Kodak Company, Rochester, NY 14650–2103
[3]Research Division, IBM Almaden Research Center, 650 Harry Road, San Jose, CA 95120–6099
[4]Central Research and Development, Organic Product Research, Dow Chemical Company, Freeport, TX 77541

Acyclic diene metathesis (ADMET) polymerization has shown to be a clean route to silicon-based unsaturated polymers. High molecular weight carbosilane and carbosiloxane polymers have been generated with Schrock's well-defined tungsten and molybdenum alkylidenes. Monomer reactivity has shown to be highly dependant on the nature of the carbon spacing between the olefin and the silicon moiety. Cyclization reactions have also been observed during bulk polymerization and in solution. In addition, chloro-functionalized silanes offer an opportunity to produce materials with a broad range of potential properties due to the reactive sites along the backbone. Further, σ, π conductive hybrid silicon polymers have been constructed which will allow the systematic study of the nondefined nature of conductivity through a σ–π system.

Olefin metathesis polymerization, which is now more than 25 years old, has led to commercial polymers based on ring opening metathesis polymerization (ROMP). The early developments in metathesis proceeded with research using classical catalyst systems, and while these systems are quite useful in producing polymers, their behavior is less well understood, and consequently well defined catalyst systems have evolved. Metathesis polymerization can be divided into two categories (Figure 1): a) ring opening metathesis polymerization and b) acyclic diene metathesis (ADMET) polymerization. The literature associated with ROMP chemistry is broad and will not be considered in the scope of this report. A brief review of ADMET polymerization, however, will prove useful in describing silicon polymers that can be made using this technique.

ADMET polymerization is based on step condensation chemistry where high molecular weight polymers are generated via the formation and removal, *in vacuo*, of a small molecule (Figure 2). The key to successful ADMET chemistry is dependent on the removal of acids from the catalyst system. This can be illustrated in Figure 3, where the adverse effect of acids is manifested in the attempted polymerization of 1,9-decadiene. In this example, the well known tungsten hexachloride catalyst system

0097–6156/96/0624–0113$12.00/0

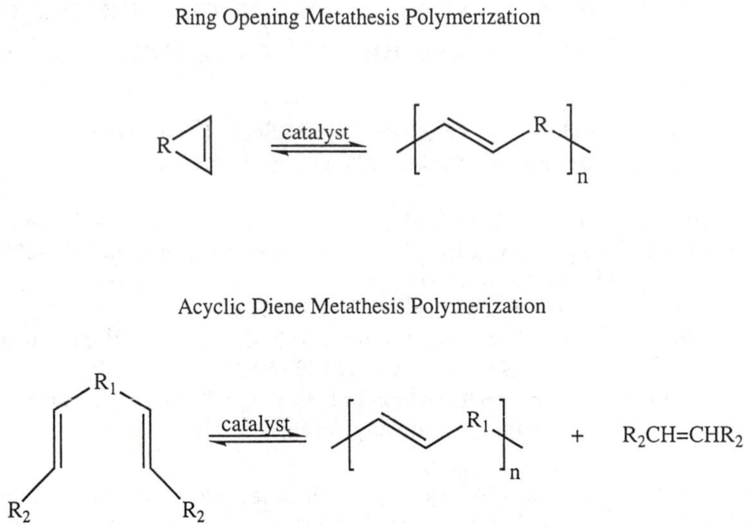

Ring Opening Metathesis Polymerization

R catalyst R n

Acyclic Diene Metathesis Polymerization

R_1 catalyst R_1 n + $R_2CH=CHR_2$

R_2 R_2

Figure 1. Classes of metathesis polymerization.

ADMET Dimerization

R + R catalyst R R

ADMET Polymerization

n R catalyst R n + $H_2C=CH_2$

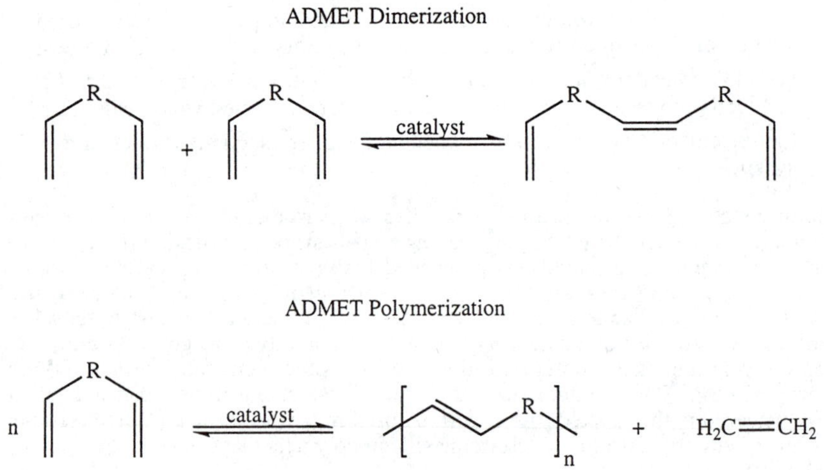

Figure 2. ADMET is a true step condensation polymerization.

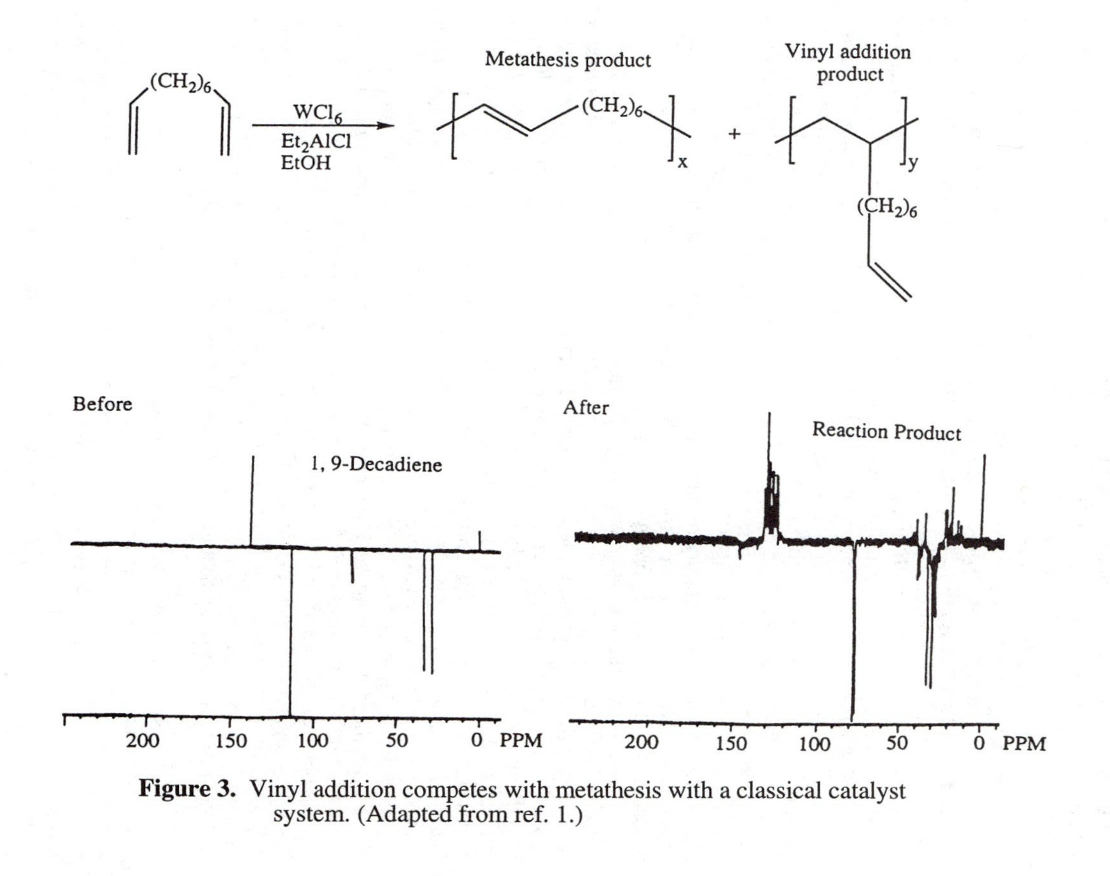

Figure 3. Vinyl addition competes with metathesis with a classical catalyst system. (Adapted from ref. 1.)

results in metathesis condensation, but also induces vinyl addition chemistry, producing a different type of repeat unit (*1*). Further, this vinyl addition reaction disrupts stoichiometric balance resulting in poorly defined polymer systems which are for the most part crosslinked. The carbon NMR spectrum shows this rather clearly, particularly in the sp^2 region where a large multiplet appears around 140 ppm. If this chemistry were perfectly clean with respect to the ADMET reaction, then one would anticipate the presence of only two lines for *trans* and *cis* sp^2 carbons. This problem was resolved by changing to an acid-free catalyst system.

Figure 4 displays this phenomenon graphically, where the catalyst system chosen is that produced by Schrock several years ago. Though the molybdenum version is shown in this figure, both molybdenum and tungsten versions are successful in this conversion, as is the recently developed Grubbs catalyst system based on ruthenium. Figure 4 illustrates that the polymer possesses only the *trans* and *cis* resonances. Polymers with number average molecular weights of a minimum of 50,000 can be made by this technique, if ethylene is completely removed (*2*).

Mechanistically, polymerization proceeds through various metallacyclobutane intermediates (Figure 5). This mechanism, which is consistent with the reacton intermediate proposed by Chauvin (*3*), and later clearly manifested by Grubbs using Tebbe's catalyst system (*4*), demonstrates the ease in which polymers can be generated using this chemistry. The polymers are perfectly difunctional in vinyl groups, an important characteristic of ADMET polymers is evident in the oligomers shown in Figure 3. The distinctive quality of ADMET polymerization derives from the fact that very clean polymer systems are produced which do not require additional purification procedures and possess only one type of repeat unit.

A wide variety of polymers can be produced using this chemistry including those which are purely hydrocarbon (*5,6*) in nature as well as systems possessing functional groups. The monomers containing functional groups can be converted to ADMET polymers if simple selection rules which govern this reaction are followed. Specifically, the negative neighboring group effect must be considered, a phenomenon which is illustrated in Figure 6. Here, the R group shown possesses Lewis basic atoms such as oxygen, sulfur, and phosphorus, and if the functional group is part of a chain which places it relatively close to the active metal center, then complexation occurs leading to a nonreactive entity. However, if the number of carbon spacers between the metal atom and the functional group is increased, then the equilibrium shifts from the stable complex to the elimination of the active catalytic methylidene carbene, which leads to propagation (polymerization). If two methylene spacers are present between the metal and the functional group, then the equilibrium preferentially is shifted towards polymer formation. This phenomenon appears to be quite general, and as a consequence has led to the synthesis of a wide variety of polymer systems including ethers (*7*), esters (*8*), carbonates (*9*), polyketones (*10*), polythioethers (*11*), and in most recent cases, polymers possessing phosphorus.

The Synthesis of Polycarbosilanes. Silicon-based polycarbosilanes do not exist naturally. A number of synthetic methods have been used to prepare them, but only now can the synthesis of polycarbosilanes, via ADMET chemistry, be realized (*12*). Our initial work began with the monomers, dimethyldivinylsilane and diphenyldivinylsilane. The attempted condensation of these monomers in a typical ADMET scheme did not lead to the formation of polymer. The explanation for the nonreactivity of these monomers can be attributed to steric interactions which preclude the formation of the required metallacyclobutane as illustrated in the polymerization cycle shown in Figure 3. This observation has precedence considering Schrock's work with vinyltrimethylsilane as illustrated in Figure 7 (*13*). Here it becomes evident that the steric interaction required to form the metallacyclobutane is inhibited by the

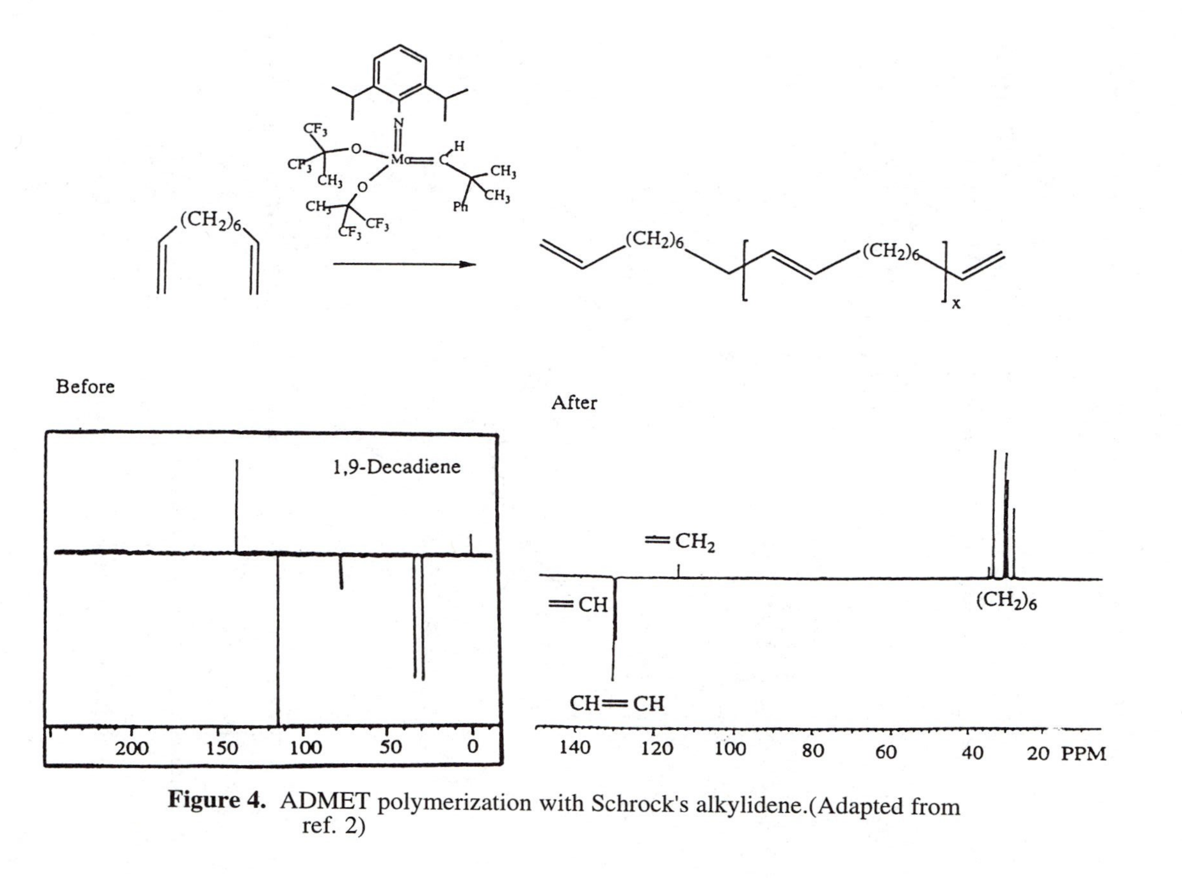

Figure 4. ADMET polymerization with Schrock's alkylidene.(Adapted from ref. 2)

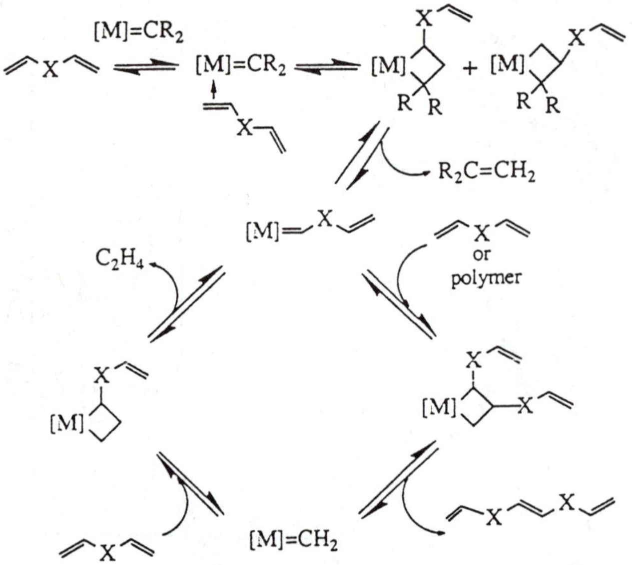

$[M]=CR_2$

X ⇌ $[M]=CR_2$ ⇌ [M] X R R + [M] X R R

X

$R_2C=CH_2$

$[M]=$ X

C_2H_4

X
or
polymer

X
[M]

X
[M] X

X

X X

$[M]=CH_2$

Figure 5. ADMET polymerization cycle.

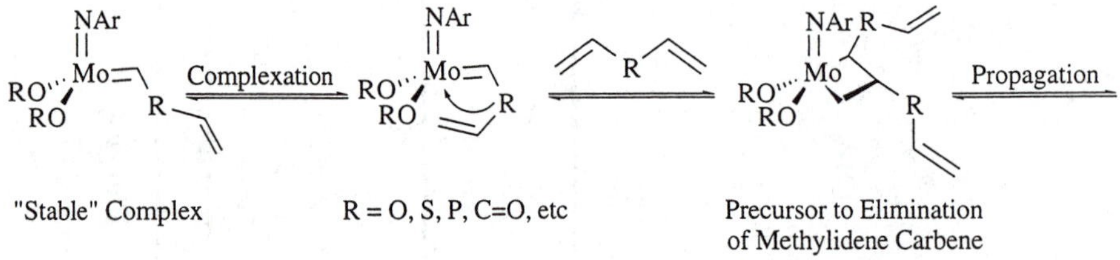

NAr
RO Mo R
RO

Complexation

NAr
RO Mo R
RO

R

NAr R
RO Mo R
RO

Propagation

"Stable" Complex

R = O, S, P, C=O, etc

Precursor to Elimination
of Methylidene Carbene

Figure 6. Schematic of the negative neighboring group effect.

presence of adjacent trimethylsilyl groups. Thus, while the exchange reaction can occur with Schrock's catalyst to form the precursor catalyst, the scheme cannot proceed to form links between the vinyltrimethylsilane olefins, and as a result, no reaction is observed. However, if one relieves the steric strain by placing methylene spacer groups between silicon and the olefin involved in metathesis, then condensation polymerization proceeds cleanly to produce a wide variety of carbosilane polymer systems. Three such examples are illustrated in Figure 8. ADMET polymerization of these monomers occurs rapidly with the release of ethylene, and high molecular weight polymers with number average molecular weights of 20,000 are observed upon reaction completion. The polymers that form are extraordinarily clean in that no other repeat unit is present. The NMR data found in Figure 9 illustrates the simplicity of this polymerization, where the NMR spectra are clean, the molecular weight is high, and in this case, the end groups are virtually nonexistent because of their low concentration. The polymers that are made by this system can exhibit a number of thermal properties depending upon the nature of the spacer itself. For example, if the spacer is aliphatic,then low glass transition temperature materials can be prepared with values as low as -60°C. However, if the spacers possess stiff carbon moieties such as aromatic rings, then the glass temperatures are in the range of 40-50°C. Further, these polymers are crystalline in nature, which is not unexpected, given the regularity of the repeat unit within the polymer.

While dimethyldivinylsilane cannot be polymerized for steric reasons, it can be copolymerized with monomers such as 1,9-decadiene. The copolymerization reaction proceeds smoothly because the steric strain described earlier is relieved since there are no adjacent trimethylsilyl groups found on the metallacyclobutane ring. In this case one of the substituents, a long alkyl chain, is less sterically hindering which allows completion of the polymerization cycle. This leads to polymers which possess dimethylsilyl co-repeat units with no two dimethylsilyl repeat units being connected to one another. Thus, the isolated presence of silyl groups in the polymer can be observed in its NMR spectrum. This provides further evidence for the steric nature in controlling ADMET chemistry.

The Synthesis of Polycarbosiloxanes. The situation for polycarbosiloxanes is similar to that observed for polycarbosilanes. Polymerizability of monomers is largely dictated by steric interactions, though one cannot exclude the electronic effect that may be present due to the silicon oxygen bond being adjacent to the metathesizing olefin. For example, bis-vinyl tetramethyldisiloxane fails to polymerize via ADMET chemistry whereas bis-allyl disiloxane does in fact condense in a rapid fashion. In the case of bis-allyldisiloxane, however, cyclization predominates rather than polymerization. The generation of the cyclic structure can be attributed to the Thorpe-Ingold effect of the dimethyl groups present in the monomer, generating a conformation which leads exclusively to intramolecular (cyclization) chemistry rather than intermolecular (polymerization) chemistry. Nonetheless, the ADMET reaction is rapid and quantitative.

Keeping these points in mind, it becomes evident that polymers can be formed by both alleviating steric strain and eliminating the possibility of intramolecular cyclization chemistry. Figure 10 illustrates examples of successful ADMET polymerizations leading to siloxane polymer structures (*14*). As previously stated these polymers possess only one type of repeat unit which is easily identified by typical characterization methods. Their thermal behavior can be varied depending upon the nature of the spacer unit, though in this case the materials are clearly more flexible than for carbosilanes with glass temperatures now closer to -100°C.

An interesting opportunity exists for combining carbosilanes and carbosiloxanes into polymer systems. It should be possible to tailor the behavior of silicon-containing polymers made by ADMET chemistry, depending upon the nature

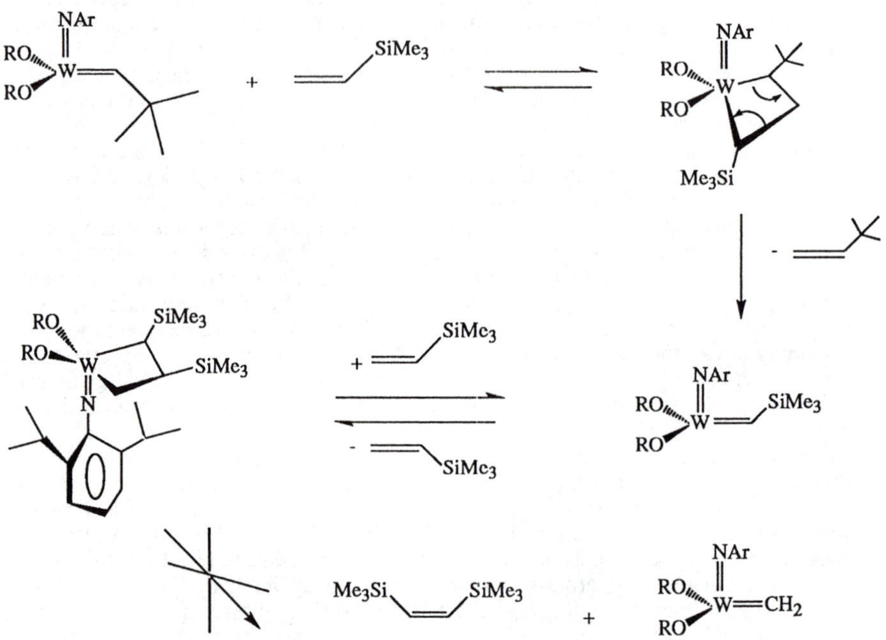

Figure 7. Schrock's vinylTMS study. R=CMe(CF_3)$_2$.

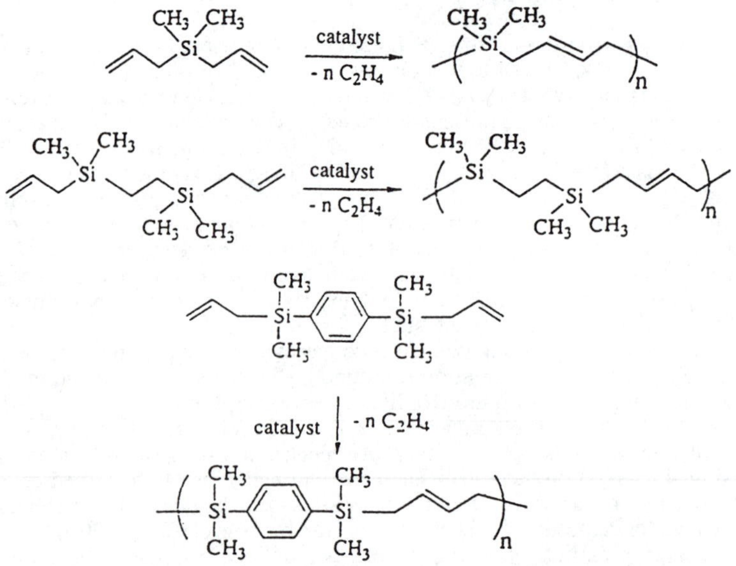

Figure 8. Polycarbo(dimethyl)silanes produced by ADMET.

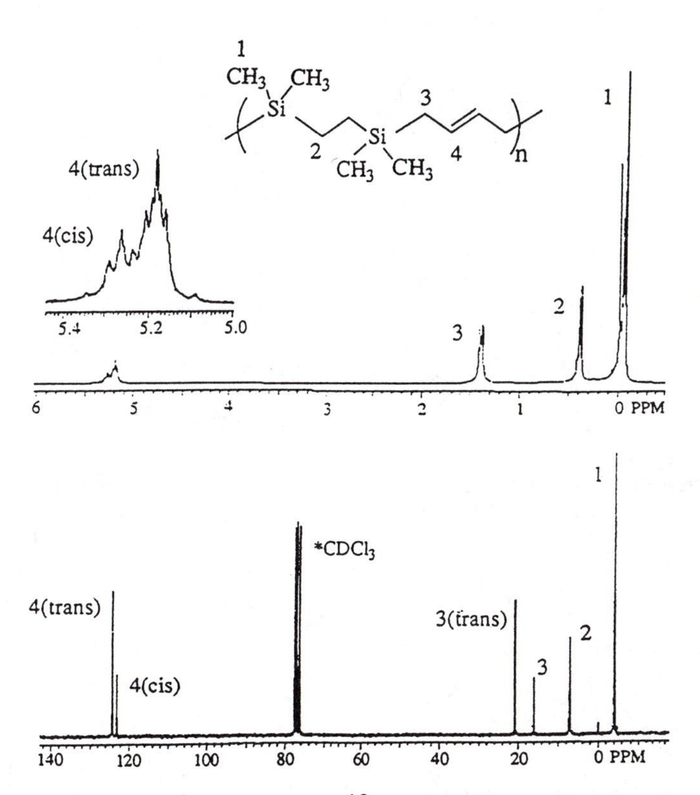

Figure 9. 200 MHz ^{1}H and 50 MHz ^{13}C NMR spectra of the above polycarbo(dimethyl)silane.(Reproduced with permission from ref. 12. Copyright 1991 American Chemical Society)

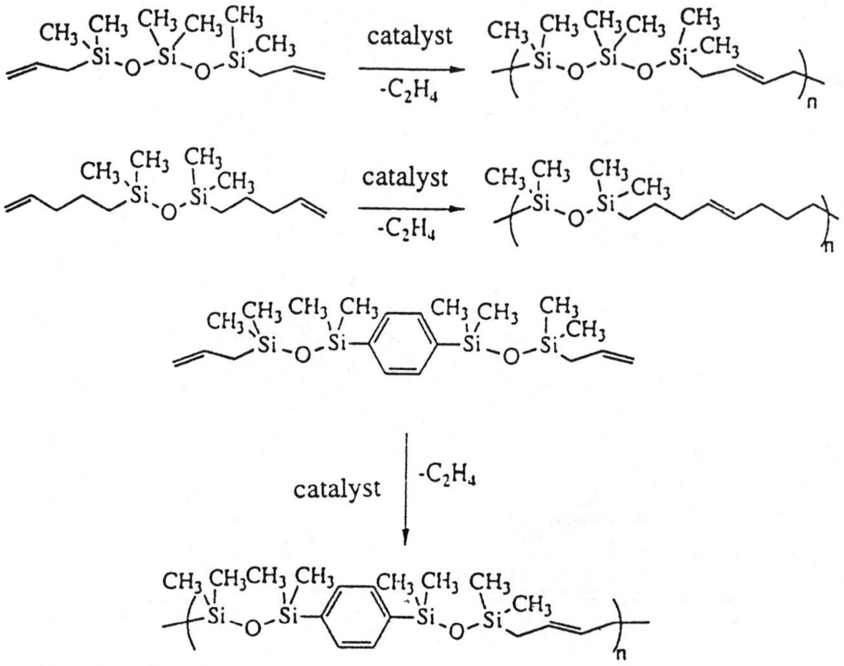

Figure 10. ADMET polymerization of carbodisiloxadienes.(Reproduced with permission from ref. 14. Copyright 1993 American Chemical Society)

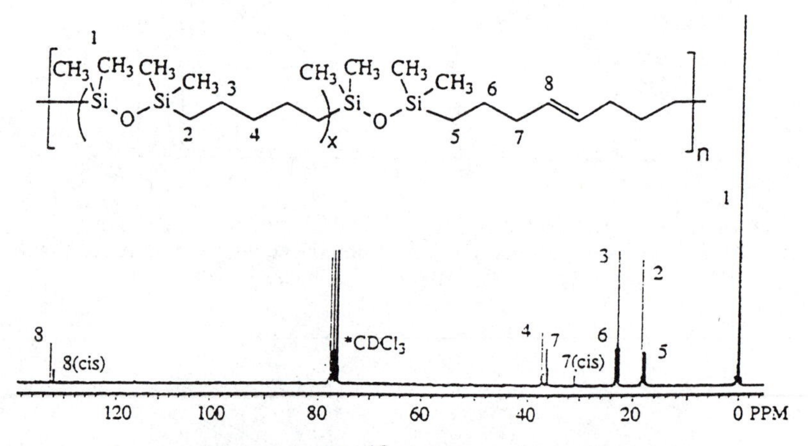

Figure 11. Quantitative 50 MHz ^{13}C NMR spectrum of the above polycarbosiloxanes.(Reproduced with permission from ref. 14. Copyright 1993 American Chemical Society)

of the spacer which is chosen in the diene monomers. Figure 11 shows the carbon NMR spectrum of a high molecular weight polycarbosiloxane with precise structure without prior purification. There are obvious advantages and opportunities to synthesizing pure, well defined polycarbosilane and polycarbosiloxane materials of this nature.

Typical of all carbosilane and carbosiloxane polymerizations is the fact that the *trans* stereochemistry predominates over the *cis* with respect to the olefin that is in the backbone. This *trans* arrangement arises from the polymerization mechanism leading to the connection of olefin units. This is observed to be true, not only for silicon polymers, but for all ADMET polymers made to this point, regardless of whether functionality is present in the backbone.

Another interesting aspect of ADMET polymerization has to do with the nature of the active chain end which is responsible for step polymerization. Once monomer concentration is sufficiently depleted, then other possible reactions can occur with respect to the chain end. One possibility is trans-metathesis which in essence is similar to exchange reactions that are found in classical step polymerizations. For example, trans-esterification in the formation of polyester naturally occurs upon depletion of monomer concentration. The same is true for polyamide chemistry. The result is the most probable distribution of molecular weights, leading to a polydispersity ratio which approaches 2.0. This polydispersity ratio value is typical for ADMET polymers, as well as for polycarbosilanes and polycarbosiloxanes.

Another typical reaction in step polymerization chemistry is the formation of small quantities of cyclics as an alternate reaction pathway. Cyclics are formed in these silicon polymers as described in a mechanism which is illustrated in Figure 12. Here, the active chain end backbites to produce rings of various sizes which can be strategically controlled by altering the medium in which the polymerization is conducted. In Figure 13, this situation is clearly illustrated for one of the simplest siloxane polymerizations that have been performed. For example, if the polymerization is carried out in bulk and under vacuum, 7 mol % of cyclics are observed with the *cis* isomer being generated in preference to the *trans* form. However, if the reactants are dissolved in a benzene solution, then cyclization occurs at a much faster rate. Dilution allows intramolecular chemistry to occur in a more preferential manner, yet the results of this cyclization are the same regardless of whether it is done in solution or bulk. The predominant cyclic structure is *cis*. Cyclization leading to the *cis* product (rather than to *trans*) occurs during postcondensation, and can be mechanistically explained as illustrated in Figure 14. Here, the *syn-cis* and *anti-cis* conformations of Schrock's catalyst system are preferred, and the metallacyclobutane rings that are generated from them lead to elimination of the *cis* cyclic rather than the *trans* one. This evidence is dramatic for identifying the factors in control of metallacyclobutane formation.

Sigma, Pi Conductive Hybrid Silicon Polymers. The facile nature of ADMET polymerization leads to other opportunities for producing clean polymer structures. This chemistry occurs at room temperature or even lower if preferred, and as a consequence, it became evident that questions might be answered regarding structure performance behavior. The objective of this section is to describe the synthesis and importance of a series of σ, π conductive polymers based on polysilanes and phenylene-vinylene. Figure 15 illustrates the chemistry at hand. The chemistry is relatively straightforward, involving the formation of a Grignard reagent which then couples with various lengths of silane units. Note in this case, the ADMET monomer possesses propenyl rather than vinyl endgroups. Propenyl groups were chosen since vinyl aromatic rings, e.g., styrenes, easily polymerize by vinyl addition chemistry. The presence of the methyl groups prevents this from occurring, and when the ADMET reaction occurs for this system, 2-butene is released rather than ethylene. These polymerizations have been attempted with values from n = 1-6, yet our first

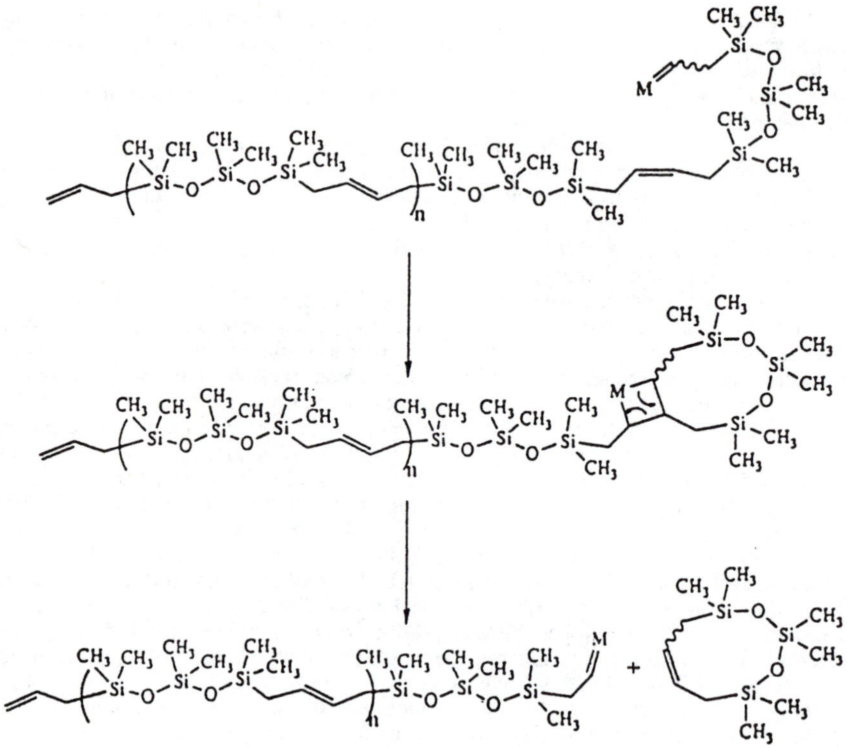

Figure 12. Backbiting mechanism for formation of cyclic dimer. M = Mo(N-2,6-C_6H_3-*i*-Pr_2)[OCMe(CF_3)$_2$]$_2$.

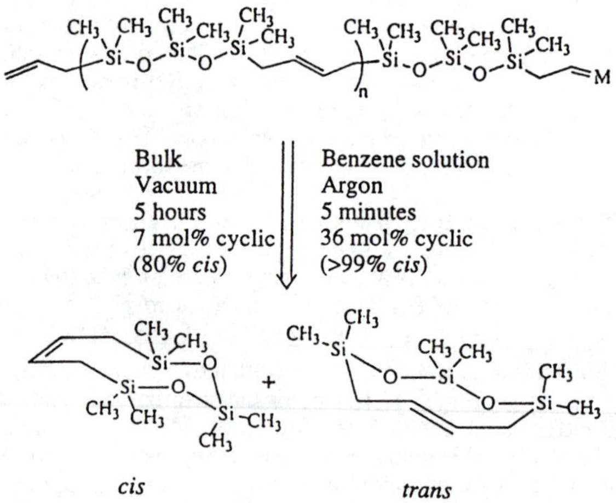

Bulk
Vacuum
5 hours
7 mol% cyclic
(80% *cis*)

Benzene solution
Argon
5 minutes
36 mol% cyclic
(>99% *cis*)

cis *trans*

Figure 13. Preparation of *cis* and *trans* cyclics formed via backbiting.(Reproduced with permission from ref. 14. Copyright 1993 American Chemical Society)

efforts to produce high polymers have not been successful, likely due to solubility problems. There is no question that ADMET condensation occurs, and in fact, a chromic shift is observed in the UV spectrum from comparing monomer and polymer. It is our intention to produce soluble forms of this polymer by proper selection of solubilizing groups at position R. Once these polymers are made, it will become

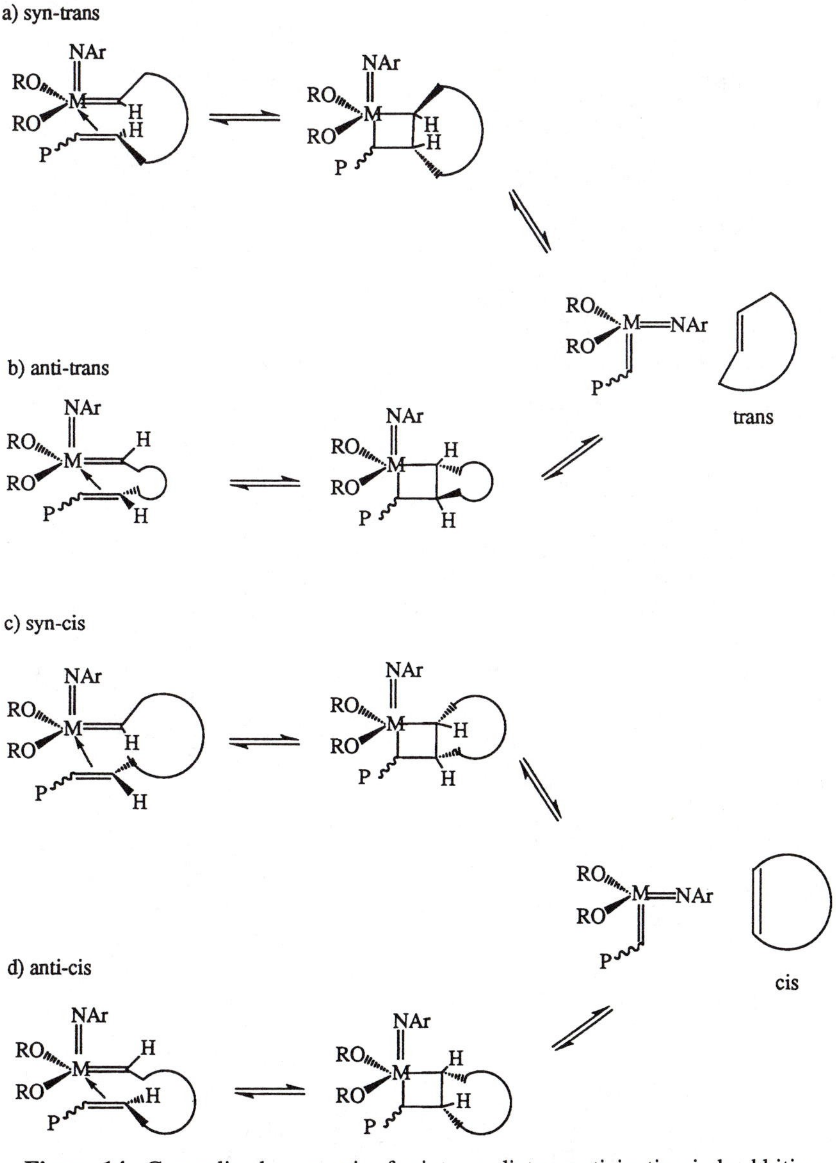

Figure 14. Generalized geometries for intermediates participating in backbiting reactions. M = Mo, R = CME(CF_3)$_2$ and P = polymer.

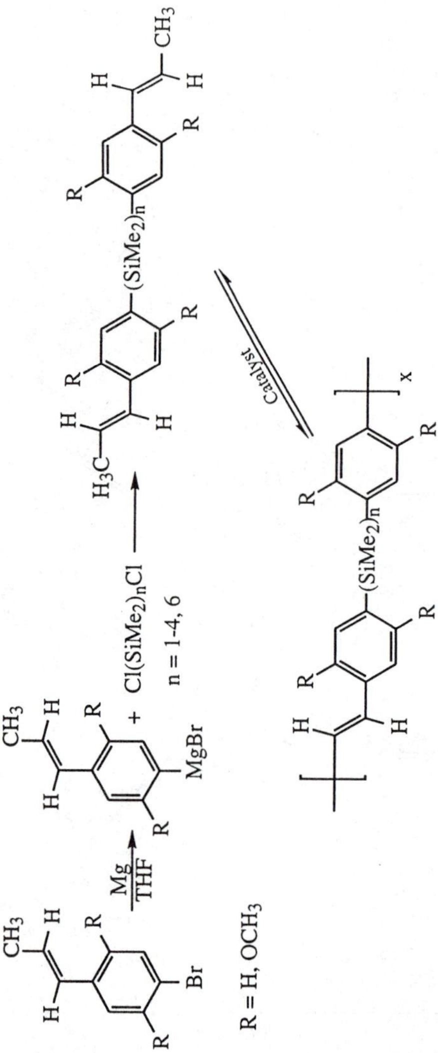

Figure 15. Synthesis and polymerization of σ–π conjugated polysilanes.

possible to study, in a systematic manner, the nature of conductive transfer from a σ-system to a π-system. This may prove to be an excellent opportunity to further elucidating the nature of conductivity of polymer backbone systems.

Silicon Polymers Possessing The Reactive Silicon Chlorine Bond. Another interesting aspect of ADMET polymerization chemistry relates to the fact that certain very reactive functional groups are completely inert to the ADMET reaction itself. For example, it has been shown that the silicon chlorine bond remains inert during ADMET chemistry in the presence of Schrock's alkylidene. As a consequence, the opportunity presents itself to form polymers at room temperature possessing reactive sites along the backbone. This has been demonstrated by synthesizing polymer backbones which possess both monochloro and dichlorosilane groups in the chain.

For example, diallylmethylchlorosilane has been synthesized using techniques described by Burns and has been condensed to form oligomers at room temperature using Schrock's molybdenum catalyst system (*15*). Further, bis-4-pentenyl-dichlorosilane has been produced and condensed as well by this chemistry (*16*). These reactions are shown in Figure 16. The NMR data for the dichloro-

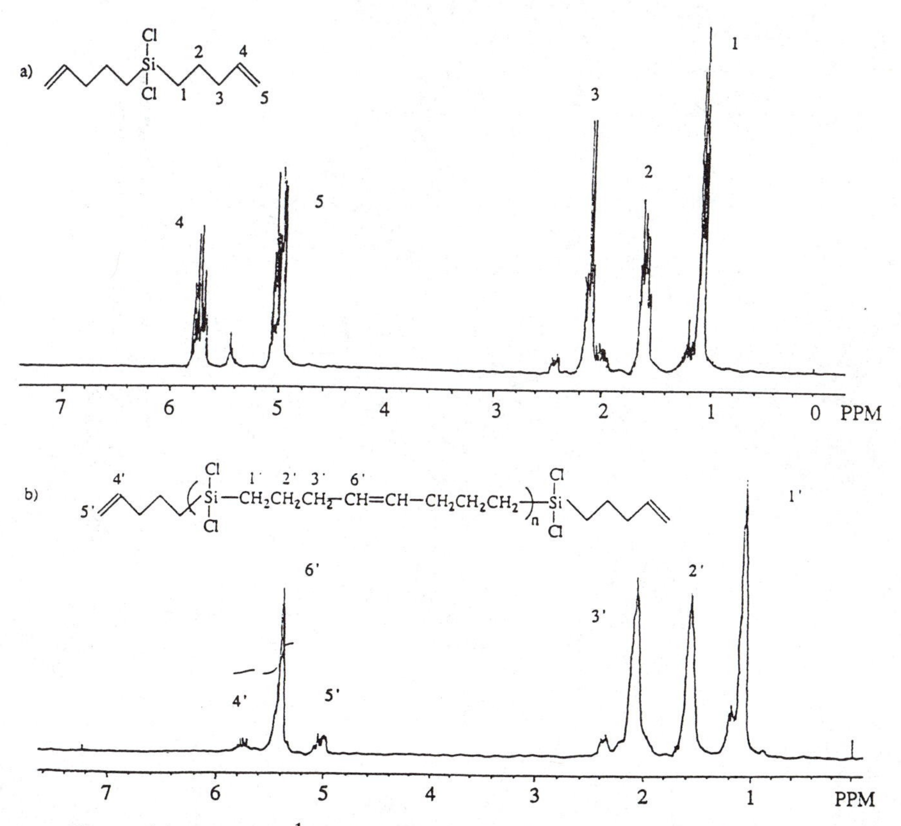

Figure 16. 300 MHz ^{1}H NMR of a dichlorsilane: a) monomer b) polymer. (Reproduced with permission from ref. 16. Copyright 1995 American Chemical Society)

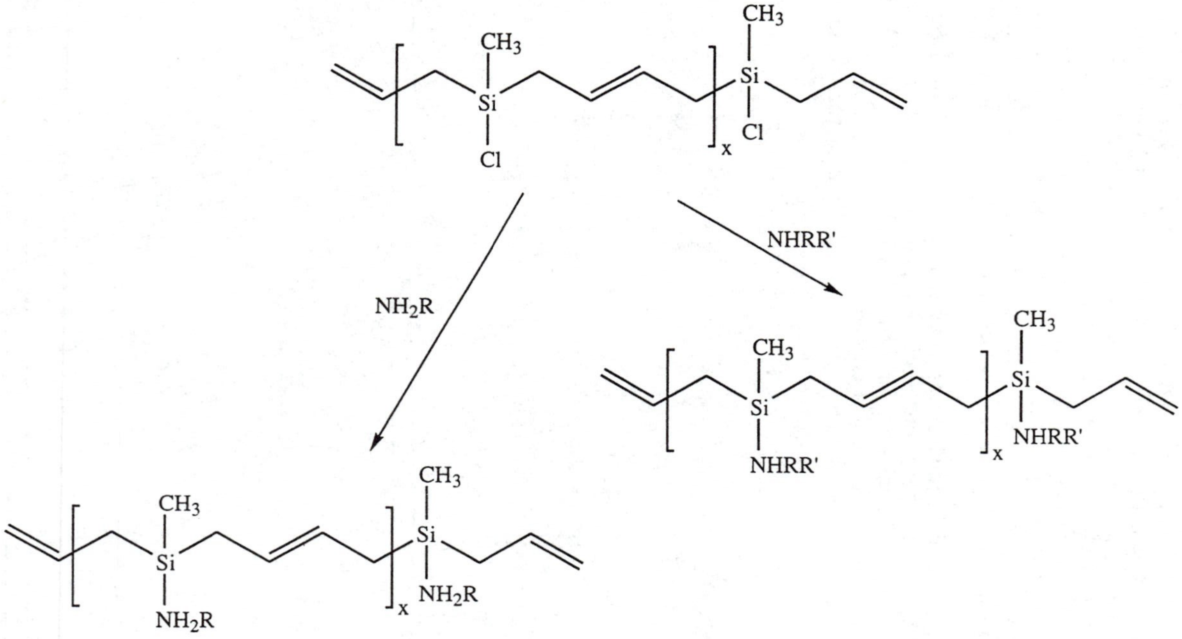

Figure 17. Grafting routes to polyphosphazene analogs. (Reproduced with permission from ref. 15. Copyright 1995 American Chemical Society)

polymerization are shown in Figure 16, illustrating that the chemistry is as clean as in other ADMET polymerization systems. These chloro-functionalized silanes offer an opportunity to produce materials having broad properties, depending upon subsequent substitution reactions that occur in a manner analogous to that done for polyphosphazenes. The physical behavior of polyphosphazines have been shown to be dependent upon the nature of the substitution that occurs at the chlorine bond (*17*). The opportunity to do the same with the silicon chlorine bond is clear, which would produce polycarbosiloxane and polycarbosilane analogs to polyphosphazenes (as shown in Figure 17). Research efforts are geared toward this goal, with further utilization of the facile ADMET reaction to produce new materials for tailored applications.

ACKNOWLEDGMENTS: We would like to acknowledge the National Science Foundation and the Army Research Office for the generous support of this research. We would also like to acknowledge Dr. Gary Burns of Dow Corning Corporation for his help in synthesis of chlorosilane monomers, and we would like to thank IBM Corporation for their contributions in preparing σ, π conjugated hybrid monomers.

Literature Cited

1. Lindmark-Hamberg, M.; Wagener, K.B. *Macromolecules* **1987**, *20*, 2949.
2. Wagener, K. B.; Boncella, J. M.; Nel, J. G. *Makromol. Chem.* **1990**, *191*, 365.
3. a) Chauvin, Y.; Herisson, J. L. *Makromol. Chem.* **1970**, *141*, 161. b) Soufflet, J. P.; Commereuc, D.; Chauvin, Y. *C. R. Hebd. Seances Acad. Sci., Ser. C* **1973**, *276*, 169.
4. Gilliom, L. R.; Grubbs, R. H. *J. Am. Chem. Soc.* **1986**, *108*, 733.
5. Tao, D.; Wagener, K. B. *Macromolecules* **1994**, *27,* 1281.
6. a) Konzelman, J.; Wagener, K. B. *Macromolecules* in press. b) Konzelman, J.; Wagener, K. B. *Polymer Preprints* **1992**, *33(1)*, 1072.
7. Brzezinska, K.; Wagener, K. B. *Macromolecules* **1991**, *24,* 5273.
8. Patton, J. T.; Wagener, K. B.; Boncella, J. M. *Macromolecules* **1992**, *25*, 3867.
9. Patton, J. T.; Wagener, K. B. *Macromolecules* **1993**, *26,* 249.
10. Patton, J. T.; Wagener, K. B. *Polymer Intl.* **1993**, *32*, 411.
11. O'Gara, J. E.; Portmess, J. D.; Wagener, K. B. *Macromolecules* **1993**, *26*, 2837.
12. Smith, D. W.; Wagener, K. B. *Macromolecules* **1991**, *24*, 6073.
13. Schrock, R. R.; DePue, R. T.; Feldman, J.; Schaverien, C. J.; Dewan, S. C; Liu, A. H. *J. Am. Chem. Soc.* **1988**, *110*, 1423.
14. Smith, D. W.; Wagener, K. B. *Macromolecules* **1993**, *26,* 6073.
15. Cummings, S. K.; Smith, D. W.; Wagener, K. B. *Macromol. Rapid Commun.* **1995**, *16*, 347.
16. Anderson, J. D.; Portmess, J. D.; Cummings, S. K.; Wagener, K. B. *Polymer Preprints* **1995**, *36(2)*, in press.
17. Allcock, H. R., Kuharcik, S. E.; Morrissey, C. T.; Ngo, D. C. *Macromolecules* **1994**, *27*, 7556, references therein.

RECEIVED December 11, 1995

Dendritic and Hyperbranched Systems

Chapter 7

Comparison of Linear, Hyperbranched, and Dendritic Macromolecules

Craig J. Hawker[1] and Jean M. J. Fréchet[2]

[1]Research Division, IBM Almaden Research Center, 650 Harry Road, San Jose, CA 95120–6099
[2]Baker Laboratory, Department of Chemistry, Cornell University, Ithaca, NY 14853–1301

A comparison of the physical properties of hyperbranched and dendritic macromolecules with linear polymers and the linear analogs of these 3-dimensional polymers is presented. It is found that thermal properties, such as glass transition temperature and degradation, are the same regardless of the macromolecular architecture but are very sensitive to the number and nature of chain end functional groups. However, other properties, such as solubility, melt viscosity, chemical reactivity, intrinsic viscosity were found to be very dependent on the macromolecular architecture.

One of the driving forces in the study of highly branched macromolecules has been the belief that these materials will have fundamentally different, and perhaps, new properties when compared to traditional linear polymers. This is not an unrealistic expectation if the structure of these three different classes of materials are examined. Traditional linear polymers are characterized by a random coil structure with only two chain ends. The contributions from these chain ends are therefore negligible at high molecular weights and the physical properties are dominated by chain entanglements and the presence, or absence, of functionalities attached to the polymer backbone. In contrast both dendrimers and hyperbranched macromolecules would be expected to adopt a globular structure, both in solution and in the solid state, which would be different to the random coil structure of linear polymers. The influence of chain entanglements and the exact nature of intermolecular interactions in such systems is therefore in question and may in fact be minor factors in determining the physical properties of these materials. Unlike linear polymers, a fundamental feature of hyperbranched macromolecules and dendrimers is the presence of a large number of chain ends. In both systems the number of chain ends is equal to the degree of polymerization plus one (for AB_2 monomers) (*1*) and it is these chain ends which would be expected to play a major

0097–6156/96/0624–0132$12.00/0

role in determining the physical properties of hyperbranched macromolecules and dendrimers. In this manuscript we would like to discuss some of the experimental work which has focused on comparing and evaluating the physical properties of dendritic and hyperbranched macromolecules with respect to the well known properties of traditional linear polymers.

Experimental

The synthesis of dendritic and hyperbranched polyesters based on 3,5-dihydroxybenzoic acid have been reported.(*2,3*)

3-Hydroxy-5-(t-butyldimethylsilyloxy)benzoic acid, 1

To a solution of 3,5-dihydroxybenzoic acid, **2**, (5.00 g, 32.0 mmol) in dry dimethylformamide (50 ml) was added t-butyldimethylsilyl chloride (9.63 g, 64.0 mmol) and imidazole (6.00 g, 90.0 mmol). The mixture was then heated at 60°C for 12 h, cooled, and evaporated to dryness. The crude silyl ester was redissolved in tetrahydrofuran (50 ml), glacial acetic acid (100 ml) was added followed by water (50 ml). The reaction mixture was then stirred at room temperature for 4 h and poured into water (500 ml). The product was extracted with ether (4 x 75 ml) and the combined extracts were washed with water (2 x 150 ml), dried, and evaporated to dryness. The crude product was purified by flash chromatography (dry loading recommended) eluting with 1:9 ether-dichloromethane gradually increasing to 3:7 ether-dichloromethane to give the desired acid, **1**, as a white solid (3.89 g, 45%). Ir. (cm^{-1}) 3600-2500, 2050, 1690, 1600 and 1010; ^1H n.m.r. (d_6-acetone) (ppm) 0.21 (s, 6H, $OSi(CH_3)_2$), 0.96 (s, 9H, $C(CH_3)_3$), 6.62 (t, 1H, J = 3 Hz, ArH), 7.06 (dd, 1H, J = 3 Hz, ArH) and 7.19 (dd, 1H, J = 3 Hz, ArH), ^{13}C n.m.r. (d_6-acetone) (ppm) -4.71, 18.28, 25.56, 110.44, 112.38, 113.02, 132.64, 157.11, 158.79 and 167.82.

Poly[3-oxy-5-(t-butyldimethylsilyloxy)benzoate], 3

To a solution of 3-hydroxy-5-(t-butyldimethylsilyloxy)benzoic acid, **2,** (1.50 g, 5.60 mmol) in dry dichloromethane (30 ml) was added DPTS (1.97 g, 6.70 mmol) and the mixture was stirred at room temperature under nitrogen for 15 minutes. DCC (1.40 g, 6.70 mmol) was then added and stirring was continued for 24 hours. The resulting precipitate was then removed by filtration and the filtrate and washings were evaporated to dryness. The crude product was purified by precipitation (twice) into 1:1 water-methanol (600 ml) and dried to give the polyester, **3**, as a white powder (1.15 g, 82 %). Ir. (cm^{-1}) 2050, 1720, 1600, 1340, 1190, 1160 and 1010; ^1H n.m.r. ($CDCl_3$) (ppm) 0.26 (s, 6H, $OSi(CH_3)_2$); 1.06 (s, 9H, $C(CH_3)_3$), 6.98 (t, 1H, J = 3 Hz, ArH), 7.54 (m, 1H, J = 3 Hz, ArH) and 7.65 (m, 1H, J = 3 Hz, ArH),

^{13}C n.m.r. ($CDCl_3$) (ppm) -4.23, 18.16, 25.57, 116.49, 119.42, 130.98, 151.55, 156.78 and 163.84.

Poly[3-oxy-5-hydroxybenzoate], 4

To a solution of poly[3-hydroxy-5-(t-butyldimethylsilyloxy)benzoate], **3,** (750 mg, 3.00 mmol equiv.) in dry tetrahydrofuran (30 ml) and acetone (10 ml) was added aqueous HCl (1N, 5.0 ml) dropwise and the reaction mixture was stirred at 50°C for 24 h. The solution was then evaporated to dryness and the residue was redissolved in tetrahydrofuran (5 ml) and precipitated into dichloromethane (400 ml). The product was purified by precipitation (twice) into dichloromethane (600 ml) and dried to give the polyester, **4,** as a white powder (0.37 g, 91 %). Ir. (cm^{-1}) 3600-2900, 1710, 1600, 1350, 1250 and 1170; ^{1}H n.m.r. (d_6-acetone) (ppm) 7.16 (br t, 1H, ArH) and 7.59 (m, 2H, ArH), ^{13}C n.m.r. (d_6-acetone) (ppm) 114.66, 114.98, 115.19, 131.90, 152.71, 159.22 and 164.33.

Results and Discussion

The first deliberate synthesis of a hyperbranched macromolecule was the preparation of a bromo-terminated hyperbranched polyphenylene, **5**, by the one-step polymerization of 3,5-dibromophenylboronic acid (Scheme 1). (*4,5*) An examination of the physical properties of **5** and its derivatives revealed that these materials possessed dramatically enhanced solubility when compared to their linear analog, poly(p-phenylene), which is essentially insoluble at molecular weights less than 1,000. Depending on the functionality at the numerous chain ends the hyperbranched polyphenylenes were soluble in wide variety of solvents at molecular weights in excess of 30,000. Significantly Kim and Webster were able to prepare a carboxylate-terminated hyperbranched polyphenylene which was soluble in water and able to act as an unimolecular micelle. Subsequently, Miller and Neenan (*6*) prepared the dendritic analog, **6,** of the hyperbranched polyphenylene, **5**, by a step-wise convergent growth approach (Figure 1). In this case the essentially monodisperse and perfectly branched dendrimers contained hydrogen end-groups which allows the contributions from the numerous bromine end-groups present in the hyperbranched macromolecules of Kim and Webster to be taken into account when making comparisons with poly(p-phenylene). Even with hydrogen end groups the solubility of **6** was significantly increased when compared to poly(p-phenylene), the magnitude of this increase was judged by Miller and Neenan to be roughly 10^5. While the comparison with linear poly(p-phenylene) is not strictly correct since it does not take into account the meta-linkages present in the dendritic or hyperbranched cases, it does strongly suggest that the highly branched, three dimensional structure of dendrimers and hyperbranched macromolecules gives rise to significantly increased solubility when compared to their linear analogs.

A more detailed study involved the synthesis of linear, hyperbranched, and dendritic macromolecules based on exactly the same monomer units or building

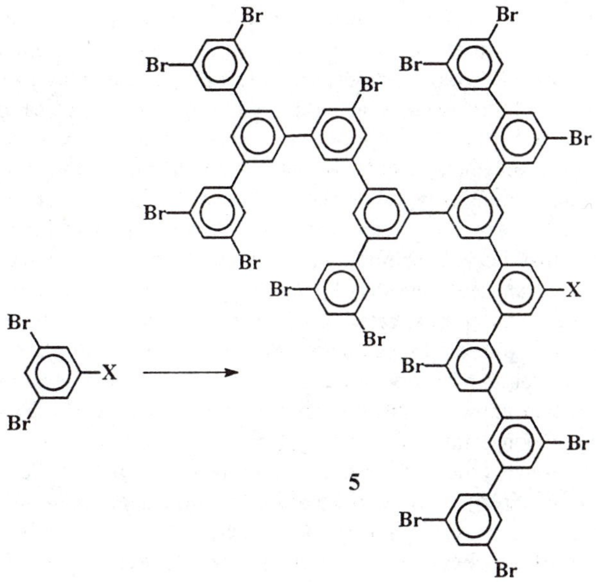

Br Br X Br Br Br Br Br Br Br Br Br Br X Br Br Br Br

5

Scheme 1.

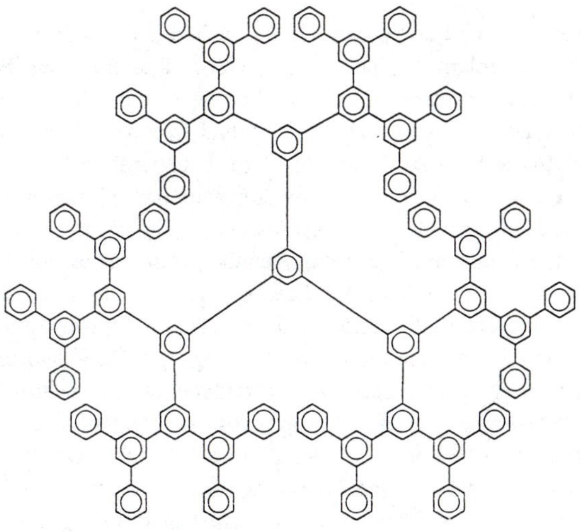

6

Figure 1: Hydrogen terminated dendritic poly(phenylene), **6**.

blocks. (*7*) In this case the repeat unit was chosen to be 3,5-dihydroxybenzoic acid which would give macromolecules with the same number of phenolic functionalities and the same all meta linkages between repeat units. In fact the macromolecules can be considered to be structural isomers which differ only in the number of branch units. A comparison of the physical properties of these materials would therefore allow the effect of architecture alone on the physical properties of polymeric materials to be evaluated. The opportunity also exists to compare these materials with normal linear polymers in order to probe the contributions from the numerous chain end groups present in hyperbranched and dendritic macromolecules.

Previously, (*3*) we have reported the thermal polymerization of 3,5-bis(trimethylsilyloxy)benzoyl chloride, **7**, which, after hydrolysis, gives the phenolic terminated hyperbranched polyester, **8** (Scheme 2). The degree of branching of **8** was determined to be approximately 60% by a combination of ^{1}H and ^{13}C nmr spectroscopy and signifies a branch point at essentially every second monomer unit. The synthesis of the dendrimers based on 3,5-dihydoxybenzoic acid employed the convergent growth approach and the trichloroethyl ester of 3,5-dihydoxybenzoic acid, **9**, as the monomer unit. (*2*) Starting from 3,5-bis(benzyloxy)benzoic acid, **10**, and using a two-step procedure involving esterification (DCC/DPTS) followed by deprotection (Zn/HOAc) a series of benzyl ether terminated polyester dendrimers, **11**, were prepared as shown in Scheme 3. Hydrogenation of **11** with Pd/C resulted in deprotection of the benzyl ether groups to give the desired phenolic terminated dendritic polyester, **12** (Scheme 4). Comparison of the structures of **8** and **12** reveals that they differ only in the degree of branching, 60% for **8** and 100% for **12**. The repeat units and the number, but not position, of the chain ends are the same for both structures.

While earlier work had compared hyperbranched and dendritic macromolecules with "normal" linear polymers, such a comparison does not take into account the numerous chain ends present in either the hyperbranched or dendritic cases. Therefore the correct analog of the dendritic and hyperbranched polyesters, **8** and **12,** is not poly(4-oxybenzoate) or poly(3-oxybenzoate), **13**, but poly[3-oxy-5-hydroxybenzoate], **4,** which contains the same number of phenolic groups as the dendrimer, **12**, of comparable molecular weight. The synthesis of **4** is however complicated by the presence of reactive phenolic groups at every repeat unit and a direct synthesis from 3,5-dihydroxy benzoic, **2**, acid is not possible since a hyperbranched structure will result. A protection-deprotection strategy was therefore developed for the synthesis of **4**. Starting from 3,5-dihydroxybenzoic acid, **2**, reaction with two equivalents of t-butyldimethylsilyl chloride followed by hydrolysis of the resulting silyl ester group gives the mono-silylated derivative, **1**, as the major product. Having protected one of the two phenolic groups, polymerization of **1** with DCC and DPTS then leads to a linear polymer, **3**, which contains no branch points and has a t-butyldimethylsiloxy group at every repeat unit. Deprotection of **3** with fluoride ion then gives the desired linear polyester, **4**, which is based on the same repeat unit and contains the same number of phenolic groups as **8** or **12** but has no branch points (Scheme 5). As detailed above the dendritic,

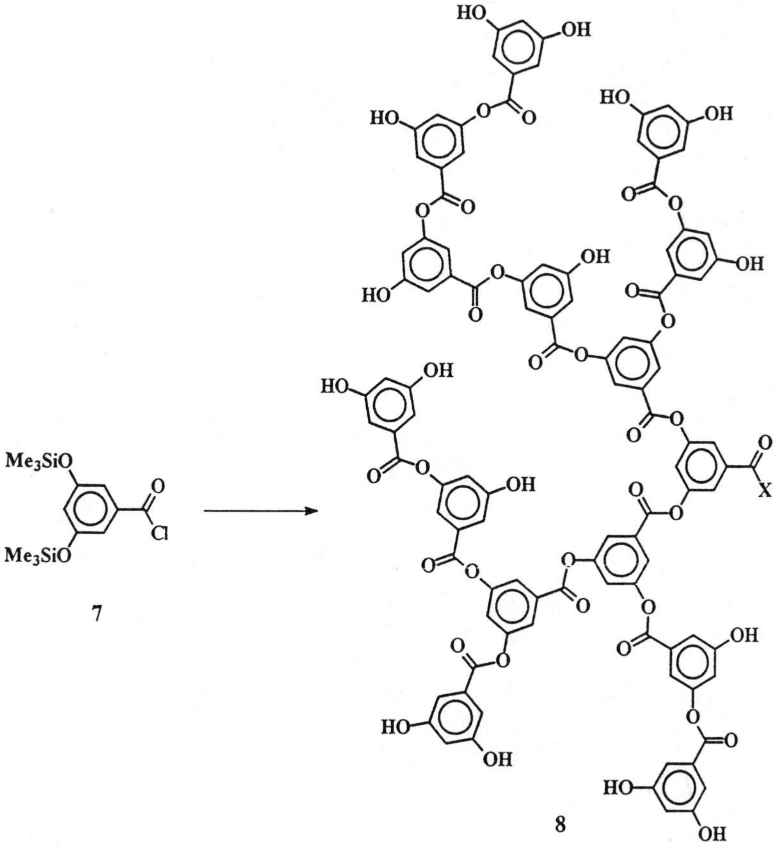

Me_3SiO
Me_3SiO
O
Cl

7

8

Scheme 2.

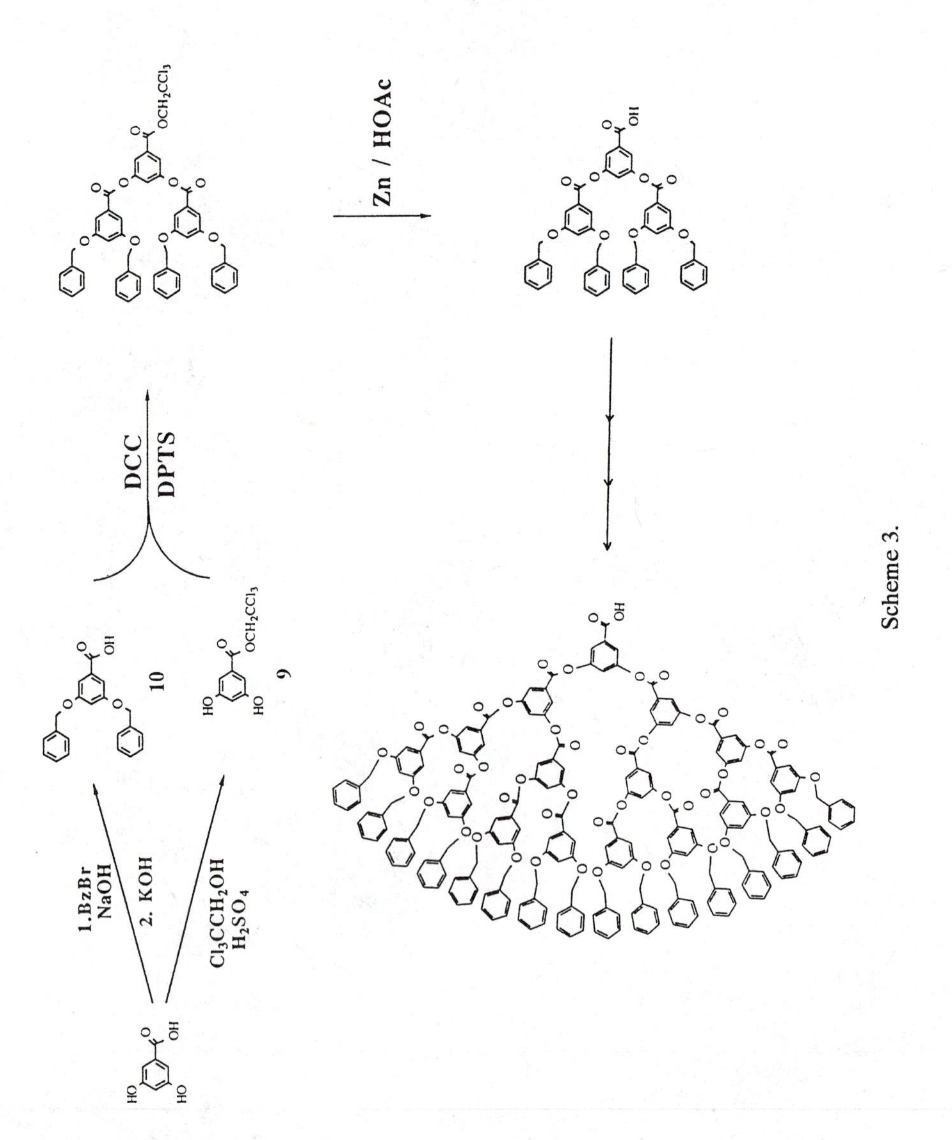

1.BzBr
NaOH
2. KOH
Cl_3CCH_2OH
H_2SO_4
10
9
DCC
DPTS
Zn / HOAc
OCH_2CCl_3
OH
HO

Scheme 3.

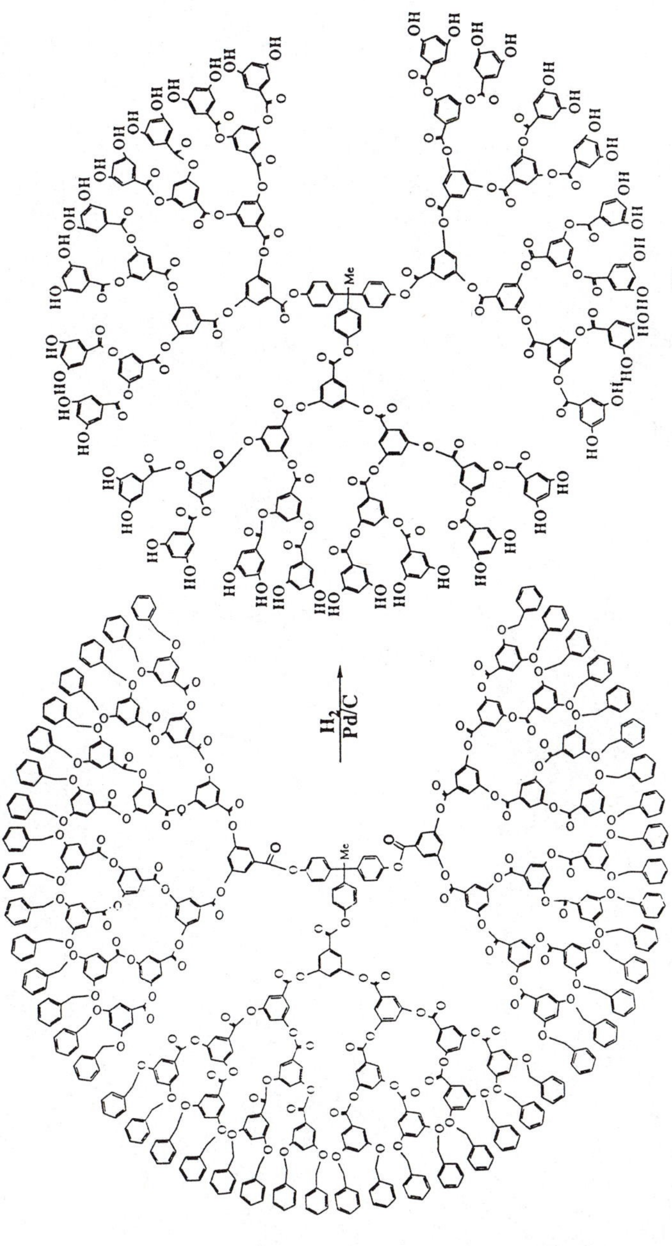

11

H_2, Pd/C

12

Scheme 4.

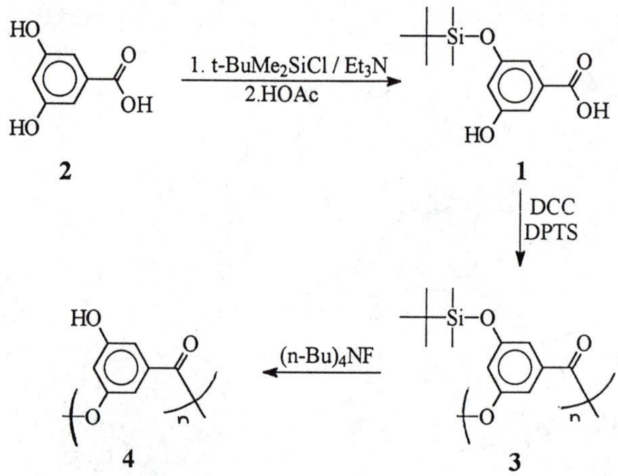

HO
O
OH
HO
2
1. t-BuMe$_2$SiCl / Et$_3$N
2.HOAc
Si–O
O
OH
HO
1
DCC
DPTS
Si–O
O
O
n
3
$(n\text{-}Bu)_4NF$
HO
O
O
n
4

Scheme 5.

hyperbranched, and linear polyesters actually constitute a form of macromolecular isomerism differing only in their degree of branching.

A comparison of the physical properties of the linear, **4**, hyperbranched, **8**, and dendritic, **12**, polyesters with themselves, and with "normal" linear polyesters such as poly(3-oxybenzoate) **13**, allows the influence of the architecture, or degree of branching, and the number and position of the chain ends to be investigated. The thermal properties of the three different polyester architectures were essentially the same, when the chain ends are phenolic the glass transition temperature of the dendrimer, **12**, is 201°C, while the hyperbranched material, **8**, has a T_g of 197°C and the linear polymer, **4**, a T_g of 204°C. Thermal degradation of all three phenolic terminated polyester architectures is essentially the same within experimental error. After a 30 minute isotherm at 250°C, no decomposition is observed up to 400°C with a 10% weight loss occurring at 440°C; rapid weight loss was then observed with 50% remaining at 575°C. Interestingly, derivatization of the chain ends resulted in significant, but similar changes in the thermal properties for each architecture. For example, the benzyl ether terminated dendritic polyester, **11**, has a glass transition temperature of 73°C while the corresponding value for the linear polyester with benzyl ether substituents is 78°C. From these results it can be concluded that thermal properties, such as T_g and decomposition temperature, are independent of the architecture or shape of the macromolecule but are heavily influenced by the nature of the chain ends. It should be noted that the linear analog, poly(3-oxybenzoate), which does not contain any phenolic functionalities has significantly different thermal properties when compared to the phenolic terminated polyesters.

In an effort to relate the increased solubility of hyperbranched and dendritic macromolecules with changes in macromolecular architecture and/or the number of chain ends the solubility behavior of the above "isomeric" polyesters, **4**, **8**, **12** was investigated. The hyperbranched macromolecules and dendrimers proved to be highly soluble in solvents that are capable of solvating the numerous chain ends. Significantly, in all cases the dendrimers proved to be more soluble that their hyperbranched analogs and both were significantly more soluble that the linear materials. The linear polymers were also found to be soluble in a reduced number of solvents when compared to their hyperbranched and dendritic analogs. For example, the phenolic terminated dendrimer, **12**, had a solubility in acetone of 1.05 g ml^{-1}, while the hyperbranched derivative, **8**, had a solubility of 0.70 g ml^{-1} and the solubility of the linear polymer, **4**, decreased to less than 0.02 g ml^{-1}. When all of these results are compared with the linear polyester, poly(3-oxybenzoate) **13**, which does not contain any "end" groups the difference becomes even more dramatic since poly(3-oxybenzoate) in insoluble in most solvents including acetone. From these results it can be concluded that the high solubility of dendrimers and hyperbranched macromolecules is a result of their highly branched architectures as well as the large number of chain ends functional groups.

The three different polymer architectures also showed different behavior in process involving interaction with a solid surface. For example the reactivity of benzyl ether terminated dendrimers was greater than that of benzyl ether terminated hyperbranched macromolecules, which in turn was more reactive than the linear polymer having benzyl ether side groups which was essentially unreactive. The reasons for these dramatic reactivity differences is not known at the present point in time but may be due to greater accessibility of the chain ends in dendrimers compared to hyperbranched compared to linear polymers, or the greater solubility of dendrimers, or it may be due to a size exclusion phenomena. In the latter scenario it is well known that dendrimers have a smaller hydrodynamic volume than hyperbranched and linear polymers of the same molecular weight.

One fundamentally important area in which dendritic macromolecules have been shown to differ from linear polymers is in their viscosity behavior. For intrinsic viscosity, it is well known that as the molecular weight of linear polymers increases their viscosity increases according to the Mark-Houwink-Sakurada equation; $[\eta] = KM^a$. However dendritic macromolecules do not obey this relationship once a threshold molecular weight is exceeded. For regular polyether dendrimers it has been shown (*8*) that a plot of log[η] vs. log(molecular weight) gives a bell shaped curve with a maximum below 5,000 a.m.u. instead of the fast rising curve expected for a classical linear polymer. This unique behavior is explained by the difference in shape between dendrimers and linear polymers. While the random coil structure of linear polymers gives the classical MHS relationship the globular, almost spherical structure of dendrimers leads to the bell shaped curve shown in Figure 2. The rational for this is that the volume of a spherical macromolecule, such as a dendrimer, increases cubically ($V = 4/3\pi r^3$) while its mass doubles at each generation and therefore increases exponentially ($Mw \sim 2^{(G-1)}$).

A recent study of the melt viscosity behavior of dendrimers has also shown behavior that differs markedly from linear polymers. (*9*) As shown in Figure 3 the normal behavior observed for linear polymers is a straight line with a slope of ca. 1.0 up to a critical molecular weight, M_c, after which a dramatic increase in the viscosity is observed and the slope increases to ca. 3.3. For a series of polyether dendrimers a plot of log(melt viscosity) vs. molecular weight is a straight line with a slope of approximately 1.1 up to molecular weights of 100,000 a.m.u. with no critical, or entanglement, molecular weight, M_c, being observed. Again this result is explained by the different macromolecular structures expected for linear and dendritic macromolecules. For linear polymers the random coil chains can entangle and the onset of these entanglements leads to the dramatic increase in viscosity observed at M_c. In contrast, the globular and highly branched architecture of dendrimers effectively prevents entanglements at all molecular weights, therefore the individual molecules do not entangle and therefore no dramatic increase in melt viscosity is observed.

A number of conclusion can be drawn from the above studies. Firstly some physical properties such as glass transition temperature and thermal degradation are independent of macromolecular architecture but are dependent on the nature of the chain ends. Other physical properties such as solubility, chemical reactivity, viscosity, etc. are dependent on macromolecular architecture and definite differences are observed between linear, hyperbranched and dendritic macromolecules based on the same building block.

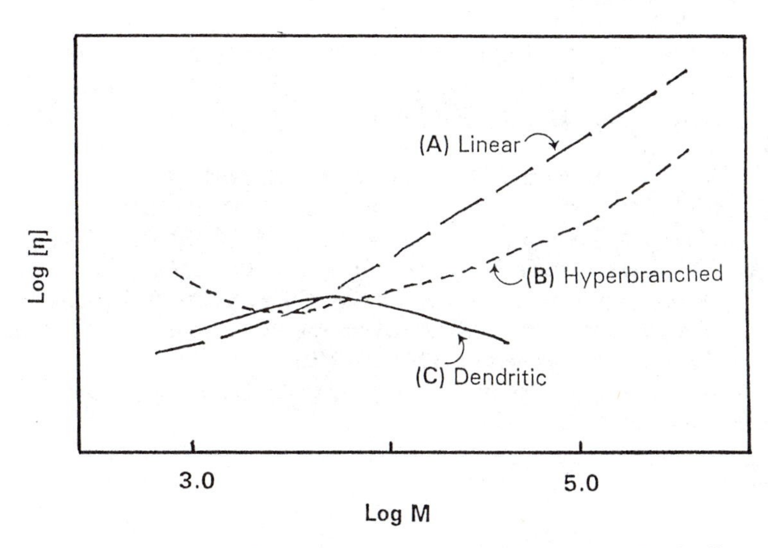

Figure 2: Plot of log(intrinsic viscosity) versus log (molecular weight) for linear (A), hyperbranched (B), and dendritic (C) macromolecules.

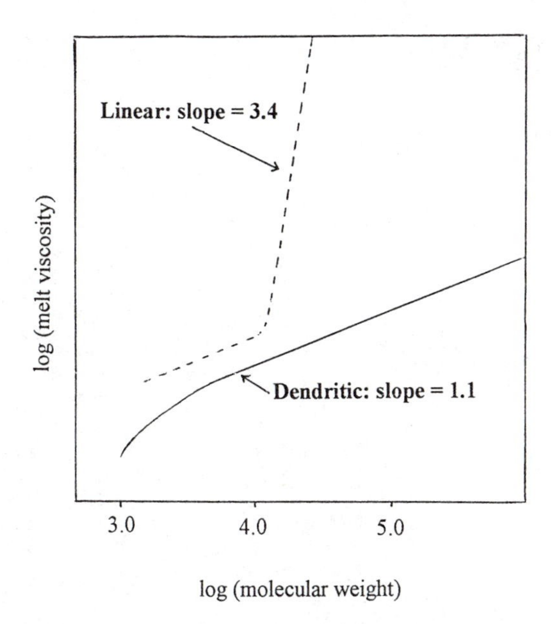

Figure 3: Plot of log(melt viscosity) versus log (molecular weight) for linear and dendritic macromolecules.

References

[1] Hawker, C.J.; Fréchet, J.M.J. *J. Am. Chem. Soc.*, **1990**, *112*, 7638.

[2] Hawker, C.J.; Fréchet, J.M.J. *J. Am. Chem. Soc.*, **1992**, *114*, 8405.

[3] Hawker, C.J.; Lee, R.; Fréchet, J. M. J. *J. Am. Chem. Soc.*, **1991**, *113*, 4583.

[4] Kim, Y.H.; Webster, O.W. *J. Am. Chem. Soc.*, **1990**, *112*, 4592.

[5] Kim, Y.H.; Webster, O.W. *Macromolecules*, **1992**, *25*, 2501.

[6] Miller,T.M.; Neenan, T.X. *Chem. Mater.*, **1990**, *2*, 346; Miller, T.M.; Neenan, T.X.; Zayas, R.; Bair, H.E. *J. Am. Chem. Soc.*, **1992**, *114*, 1018.

[7] Wooley, K.L.; Fréchet, J.M.J.; Hawker, C.J. *Polymer*, **1994**, *35*, 4489.

[8] Mourey, T.H.; Turner, S.R.; Rubenstein, M.; Fréchet, J.M.J.; Hawker, C.J.; Wooley, K.L. *Macromolecules*, **1992**, *25*, 2401.

[9] Hawker, C.J.; Farrington, P.; Mackay, M.; Fréchet, J.M.J.; Wooley, K.L. *J. Am. Chem. Soc.*, **1995**, *117*, 6123.

RECEIVED December 6, 1995

Chapter 8

Dendritic Structures with Polyfunctional Cores

A.-Dieter Schlüter, Wilhelm Claussen, Birol Karakaya, and W. Lamer

Institut für Organische Chemie, Freie Universität Berlin, Takustrasse 3, D–14195 Berlin, Germany

The synthesis of hydroxy functionalized poly(para-phenylene)s and their grafting with Fréchet-type dendritic fragments of the first two generations are described.

This article describes an approach to hybridize two recently developed, very successful concepts of macromolecular chemistry, the rod-like polymers and the dendrimers. The idea behind this hybridization is to make dendritic structures available with non-spherical, but still defined and predictable shapes in solution. In dendrimer synthesis large dendritic structures are built from small cores with a few functional groups. This can be done by employing either a divergent or a convergent strategy (*1*). In convergent strategy pre-formed dendritic fragments are reacted with the core's functional groups until complete coverage is reached (*2*). If dendritic fragments (dendrons) could be attached to rod-like polymers with functional groups in the periphery instead of to small, "dot-like" core molecules, a new kind of macromolecule could be formed. The molecular architecture of this macromolecule would then be characterized by a rigid backbone which is wrapped about by wedges that increasingly branch as they go from the inner to the outer regions. Depending upon the backbone stiffness, the degree of coverage, and the fragment size, the envelope of these macromolecules would no longer be a sphere, as is postulated for the dendrimers known today, but a cylinder.

Synthesis of Rigid Backbones with Hydroxy Anchor Groups

General Remarks. The two kinds of polymeric cores used in this project, poly([1.1.1]propellane)s and the poly(*para*-phenylene)s (*3,4*), differ in their mode of synthesis, the availability of model compounds, and the lengths of backbones, as well as many other aspects (*5,6*). They complement each other very well and their investigation should result in a conclusive picture of the whole matter. Since the general theme of the present volume is step-growth polymerization, the PPP derivatives, **1a** and **2a**, which are synthesized by Suzuki polycondensation (*4*), play an important role here. Why not use a flexible polymeric core, which, in principle, should be feasible and has been attempted (*7*)? Dendrimerization with a rod-like backbone instead of a flexible one seems to have two advantages, which may lead to the achievement of a high degree of coverage with dendritic fragments. The first advantage has to do with the approachability of functional groups. Since flexible polymers form random coils, many of the functional groups are "buried" in the inner

0097–6156/96/0624–0145$12.00/0

part and are not so exposed to chemical modification as those on rod-like polymers. The second advantage is related to entropy. If flexible polymers are grafted with bulky substituents, the backbone conformation transforms from a random coil to a stretched, more or less linear one (*8*). This is costly in entropy and, therefore, contraproductive to the achievement of a high degree of coverage.

1a (8.7 Å)

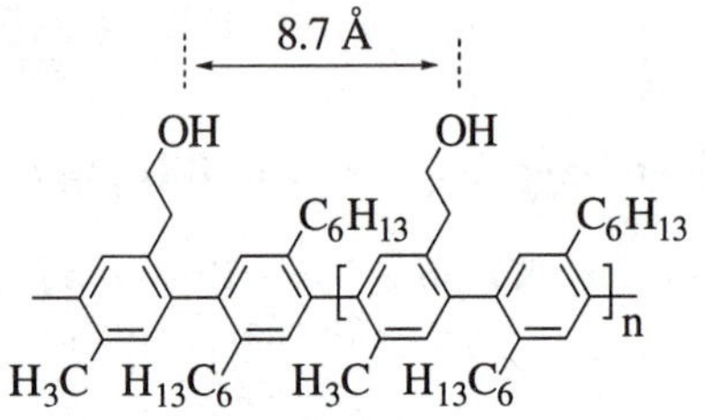

2a (13.0 Å)

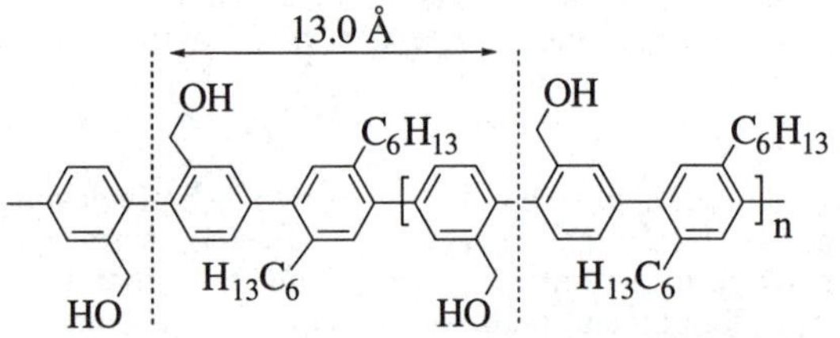

Both PPPs used in our study carry hydroxy functions, a feature which was deliberately designed in order to be able to attach Fréchet-type dendrons with benzylic bromide functions in the focal point (2). Not only are these dendrons, specifically those of lower generations, easy to make, but the coupling chemistry (Williamson ether synthesis) is also well understood. Together with the different anchor group density in polymers **1a** and **2a** these features were considered important prerequisites for the realization of a complete polymer analogous derivatization (dendrimerization), which is notoriously difficult to achieve.

Monomer synthesis. Suzuki polycondensation is highly compatible towards functional groups (*4*). Important exceptions here are the hydroxy and the amine functions. The AA-type monomers **3** and **4**, which contain hydroxy functions, were therefore synthesized with a methoxymethyl (MOM) protective group which can be cleaved off under acidic conditions (*9*). The syntheses involve standard procedures and these compounds are available as analytically pure materials on the 25 g and 5 g scale, respectively. The known synthesis of BB-type monomer **5** was easily carried out on the 10 g scale (*10*).

Monomer **5** deserves some comment, because there occasionally seems to be some "confusion" whether it has been obtained in polycondensation grade purity. The reason for this confusion is the appearance of the ^{1}H NMR spectra of **5** which differ quite markedly depending upon the history of the sample. There is a subtle balance between the water content of **5** and the formation of self-condensation products of the boronic acid function, which gives, for example, boroxines (*11*). With the increasing degree to which monomer **5** is dried, new signals become apparent in the spectrum. Figure 1 shows the low field regions of the NMR-spectra of one and the same preparation of **5** but at four different drying stages. The water content increases from spectrum a to d. Spectrum c is easiest to understand. It shows singlets for the aromatic protons at $\delta = 7.10$ and the acid protons at $\delta = 7.75$. The signal of water contained in this sample at $\delta = 3.45$ is not shown. With increasing water content (spectrum d) exchange between the protons of water and of **5** is observed. The signals for the boronic acid and water merge to a broad signal at $\delta = 5.1$ ppm. If one goes in the opposite direction, thus if the sample is dried more rigorously, small signals at $\delta = 7.2$ and $\delta = 7.6 - 7.7$ grow into the spectra. These signals do not reflect what one typically calls an impurity but are caused instead by the aromatic protons of condensed boronic acids. The remaining proton signal of the boronic acid functions appears with reduced intensity and higher line width, the latter indicating perhaps higher correlation times because of the increased size of the condensate.

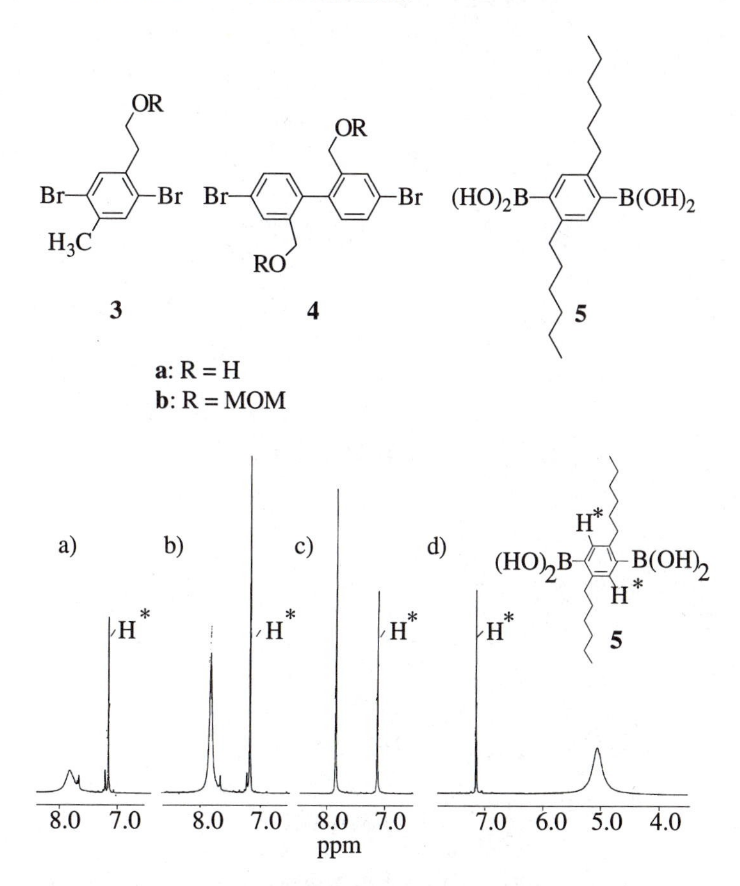

Figure 1. Low field regions of ^{1}H NMR spectra of 5 with increasing water content (a → d). The spectra were recorded in d_6-DMSO at room temperature.

If signals of boroxines and related structures appear in the spectrum of **5**, a reliable determination of the actual content of **5** is difficult. Therefore it is recommended to hydrolize the condensates in the mixture with acid until a situation like in spectrum c (or d) is achieved. Simple integration provides the desired value, which then can be used to match the stoichiometric balance required for the step-growth polymerization.

Polycondensation and deprotection. The polycondensations of dibromides **3b** and **4b** with diboronic acid **5** were done according to standard procedures using 1.0 mol-% of palladium tetrakis(triphenylphosphin) as catalyst precursor (*10*) (Figure 2).

Polymers **1b** and **2b** were obtained as colorless materials in yields exceeding 95%. Size exclusion chromatography in THF versus polystyrene standard gave the following molecular weight values which are representative for unfractionated samples. **1b**: Mn = 9,700 (Pn = 22), Mw = 15,500 (Pw = 35), D = 1.6; **2b**: Mn = 12,000 (Pn = 22), Mw = 20,000 (Pw = 36), D = 1.7. The polymers were characterized further by ^{1}H- and ^{13}C NMR spectroscopy and elemental analysis. All data are in agreement with the proposed structures. (For a recent discussion of possible side reactions of the Suzuki cross-coupling see reference 12.)

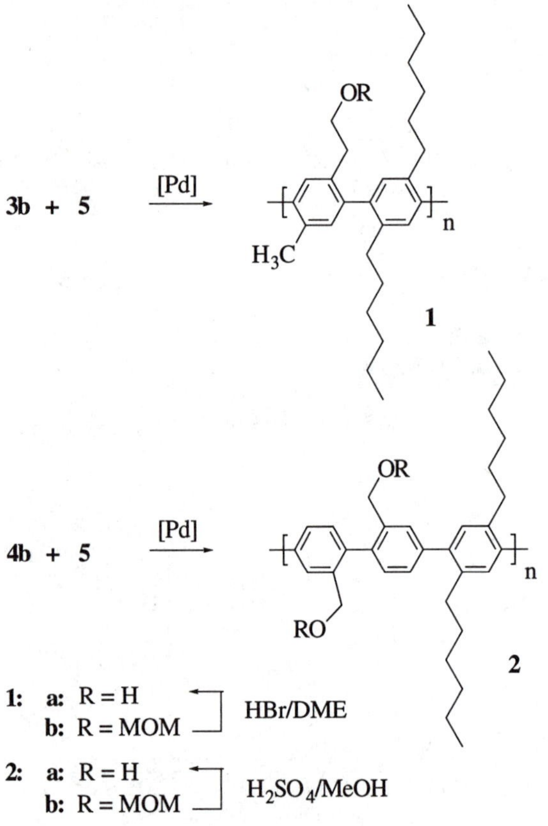

3b + 5 →[Pd] 1

4b + 5 →[Pd] 2

1: a: R = H; b: R = MOM (HBr/DME)

2: a: R = H; b: R = MOM (H_2SO_4/MeOH)

Figure 2. Synthesis of polymers **1b** and **3b**, and their deprotection to **1a** and **2a**.

The conditions for the deprotection had to be individually developed for each polymer. Best results were obtained for **1b** with hydrogen bromide in dimethoxyethane (DME) and for **2b** with sulfuric acid in methanol. As expected, relatively high dilution conditions are required to bring about complete deprotection. For example, when **1b** was used in less than roughly 0.01 molar DME solutions, the MOM group was completely removed. If, however, the same reaction was done in a 0.1 molar solution, deprotection could not be driven beyond approximately 90%, not even with extended reaction time. Figure 3 compares the ^{1}H NMR spectra of **1b** and its deprotected analogue **1a** to demonstrate the degree of deprotection achieved. As can be seen, the MOM signals in **1a** have virtually disappeared. As a result of the removal of the MOM group there is a change in the line shape of the CH_2-$\mathit{CH_2}$-O group signal at δ = 3.7 ppm which is presumably due to atropic isomerism, a typical phenomenon for substituted oligo- and polyphenylenes (*13*).

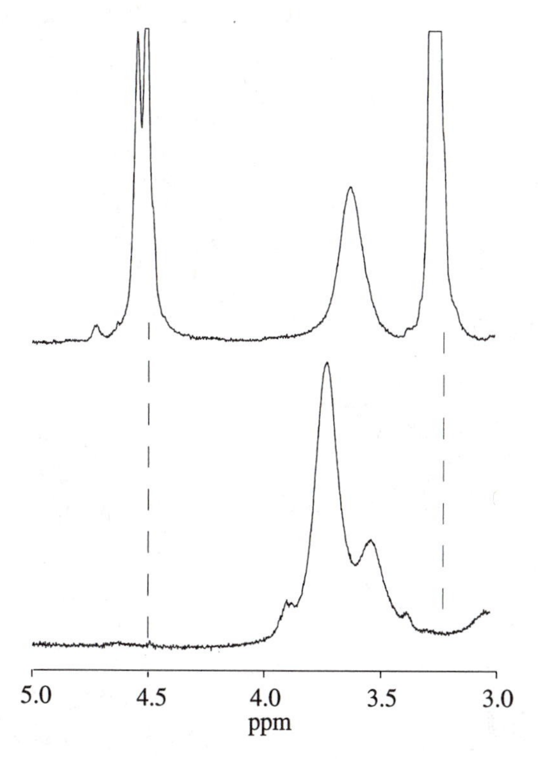

Figure 3. Identical sections of the ^{1}H NMR spectra of protected polymer **1b** (top) and deprotected polymer **1a** (bottom). The signals of the MOM protective group at δ = 3.25 and δ = 4.5 ppm (top) have virtually disappeared in the bottom spectrum indicating a complete deprotection. The signal at approx. δ = 3.7 ppm (top) undergoes a change in line shape and shift (bottom), which is presumably due to atropic isomerism.

On the Dendrimerization of PPPs

The first dendrimerization experiments were done with polymer **2a** and Fréchet-type dendrons of the first and second generation, **(G-1)Br** and **(G-2)Br** (*2*). According to the Williamson procedure ether syntheses are done by deprotonating alcohol to alkoxide and by using the latter more powerful nucleophile in the following displacement reaction. This very successful protocol in low molecular weight chemistry cannot be automatically transferred to the polyalcohols **1a** and **2a**. Attempted complete deprotonation of **2a** resulted in the precipitation of the corresponding polyalkoxide even in the presence of highly polar, aprotic solvents like dimethylformamide (DMF). The precipitated deprotonated **2a** did not react with the dendrons to any significant extent even under relatively harsh reaction conditions. The dendrimerized alkoxides, which are certainly formed at the surface of the precipitate, unfortunately did not "pull" the backbones into solution so that further dendrimerization could occur. To circumvent this problem, a solution of polymer **2a** and the respective dendron in DMF/toluene (1:1) was treated with the stoichiometrically required amount of the base sodium hydride in very small portions during the course of several hours (Figure 4).

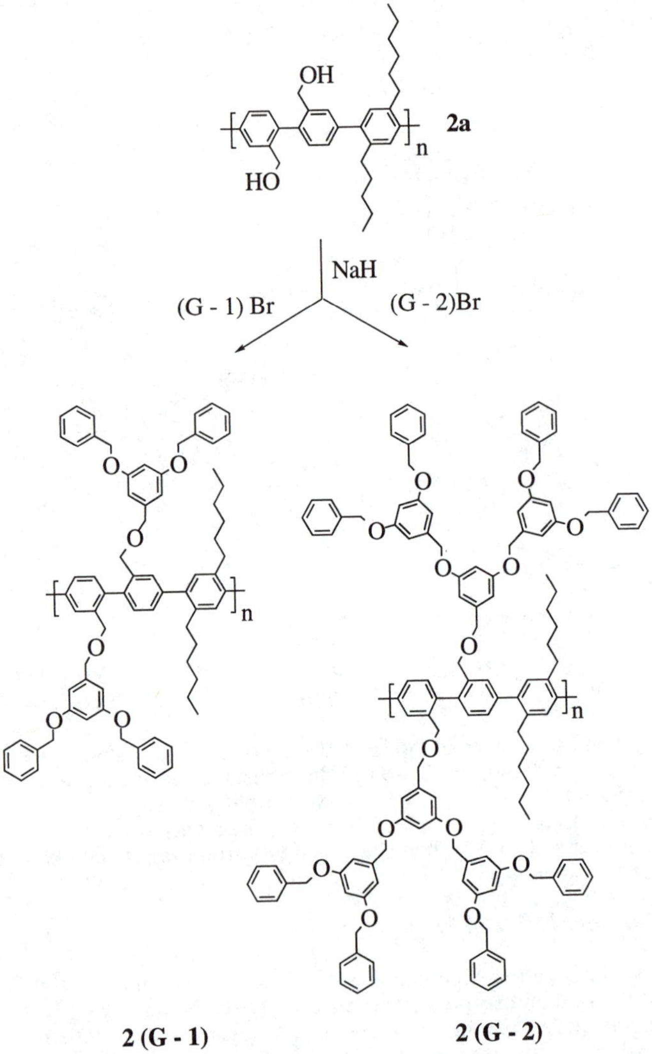

Figure 4. Dendrimerization of polymer **2a** with Fréchet type dendrons of generation 1 and 2. Whereas 2(G-1) is covered completely, 2(G-2) has approx. 40% free hydroxy functions.

The polymer stayed in solution because only a few hydroxy functions were deprotonated and immediately dendrimerized by the dendrons simultaneously present. This procedure was repeated until all hydroxy functions were thought to be covered. After the reaction was finished, excess dendrons were removed by repeated precipitation/dissolution cycles, which turned out to be quite a tedious procedure, specifically for the second generation dendrons.

What is the maximum achievable coverage? For the first generation dendron it was higher than 95% by ^{1}H and ^{13}C NMR, but could not be driven beyond 50-60% for the second one. The determination of this value rests upon ^{1}H NMR integration, which is subject to some uncertainty, because of the considerable line widths in the spectra. This result was considered quite disappointing, specifically in regard to the goal to completely attach even larger than second generation dendrons to the rigid backbones. The question was then whether this was due to the specific coupling reaction used or to an unfavorable relation between the space demand of the wedge and the space available for it at the backbone.

The Dimensions of a Fréchet-type (G-2) Wedge from a Single Crystal X-ray Study and its Implication on the Maximum Coverage. The average distances between the hydroxy functional groups of a number of rigid rod polymers were previously calculated to lie between 3.3 and 8.7 Å (*14*) For polymer **1a** this distance is 8.7 Å. Polymer **2a** has two hydroxy groups every 13.0 Å. Due to the few conformational constraints that the repeating units have in regard to rotation, wedges in the side chain should be able to occupy the entire space around the backbone. To determine the minimum amount of space required for a wedge, single crystals of (G-2) ester **6** were grown, for which the crystal structure was solved. (Saenger, W.; Geßler, K. unpublished.) The ORTEP plot (Figure 5a) shows that **6** has a remarkably flat structure with significant deviations from planarity only at two terminal benzene rings. The wedge fits into a 45% slice of a cylinder with a height of 4.0 Å (half of the a-axis of the unit cell) and a radius of approx. 14 Å (Figure 5b). Disregarding solvent/wedge interactions and assuming a strictly linear chain, the space demand of a G-2 wedge should therefore pose no problems even where the anchor groups have the minimum distance of 3.3 Å (Figure 5c). As a result of this, we are inclined to believe that the kind of coupling reaction used here is responsible for the limitation of the achievable coverage. There does not seem to be a spatial restriction even for the attachment of (G-3) fragments to most of the backbones, certainly not for polymer **1a** where the anchor groups are separated from one another by 8.7 Å.

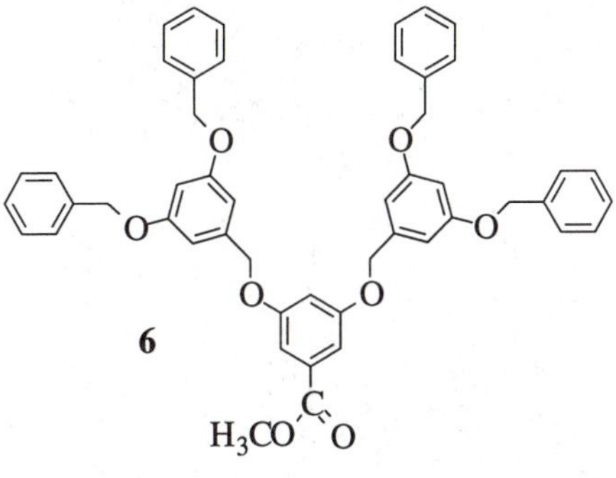

6

H_3CO C O

Modified Attachment Chemistry. Two modifications were introduced to broaden the scope of the coupling reactions and to make the flexibility of coupling conditions as great as possible: (a) Wedges with isocyanate groups at the focal point like **7d** (Figure 6), were prepared (*14*) and anchored to the hydroxylated polymers **1a** and **2a** and (b) the PPP derivative **10b** was synthesized with iodomethyl side chains (Figure 7). The use of which reverses the nucleophilic and electrophilic centers in Williamson synthesis. Dendrons with hydroxy groups in the focal point can be advantageously used in the dendrimerization reaction, since their deprotonation should be far less critical than for hydroxylated polymers. Additionally, because the leaving group (iodide) is benzylic, the displacement reaction should proceed with high conversion and without competitive elimination.

The attachment of **7d** to polymer **1a** to furnish the new polymer **8** was tested. Initial experiments show that the coverage can be driven to approximately 80% which is much further than in the case of polymer **2(G-2)**. Considering the fact that the

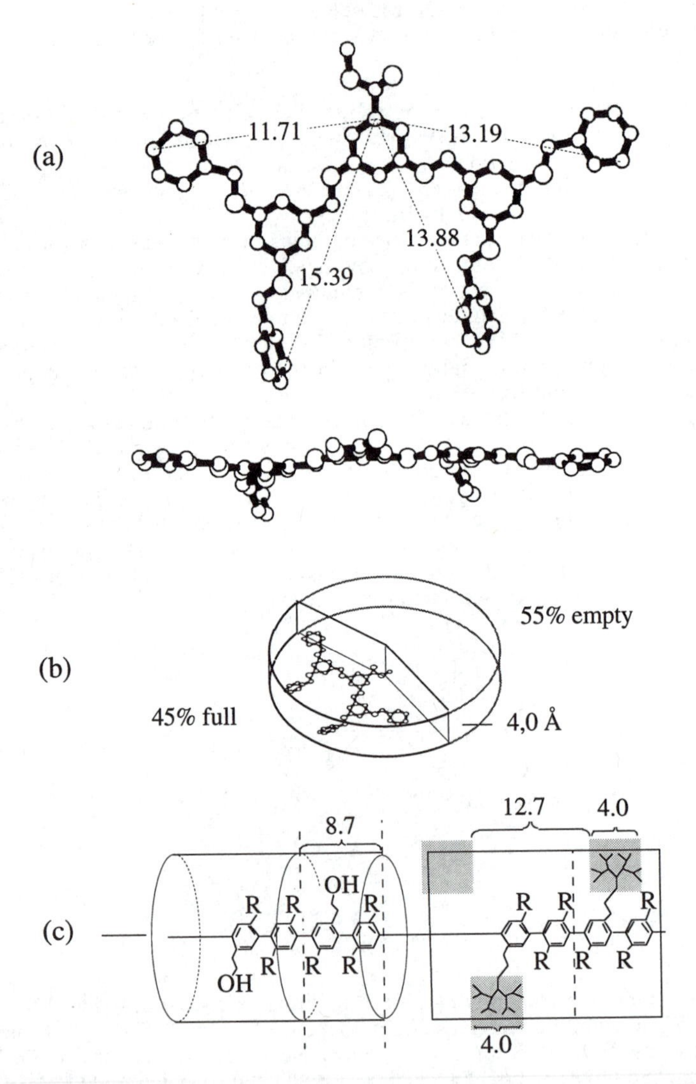

Figure 5. (a) Two perspectives of the crystal structure of **1** (ORTEP), (b) illustration of the space demand of **1**, and (c) oblique and sectional view of the cylindrical slices available for wedges in the case of a polymer with the shortest distance between two anchor groups (3.3 Å). The space demand of a G-2 wedge is indicated in grey.

dendron was used stoichiometrically and not in excess in these experiments, the final coverage has certainly not been reached yet and a 100% coverage seems to be within reach. The alternative route using polymer **10b** as a precursor for **11** also looks promising. With **(G-2)OH** a coverage of 90% has already been realized in the initial experiments (Karakaya, B; Claussen, W.; Klopsch, R: Schlüter, A.-D., to be published).

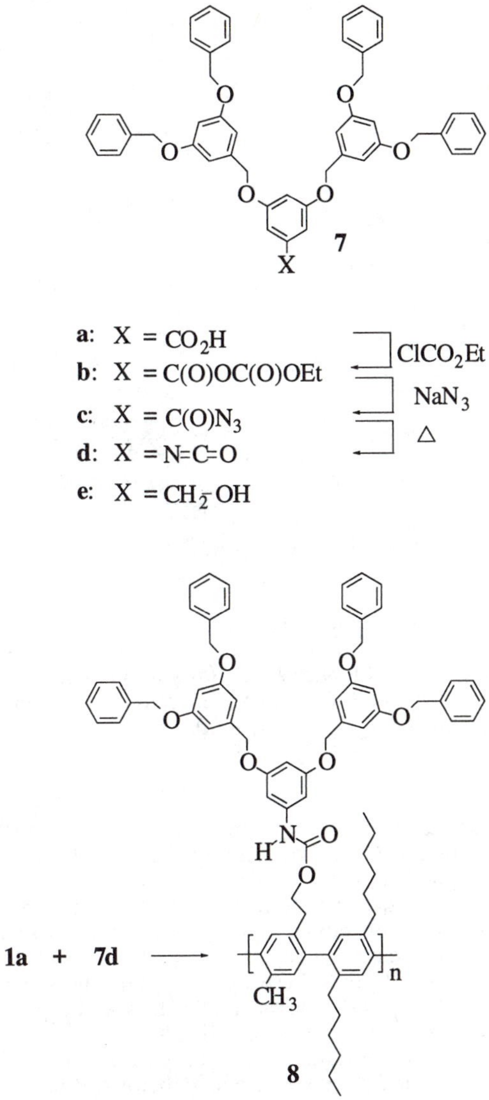

7

a: X = CO_2H
b: X = C(O)OC(O)OEt
c: X = $C(O)N_3$
d: X = N=C=O
e: X = CH_2-OH

$ClCO_2Et$
NaN_3
Δ

1a + **7d** → **8**

Figure 6. Synthesis of the second generation dendritic wedges, **7d**, with an isocyanate function at the focal point (top) and its reaction with polymer **1a** to furnish the dendrimerized polymer **8** (bottom).

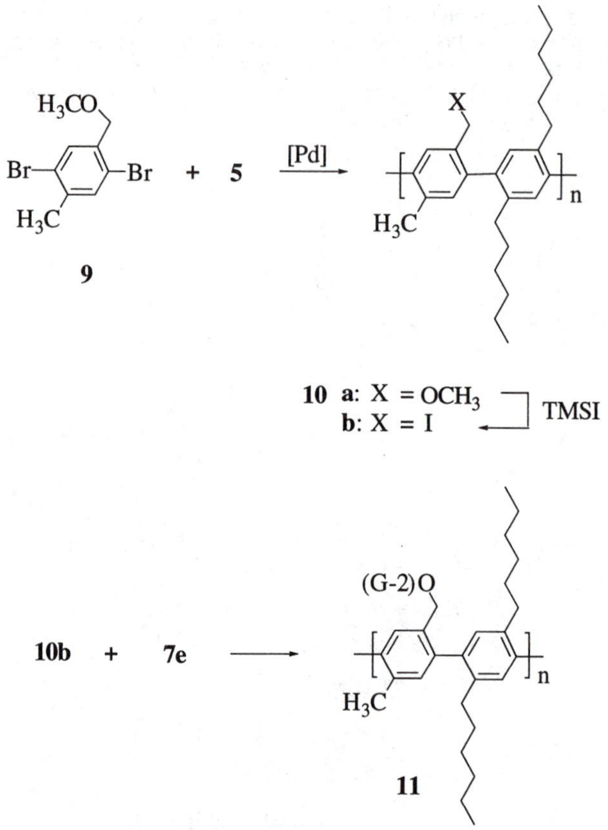

Figure 7. Synthesis of the PPP derivative **10b** with iodomethyl side chains (top) and its dendrimerization with Fréchet-type dendrons of the second generation, **7e**, to give the corresponding polymer **11** (bottom).

Outlook

Despite the preliminary nature of the experiments described just now, it becomes clear that the key to the successful synthesis of dendritic "macrocylinders" lies in the kind of attachment chemistry applied. The achievement of a 100% coverage with a second generation dendron is only a question of time and the first experiments with wedges of higher generation are already underway. Once the synthesis of rigid rod polymers with higher generation dendrons has been accomplished, it will be of prime importance to actually prove experimentally that these hybrids attain the shape of a cylinder in solution. Besides their beauty, such macrocylinders may find application, for example, in the area of vesicles and membranes, specifically when they can be made with a hydrophobic surface in one part and a hydrophilic in the other. There is, however, considerable synthetic and analytical mileage to go before getting close to such goals.

Acknowledgment
Financial support of the work by the Deutsche Forschungsgemeinschaft and the Fonds der Chemischen Industrie is gratefully acknowledged.

Literature Cited

1. Tomalia, D. A. *Adv. Mater*. **1994**, *6*, 529, and literature cited therein.
2. Fréchet, J. M. J.; Wooley, K. L.; Hawker, C. J. *J. Am. Chem. Soc*. **1991**, *113*, 4252.
3. Schlüter, A.-D.; Bothe, H.; Gosau, J.-M. *Makromol. Chem*. **1991**, *192*, 2497.
4. Schlüter, A.-D.; Wegner, G. *Acta Polymer*. **1993**, *44*, 59.
5. Freudenberger, R.; Claussen, W.; Schlüter, A.-D.; Wallmeier, H. *Polymer* **1994**, *35*, 4496.
6. Claussen, W.; Schulte, N.; Schlüter, A.-D. *Makromol. Rapid Commun*. **1995**, *16*, 89.
7. For example see: Tomalia, D. A.; Naylor, A. M.; Goddard III, W. A. *Angew. Chem. Int. Ed. Engl.* **1990**, *29*, 138.
8. Wintermantel, M.; Schmidt, M.; Tsukuhara, Y.; Kajiwara, K.; Kolijiva, S. Makromol. Rapid Commun. **1994**, *15*, 2791
9. Greene, T. W.; Wuts, P. G. M. *Protective Groups in Organic Synthesis*; Wiley: New York, 1991.
10. Rehahn, M.; Schlüter, A.-D.; Wegner, G. *Makromol. Chem.* **1990**, *191*, 1991.
11. *Boron and Oxygen, Gmelin Handbook of Inorganic and Organometallic Chemistry*, Springer: Berlin, 1987; B 3rd Supplement , Vol. 2. p. 147.
12. Novak, B. M.; Wallow, T. I.; Goodson, F.; Loos, K. *Am. Chem. Soc. Polym. Div., Polym Prepr*. **1995**, *36(1)*, 693.
13. For example, see: Katz, H. E. *J. Org. Chem.* **1987**, *52*, 3932.
14. Schlüter, A.-D., *Am. Chem. Soc. Polym. Div., Polym. Prepr.* **1995**, *36(1)*, 745.

RECEIVED December 4, 1995

Chapter 9

Syntheses of Telechelic, Star-Shaped, and Hyperbranched Aromatic Polyesters

Hans R. Kricheldorf, Olaf Stöber, Gerd Löhden, Thomas Stuckenbrock, and Dierk Lübbers

Institut für Technische und Makromolekulare Chemie, Universität Hamburg, Bundesstrasse 45, D–20146 Hamburg, Germany

The Syntheses of numerous telechelic, star-shaped and hyperbranched polyesters were studied on the basis of two difunctional monomers namely β-(4'-hydroxyphenyl)propionic acid (Phla) and 3-hydroxybenzoic acid (3-Hybe). Telechelic oligomers with two different or two identical endgroups were obtained by cocondensation with chain stoppers such as 4-tert.-butylbenzoic acid or bisphenol-P. Most oligoesters of Phla were highly crystalline and difficult to dissolve. Soluble star-shaped oligoesters were isolated from cocondensations of tetrahydroxyspirobisindane. Hyperbranched copolyesters with a variable degree of branching were prepared by copolycondensation of acetylated 3-Hybe and 3,5-dihydroxybenzoic acid or 5-hydroxyisophthalic acid. In addition to monomers with free carboxyl group monomer with silylated carbonxyl groups were used. The silylation avoids the presence of acidic protons, and thus, reduces the risk of side reactions resulting in crosslinks. The properties of copolyesters were studied with regard to the degree of branching. Furthermore, a novel "one-pot procedure" was developped which allows a facile synthesis of hyperbranched copoly(ester-amide)s from combinations of 3-Hybe and 3,5-diaminobenzoic acid or 3,5-dihydroxybenzoic acid and 3-aminobenzoic acid. Also these copolyesteramides are completely amorphous. Star-shaped and hyperbranched copolyesters and copoly(ester-amide)s were obtained by copolycondensation of trifunctional monomers and small amounts of difunctional or tetrafunctional (a_n-type) star-centers. The ^{1}H- and ^{13}C NMR spectra of all polyesters are discussed.

0097–6156/96/0624–0156$14.50/0

Regardless of details of the chemical structure it makes sense to subdivide monomers useful for step-growth polymerizations into two large groups:

1) a-b-type monomers containing two different functional groups (a,b) and
2) a-a/b-b-type monomers, each of which contains two identical functional groups.

Although, text books usually do not differentiate between both groups (1-9), the synthetic potential of the "a-b monomers" is by far higher than that of the "a-a/b-b monomers" as illustrated in the scheme of Figure 1. Polycondensations and polyadditions of "a-a/b-b monomers" normally yield mixtures of oligomers or polymers with three different combinations of endgroups. Telechelic materials with uniform endgroups are difficult to prepare. In contrast, a clean polycondensation of polyaddition of "a-b monomers" automatically yield telechelic oligomers and polymers with two different functional endgroups. Cocondensation with a small amount of an "a-a monomer" yields telechelics having two "a-endgroups", whereas cocondensation with a "b-b monomer" yield exclusively "b-terminated" telechelics. Copolycondensations of "a-b monomers" with "a_n monomers" may result in the formation of stars-haped polymers containing n-stararms. Copolycondensations of "a-b" and "a_n-b monomers" finally results in the formation of more or less hyperbranched polycondensates. In contrast, "a-a/b-b monomers" will automatically produce crosslinked materials, when polycondensed in the presence of any kind of multifunctional comonomers.

It was,and still is, a purpose of our work to illustrate the synthetic potential of "a-b monomers" in the field of aromatic polyethers, polyesters and polyamides (concentrating on polyesters in the present contribution). The preparation of star-shaped and hyperbranched polycondensates is plagued by side-reactions resulting in crosslinks, and thus, clean step-growth processes are a basic requirement for a successful synthesis. In this connection the potential of silicon mediated polycondensations should be explored, because polycondensations of silylated monomers may be a cleaner process than that of the corresponding nonsilylated (protonated) monomers, for instance, because proton catalyzed side reactions, such as the Fries-rearrangement, are avoided.

Telechelic Oligoesters

The syntheses of telechelic, star-shaped or hyperbranched oligo- and polyesters reported in this work are based on two "a-b monomers": 3-hydroxybenzoic acid (3-Hybe) and 3-(4'-hydroxyphenyl)propionic acid (phloretic acid, Phla). These monomers were selected for the following reasons. Polyesters of 3-Hybe melt below 200°C, and they are soluble in various organic solvents (10). Polycondensations of 2-Hydroxybenzoic acid derivatives show a high tendency to yield cyclic oligomers and substituted and unsubstituted 4-hydroxybenzoic acids, 6-hydroxynaphthalene-2-carboxylic acid or 4-'-hydroxybiphenyl-4-carboxylic acid yield oligo- and polyesters which are infusible and insoluble in all common solvents. Also poly(Phla) is highly crystalline and insoluble in common solvents, but the oligoesters may be soluble and meltable, depending on their endgroups and comonomers. Despite problems with the solubility of poly(Phla), homo- and copolyesters of Phla are interesting materials for the following reasons. Ester groups of Phla are more electrophilic, and thus, more sensitive to hydrolysis than ester groups of hydroxybenzoic acids. Hence polyesters containing Phla may be considered to be biodegradable, in as much, as Phla itself belongs to the metabolism of several plants and micro organisms, and thus, is a biocompatible degradation product. Furthermore, copolyesters containing 4-hydroxy acid may form liquid-crystalline (nematic) melts (11).

The polycondensation of Phla requires its acetylation. In contrast to fully aromatic hydroxy acids, the acetylation of Phla is affected by acid catalyzed oligomerization, which renders the isolation of the pure monomer more cumbersome. Silylation of the crude reaction mixture allows an efficient purification and isolation of the silylated monomer (1) by distillation in vacuum (eqs. 1,2). Furthermore, the trimethylsilyl ester can be used as monomer without saponification. As revealed quite recently (12,13) the silyl ester group is the only ester group which undergoes polycondensation with acetylated phenols even in the absence of transesterification catalysts.

A first series of polycondensations were conducted in such a way that 4-tert.-butylbenzoic acid was added as a "chain-stopper" in various molar ratios (Table 1). These polycondensations were conducted in bulk at a maximum reaction temperature of 270°C, because preliminary studies had evidenced, that higher temperatures might cause side reactions with cross-linking. The ^{13}C NMR CP/MAS spectrum of sample No. 1, Table 1 indicated the successful incorporation of 4-tert.-butylbenzoic acid, and thus, the formation of telechelics with two different chain ends (eq. 3 and formula 2). Unfortunately, all oligoesters of this series proved to be insoluble in all common solvents due to a high degree of crystallinity (compare Figure 5). Therefore, a more detailed characterization in solution was not feasible. An analogous series of polycondensations was synthesized with acetylated 4-tert-butylphenol as chain-stopper (eq. 4 and structure 3). In this case the polyesters 3 prepared with a monomer/chain-stopper ratio of 10/1 and 20/1 were soluble (Table 1). The ^{1}H NMR spectra proved the incorporation of 4-tert.-butylphenol and allowed the calculation of DP's (Figure 2). The solubility of oligoesters of structure 3 in contrast to those of structure 2 illustrates a significant influence of the endgroups on their properties.

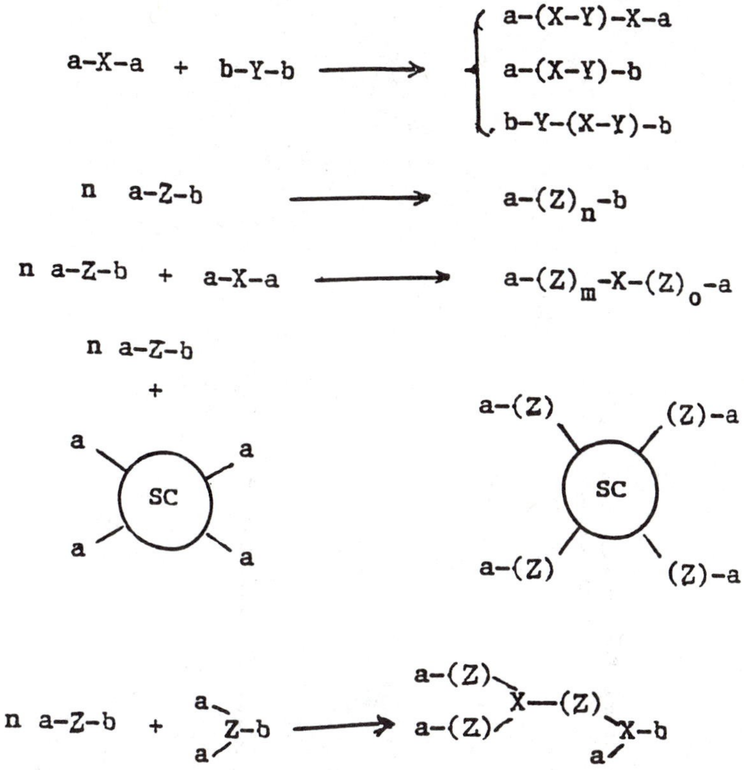

Figure 1: Schematic polycondensations of a-a + b-b monomers.

Table 1: Yields and properties of telechelic polyesters prepared from 1 and 4-tert.-butylbenzoic acid (2) or 1- and 4-tert.-butylphenol (3) in bulk at 8 h

Poly-ester	Mon. (1) / Chain stopper	Yield (%)	$\eta_{inh.}$ [a] (dl/g)	DP [b] (^{1}NMR)	T_m [c] (°C)
2	10/1	92	-	-	247, 266
2	20/1	93	-	-	236, 266
2	40/1	94	-	-	248, 267
3	10/1	94	0.10	12	240
3	20/1	93	0.14	23	237, 267
3	40/1	96	-	-	244, 267

a) measured at 20°C with c = 2 g/l in CH_2Cl_2/trifluoroacetic acid (volume ratio 4:1)
b) from ^{1}H NMR endgroup analyses
c) DSC measurements with a heating rate of 20°C/min (1rst heating)

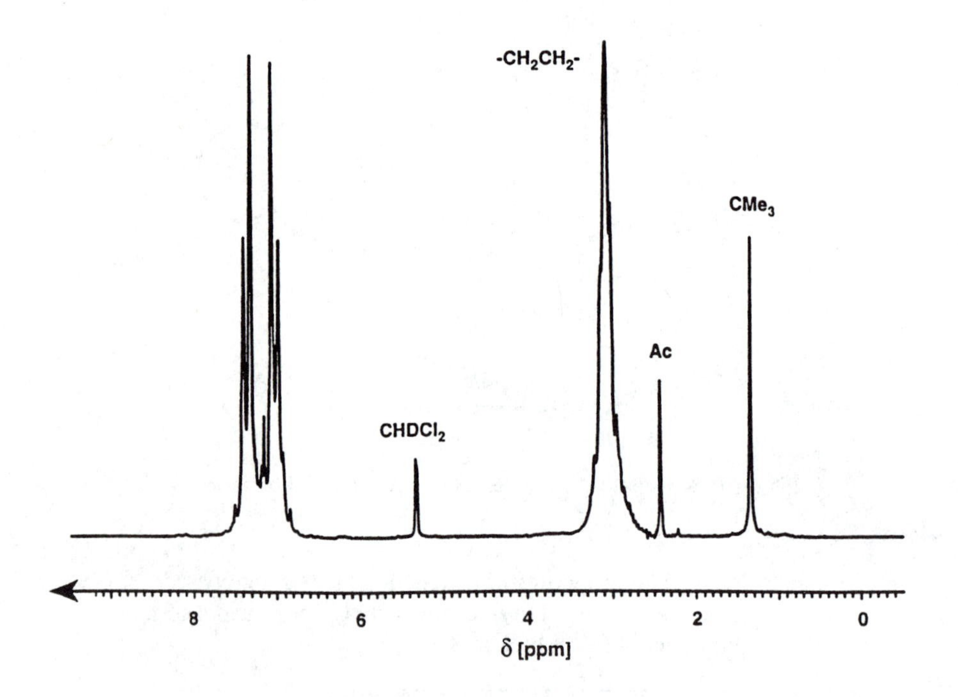

Figure 2: 100 MHz ^{1}H NMR spectrum of oligo(Phla) terminated with 4-tert.butylphenol (No 1, Table 1).

$$HO-C_6H_4-CH_2-CH_2-CO_2H \xrightarrow{+Ac_2O} AcO-C_6H_4-CH_2-CH_2-CO_2H \quad (1)$$

$$+ \; Ac\left(O-C_6H_4-CH_2-CH_2-CO-\right)-OH \quad (2)$$

$$Me_3C-C_6H_4-CO_2SiMe_3 \;+\; AcO-C_6H_4-CH_2-CH_2-CO_2SiMe_3 \;\; (\underline{1}) \quad (3)$$

$$\xrightarrow{-\,AcOSiMe_3}$$

$$Me_3C-C_6H_4-CO-\left(-O-C_6H_4-CH_2-CH_2-CO-\right)_n-OH \quad \underline{2}$$

$$Me_3C-C_6H_4-CO-Ac + n\,\underline{1} \longrightarrow Me_3C-C_6H_4-O-\left(-CO-CH_2-CH_2-C_6H_4-O-\right)_n-Ac \quad \underline{3} \quad (4)$$

$$n\,\underline{1} \;+\; AcO-C_6H_4-C(CH_3)_2-C_6H_4-C(CH_3)_2-C_6H_4-OAc \quad \underline{4}$$

$$\underline{5}: \; Ac\left(-O-C_6H_4-CH_2-CH_2-CO-\right)_m-O-C_6H_4-C(CH_3)_2-C_6H_4-C(CH_3)_2-C_6H_4-O-\left(-CO-CH_2-CH_2-C_6H_4-O-\right)_n Ac$$

A third series of polycondensations was conducted with acetylated bisphenol-P (4) as a "chain-stopper", to obtain telechelic poly(Phla) with two identical functional endgroups (5 in eq. 5). Bisphenol-P was selected as a chain-stopper, because its acetate is not volatile, and because the four methyl groups allow again an easy ^{1}H NMR spectroscopic detection of this building block in the isolated polyesters (Figure 3). Again both ^{1}H NMR spectra and inherent viscosities (Table 2) indicate the successful incorporation of the chain-stopper. In this case soluble oligoesters were obtained up to a M/I ratio of 80.

Stars-haped Polyesters

All star-shaped polyesters described in this work were prepared in such a way that a suitable "star-center" was directly polycondensed with the difunctional monomers. The star-centers were selected, so that they contained isolated methyl groups which yield sharp singlet signals in the ^{1}H NMR spectra to allow an easy identification and quantification (analogous to bisphenol-P).

A series of three armed stars (7) was obtained by polycondensation of the acetylated triphenol 6 with either 3-acetoxybenzoic acid or its trimethylsilyl ester (eq. 6). When these polycondensations were conducted at the same maximum reaction temperature (270 °C), it was found that the monomer with the free carboxyl group yields the higher molecular weights (viscosities) (Table 3). Since the maximum molecular weights are limited by the monomer/starcenter ratio, the lower viscosities obtained from the silylated monomer mean that the conversion was not complete. This conclusion was confirmed by the ^{1}H NMR spectroscopic detection of a CO_2SiMe_3 signal at 0.3 ppm, when the crude reaction mixture was measured at the end of the reaction time. As already discussed previously the silylated carboxyl group is less reactive than the free carboxyl group when polycondensations are conducted at identical reaction conditions. However, the silylated carboxyl group allows in most cases higher reaction temperatures which compensate its lower reactivity. Anyway, the data compiled in Table 3 show increasing viscosities with higher M/I ratio, and thus, a successful control of the molecular weight by incorporation of the star-center 6. Furthermore, the ^{1}H NMR spectra allowed again the calculation of apparent DP's. In this case the ^{1}H NMR of signal of the CH_3 "endgroup" was useless, because it was overlapping with the stronger signal of the acetate endgroups. However, the signal of the aromatic protons of 6 were well separated from those of the repeating unit, and thus, allowed an easy calculation of the DP.

A second series of star-shaped polyesters (10) was prepared by polyconden-

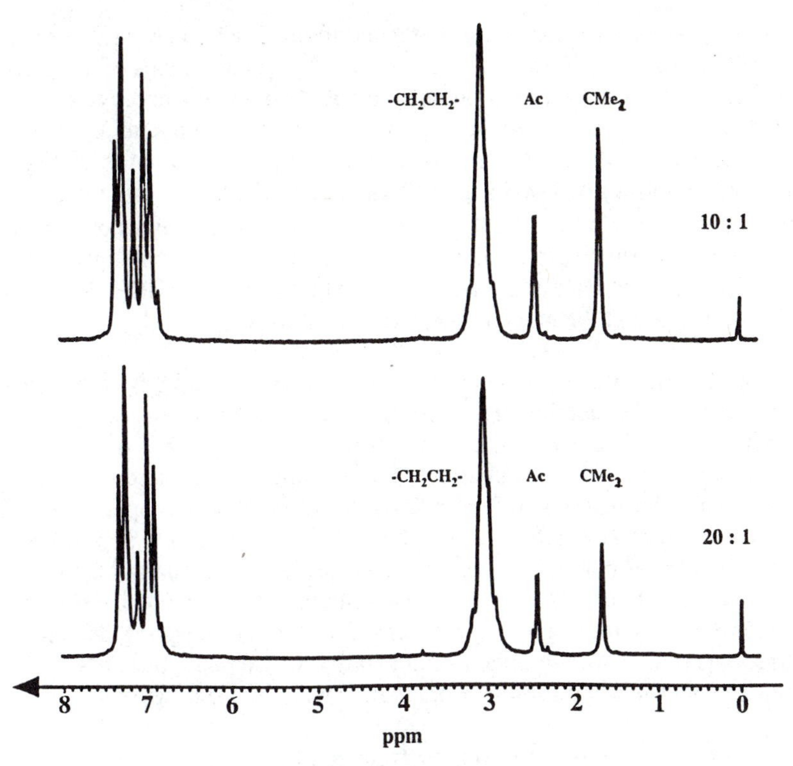

Figure 3: 100 MHz ^{1}H NMR spectra of oligo(Phla) 3 containing bisphenol-P (Nos. 1,2, Table 1).

Table 2: Yields and properties of the telechelic polyesters 5 prepared from monomer 1 and acetylated bisphenol-P (4) in bulk at 270°C/4 h

Poly-ester	Mon. (1) / Bisphenol-P(4)	Yield (%)	ηinh.[a] (dl/g)	DP[b] (^{1}NMR)	T_g[c] (°C)	T_m[c] (°C)
1	10/1	95	0.10	12	47	233
2	20/1	96	0.14	24	53	235
3	40/1	96	0.22	44	58	broad
4	80/1	94	0.26	90	-	257
5	160/1	94	-	-	-	230/267

a) measured at 20°C with c = 2 g/l in CH_2Cl_2/trifluoroacetic acid (volume ratio 4:1)
b) ^{1}H NMR endgroup analyses (see Figure 3)
c) DSC measurements with a heating rate of 20°C/min

sation of the Phla monomer 1 with the tetrafunctional "star-center" 8 (14). In order to find out, if transesterification of the acetate groups of 8 is sterically hindered, a model reaction with silylated 4-methylbenzoic acid was conducted (eq. 7). The ^{1}H NMR spectrum of model compound 9 proved that this condensation reaction was successful. The results of the analogous polycondensations of 1 and 8 are summarized in Table 4. In preliminary experiments not listed in Table 4, it was found that reaction times of $\geq$ 8 h are required to obtain nearly quantitative conversions. Correspondingly, both the inherent viscosities and the DP's determined from the ^{1}H NMR spectra (Figure 4) parallel the monomer/star-center ratio.

One reason why the polycondensation of monomers 1 and 8 is particularly interesting, is the fact the oligo- and polyesters of Phla possess a high tendency to crystallize, whereas the spirocyclic "star-center" 8 is unfavorable for crystallization of the oligoester 10. The results is illustrated by the WAXD powder patterns of Figure 5. At low DP's the star-shaped poly(Phla) is mainly amorphous, whereas at higher DP's the crystallinity approaches that of the linear homopolyester, which possesses an unusually high crystallinity around 70 - 80%. A comparison of the polyesters obtained from monomer 1 and 4-tert-butylphenol, bisphenol-P or star-center 8 demonstrates that the chance to obtain soluble polyesters with higher DP's increases with the steric influence of chain-stopper and star-centers.

Hyperbranched Poly(3-hydroxybenzoic acid)

The copolycondensation of a difunctional ("a-b") monomer and a trifunctional ("a_2-b") monomer is of particular interest, because it enables the preparation of randomly branched (hyperbranched) copolymers with a variable degree of branching (DB). Furthermore, if it is possible to vary the nature of the endgroups, the influence of the endgroups on the properties of the polyesters may also be studied. For this purpose four different series of hyperbranched copolyesters of 3-Hybe, were synthesized (eqs. 8-11). The polycondensations according to eqs. (8) and (9) represent the earliest syntheses of randomly branched polyesters and one of the earliest synthesis of randomly branched polycondensates 10). The details of these polycondensations were described previously (10,15), and thus, should not be reported here agian. However, it should be emphasized that nearly random sequences of both comonomers were obtained in all four series, as illustrated by the ^{13}C NMR spectrum (Figure 6). Furthermore, the syntheses and properties of the hyperbranched homopolyesters of 3,5-bishydroxybenzoic acid and 5-hydroxyisophthalic acid were described by three research groups in much detail (16-21), and thus, will not be repeated here. Using 3,5-bistrimethylsiloxybenzoyl chloride, 3,5-bisacetoxybenzoic acid or its trimethylsilyl ester the hyperbranched homopolyesters of structures 9 and 10 were obtained

Table 3: Yields and properties of the star-shaped polyesters 7 prepared from silylated 3-acetoxybenzoic acid and "star-center" 6 in bulk at 270°C/5 h

Poly-ester	Mon. (1) / Star-center 6	Yield (%)	$\eta_{inh.}$ [a] (dl/g)	T_g [b] (°C)
1	30/1	94	0.08	95
2	60/1	96	0.11	111
3	90/1	96	0.16	128
4	120/1	97	0.22	137

a) measured at 20°C with c = 2 g/l in CH_2Cl_2
b) DSC measurements with a heating rate of 20°C/min

Table 4: Yields and properties of the star-shaped polyesters 10 prepared from monomer 1 and star-center 8 in bulk at 270°C/8 h

Poly-ester	Mon. (1) / Star-center 8	Yield (%)	$\eta_{inh.}$ [a] (dl/g)	DP [b] (^{1}NMR)	T_g [c] (°C)	T_m [c] (°C)
1	10/1	94	0.10	12/1	53	231
2	20/1	93	0.14	21/1	55	233
3	40/1	95	0.21	38/1	57	241
4	80/1	93	0.36	70/1	62	242
5	160/1	96	-	-	64	243

a) measured at 20°C with c = 2 g/l in CH_2Cl_2/trifluoroacetic acid (volume ratio 4:1)
b) ^{1}H NMR endgroup analyses
c) DSC measurements with a heating rate of 20°C/min (1st heating)

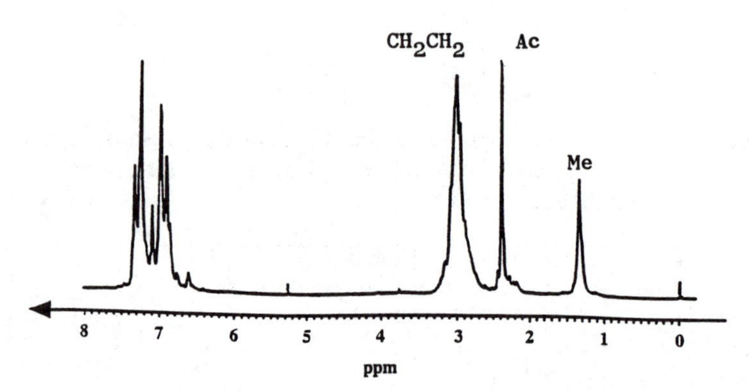

Figure 4: 100 MHz ^{1}H NMR spectrum of starshaped poly-(Phla) derived from "star center" 8.

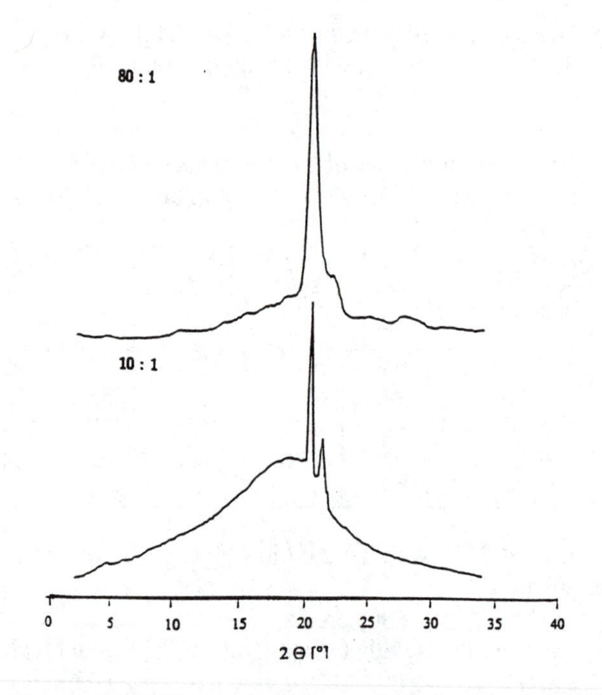

Figure 5: WAXD powder patterns of starshaped poly(Phla) derived from "star center" 8 with variation of the 1/8 ratio.

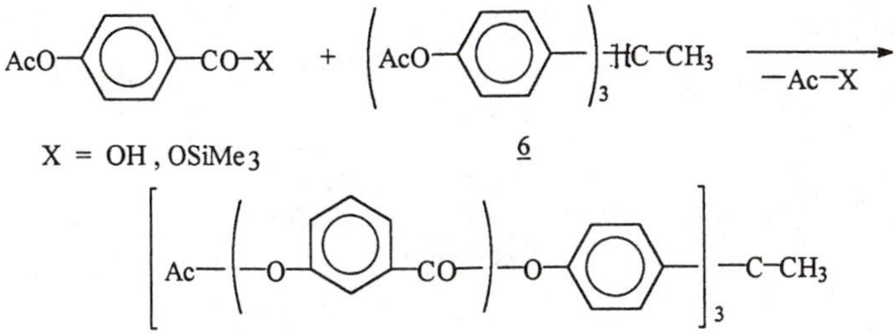

AcO–C₆H₄–CO–X + (AcO–C₆H₄–)₃C–CH₃ → (−Ac–X) (6)

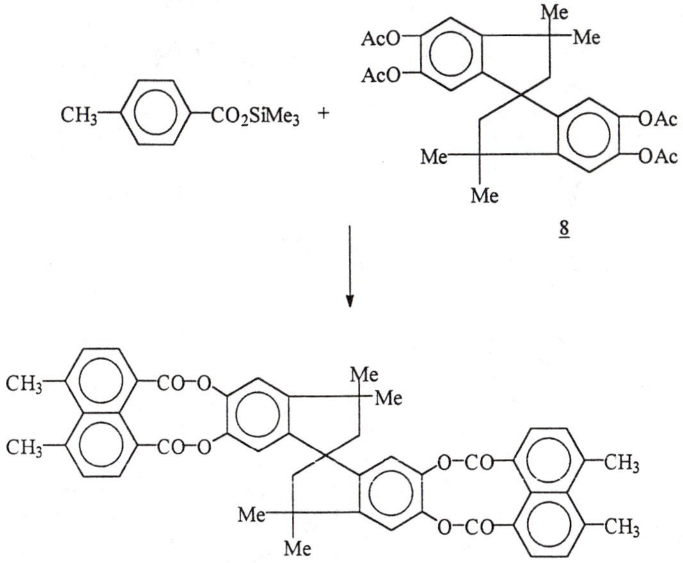

X = OH, $OSiMe_3$ — 6

[Ac–(–O–C₆H₄–CO–)–O–C₆H₄–]₃–C–CH₃

7

CH_3–C₆H₄–CO_2SiMe_3 + 8 → 9 (7)

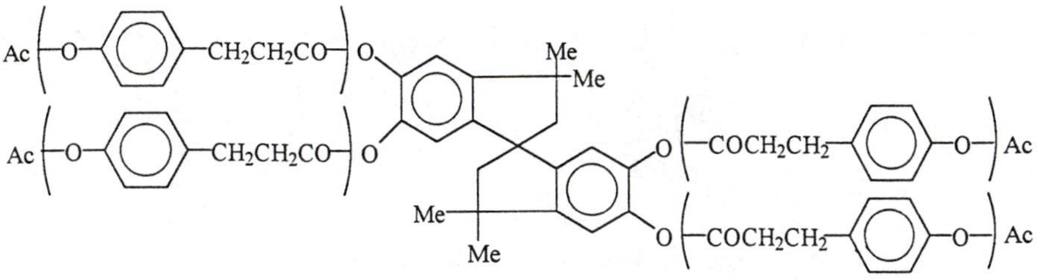

10

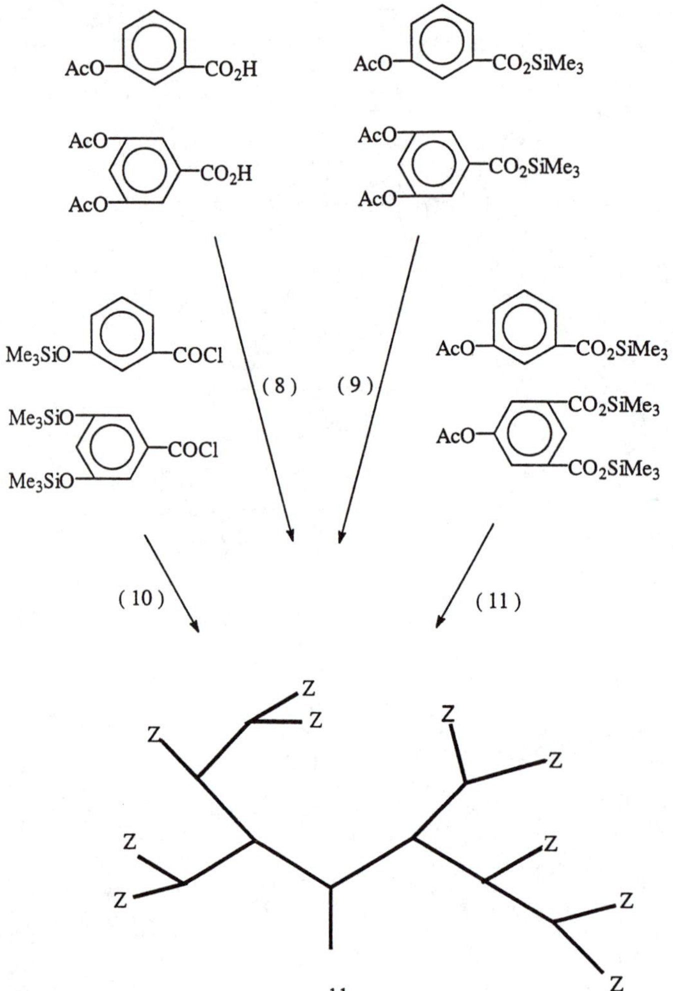

AcO–C₆H₄–CO_2H AcO–C₆H₄–CO_2SiMe_3

$(AcO)_2C_6H_3$–CO_2H $(AcO)_2C_6H_3$–CO_2SiMe_3

Me_3SiO–C₆H₄–COCl AcO–C₆H₄–CO_2SiMe_3

$(Me_3SiO)_2C_6H_3$–COCl AcO–$C_6H_3(CO_2SiMe_3)_2$

(8) (9) (10) (11)

<u>11</u>

Z = OH, OAc, $OSiMe_3$, CO_2H, CO_2SiMe_3

$$DB = \frac{N_B + N_T}{N_L \; N_B + N_T} \qquad (12)$$

N_L, N_B, N_T = molar fractions of linear, branching and terminal units in a branched polyester

(16-21). It was found that regardless of the synthetic method all these hyperbranched polyesters possess a degree of branching (DB) in the range of 0.55-0.60 (as defined by eq. 12), which means that somewhat less than every second repeating unit is a branching point. Obviously, this DP is a direct consequence of the steric demands of the "side chains", and does not reflect different reactivities of functional groups. The DB was calculated via eq. (12) on the basis of high resolution ^{1}H NMR spectra which display a separate endgroup signal (x in Figures 7 and 8) of the 3,5-acetoxybenzoyl units. This endgroup signal disappears in copolyesters containing higher molar fractions of 3-oxybenzoyl units, because the 3,5-bisoxybenzoyl moieties then play the role of linear or branching units (Figures 8 and 9).

In the case of the copolyesters 11, Z=OAc, the influence of the number of branching points on the crystallization was studied. Linear poly(3-Hybe) crystallizes spontaneously (i.e. without artifical nucleation) from its solution in aromatic solvents (10). After period of 3-4 days nearly 100% of the dissolved polyester has precipitated. This experiment was repeated with hyperbranched poly(3-Hybe) and the number of branching units was systematically varied. As demonstrated by Figure 10, the fraction of poly(3-Hybe) capable of crystallization decreases continously with increasing DB. At a molar ratio of one branching unit per ten 3-Hybe units crystallization is completely suppressed. It seems that linear chain segments with ≥ 20 repeating units are required to form crystallites.

Another interesting relationship was revealed when the Tg's were plotted versus the molar fraction of branching units. This plot (Figure 11) includes three variations of the endgroups (curves A, B and C). A high DB severely hinders the segmental motion, and thus, all hyperbranched homopolyesters (12 and 13) possess higher Tg's than linear poly(3-Hybe). In the case of copolyester 11, Z=CO_2H the T_g's display a continous decrease from the value of the hyperbranched homopolyester to that of linear poly(3-Hybe) (curve A). This curve in principle agrees with the Fox equation (22). In contrast the Tg's of copolyesters 11, Z= OH (curve B) and Z=OAc (curve C) show a minimum around one branching unit per 10 repeating units. Interestingly this result fits well in with the degree of branching limiting the crystallization as demonstrated in Figure 10. Obviously a minimum segment length around twenty 3-Hybe units in required, for the chain segments to, pack in dense more or less parallel fashion. Such a denser parallel chain packing reduces the segmented mobility and favors the crystallization. Thus random branching exerts two contradictory effects on the segmental mobility. A low to moderate DB creates more free volume and a higher mobility, when compared to the perfectly linear homopolymer, but at high DB's the steric hindrance of the segmental motion prevails. Strong electronic interactions

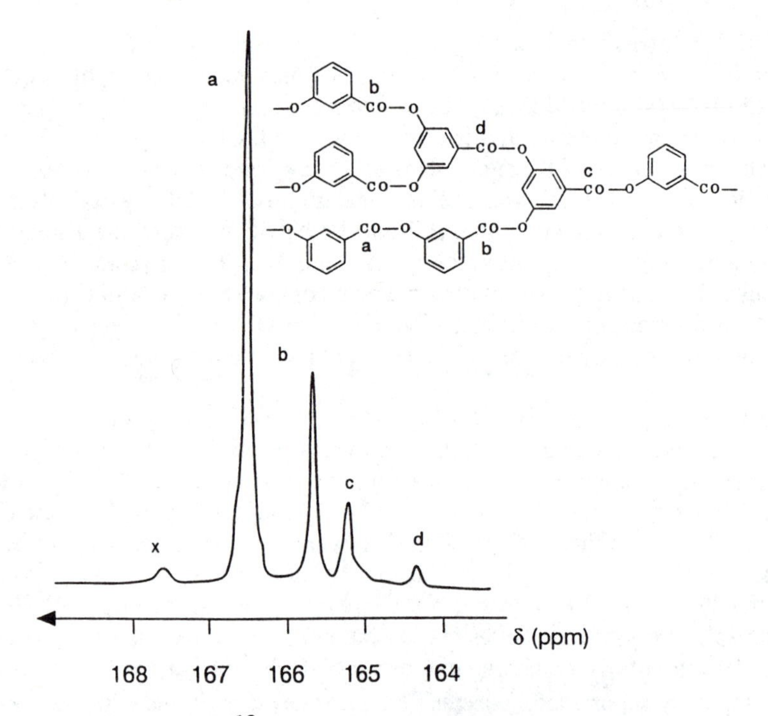

Figure 6: 75.4 MHz ^{13}C NMR spectrum (CO-signals) of a copolyester prepared by cocondensation of silylated 3-acetoxybenzoic acid and 3,5-bisacetoxybenzoic acid (molar ratio 3:1) in bulk at 280°C. (Reproduced with permission. Copyright 1995 Huethig.)

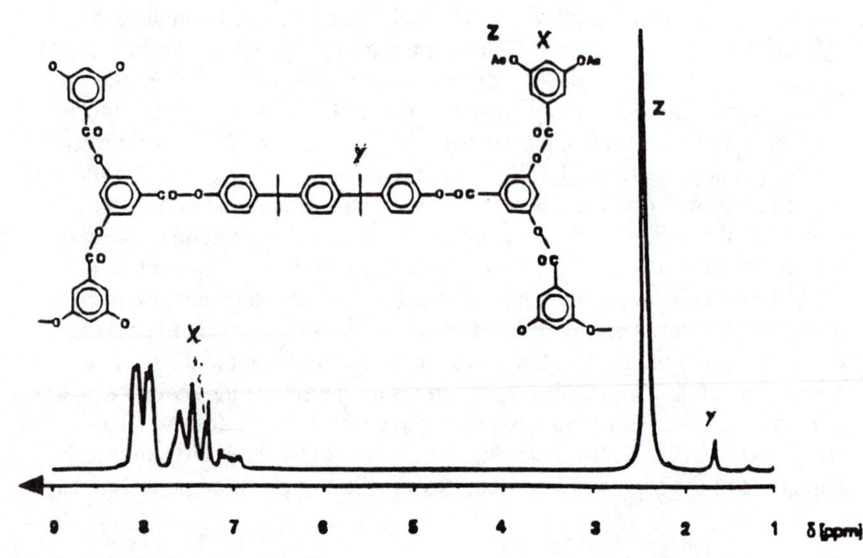

Figure 7: 100 MHz ^1H NMR spectrum of the star-shaped, hyperbranched polyester 15 (in $CDCl_3$).

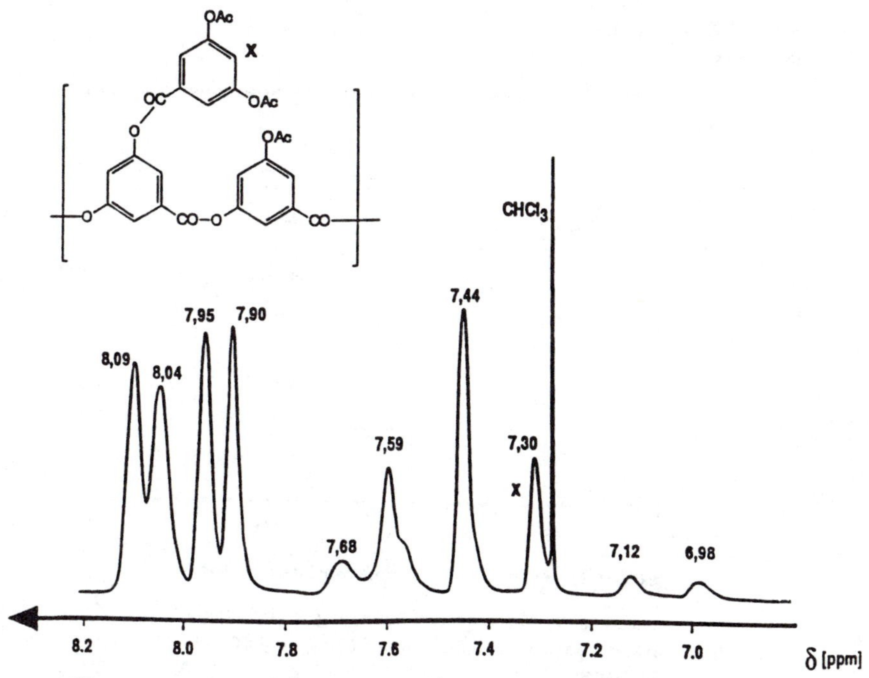

Figure 8: 360 MHz ^{1}H NMR spectrum of polyester 15 (in $CDCl_2$).
(Reproduced with permission. Copyright 1995 Huethig.)

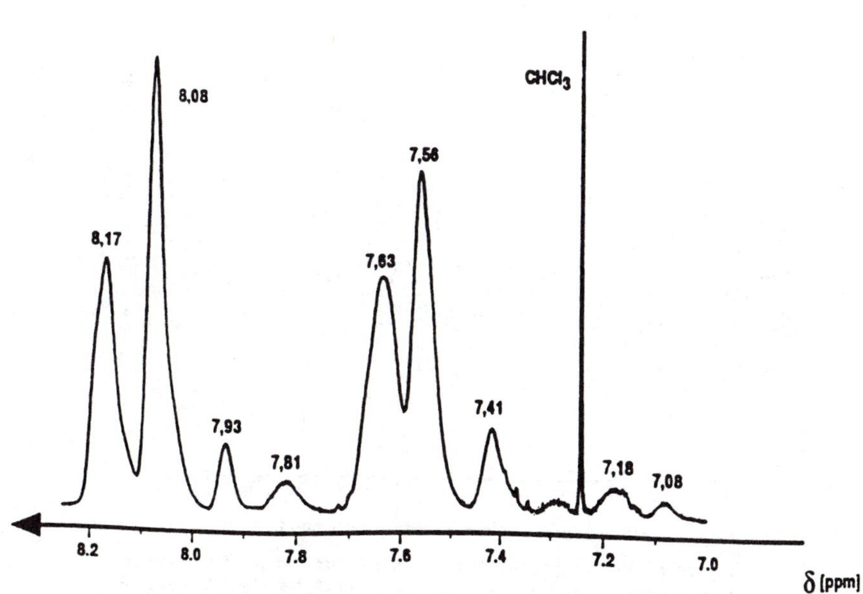

Figure 9: 360 MHz ^{1}H NMR spectrum of a 3:1 copolyester of structure 11 (2=OAc).
(Reproduced with permission. Copyright 1995 Huethig.)

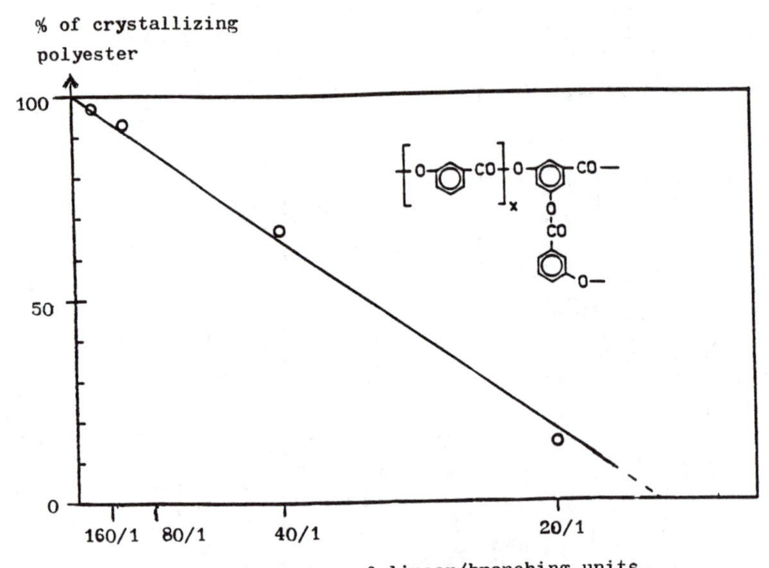

Figure 10: Influence of the degree of branching on the weight fraction of poly(3-Hybe) crystallizing from its solution in benzene/pyridine (1:1 mixture).

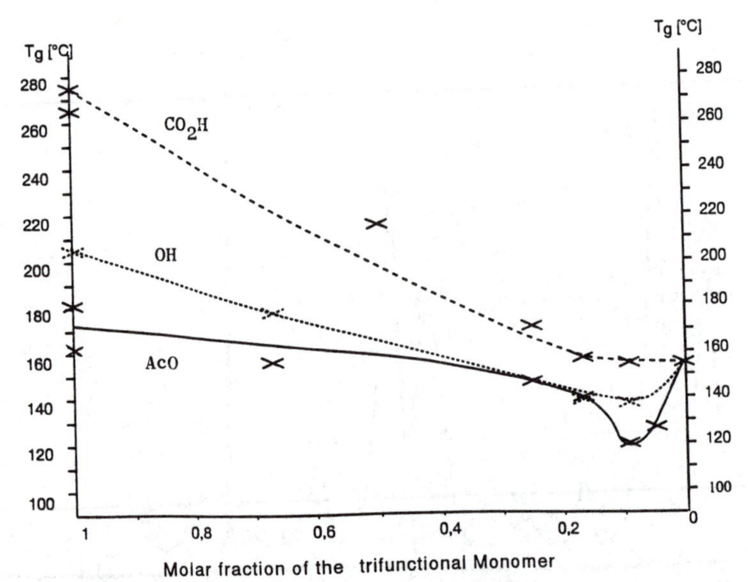

Figure 11: Plot of glass-transition temperatures versus the molar fraction of branching units in hyperbranched poly(3-Hybe): A) structure 11, CO_2H endgroups, B) 11, OH-endgroups, C) 11, OAc endgroups. (Reproduced with permission. Copyright 1995 Huethig.)

(such as H-bonds) between endgroups and chain segments reduce the mobility at any DB.

Star-shaped and Triblock, Hyperbranched Polyesters

The synthetic strategy used for the preparation of star-shaped polyesters with linear star arms (structures 7 and 10) can also be applied to the synthesis of star-shaped polyesters with hyperbranched star arms. For this purpose silylated 3,5-bisacetoxybenzoic acid was polycondensed with the di-, tri- or tetrafunctional phenolacetates 3, 6 or 14. In all cases both viscosity and GPC measurements confirmed that the molecular weights varied with the feed ratio monomer/"star-center". In the case of structure 15 ^{1}H NMR spectroscopy also allowed the determination of the DP which also paralleled the feed ratio (Figure 7). The results obtained from hyperbranched polyesters of structure 15 are summarized in Table 5 (19). Unfortunately, the "star-center" 6 turned out to be unfavorable for ^{1}H NMR spectroscopic determination, because all its ^{1}H NMR signals were obscured of DP's by the signals of the 3-Hybe units and acetate endgroups. In the case of "star-center" 14 the tert.butyl groups was split of as isobutylene in the course of the polycondensation (20).

Lengthening of the central unit in structure 15 should result in A-B-A-triblock copolymers with hyperbranched A-blocks. This goal was approached in the following way. Telechelic oligo(ether-ketone)s (16) were prepared from silylated bisphenol-P according to eq. (13) as described previously (23). When such an oligo(ether-ketone) was heated with an excess of acetyl chloride in an inert high boiling solvent (e.g. chloronaphthalene) a complete acetylation of the endgroups took place. The structure of the resulting oligo(ether-ketone) (17) was confirmed by ^{1}H NMR spectroscopy. As demonstrated in Figure 12 a 360 MHz ^{1}H NMR spectrum indicates the presence of the bisphenol-P unit and the acetate endgroups and allows the determination of the DP (15 in the case of Figure 12). The melting temperature (T_m) of 17 is 290-295°C (by DSC), and thus, its polycondensation with silylated 3,5-bisacetoxybenzoic acid (eq. 14) occurs in the molten state of all components. The resulting triblock copolymer (18) was almost insoluble in THF in contrast to 15, but soluble in chloroform containing 10 vol% of trifluoroacetic acid. These properties indicate the covalent connection of hyperbranched A-blocks and oligo(ether ketone) blocks, on the one hand, and prove the absence of crosslinks, on the other hand. The ^{1}H NMR spectrum of 18 is almost identical with that of 15 (Figure 7), because the signals of the ether ketone segments are hidden by the signals of the polyester blocks.

This work can be extended to hyperbranched triblock copolymers with an

Table 5: Copolycondensations[a)] or Trimethylsilyl 3,5-Diacetoxybenzoate with Acetylated Bisphenol-P

expt no.	Monomer/init[b)]	Yield (%)	ηinh.[c)] (dl/g)	T_g (°C)	DP[d)]	Mn	elem.anal.	%C	%H
1	20	63	0.18	144	45±5	~ 8 x 10^3	calcd.	62.56	3.74
							found	60.25	3.48
2	40	72	0.24	149	65±5	~ 12 x 10^3	calcd.	61.67	3.60
							found	59.66	3.53
3	80	75	0.27	150	100±5	~ 18 x 10^3	calcd.	61.19	3.50
							found	60.00	3.57
4	150	75	0.37	152	180±5	~ 32 x 10^3	calcd.	60.96	3.45
							found	60.12	3.50
5	250	76	0.38	154	—	(~ 40 x 10^3)	calcd.	60.85	3.43
							found	60.24	3.60

a) 1 h/200°C + 3 h/280°C + 0.5 h/280°C with vacuum
b) molar feed ratio of monomer and 4
c) measured at 20°C with c = 2 g/l in CH_2Cl_2/trifluoroacetic acid (volume ratio 4:1)
d) average degree of polymerization as determined from the ratio of bisphenol-P and acetate groups

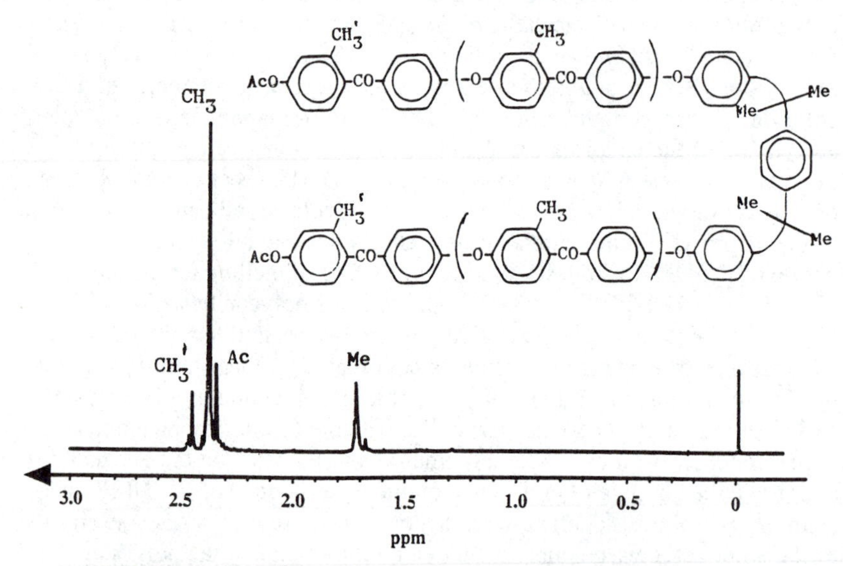

Figure 12: 360 MHz ^{1}H NMR spectrum of the acetylated oligo-(ehter-ketone) 17 (DP ~ 15).

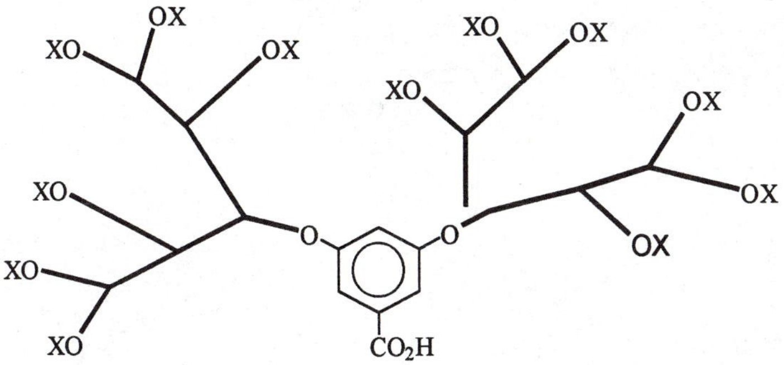

OX XO OX XO OX XO OX XO OX XO OX XO O O OX CO_2H

12, X = H, Ac, $SiMe_3$

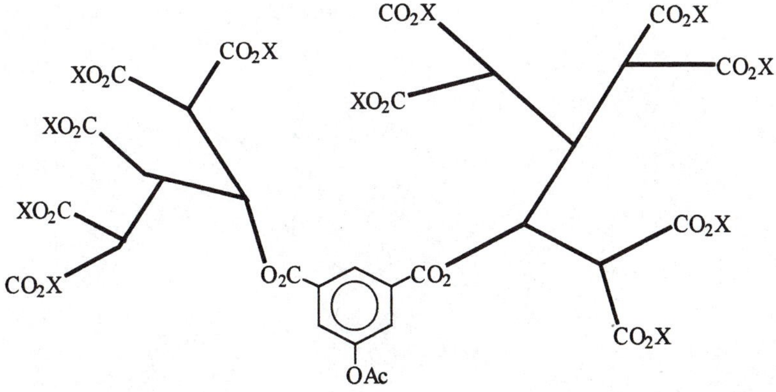

CO_2X CO_2X CO_2X XO_2C CO_2X XO_2C XO_2C XO_2C CO_2X CO_2X O_2C CO_2 CO_2X OAc

13, X = H, C_2H_5, $SiMe_3$

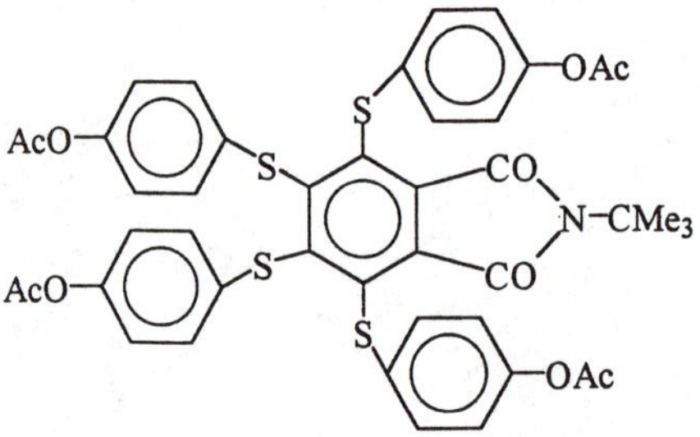

AcO, OAc, AcO, OAc, S, S, S, S, CO, N–CMe$_3$, CO

14

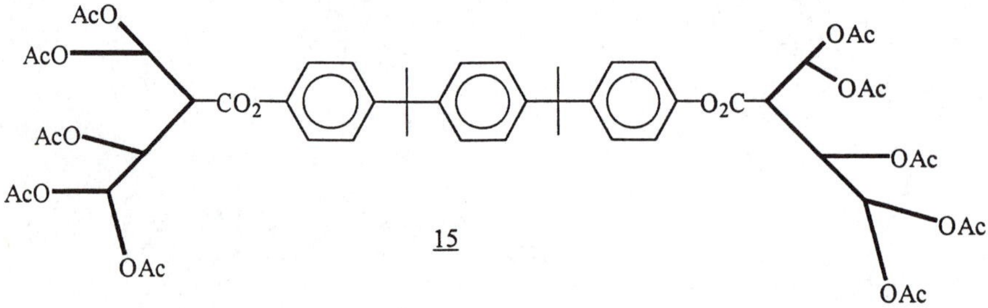

AcO, AcO, AcO, AcO, OAc, CO$_2$, O$_2$C, OAc, OAc, OAc, OAc, OAc

15

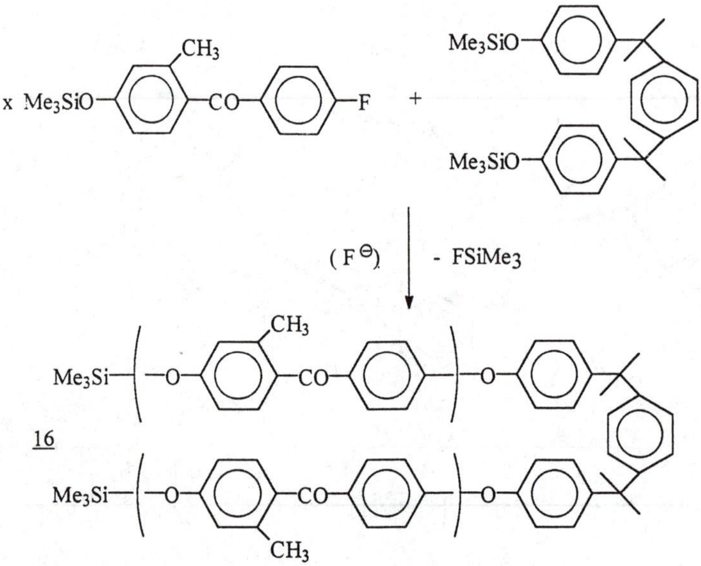

x Me_3SiO–(C$_6$H$_3$CH$_3$)–CO–(C$_6$H$_4$)–F + Me_3SiO, Me_3SiO

(F$^{\ominus}$) - $FSiMe_3$ (13)

Me_3Si, CH$_3$, CO, O, CH$_3$

16

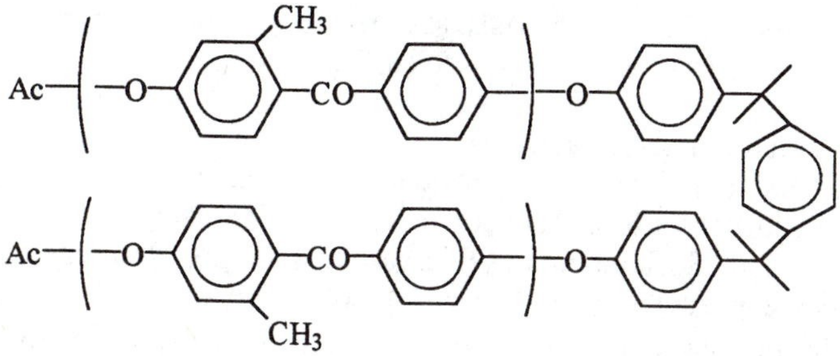

Ac—(—O—C6H3(CH3)—CO—C6H4—)—O—C6H4—C(CH3)2—C6H4—C(CH3)2—C6H4—O—(—C6H4—OC—C6H3(CH3)—O—)—Ac

17

AcO, AcO (3,5-diacetoxy)—C6H3—CO_2SiMe_3

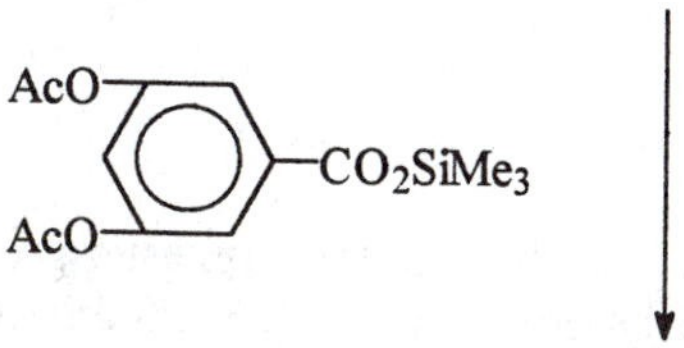

(14)

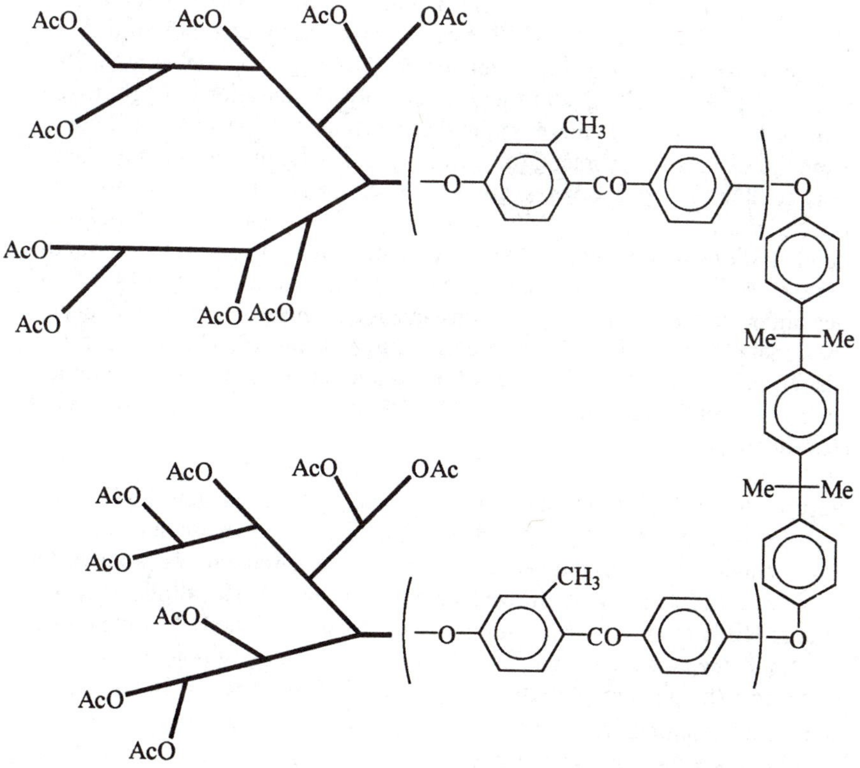

18

oligo(ether-sulfone) B-block, because telechelic oligo(ester sulfone)s can also be obtained by a route analogous to eq. (13) (24).

Hyperbranched Poly(ester-amide)s

Star-shaped and hyperbranched aromatic poly(ester-amide)s can be synthesized by polycondensation of "dimeric monomers" consisting of a hydroxy and an amino acid. Such monomers (e.g. formulas 19, 20, 21 and 22) are easy to prepare by acylation of silylated aminobenzoic acids with acetoxybenzoylchlorides (eq. 15). At moderate temperatures (< 100°C) the trimethylsilylgroup protects the carboxyl group of the aminobenzoic acid against acylation, whereas the aminogroup is activated by the silylation. In consequence, a highly selective and almost quantitative acylation of the amino group takes place, and the resulting silylated monomers (e.g. 19, 21a, 22a) can be used for polycondensations in an "one-pot procedure" (25-26).

An alternative approach is based on the hydrolysis of the silyl ester groups, and isolation of the monomers with free carboxyl groups (eq. 16, and formulas 20, 21b, 22b). The polycondensations of these silicon free monomers yield in principle the same poly(ester-amide)s as the silylated monomers (formula 23). However, the NMR spectroscopic characterization of the poly(ester-amide)s derived from monomers 21 or 21b revealed slight differences. The ^{13}C NMR spectra suggest that polycondensations of 21b involve more side reactions. Furthermore, higher molecular weights were obtained from the silylated monomer 21a. Unfortunately, all poly(ester-amide)s are only soluble in acidic solvents, and thus, GPC measurements were not conducted. Both ^{1}H and ^{13}C NMR spectroscopy also revealed that all polycondensations of silylated or nonsilylated monomers involve a low to moderate extent of ester-amide interchange reactions. The formation of acetamide endgroups is a typical consequence of this process as evidenced by signal y' in the ^{1}H NMR spectra of Figures 14 and 15. The IR-spectra displayed the expected CO-bands of ester- and amide groups along with a broad band in the range of 3000-3500 cm^{-1} resulting from NH-vibrations of disordered H-bonds (Figure 13).

Polycondensations of monomers 19a, 21a and 22a with the tetrafunctional "star-centers" 24, 25 or 26 yielded star-shaped and hyperbranched poly-(ester-amide)s (15). As indicated by viscosity measurements variation of the monomer/star-center ratio allowed a control of the DP. The aliphatic protons of the "star-centers" also allow a determination of the DP by means of ^{1}H-NMR spectroscopy (Figures 14 and 15). In the case of polyesters derived from the piperazine derivative 20 the ^{1}H NMR spectra need to be conducted at temperatures ≥ 80°C, because lower temperatures yield a broad CH_2-signal due to cis-/trans-isomerism and slow rotation around the CO-N-bond (27).

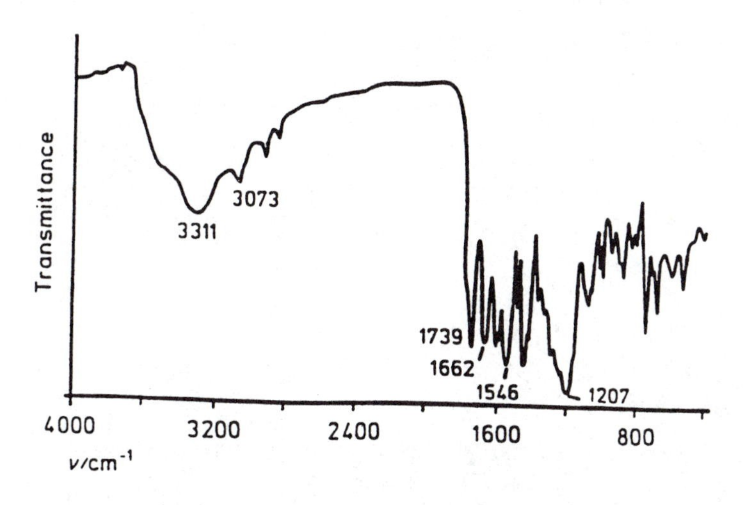

Figure 13: IR-spectrum (KB pellets) of the hyperbranched poly(ester amide) 23 prepared by polycondensation of monomer 21a in bulk at 270°C.

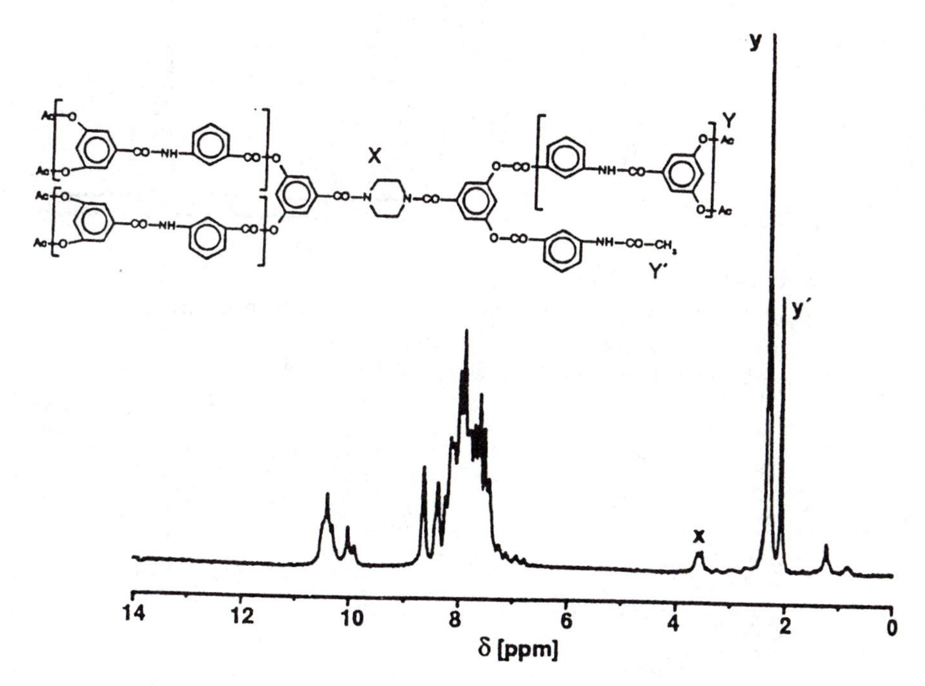

Figure 14: 100 MHz ^{1}H NMR spectrum of the star-shaped hyperbranched poly(ester-amide) prepared from 22a and 24.
(Reproduced with permission. Copyright 1995 Marcel Dekker.)

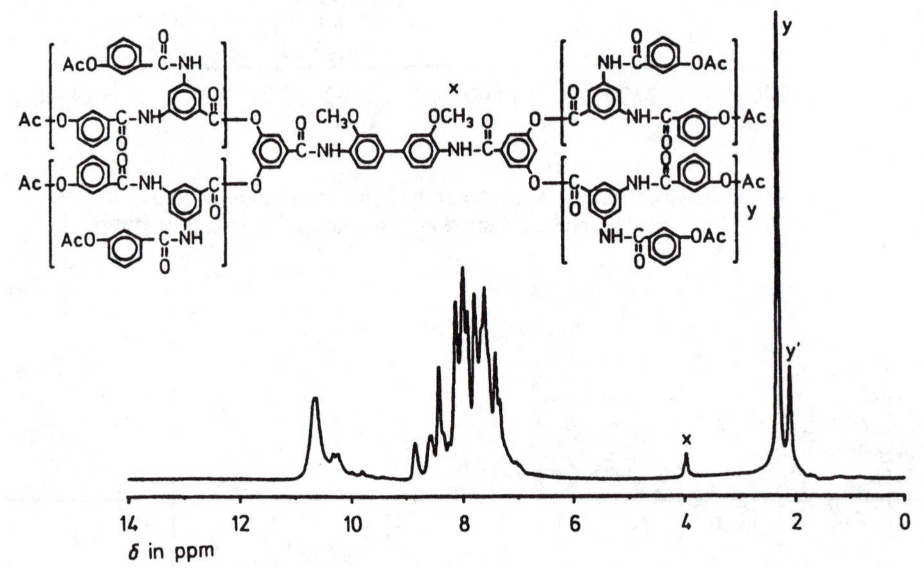

Figure 15: 100 MHz ^1H NMR spectrum of a star-shaped and hyper-branched poly(ester-amide) prepared from 21a and 25.
(Reproduced with permission. Copyright 1995 Huethig.)

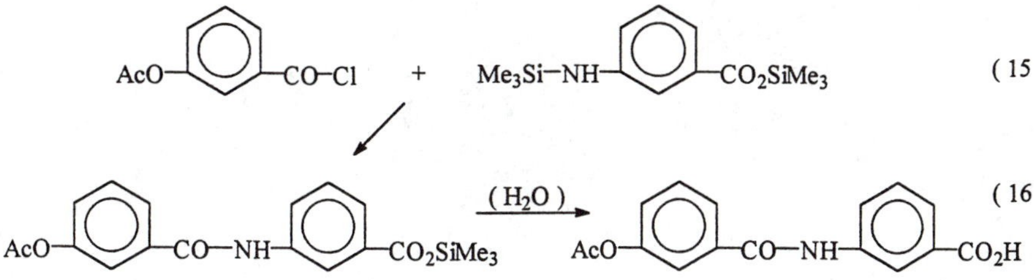

(15)

(16)

19

20

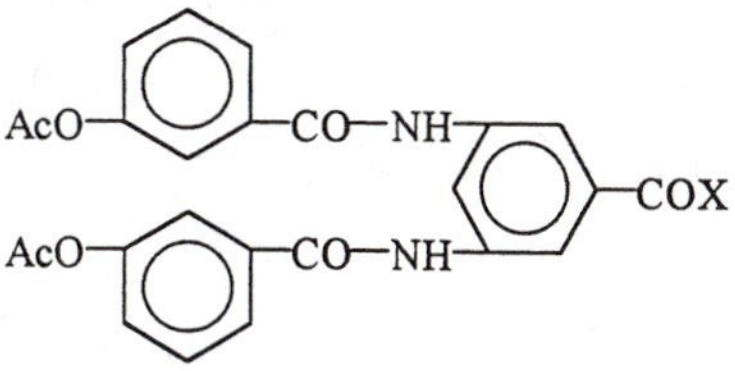

21 a : X = $SiMe_3$
b : X = H

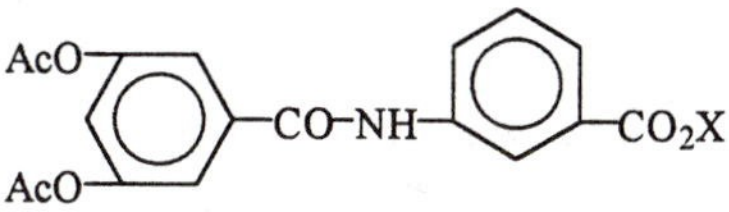

22 a : X = $SiMe_3$
b : X = H

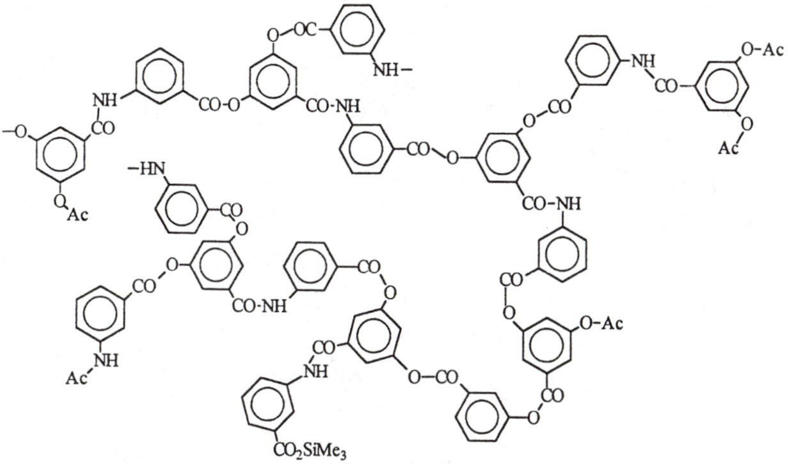

23

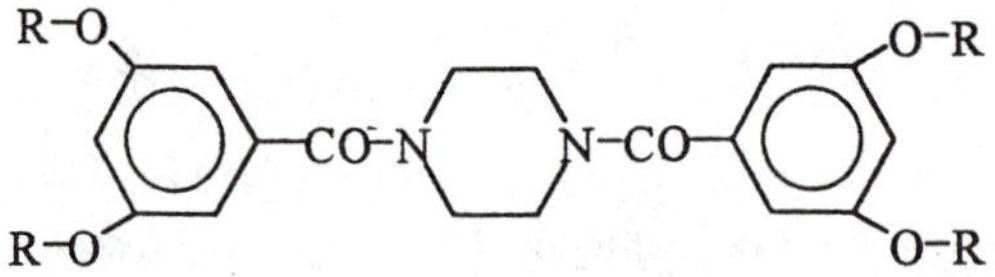

R–O O–R
–CO–N N–CO–
R–O O–R

24

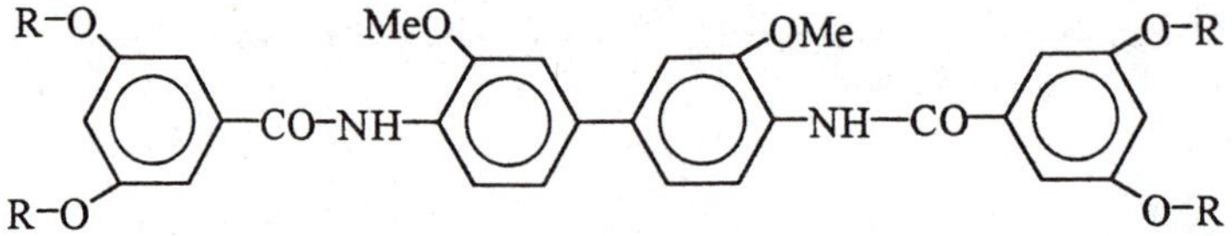

R–O MeO OMe O–R
–CO–NH– –NH–CO–
R–O O–R

25

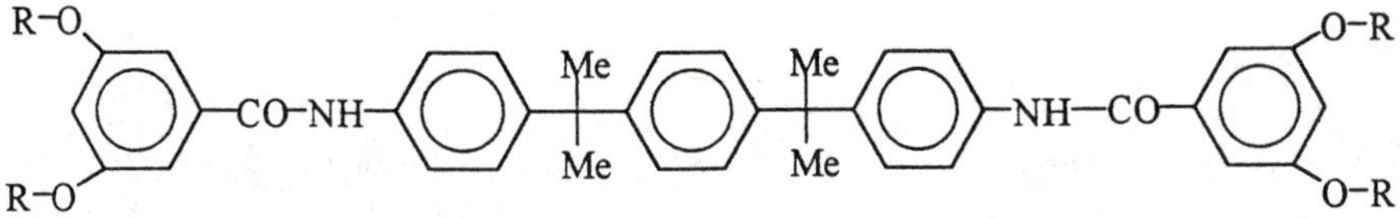

R–O Me Me O–R
–CO–NH– –NH–CO–
Me Me
R–O O–R

26

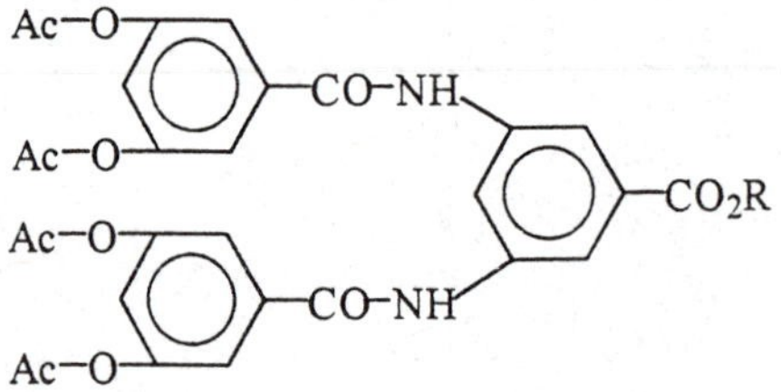

Ac–O
–CO–NH
Ac–O
–CO_2R 27 a, b
Ac–O
–CO–NH
Ac–O

R = H, $SiMe_3$

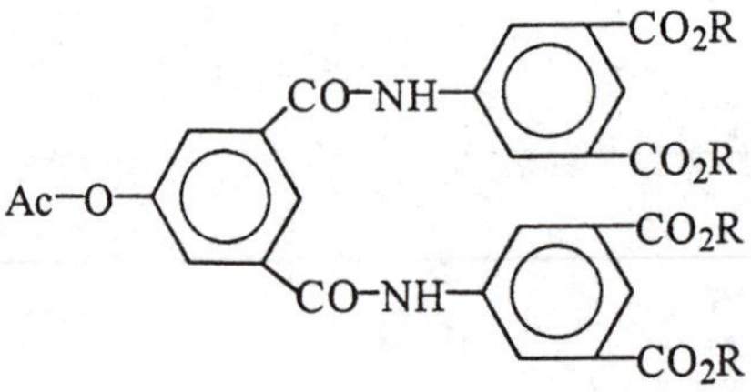

CO_2R
CO–NH–
CO_2R
Ac–O– 28 a, b
CO_2R
CO–NH–
CO_2R

The synthetic strategy described above is versatile and allows a broader variation of the chemical structure of poly(ester-amide)s. Polymers derived from hyperbranched monomers such a 27a,b or 28a,b are currently under investigation.

EXPERIMENTAL

Materials: 3-Hydroxy-, 3,5-dihydroxy, 3-amino- and 3,5-diaminobenzoic acid were gifts of BAYER AG (Leverkusen, FRG) and used without further purification. Bisphenol-P and bisaniline-P were purchased from Kennedy and Klim (Little Silver, N.J., USA). Piperazine, 1,1,1-tris(4'-hydroxy-phenyl)ethane, phloretic acid tetrahydroxy spirobisindane were purchased from Aldrich Co. (Milwaukee, Wisc., USA). All hydroxy acids and phenols were acetylated with an excess of acetic acid and a catalytic amount of pyridine in refluxing toluene. All silylations were conducted with chloro-trimethylsilane and triethylamine in refluxing toluene or dioxane.

Polycondensations

A) Telechelic poly(phloretic acid)s (Tables 1, 2 and 4).
Silylated 3-(4'-acetoxyphenyl) propionic acid (1, 25 mmol) and a chain-stopper or star-center were weighed into a cylindrical glass reactor equipped with stirrer, gas-inlet and gas-outlet tubes. The reactor was placed into metal bath preheated to 150°C, and the temperature was raised to 270°C within 0.5 h. This temperature was maintained for 4 or 8 h (+ 0,5 h in vacuo). The liberated chlorotrimethylsilane was removed with a slow stream of nitrogen. The cold polyester was dissolved in CH_2Cl_2/tri-fluoroacetic acid (volume ratio 4:1) and precipitated into cold methanol.

B) Star-shaped polyesters 7 (Table 3).
Trimethylsilyl 3-acetoxybenzoate (60 mmol) and 1,1,1-tris(4'-acetoxy-phenyl)ethane (2,1, 0.67 or 0.5 mmol) were weighed into a cylindrical glass reactor and reacted and worked up as described above.

C) Hyperbranched polyesters 11 and poly(esteramide)s.
The syntheses of theses hyperbranched copolymers have been described in refs. 20, 21, 25, 26.

MEASUREMENTS

The inherent viscosities were measured with an automated Ubbelohde vis-cometer thermostated at 20°C.
The IR spectra were recorded from KBr pellets with a Nicolet SXB-20 FT IR-spectrometer.

The DSC measurements were conducted with a Perkin Elmer DSC-4 using aluminum pans under nitrogen.
The WAXD powder patterns were recorded with a Siemens D-500 diffractometer using Ni-filtered CuK_{α}-radiation.
The 100 MHz ^{1}H NMR spectra were recorded with a Bruker AC-100 FT-spectrometer in 5 mm o.d. sample tubes. Internal TMS served as standard.
The 360 MHz ^{1}H NMR-spectra were recorded with a Bruker AM-360 FT spectrometer.

Acknowledgment: We thank the Deutsche Forschungsgemeinschaft for financial support.

REFERENCES

1) Elias, H.G.; "Macromolecules" 5th ed., Hüthig & Wepf, Basel, New York 1990
2) Cowie, J.M.G.; "Polymers: Chemistry & Physics of Modern Materials", Intertext Books Ltd., Aylesbury, U.K. 1973
3) Billmeyer, F.W.; "Textbook of Polymer Science" 2nd ed., Wiley-Interscience, New York 1970
4) Bowie, F.A., Winslow, F.H.; "Macromolecules. An Introduction to Polymer Science", Academic Press, New York 1979
5) Campbell, J.; "Introduction to Synthetic Polymers", Oxford University Press, Oxford, New York 1994
6) Vollmert, B.; "Grundriss der Makromolekularen Chemie", 2nd ed., B. Vollmert Press, Karlsruhe (FRG) 1982
7) Batzer, H., Lohse, F.; "Einführung in die makromolekulare Chemie", 2nd ed, Hüthig & Wepf Verlag, Basel, Heidelberg 1976
8) Rempp, P.; "Polymer Synthesis", Hüthig & Wepf Verlag, Basel Heidelberg 1986
9) Lehnere, M.D., Gehrke, K., Nordmeier, E.H.; "Makromolekulare Chemie", Birkhäuser Verlag 1993
10) Kricheldorf, H.R., Zang, Qu-Zb, Schwarz, G.; Polymer 1982 23 1821
11) Kricheldorf, H.R., Conradi, A.; J.Polym.Sci.Part A, Polym.Chem. 1987 25 489
12) Kricheldorf, H.R., Schwarz, G., Ruhser, F.; Macromolecules 1991 24 3485
13) Kricheldorf, H.R., Lübbers, D.; Makromol.Chem. Rapid Commun. 1991 12 691
14) Kricheldorf, H.R., Stuckenbrock, T.; Manuscript in preparation
15) Kricheldorf, H.R., Stöber, O., Lübbers, D.; Macromol.Chem. Phys. 196 3549 (1995)
16) Hawker, C.J., Lee, R., Fréchet, J.; J.Am.Chem.Soc. 1991 113 4583
17) Voit, B.I., Turnere, S.R.; Polym.Prepr. (Am.Chem.Soc.Polym. Chem. Div.) 1992 34 184

18) Turner, S.R., Voit, B.I., Mourey, T.H.; Macromolecules 1993 26 4617
19) Walter, F., Turner, S.R., Voit, B.I.; Polym.Prepr. (Am.Chem. Soc., Div.Polym.Chem.) 1993 34 79
20) Kricheldorf, H.R., Stöber, O., Lübbers, D.; Macromolecules 1995 28 2118
21) Kricheldorf, H.R., Stöber, O.; Macromol.Rapid Commun. 1994 15 87
22) Pochan, J.M., Beatty, C.L., Dochan, D.F.; Polymer 1979 20 879
23) Kricheldorf, H.R., Adebahr, T., Majidi A.M.; Macromolecules 1995 28 2112
24) Kricheldorf, H.R., Adebahr, T.; Makromol.Chem. 1993 194 2103
25) Kricheldorf, H.R., Löhden, G.; Macromol.Chem.Phys. 1993 196 1839
26) Kricheldorf, H.R., Löhden, G.; J.M.S.-Pure & Appl.Chem. A 32 1915 (1995)
27) Kricheldorf, H.R., Rieth, K.-H.; J.Polym.Sci., Polym. Letters Ed. 1978 16 379

RECEIVED January 25, 1996

Chapter 10

Design, Synthesis, and Properties of Dendritic Macromolecules

Craig J. Hawker and Wayne Devonport

Research Division, IBM Almaden Research Center, 650 Harry Road, San Jose, CA 95120–6099

The fundamental requirements for the synthesis of dendrimers and hyperbranched macromolecules is examined. Examples of the divergent and convergent approaches are presented and a comparison of both methods with respect to each other and also the one-step procedure for hyperbranched macromolecules is made. The structural similarities and differences between dendrimers and hyperbranched macromolecules are described and the effect of this on the physical properties of these novel three-dimensional materials is discussed. Finally, a comparison of these materials with linear polymers is examined.

As their name implies, dendritic macromolecules have a highly branched three-dimensional structure which in many ways mimic the structure of trees. From a central point the macromolecule branches outwards with the number of chain ends or terminal groups increasing as the molecular weight increases. It is the number and nature of these branch points which gives rise to the three-dimensional structure of dendritic macromolecules and to the primary difference between the three broad types of dendritic macromolecules which are recognized today. If the macromolecule has highly regular branching, which typically follows a strict geometric pattern, it is termed a dendrimer and these materials are frequently monodisperse and are prepared in a multi-step synthesis with purifications at each step. In contrast, if the branching is random and irregular, with significant amounts of failure sequences being present in the macromolecule, these materials are termed hyperbranched macromolecules. A consequence of this irregular structure is that these materials are typically as polydisperse as other step-growth polymers. The final type of dendritic macromolecule are hybrid dendritic-linear polymers in which a combination of dendritic and linear segments are covalently attached. While some extremely interesting and unusual macromolecular architectures are obtained in this

0097–6156/96/0624–0186$12.00/0

hybrid dendritic-linear area an analysis of these materials is beyond the scope of this review.

Synthesis of Dendrimers

One of the fundamental differences between dendrimers and hyperbranched macromolecules is that the former are prepared in a multi-step synthesis with purifications at each step while the latter are prepared in a one-step procedure. For dendrimers, two synthetic strategies have been developed for their preparation. The initial approach, which was independently conceived and demonstrated by Tomalia (1) and Newkome (2), has come to be known as the divergent, or starburst methodology. In this approach, growth is started at the central core of the dendrimer and proceeds radially outwards with the stepwise addition of successive layers of monomer building blocks. The synthesis of poly(amidoamine) (PAMAM) dendrimers by Tomalia demonstrates many of the important features that must be considered in any divergent synthesis. Scheme 1 shows the synthesis starts with reaction of a polyfunctional core molecule, in this case ammonia, with methyl acrylate to give the triester, **1**. Regeneration of the reactive NH groups is then accomplished by exhaustive amidation of the ester functionalities with 1,2-diaminoethane. This leads to a doubling in the number of reactive NH groups from three for ammonia, to six for the first generation dendrimer, **2**. Interestingly, while Tomalia used addition chemistry to regenerate his reactive chain ends, a number of other workers in the field have used a protection/deprotection strategy. Larger dendrimers result from the successive Michael additions and amidations, with purification procedures at each stage. Successful dendrimer synthesis using the Tomalia route can be acheived by attaining the following prerequisites: i) high yielding reactions must be used for each of the generation growth steps, this becomes especially crucial at higher generations when the number of reactive chain ends becomes extremely large ii) monomer units should be carefully selected to be readily available and, if possible, they should be symmetrical so that each of the reactive functional groups has the same reactivity iii) purification of the dendrimers must allow complete removal of excess monomers and/or reagents as these can lead to unwanted side reactions or growth in the next step of the synthesis.

Although the divergent growth approach has proven to be extremely successful for the preparation of dendrimers, it does have a number of drawbacks and limitations with regard to ideal dendrimer growth. This is especially true if only a single reactive functionality on the dendrimer is required or if accurate control over the placement of end groups at the periphery of the globular dendrimer is desired. To overcome these limitations an alternative synthetic strategy was developed for the synthesis of dendrimers. This approach was termed the convergent growth methodology and again it relies on a multi-step strategy involving step-growth or condensation chemistry (3,4).

Convergent methodology relies on the disconnection approach used in traditional organic synthesis coupled with the fractal and highly symmetrical nature of dendrimers. Disconnection of a dendrimer therefore eventually leads to the chain

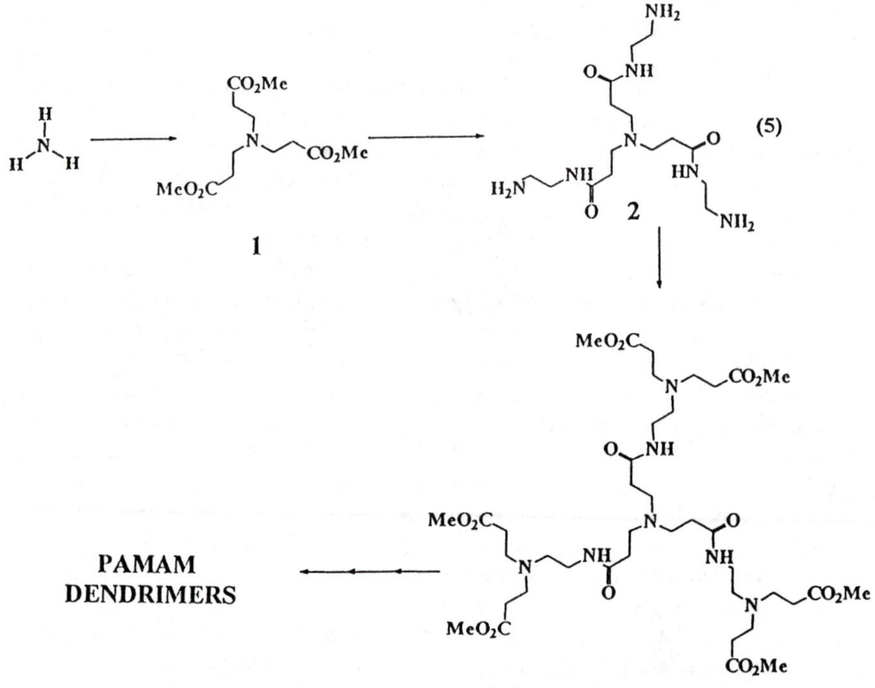

$$\text{NH}_3 \longrightarrow \mathbf{1} \longrightarrow \mathbf{2} \quad (5)$$

1: $N(CH_2CH_2CO_2Me)_3$

2: $N(CH_2CH_2CONHCH_2CH_2NH_2)_3$

PAMAM DENDRIMERS

Scheme 1. Synthesis of poly(amidoamine) dendrimers.

ends as the starting point in the synthesis, which is exactly the reverse of the divergent growth approach where synthesis starts at the central core. Convergent methodology is demonstrated in Scheme 2. Growth begins with the chain ends or "surface" functional group, S, and coupling with an AB_x building block, or monomer unit, **3**, leads to the next generation dendron, **4**. It should be noted that the value of x, which is the multiplicity of the branch point, has to be two or greater for dendrimer synthesis. Activation of the single latent functionality, P, which is present in **4** gives the reactive functional group, R, which can again be coupled with the monomer unit to give the next generation dendron, **5**. The dendritic structure can already be noted in this second generation dendron, **5**, where there are four chain ends and two inner layers of building blocks. Repetition of this two step procedure then leads to larger and larger dendrimers and, if desired, the final reaction step may be coupling of several reactive dendrons to a polyfunctional core, C, to give a dendritic macromolecule similar to that which might be obtained from a divergent strategy.

A special feature of the convergent growth approach to dendrimers should also be noted, while the number of chain ends doubles at each generation growth step only a single functional group is present at all times at the focal point of the growing dendrimer. An important consequence of this single focal point group is that generation growth requires only a single activation reaction and usually only two coupling steps. Therefore, the probability of side reactions and failure sequences is not only low but the dramatic difference in the size and polarity between the desired dendrimers and unreacted or partially reacted materials simplifies purification.

As has been shown above, while both the convergent and divergent methodologies involve a step-wise growth approach and, depending on the choice of building blocks, can give the same final dendritic structure there are some fundamental differences between the two approaches. These differences results in an almost complimentary relationship between the two approaches and a careful examination of the desired dendrimer should be undertaken before planning any dendrimer synthesis, since one approach may be more viable than the other. In general, the more controlled nature of the convergent approach is better suited for the synthesis of very regular or precisely functionalized dendrimers where the number and nature of the chain ends, "interior" building blocks, and the focal point group need to be controlled. In contrast, for very large dendrimers (MW > 100,000) the divergent approach is the methodology of choice due to the ability to use large excess of reagents and the minimization of steric hindrance to reaction.

Irrespective of these differences, both approaches have proved to be extremely successful for the preparation of dendritic macromolecules. By the use of either the divergent or convergent growth approaches a number of different groups have prepared dendrimers based on a wide variety of functional groups i.e. dendritic poly(amides) (5), poly(etherketones) (6), poly(amines) (7), poly(phenylenes) (8), poly(silanes) (9), poly(phosphonium salts) (10), poly(esters) (11), poly(phenylacetylenes) (12), poly(alkanes) (13), *etc*. Interesting variations on this theme have been the preparation of optically active dendrimers (14), dendritic poly(radicals) (15), and dendrimers based on co-ordination chemistry with

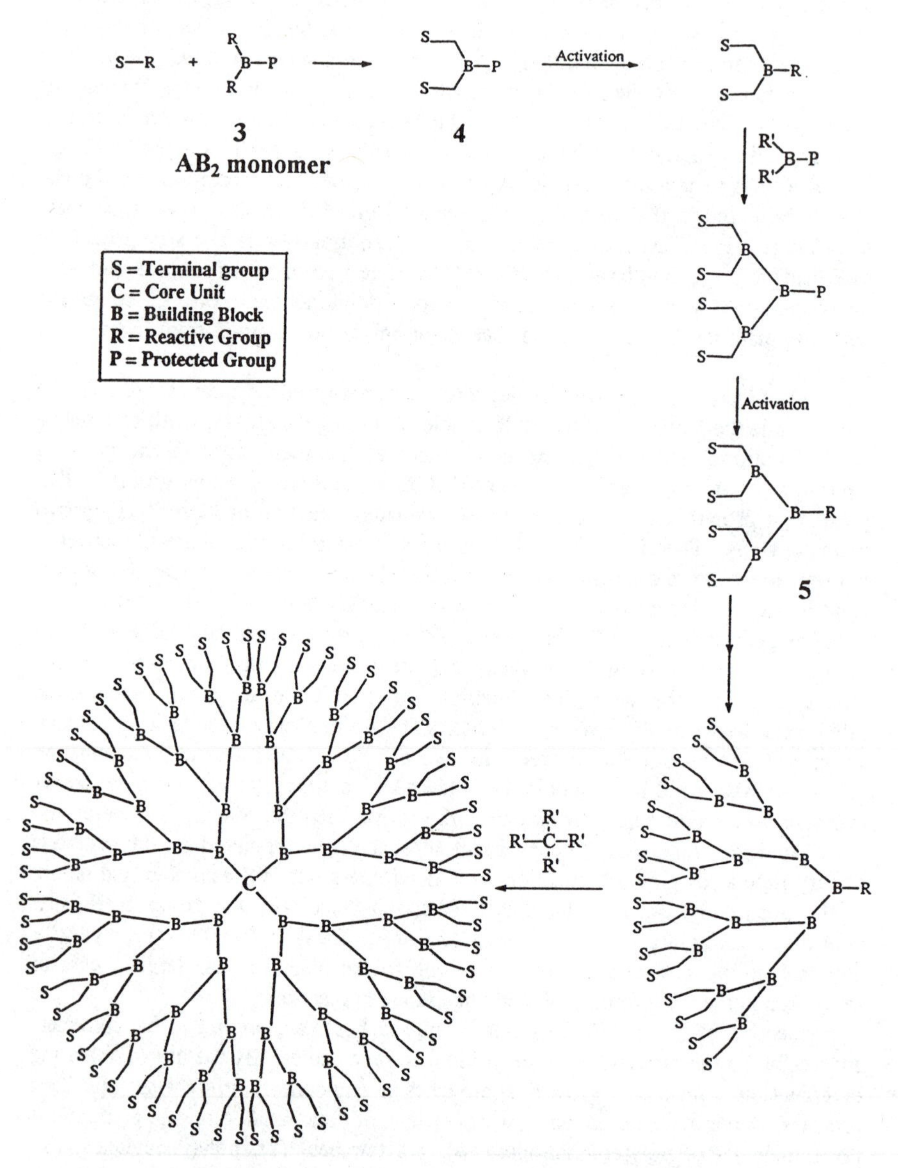

Scheme 2. Synthesis of dendrimers using convergent growth methodology.

transitional metals atoms as the branch points (16). From these studies it is obvious that the range of structures and building blocks that can be used to construct dendrimers is limited only by synthetic requirements and the researchers imagination.

Synthesis of Hyperbranched Macromolecules

The one-step synthetic strategy used to prepare hyperbranched macromolecules places some additional conditions on the structure of the AB_x monomer unit and the chemistry employed for polymerization when compared to dendrimers. Unlike dendrimers which are purified at each step and can therefore tolerate low yields or the presence of unreacted side products the requirements for hyperbranched systems are much more rigorous. Firstly, the reactive groups, A and B, should react with each other only after a suitable activation step (*e.g.* removal of a protecting group *via* a chemical, thermal, or photochemical process) or in the presence of a catalyst, otherwise the synthesis and handling of the monomer unit can be extremely complicated. Also the mutual reactivity of the functional groups, A and B, should be very high in order to achieve growth and polymer formation. This is similar to normal step growth polymerization requirements, though it should be noted that this same requirement need not be satisfied for dendrimers. Side reactions should also be kept to an absolute minimum in order to prevent crosslinking of the growing hyperbranched macromolecule which is aggravated by the presence of a large number of reactive B groups at the chain ends. Deactivation of the single reactive A group at the focal point of the growing hyperbranched macromolecule should also be kept to an absolute minimum since it would be expected to severely disturb the growth process and may lead to low molecular weight materials.

The one-step synthetic strategy used for the production of hyperbranched macromolecules results in significant structural differences between these two related families of globular macromolecules. As shown in Scheme 3, the one-step polymerization of an AB_2 monomer unit does not result in perfect growth, instead growth is uncontrolled and leads to a complex polydisperse product, **6**. This hyperbranched product, **6**, contains regions that resemble both dendrimers and linear polymers. In fact if the structure is analyzed in detail only three different types of individual sub-units can be identified. These have come to be know as dendritic, linear, and terminal sub-units and differ in the number of B functionalities that have undergone reactions to form polymeric linkages, C. Dendritic units are obtained by reaction of both B functionalities and resemble the internal building blocks present in dendrimers and contribute to branching. Alternatively, if only one of the two B functionalities react, a linear unit, analogous to that which would be found in a true linear polymer, is obtained. It is these failure sequences which constitute that main differences between hyperbranched macromolecules and dendrimers. Finally, if neither of the B functionalities react, a terminal unit is obtained which can be consider to be the equivalent of an outer layer, or chain end unit of a dendrimer. Therefore, the structure of hyperbranched macromolecules are intermediate between those of "perfect" dendrimers and linear polymers. These intermediate structures

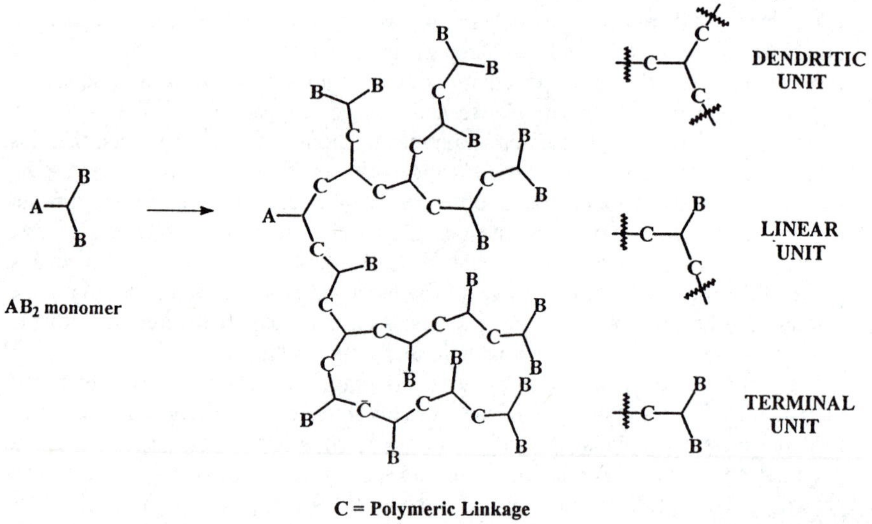

AB_2 monomer

DENDRITIC UNIT

LINEAR UNIT

TERMINAL UNIT

C = Polymeric Linkage

6

Scheme 3. Hyperbranched polymer architectures.

can be described by their "degree of branching", or DB. This has been defined by the formula:

$$DB = \frac{\text{number of terminal units} + \text{number of dendritic units}}{\text{total number of units}}$$

From this definition, a linear polymer has a degree of branching of 0.0, while a perfectly branched dendrimer has a degree of branching of 1.0 and a hyperbranched macromolecule has a DB between 0.0 and 1.0. For hyperbranched macromolecules the degree of branching can be determined by either NMR spectroscopy (17,18) or a degradative technique (19). Where DB has been determined, the value is typically between 0.45 and 0.60 though values as high as 0.85 and as low as 0.15 have been reported.

Concerted efforts to examine the synthesis and polymerization of such AB_x monomers was however not attempted until 1989 when Kim and Webster (17) reported their preparation of hyperbranched polyphenylenes from the AB_2 monomers, 3,5-dibromophenyl boronic acid, 7, or 3,5-dibromophenyl magnesium bromide, **8** (Figure 1). In a similar way to dendrimers, the field of hyperbranched macromolecules has been extended considerably since this initial report. Hyperbranched polyesters based on fully aromatic, fully aliphatic, or partially aliphatic/aromatic systems have been extensively studied by a number of groups (18-20). Similarly, hyperbranched poly(etherketones) (21), poly(ethers) (22), poly(urethanes) (23), poly(siloxanes) (24), *etc.* have been prepared using classical step growth chemistry.

Physical Properties

In contrast to the high level of maturity that the synthesis of dendrimers and hyperbranched macromolecules have reached the study of their physical properties is still only in its infancy. The results that have been reported have sparked considerable interest and are helping to build a general understanding on the behavior of these novel globular macromolecules. In general, it has been shown that thermal properties, such as glass transition temperature and thermal decomposition, of dendrimers and hyperbranched macromolecules are both essentially the same as their linear analogs (25). However, the solubility of dendrimers is significantly improved when compared to their linear analogs. Miller and Neenan have reported solubility enhancements of 10^5 for dendritic poly(phenylenes) when compared to linear poly(phenylene) (4). Similarly, hyperbranched macromolecules have enhanced solubilities when compared to linear analogs, though the magnitude of enhancement is slightly less than that for the dendrimer case. Presumably the solubility differences arise due to the highly branched, globular nature of dendrimers and hyperbranched macromolecules. Another consequence of this novel structure can be found in studies concerning both the intrinsic (26) and melt viscosity (27) of dendritic and hyperbranched macromolecules. A plot of log[intrinsic viscosity] vs log[molecular weight] gives a different curve for all three architectures (Figure 2). For dendrimers,

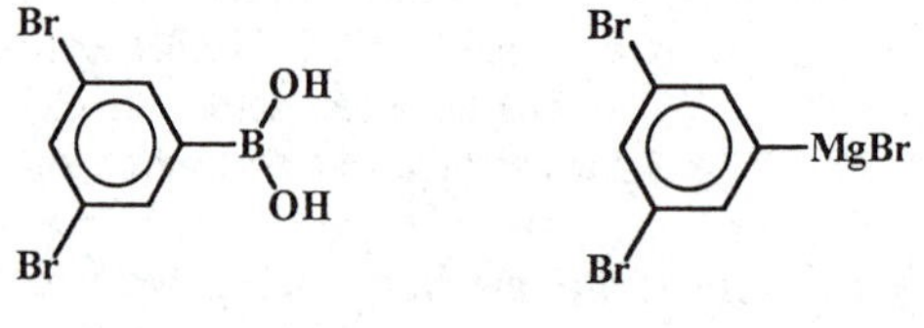

Br Br
OH
B —MgBr
OH
Br Br

7 8

Figure 1. Building blocks for the synthesis of polyphenylenes.

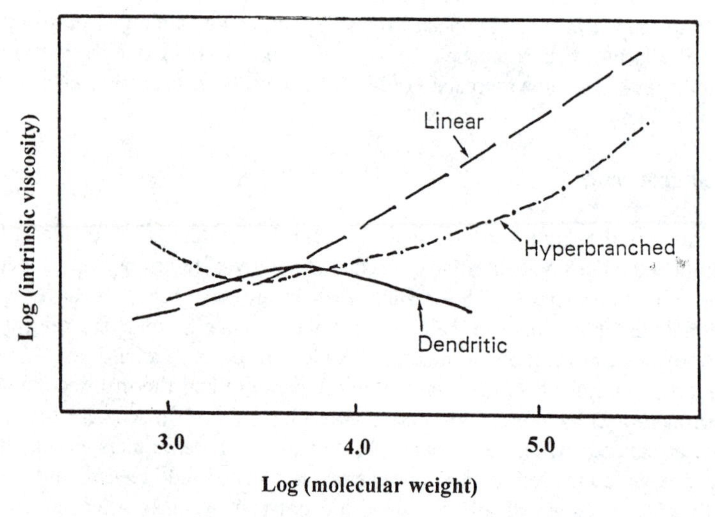

Figure 2. A plot of log[intrinsic viscosity] versus log[molecular weight].

a totally unique bell shaped relationship with a maximum below 5,000 amu is observed which does not obey standard theory. In contrast, hyperbranched macromolecules do follow the Mark-Houwink-Sakurada relationship, albeit with extremely low values of "a" when compared to linear polymers. Similarly, melt viscosity studies on dendrimers do not show the expected dramatic increase in viscosity on increasing molecular weight, instead, a linear relationship between viscosity and molecular weight is observed up to molecular weights of 100,000 amu. Hyperbranched macromolecules have also been shown to have very low melt viscosities when compared to linear polymers. This unique viscosity behavior can be explained by either standard geometrical considerations, if it is assumed that dendrimers have a spherical shape, or by a significant reduction, or lack, of entanglements for dendrimers and hyperbranched macromolecules. Other interesting physical properties found for dendrimers and hyperbranched macromolecules are enhanced compatibility with other polymers (28), high chemical reactivity and catalytic activity of chain end functional groups (29), as host-guest systems for drug delivery or molecular inclusion studies (30), *etc.*

Conclusions

In the short space of 10 years the field of dendritic macromolecules has progressed at an extremely rapid pace with these globular materials effectively constituting a new branch of polymer science. While the synthesis of these materials draws heavily from standard step growth literature, the degree of control present in dendrimers and, to a lesser extent, in hyperbranched systems is unparalleled. This key feature, coupled with their globular, three-dimensional structure allows an extremely large range of novel macromolecular architectures to be prepared. From these materials unique structure property relationships can be defined and in many instances either enhanced or new physical properties are found for the pure dendritic macromolecules or for blends of these materials with commodity linear polymers.

Literature Cited

[1] Tomalia, D.A.; Baker, H.; Dewald, J.; Hall, M.; Kallos, G.; Martin, R.; Ryder, J.; Smith, P. *Polym. J.*, **1985**, *17*, 117.

[2] Newkome, G.R.; Yao, Z.; Baker, G.R.; Gupta, V.K. *J. Org. Chem.*, **1985**, *50*, 2003.

[3] Hawker, C.J.; Fréchet, J.M.J. *J. Am. Chem. Soc.*, **1990**, *112*, 7638.

[4] Miller,T.M.; Neenan, T.X. *Chem. Mater.*, **1990**, *2*, 346.

[5] Uhrich, K.E.; Fréchet, J.M.J. *J. Chem. Soc. Perkin Trans.* I, **1992**, 1623.

[6] Morikawa, A.; Kakimoto, M.; Imai, Y. *Macromolecules*, **1993**, *26*, 6324.

[7] de Brabander-van den Berg, E.M.M.; Meijer, E.W. *Angew. Chem. Int. Ed. Engl.*, **1993**, *32*, 1308.

[8] Miller, T.M.; Neenan, T.X.; Zayas, R.; Bair, H.E. *J. Am. Chem. Soc.*, **1992**, *114*, 1018.

[9] van der Made, A.W.; van Leewen, P.W.N.M.; de Wilde, J.C.; Brandes, R.A.C. *Adv. Mater.*, **1993**, *5*, 466.
[10] Rengan, K.; Engel, R. *J. Chem. Soc., Perkin Trans. 1*, **1991**, 987.
[11] Hawker, C.J.; Fréchet, J.M.J. *J. Am. Chem. Soc.*, **1992**, *114*, 8405.
[12] Xu, Z.F.; Kahr, M.; Walker, K.L.; Wilkins, C.L.; Moore, J.S. *J. Am. Chem. Soc.*, **1994**, *116*, 4537.
[13] Newkome, G.R.; Moorefiled, C.N.; Baker, G.R.; Saunders, M.J.; Grossman, S.H. *Angew. Chem. Int. Ed. Engl.*, **1991**, *30*, 1178.
[14] Seebach, D.; Lapierre, J.M.; Skobridis, K.; Greiveldinger, G. *Angew. Chem. Int. Ed. Engl.*, **1994**, *33*, 440.
[15] Rajca, A.; Utamapanya, S. *J. Am. Chem. Soc.*, **1993**, *115*, 10688.
[16] Campagna, S.; Denti, G.; Serroni, S.; Ciano, M.; Juris, A.; Balzani, V. *Inorg. Chem.*, **1992**, *31*, 2982.
[17] Kim, Y.H.; Webster, O.W. *J. Am. Chem. Soc.*, **1990**, *112*, 4592.
[18] Hawker, C.J.; Lee, R.; Fréchet, J.M.J. *J. Am. Chem. Soc.*, **1991**, *113*, 4583.
[19] Kambouris, P.; Hawker, C.J. *J. Chem. Soc., Perkin Trans.* 1, **1993**, 2717.
[20] Johansson, M.; Malmstrom, E.; Hult, A. *J. Polym. Sci., Polym. Chem.*, **1993**, *31*, 619.
[21] Chu, F.; Hawker, C.J. *Poly. Bull.*, **1993**, *30*, 265.
[22] Uhrich, K.E.; Hawker, C.J.; Fréchet, J.M.J.; Turner, S.R. *Macromolecules*, **1992**, *25*, 4583.
[23] Spindler, R.; Fréchet, J. M. J. *Macromolecules*, **1993**, *26*, 4809.
[24] Mathias, L.J.; Carothers, T.W. *J. Am. Chem. Soc.*, **1991**, *113*, 4043.
[25] Wooley, K.L.; Fréchet, J.M.J.; Hawker, C.J. *Polymer*, **1994**, *35*, 4489.
[26] Mourey, T.H.; Turner, S.R.; Rubenstein, M.; Fréchet, J.M.J.; Hawker, C.J.; Wooley, K.L. *Macromolecules*, **1992**, *25*, 2401.
[27] Hawker, C.J.; Farrington, P.; Mackay, M.; Fréchet, J.M.J.; Wooley, K.L. *J. Am. Chem. Soc.*, **1995**, *117*, 6123.
[28] Kim, Y.H.; Webster, O.W. *Macromolecules*, **1992**, *25*, 2501.
[29] Wooley, K.L.; Fréchet, J.M.J.; Hawker, C.J. *Polymer*, **1994**, *35*, 4489.
[30] Jansen, J.F.; de Brabander van den Berg, E.M.; Meijer, E.W. *Science*, **1994**, *266*, 1226.

RECEIVED December 6, 1995

Chapter 11

Adamantane-Containing Polymers

Lon J. Mathias, Jennifer J. Jensen, Veronica T. Reichert, Charles M. Lewis, and Gordon L. Tullos

Department of Polymer Science, University of Southern Mississippi, Southern Station 10076, Hattiesburg, MS 39406–0076

Adamantane is a rigid ring system comprised of three fused chair conformation cyclohexane rings.[1] Its excellent thermal stability, bulkiness, and tetrahedral geometry lead to improved physical properties such as stiffness, glass transition temperature (T_g), and solubility.[2] Star polymers include polyaramids[3,4] and polybenzoxazoles[5] based on adamantane and biadamantane. All hydrocarbon three-dimensional networks use acetylene[6], phenylacetylene[7], or diphenylacetylene[7] groups attached at the bridgehead positions with thermal polymerization or nickel catalyzed reactions of iodophenyl groups attached directly to the bridgehead positions.[4] Pendent adamantane groups were incorporated into acrylates[8], phenolics[9], poly(phenylenes)[10], and poly(ether ether ketones).[11] Each type shows a large increase in T_g and thermal properties over the linear, unsubstituted polymer.

Star Polymers

Aramids are high performance rigid rod materials such as Kevlar® and Nomex®. Benzoxazoles are even more rigid materials and are noted for their excellent thermooxidative stability and chemical resistance.[5] Both types of polymers are processed from lyotropic solutions into fibers or films which exhibit high moduli and tensile strengths due to the high degrees of chain orientation.[5] Several researchers have looked at incorporating the rigid rod materials as arms on a star polymer, the intent being to improve properties along the transverse direction of the fibers and films.[5] The use of adamantane as a rigid core enforces a tetrahedral arrangement of the aramid and benzoxazole arms. These star polymers exhibit comparable thermal stability and decreased viscosity over the linear polymer. The aramids are interesting in that they don't behave as rigid rods in solution and are more soluble than linear aramids of comparable molecular weight.[3] The impetus for examining adamantane as the core for rigid rod star polymers is to study its effect on polymer viscosity,

0097–6156/96/0624–0197$12.00/0

thermal behavior, and ultimate physical properties. Some examples of aramid star polymers were synthesized in our lab (**1**)[3] or elsewhere (**2**)[4]. In these and subsequent figures, 4-armed species are represented by the structures drawn, ie, each bridgehead contains one arm. The star polymers shown were synthesized using either 1,3,5,7-tetrakis(4-iodophenyl)adamantane or 1,3,5,7-tetrakis(4-aminophenyl)adamantane as the core, and 4-iodoaniline as the monomer. Polymerization occurs via a palladium catalyzed carbonylation reaction as described by Perry and coworkers.[12]

Incorporation of adamantane was confirmed by FTIR, solid state ^{13}C NMR, and solution ^{13}C NMR. It was found that the amine core star polymers were higher molecular weight than the iodo core star polymers. According to previous observations, an excess in amine groups, which is the case for the amine core star polymers, tends to favor higher molecular weight polymers, possibly due to a decrease in side reactions.[12] A linear polybenzamide (PBA) of comparable molecular weight was also synthesized for comparison. Plots of inherent viscosity and reduced viscosity versus concentration for the star polymers gave positive slopes which indicate the star polymers do not behave as rigid rods in sulfuric acid solutions, in contrast to linear aramids. All three of the polymers were soluble in DMAc-5% LiCl as well as in concentrated sulfuric acid. Films of star$\mathbf{NH_2}$ were cast from dilute solutions in DMAc-5% LiCl; concentrated solutions formed gels, probably due to strong interaction between the arms resulting in increased entanglements. Star**I** and the linear PBA did not form films or gels, probably due to lower molecular weights and the absence of entanglement. Table I gives the properties of aramid star and linear polymers studied.

Table I. Properties of Aramid Star and Linear Polymers

Polymer	$[n]$ (dL/g)[a]	T_d(°C)[b]	transition (°C)
linear	0.24	450	none
star**I**	0.27	450	355
star$\mathbf{NH_2}$	0.74	445	none

[a] 0.25 g/dL of H_2SO_4, 30 °C. [b] 10% weight loss in air.

Compound **3** is an example of a star polymer with benzoxazole arms.[5] The three dimensional rigid-rod polymer was obtained from the polycondensation of 4-[5-amino-6-hydroxybenzoxazol-2-yl]benzoic acid (ABA) with 1,3,5,7-tetrakis(4-carboxylatophenyl)adamantane (TCBA) in polyphosphoric acid (PPA). Polymer structures were confirmed by FTIR and elemental analysis, although the extent of attachment of PBO arms to the adamantane core was not confirmed. In addition, stir opalescence was observed indicating lyotropic-like behavior. However, the more compact star structure resulted in intrinsic viscosities that were significantly less than linear PBO's obtained under identical conditions. The onset for weight loss under TGA in air occurred at 500 °C for both linear and star polymers and no transitions were observed by DSC.

The impetus of the star benzoxazole research was to utilize two and three dimensional cores for rigid-rod polymers in hopes of increasing compressive strength over the linear polymers.[13] Fibers were prepared via a dry-jet wet spinning process at 90 °C using various spin-draw ratios. The mechanical properties of drawn fibers indicated a slight increase in compressive strength of the 3D rigid-rod polymers based on adamantane over the linear polymer but a decrease in tensile strength and modulus. Morphology studies obtained from wide angle x-ray diffraction (WAXD) measurements revealed more disorder in the 3D samples compared to the linear and 2D samples. Perhaps the inability of the rigid-rod arms to pack laterally gives rise to the decrease in mechanical properties observed in the 3D rigid-rod polymers as compared to linear rigid-rod polymers of comparable molecular weight.

The last star step-growth polymers to be discussed are hyperbranched aramids.[3] The first hyperbranched aramid (**4**) was synthesized using 1,3,5,7-tetrakis(4-iodophenyl)adamantane as the core, 3,5-Dibromoaniline was polymerized from this using the palladium catalyzed carbonylation reaction discussed previously. A two-step carbonylation process was used in which higher pressures of CO were used at the beginning of the reaction to insure complete displacement of the very reactive iodo groups and insertion of the carbonyl to form the first amide linkage at the core. Polymerization (and formation of hyper-branches) through the less reactive bromo groups was achieved by lowering the CO pressure. For comparison, a hyper-branched aramid (**5**) was synthesized from 3,5-dibromoaniline using the carbonylation reaction.

Polymer **5** is structurally similar to the branched aramids reported, except that **5** (as well as **4**) should have bromine end-groups instead of acid or amine moieties.[14] The previously reported branched aramid systems were soluble in amide solvents such as DMF, DMAc, and NMP. Polymer **5** was found to be insoluble in these solvents as well as insoluble in DMAc-5% LiCl and sulfuric acid. This may be due to cross-linking through an unknown coupling of the reactive intermediates and end-groups. Solid-state NMR was used to confirm the overall product structure. Aramid carbonyls were present at ca 165 ppm indicating reaction had occurred. Peaks for residual DMAc and/or CH_2Cl_2 were present in the NMR spectra even after extensive extraction and vacuum drying, implying an inherent tenacity for solvent retention in these star polymers.

Polymer **4** was partially soluble in DMSO and amide solvents such as DMF and DMAc. Solution and solid-state NMR were used to confirm the product structure. Peaks due to the adamantane moiety and aramid carbonyls were present indicating reaction. Again, residual solvent peaks were observed in the solid-state NMR spectra, even after extensive extraction and vacuum drying. It appears that hyperbranched structures and adamantane incorporation disrupts crystalline packing and leads to a more open and molecularly-porous branched structure that is capable of taking up and holding solvent.

Thermal analysis of these two highly branched polymers showed no transitions below 500 °C by DSC, although gradual changes in the base line occurred above 310 and 340 °C, respectively, corresponding to the onset of weight loss seen in the TGA scans. Thermal decomposition began ca 80 °C lower and was more rapid above 350 °C, than for linear and star-branched polymers discussed previously. Perhaps the presence of one unreacted aryl bromide group per repeat unit (on average) may thermolyze or promote backbone degradation, leading to decreased thermal stability.

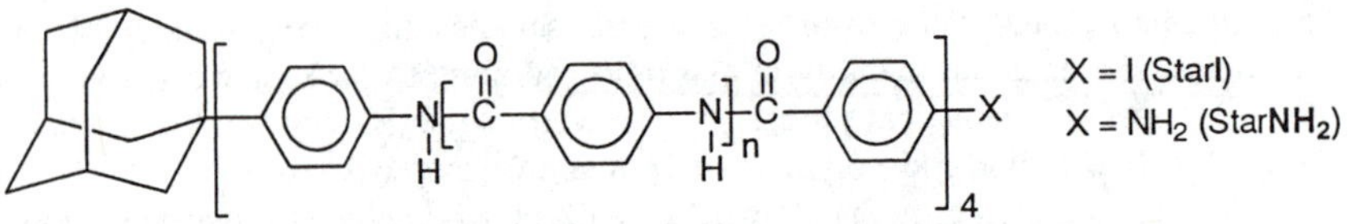

X = I (StarI)
X = NH_2 (StarNH_2)

Figure 1. Structures of Star I and Star NH_2.

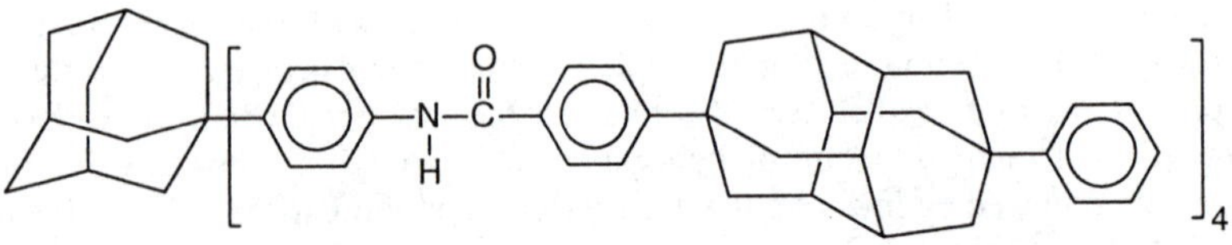

Figure 2. Star aramid structure.

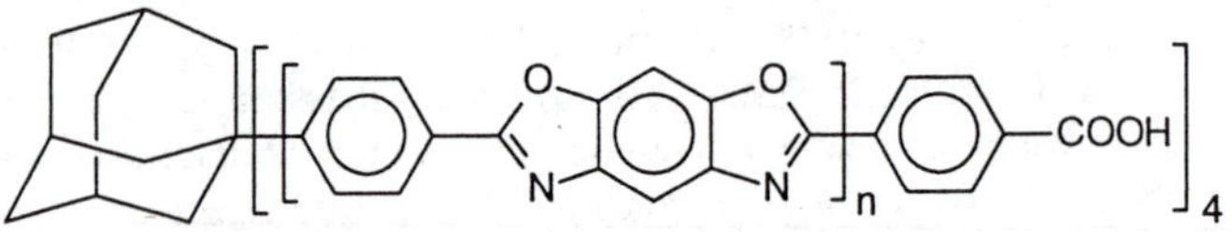

Figure 3. Star poly(benzoxazole) general structure.

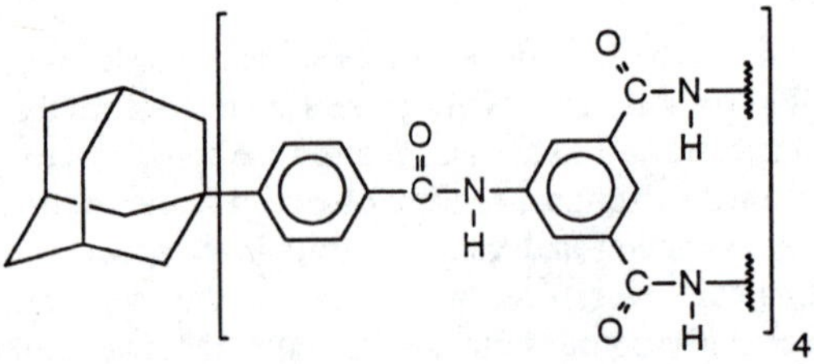

Figure 4. Hyperbranched aramid (adamantane core).

In all cases, it appears that the adamantane core may enhance solubility which promotes the formation of high molecular weight polymers without sacrificing thermal stability. However, in the case of highly branched aramids, the presence of adamantane units as defects to prevent close packing did not maintain solubility (probably due to crosslinking) and actually resulted in a decrease in the thermal stability of the polymer obtained. Lastly, adamantane has been demonstrated as a good candidate for step-growth polymerization of star polymers.

Three Dimensional Hydrocarbon Networks

There has been a lot of research recently devoted to the formation of all-hydrocarbon materials. They offer advantages over heteroatom containing materials in long term thermal and environmental stability, in possessing low dielectric constant, and having low water absorption. Also, the strong carbon-carbon bonds should lead to improved moduli, strength, and toughness. Adamantane is an ideal choice for use in such materials in that it is all hydrocarbon and its tetrahedral geometry allows for growth of hydrocarbon moieties in three dimensions. Adamantane has been functionalized with terminal acetylene[6] (7), phenylacetylene[7], and diphenylacetylene[7] (**6**) groups. Upon thermal cure, these materials undergo dimerization, oligomerization, and polymerization. The cured products exhibit excellent thermal stability and potential for forming "diamond" like structures.

Structure **6** gives some examples of hydrocarbon networks synthesized in our lab. Utilizing 1,3,5,7-tetrakis(4-iodophenyl)adamantane as the starting material, both 1,3,5,7-tetrakis(4-ethynylphenyl)adamantane (TEPA) and 1,3,5,7-tetrakis(4-phenylethynylphenyl)adamantane (TPEPA) were synthesized. The acetylene groups in TEPA were found to be quite stable, and no special precautions were taken with storage and handling of this compound. FTIR, elemental analysis, and solid state ^{13}C NMR confirmed the structures. TEPA was cured for 1 h at 200 °C, 1 h at 250 °C and 30 min at 300 °C. Disubstituted acetylenes require higher curing temperatures; therefore, TPEPA was heated at 300 °C for 30 min, 350 °C for 30 min, and 400 °C for 1 h.

Curing was followed by FTIR and solid state ^{13}C NMR, with disappearance of the acetylene stretches in FTIR and chemical shifts in NMR spectra indicate complete reaction. TGA in air of the cured materials revealed onset of decomposition at 450°C and 490 °C, respectively. This project also involved investigation of the cure mechanism of the acetylene groups. It is thought that identification of the cure products may give insight into the structure-property relationship for these cured polymers. Terminal acetylenes can cure by cyclotrimerization, biradical mechanisms, Glaser coupling, and Straus coupling.[15] Solid state ^{13}C NMR of partially cured TEPA revealed the formation of an acetylene-containing intermediate, while the fully cured product appears to contain only adamantane and aromatic structures. The cure products of disubstituted acetylenes have not been clearly identified yet but are thought to consist of a complex mixture of components. Solid state ^{13}C NMR spectra of the fully cured TPEPA consisted of resonances due to aromatic structures and adamantane. Dipolar dephasing experiments revealed non-protonated carbons on adamantane and the phenyl attached to it, and overlapping peaks due to non-protonated carbons resulting from cure. These data suggest that an all-aromatic and/or extended polyacetylene structure formed that contains adamantane as the core.

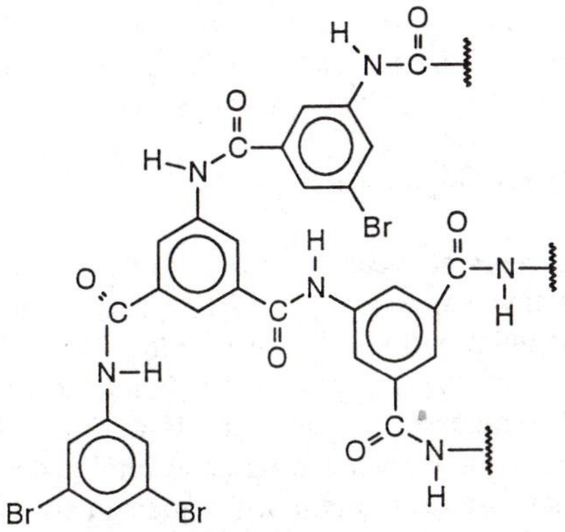

Figure 5. Hyperbranched aramid with random branching.

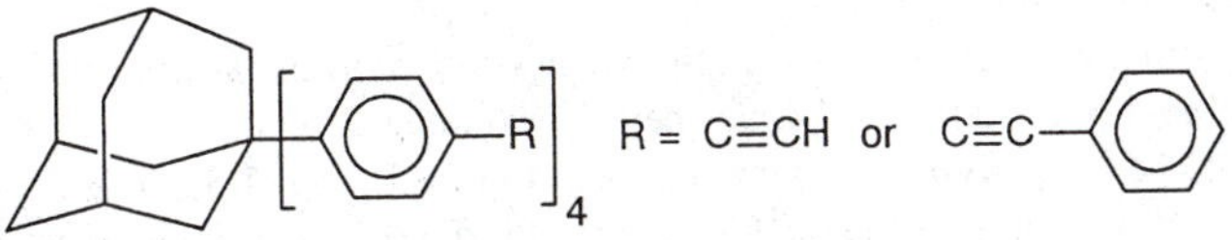

R = C≡CH or C≡C–Ph

Figure 6. 1,3,5,7-tetrakis(4-ethynylphenyl)adamantane and 1,3,5,7-tetrakis(4-phenylethynylphenyl)adamantane.

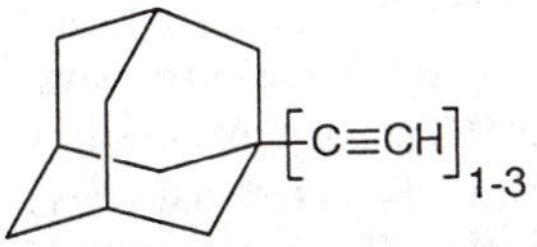

Figure 7. Multi-ethynyl substituted adamantane

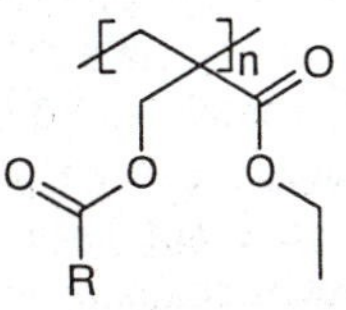

R = H, CH_3, C_6H_5, t-butyl, adamantyl

Figure 8. Ester derivatives of ethyl α-hydroxymethylacrylate.

Compound **7** is an example of adamantane with directly attached acetylene groups.[6] The monomer was prepared in three steps from adamantane. Complete cure was accomplished by heating above 320 °C. The onset of decomposition for the cured materials was 475°C in air and helium. Thermal polymerization gave a complex mixture of oligomers with number-average molecular weights ranging from 184 to 10,000 as measured by reverse phase HPLC. ^{13}C NMR spectra revealed formation of a linear polyene structure containing adamantane with terminal acetylene groups.

The exact mechanism of acetylene group thermal polymerization is still not clear. It is evident that the adamantane ring stays intact upon polymerization and contributes to the high thermal stability of the cured hydrocarbon networks. We hope to use ^{13}C labeled monomers to further investigate cure mechanism details.

Polymers with Adamantyl Pendent Groups

The bulky rigid structure of adamantane greatly increases T_g's when incorporated as a pendent group onto polymer backbones. Intuitively, the attachment of a large group pendent from the polymer should reduce chain mobility and raise T_g; however, what is surprising with the adamantyl moiety is the magnitude of its effect. Adamantane also maintains thermooxidative stability because, unlike a typical linkage to a tertiary carbon, (for example a t-butyl ester), elimination cannot easily occur at the bridgehead position which is the point of attachment. Our group is currently examining a whole range of adamantane-substituted polymers in order to understand how changes in molecular mobility and packing behavior affect thermal and mechanical properties

As a basis for reference, a series of chain-growth polymers based on ester derivatives of ethyl α-hydroxymethylacrylate were synthesized which illustrate the adamantyl effect. Polymers of structure **8** are examples of acrylates with pendant ester groups ranging from hydrogen (formate) to adamantane (adamantate).[8] Table II shows the series of polymers including number-average molecular weights and T_g. As the size of the pendent group increased from methyl (acetate), to t-butyl, to phenyl (benzoate), and finally to adamantane, the T_g values increased with values of 49, 100, 130, and 214 °C, respectively.

Table II. Molecular weights (SEC determined) and T_g's of ester derivatives of ethyl α-hydroxymethyl acrylate

Polymer	Mn (/1000)	T_g
acetate	610	49
acetate - t-butyl ester	225	100
benzoate	112	130
adamantate	319	214

Recent work in our group has centered on step-growth polymerization using several new monomers with pendant adamantyl groups. The monomers were synthesized by alkylating phenolics and halobenzenes with 1-bromoadamantane. The monomers were synthesized in high yields and the phenolic starting materials require no catalyst for alkylation.

The first example is 4-(1-adamantyl)phenol which is obtained by simply heating 1-bromoadamantane in phenol. The monomer, along with formaldehyde and acid catalyst, were used in the step-growth polymerization to novolacs. Novolacs are thermoplastic resins with molecular weights up to 2000 and T_g's of 45 - 70 °C.[16,17] Their synthesis is usually carried out at a molar ratio of 1 phenol to 0.75 - 0.85 formaldehyde, resulting in a linear or slightly branched product.[16,17] This study explored the use of variations and amounts of formaldehyde and reaction times in the synthesis of adamantyl-substituted phenolic polymers (**9**).[9] The polymers showed number- average molecular weights of ca 3000. Although the molecular weights are low, they seem to correlate well with the molecular weight limits suggested in the literature.[16,17] The T_g's observed by DSC ranged from 175 to 230 °C compared to 45-70 °C for the unsubstituted analogs. The polymers exhibited a 10% weight loss at 400 °C in nitrogen as measured by TGA. A summary of pendent adamantyl step-growth polymers is given in Table III.

Another pendent adamantyl step-growth polymer exhibiting large increases in T_g was synthesized from adamantyl-substituted resorcinol and 4,4'-difluorobenzophenone to form a poly(ether ether ketone) (PEEK).[11] The monomer was obtained by heating resorcinol with 1-bromoadamantane. Polymer **10** had a number-average molecular weight of 55,000 (SEC values relative to PSt), so the apparent steric hindrance of the adamantyl group did not prevent polymerization or decrease the rate of polymerization relative to that of unsubstituted monomers. The polymer had a T_g of 235 °C, which is a 115 °C increase over the unsubstituted analog. The polymer was readily soluble in common organic solvents and exhibited a 5% weight-loss at > 490 °C.

A series of PEEK copolymers were synthesized using varying compositions of adamantyl-substituted resorcinol and hydroquinone. Glass transition temperatures varied linearly with composition, suggesting random copolymer formation. Thin films were easily formed by solution casting or melt pressing at 250 - 270 °C and were off-white to tan in color. Films cast from 100% adamantyl-substituted resorcinol were transparent and strong, but brittle. Films melt-pressed with copolymers containing 50% adamantyl-substituted resorcinol were transparent, tough and flexible.

The last step-growth polymerization to be discussed involves coupling of aryl halides using a $NiCl_2$/Zn/triphenylphosphine system to form a polyphenylene. Many polyphenylenes, especially poly-p-phenylene (PPP), are difficult to synthesize due to lack of solubility of the polymer or difficulty in achieving the conversion necessary to develop high molecular weight.[18] The impetus of this research is to explore the synthesis of various meta substituted polyphenylenes, the goal being to solubilize the growing polymers in hopes of achieving high molecular weight. The monomer was obtained in high yield by the alkylation of 1,3-dichlorobenzene with 1-bromoadamantane in the presence of a Lewis acid to form 1,3-dichloro-5-(1-admantyl)benzene. This particular reaction is unique in that greater than 90% substitution occurs in the meta position contrary to typical electrophilic aromatic substitution for the ortho- and para- directing chlorine substituents. Unfortunetly, the adamantyl substituted poly-m-phenylene (**11**) precipitated from the DMAc or NMP

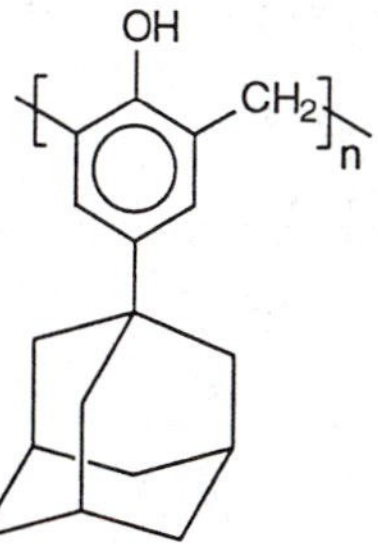

Figure 9. Adamantyl-substituted phenolic polymer.

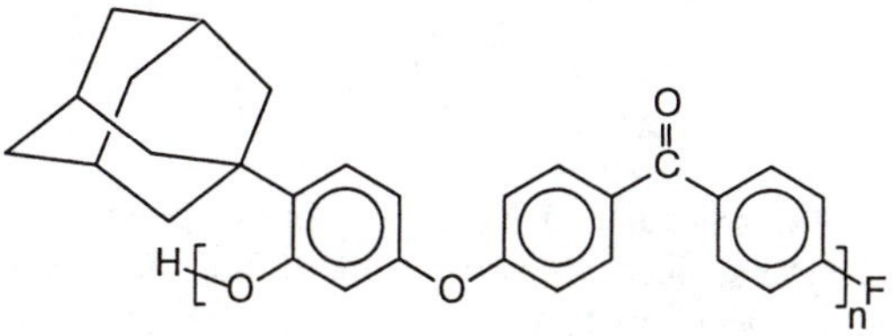

Figure 10. Adamantyl-substituted poly(ether ether ketone).

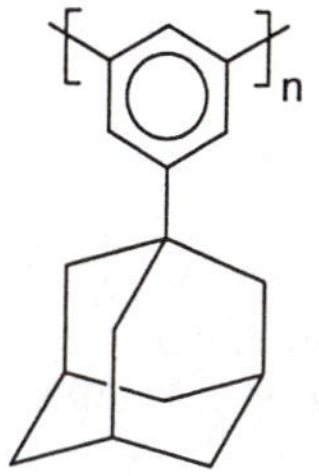

Figure 11. Adamantyl-substituted poly(phenylene).

solvent during polymerization; ie, the polar aprotic solvents necessary to generate the nickel catalyst *in situ* are not good solvents for the all-hydrocarbon polymer. The polymer was soluble in chloroform and THF and had a number-average molecular weight (SEC values relative to polystyrene) of only ca 2000. No thermal transitions were observed before the onset of decomposition at 350 °C.

Table III. Summary of pendent adamantyl polymers

Polymer	Mn[a]	T_d (°C)	T_g (°C)	T_g increase[e] (°C)
phenolic	3000	400[b]	175 - 230	160 - 185
PEEK	55,000	490[c]	235	115
poly-m-phenylene	2000	350	nd[d]	na

[a] relative to polystyrene. [b] 10% weight loss (nitrogen). [c] 5% weight loss (air).
[d] none detected before onset of decomposition.
[e] increase over linear unsubstituted analogs.

Conclusions

The incorporation of adamantane into various types of polymer shows some promise for thermal property enhancement without sacrificing processability. The overall approach is twofold; on the one hand, to use the tetrahedral geometry of adamantane to enforce three dimensional structures, and on the other to use bulky pendant adamantanes to promote solubility, maintain thermal stability, and raise glass transition temperatures. We believe both approaches offer valuable methods for modifying physical and mechanical properties of commercially important polymers, and we hope to continue to explore such opportunities.

Acknowledgements

This work was supported in part by grants from the Office of Naval Research, donors to the Petroleum Research Fund (administered by the American Chemical Society) and the National Science Foundation (DMR-9111903)

References

1. Fort, R. F. In *Adamantane: The Chemistry of Diamond Molecules*; Gassman, P. G., Ed.; Studies in Organic Chemistry; Vol. 5, Marcel Dekker: New York, 1976, Chap. 1, p. 1.
2. Khardin, A. P.; Radchenko, S. S. *Russ. Chem. Rev.* **1982**, *51*, 272.
3. Reichert, V. R.; Mathias L. J. *Macromolecules*, **1994**, *27*, 7024.
4. Ortiz, R. Ph.D. Dissertation, University of California, Los Angeles, 1993.
5. Dotrong, M.; Dotrong, M. H.; Moore, G. J.; Evers, R. C. *Polym. Prepr. (Am. Chem. Soc. Div. Polym. Sci.)*, **1994** *35(2)*, 673.

6. Archibald, T. G.; Malik, A. A.; Baum, K. *Macromolecules*, **1991**, *24*, 5261.
7. Reichert, V. R.; Mathias L. J. *Macromolecules*, **1994**, *27*, 7030.
8. Avci, D.; Kusefoglu, S. H.; Thompson, R. D.; Mathias, L. J. *Polym. Prepr. (Am. Chem. Soc., Div. Polym. Sci.)*, **1994**, *35(2)*, 655.
9. Jensen, J. J., Grimsley M., and Mathias, L. J. *J. Polym. Sci., Polym. Chem. Ed.*, in press.
10. Mathias L. J.; Tullos, G. L. submitted for publication.
11. Mathias L. J.; Lewis, C. M. submitted for publication.
12. Turner, R. S.; Perry, R. J.; Blevins R. W. *Macromolecules* **1992**, *25*, 4819. Perry, R. J.; Wilson, B. D. *Macromolecules* **1993**, *26*, 1503. Perry, R. J.; Turner, R. S.; Blevins, R. W. *Macromolecules* **1993**, *5*, 4. Perry, R. J. *CHEMTECH* **1994**, *24*, 18.
13. Dotrong, M.; Dotrong, M. H.; Moore G. J.; Evers, R. C. *Polym. Prepr. (Am. Chem. Soc., Div. Polym. Sci.)* **1994**, *35(2)*, 673.
14. Kim, Y. H. *Polym. Prepr. (Am Chem. Soc. Div. Polym. Chem)* **1993**, *34*(1), 56. Kim, Y. H. *J. Am. Chem. Soc.* 1992, *114*, 4947.
15. Swanson, S. A.; Fleming, W. W.; Hofer, D. C. *Macromolecules*, **1992**, *25*, 582.
16. Kopf, P. W.; Little, A. D. *Encyclopedia of Polymer Science and Engineering*; Vol. 11, John Wiley and Sons, Inc., New York, 1988, pp. 51-52, 70.
17. Kopf A.; Pilato L. A. *Phenolic Resins: Chemistry, Applications and Performance, Future Directions*, Springer-Verlag, Berlin Heidelberg, 1985 chap. 4, p. 63, chap. 5, p. 91.
18. Chaturvedi, V.; Tanaka, S.; Kaeriyama, K. *Macromolecules* **1993**, *26*, 2607.

RECEIVED December 4, 1995

Poly(aryl ether) Synthesis

Chapter 12

Poly(aryl ether) Synthesis

Jeff W. Labadie[1], James L. Hedrick[2], and Mitsuru Ueda[3]

[1]Argonaut Technologies, Inc., 887 Industrial Road, San Carlos, CA 94070
[2]Research Division, IBM Almaden Research Center, 650 Harry Road, San Jose, CA 95120–6099
[3]Department of Materials Science and Engineering, Yamegata University, 4–3–16 Jonan, Yonezawa, Yamagata 992, Japan

Poly(aryl ethers) are an important class of commercial polymers and are a member of the family of materials referred to as engineering thermoplastics *(1)*. Commercial examples include Amoco's poly(aryl ether sulfone) (Udel), ICI's poly(ether ether ketone) (PEEK), and General Electric's Ultem poly(ether imide). They display an attractive balance of properties such as relatively low cost, good processability, excellent chemical resistance, high thermal stability and good mechanical properties. Since the initial report of their synthesis by nucleophilic aromatic displacement polymerization of activated aryl dihalo compounds with bisphenolates *(2)*, significant effort has been devoted towards these polymer systems *(3)*. It is the purpose of this article to review many of the latest developments in the field of poly(aryl ether) synthesis, including mechanistic results, new activating groups and polymer structures, and alternative synthetic routes.

The most commonly used synthetic route to poly(aryl ether)s involves generation of an ether linkage by nucleophilic aromatic substitution (S_NAr) as the polymer-forming reaction (Scheme I). Early work focused primarily on sulfones and ketones as activating groups for the displacement of halides *(2)*, and nitro groups *(4)*. The role of heterocycles as activating groups was first reported for a 1,3,4-oxadiazole group *(2a)*, and later for the nitro-displacement polymerization of bis(nitrophthalimides) to afford poly(ether imides) *(4b)*. The nature of the activating group, leaving group, bisphenolate, and solvent all play an important role in the polymer forming reaction, the details of which will be discussed further in the following section.

Reaction Conditions and Mechanism

The primary mechanism for formation of aryl ether linkages involves nucleophilic aromatic substitution of an activated leaving group by phenolate. Polar aprotic solvents, e.g., dimethylsulfoxide (DMSO), N-methylpyrrolidone (NMP) and dimethylacetamide (DMAC) are required to effect the reaction. The use of dimethylproylene urea has been reported as an alternative solvent

0097–6156/96/0624–0210$12.00/0

which affords increased polymer molecular weights and yields with less reactive aryl fluorides, and it was found to dissolve poorly soluble rigid heterocyclic-based poly(aryl ethers) *(5)*. Special reaction conditions are required for the crystalline ketone based poly(aryl ethers), where diphenylsulfone is used at reaction temperatures above the melting point of the polymer to avoid premature crystallization *(2c)*. Potassium carbonate is the preferred base for bisphenolate generation as the reaction can be carried out in the reactor, in the presence of the dihalide, using toluene to dehydrate the system *(6)*.

The key characteristics of effective activating groups for an S_NAr mechanism are high electron affinity and the presence of a site of unsaturation which can stabilize the negative charge developed along the reaction coordinate through resonance to a hetero atom. This step involves the formation of a Meisenheimer complex (I) which lowers the activation energy of the displacement (Scheme II) *(7)*. Formation of the Meisenheimer complex is the slow rate-determining step, as reflected in the greater reactivity of aryl fluorides relative to chlorides and bromides. The higher reactivity of the fluoride is attributed to its small size and higher electronegativity relative to chloride and bromide. Recently, aryl triflates were shown to be effective leaving groups, displaying a higher reactivity than fluoride *(8)*. Aryl triflate displacement was complicated by competitive S-O cleavage which afforded sulfonate exchange to the phenolate. This exchange limited the phenols which could be used to those which can form activated sulfonates themselves. Nitro groups are readily displaced, but the resulting nitrite ion is reactive and undergoes side reactions, such as those observed with imides if conditions are not carefully controlled *(9)*. In general, fluoride is the most desirable leaving group owing to its reactivity. A complication observed with fluoride is that its nucleophilic nature in polar aprotic solvents leads to a back reaction where fluoride cleaves the aryl ether bond *(7a)*. The generation of high polymer is assisted by the precipitation of fluoride salts from the polymerization to limiting the effect of the back reaction. The aryl ether bonds are also cleaved by phenoxide, which leads to ether interchange in the polymerization *(2b,7a)*.

Although activated aryl fluorides are the most common substrates for poly(aryl ether) synthesis, there is a great deal of interest in the use of aryl chlorides due to their reduced cost. Hergenrother and co-workers showed that a variety of dichlorobenzophenone, phenylsulfone, and bis(benzoyl)benzene monomers afforded high polymer with a variety of bisphenols *(10)*. Percec *(11-13)* and Mohanty *(14)* found that in certain cases the polycondensation of a bis(aryl chloride) with a hydroquinone afforded low molecular weight polymer due to a reductive dehalogenation reaction. The degree to which this occurred was dependent on the oxidation potential of the bisphenolate, the solvent, and the polymerization temperature. The results are explained by a competition between a polar and single electron transfer (SET) pathway (Scheme III), where the latter is based on an $S_{RN}1$ mechanism *(15)*. In the SET pathway, an electron is transferred from the bisphenolate (electron donor) to the aryl halide (electron acceptor) to form a radical anion-radical pair, II. This pair may collapse to form the Meisenheimer complex, I, and proceed to the aryl ether, or separate and eliminate the halide to give a phenyl radical. The phenyl radical can abstract a hydrogen from the solvent to give the dehalognated chain-end. It is also proposed that a aryl ether can form via the SET pathway rather than strictly the

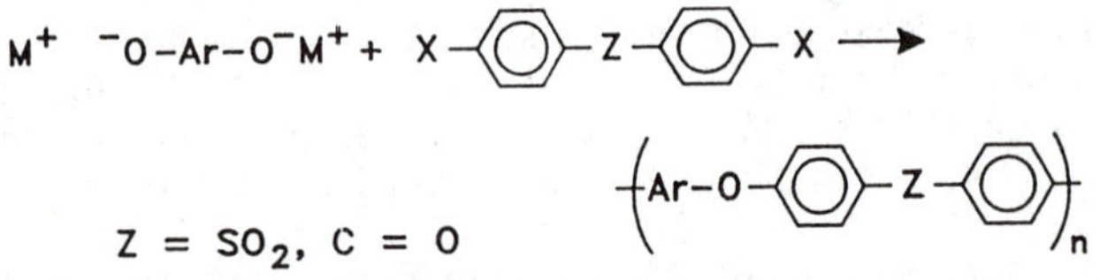

$M^+ \; ^-O-Ar-O^-M^+ + X-\langle\bigcirc\rangle-Z-\langle\bigcirc\rangle-X \longrightarrow$

$Z = SO_2, \; C = O$ $\quad -(Ar-O-\langle\bigcirc\rangle-Z-\langle\bigcirc\rangle)_n-$

Scheme I.

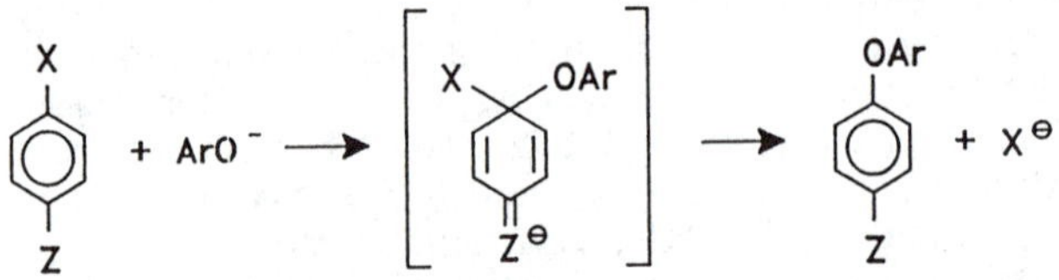

X, Z, + ArO⁻ → [X, OAr, Z⊖] I → OAr, Z + X⊖

Scheme II.

−CO, X, ⁻OArO⁻ ⇌ −CO, X, ⁻OArO⁻ ⇌ −CO, X, •OArO⁻ II

−CO, X, OArO⁻ I

−CO, OArO⁻ + X⁻

−CO, X + •OArO⁻

−CO, • + X⁻

−CO, H

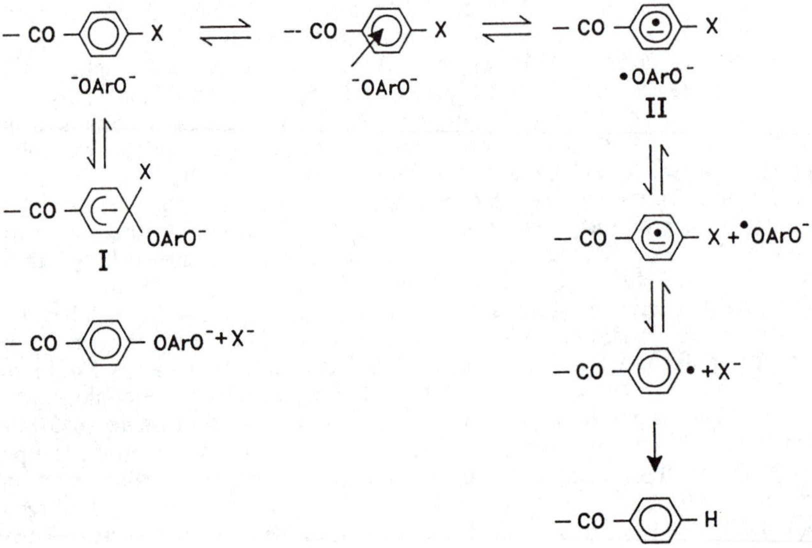

Scheme III.

polar pathway *(14)*. Since the SET relies on the bisphenolate acting as the electron donor, the reductive halogenation side reaction is limited to electron-rich bisphenolates, e.g., that derived from hydroquinone. The balance between polar and SET pathways is also dependent on aryl halide structure, with the polar pathway favored to a greater extent with more electronegative leaving groups and with sulfone compared to ketone activating groups *(13)*.

The suppression of the SET pathway can be affected by addition of a radical scavenger, e.g., tetraphenylhydrazine, to the polymerization *(14)*. Addition of 0.01 mole% of tetraphenylhydrazine to the polymerization of 1,3-bis(p-chlorobenzoyl)benzene and hydroquinone resulted in high polymer with an inherent viscosity equivalent to those polymers prepared with the difluoro monomer. In addition, the reductive dehalogenation side reaction can be avoided by the use of diphenylsulfone rather than NMP or DMAC, due to the absence of labile hydrogen atoms towards hydrogen abstraction *(16)*.

Poly(Aryl Ether Ketones)

Since the first reports of poly(aryl ether ketones) prepared via nucleophilic aromatic substitution nearly three decades ago, various synthetic strategies have been pursued to prepare semicrystalline materials under less stringent polymerization conditions (i.e., common organic solvents at mild temperatures). Of particular interest is poly(aryl ether ether ketone), PEEK, which is a highly crystalline polymer with high solvent resistance and a T_g of 145 °C and a T_m of 340 °C *(1,17-19)*. The polymerization of PEEK is mediated in diphenylsulfone at temperatures in the proximity of the Tm.

To circumvent the use of high polymerization temperatures and diphenylsulfone as solvent, the synthesis of soluble precursor polymers of PEEK has been investigated. This strategy was applied to ketimine derivatives of 4,4′-difluorobenzophenone *(20-22)*. The ketimine group activates fluoride displacement, allowing polymerization with hydroquinone and, since the substituent on the imine nitrogen retards crystallization, conventional polyether synthetic conditions (DMAC or NMP, 160 – 180 °C) can be used without premature precipitation of the polymer. Likewise, an amorphous and soluble poly(ketal ketone) has been reported by nucleophilic displacement polycondensation of an acetal monomer, 2,2-bis(4-hydroxyphenyl)-1,3-dioxoline, with 4,4′-difluorobenzophenone at 150 – 220 °C in an aprotic dipolar solvent *(23)*. Each of the above amorphous/soluble polymer systems could be quantitatively hydrolyzed to produce the parent ketone structure.

Sogah *(24)* and McGrath *(25)* have reported tert-butyl and phenyl-substituted poly(aryl ether ketones) respectively, which are prepared by nucleophilic substitution reaction of the corresponding substituted hydroquinones and 4,4′-difluorobenzophenone. The resulting polymers were amorphous and highly soluble in common organic solvents as a result of the bulky substituents which suppressed crystallization. The bulky substituents were cleaved with acid in a reverse Friedel-Crafts alkylation reaction to produce the semi-crystalline PEEK *(24)*.

A novel approach to semicrystalline poly(aryl ether ketones) involves the reaction of 4,4′-dichlorobenzophenone with sodium carbonate in the presence of a silica/copper catalyst in diphenylsulfone at 280 – 320 °C *(26)*. Under these conditions, the sodium carbonate/catalyst combination behaves as a water-equivalent, inserting an ether linkage between benzophenone units. The proposed mechanism involves formation of a silyl ether by the reaction of 4,4′-dichlorobenzophenone with a silanol on the silica surface as a key intermediate. This process was also applied to the synthesis of amorphous poly(aryl ether ketones) and polysulfones.

In addition to new routes to crystalline poly(aryl ether ketones), a number of new monomers and polymerization methods have been reported for the synthesis of amorphous poly(aryl ether ketones). The driving force for much of this research has been the development of processable high T_g materials with tough, ductile mechanical properties. An important example of this research is the preparation of high molecular weight poly(aryl ketones) derived from 1,3-bis(4-chlorobenzoyl)benzene reported by Hergenrother and co-workers *(27)*. The resulting poly(aryl ether ketones) were high T_g materials with exceptionally tough mechanical properties. Likewise, 1,3-bis(4-chlorobenzoyl)benzene and related ketone-containing monomers were polymerized with the bulky bisphenol 9,9′-bis(3,5-diphenyl-4-hydroxyphenyl)fluorene, to produce a series of high T_g poly(aryl ether ketones) *(28)*. Other new bis(fluorobenzoyl) monomers which have been developed include naphthalene- *(29)*, biphenyl- *(30)*, and indan-based *(31)* compounds, affording high T_g, processable amorphous poly(aryl ether ketones) in polymerizations with common bisphenols.

Hay has reported the synthesis of poly(aryl ether ketones) containing o-dibenzoylbenzene moieties by the polymerization of 1,2-bis(4-fluorobenzoyl)benzene with various bisphenolates in DMAC *(32)*. Transformation of the o-dibenzoyl(benzene) moiety in the polymer chain to a heterocycle by cyclization with small molecules was developed as a means of increasing the polymer chain stiffness and solvent resistance. It was demonstrated that reaction of the polymer with hydrazine monohydrate in the presence of a mild acid in chlorobenzene converted the poly(aryl ether ketone) to a poly(aryl ether phthalazine), a new class of heterocyclic-containing polyether (Scheme V) *(33)*. Likewise, reaction of the polymer with benzylamine in a basic medium led to amorphous, thermally stable poly(aryl ether isoquinolines) (Scheme V) *(34)*. Another example of the use of the o-benzoyl cyclization strategy is the intramolecular ring closure of poly(aryl ketones) containing 2,2′-dibenzoylbiphenyl units to form poly(aryl ether phenanthrenes) *(35-37)*.

Trans-etherification involving cleavage of aryl ether bonds by phenoxide has been observed to be a significant side reaction in poly(aryl ether) synthesis and leads to ether interchange *(7a)*. This process has been exploited in the synthesis of poly(aryl ether ketones) from polymers with related structures. High molecular weight poly(aryl ether ketone) was found to react with 4,4′-difluorobenzophenone in the presence of potassium carbonate in benzophenone or diphenyl sulfone at 300 °C, to afford low molecular weight poly(aryl ether ketone) with fluorophenyl end groups *(38)*. High temperature polycondensation of 4,4′-dihydroxydiphenyl sulfone with 4,4′-bis(4-fluorobenzoyl)biphenyl using diphenyl sulfone as solvent and sodium carbonate as base, led to formation of the expected alternating polymer sequence. Replacement of sodium carbonate by potassium gave the random sequence polymer *(39)*. The ether linkages, being activated by electron-withdrawing groups on both adjacent aromatic rings, are susceptible to trans-etherification as a result of nucleophilic cleavage by fluoride ion as an intermediate step. Potassium fluoride has greater solubility in dipolar aprotic solvents which promotes fluoride-catalyzed trans-etherification.

Generally, the synthesis of poly(aryl ether ketones) by nucleophilic displacement requires the use of expensive fluoroarylketone monomers. Alternatively, potassium fluoride or a mixture of potassium fluoride and the phase transfer catalyst N-neopentyl-4-(dialkylamino)pyridinium chloride have been shown to be very effective in promoting the polymer-forming nucleophilic displacement reaction in poly(aryl ether ketone) synthesis from dichloroaryl ketone monomers *(40)*. This reaction is based on phase-transfer catalyzed chloride/fluoride exchange.

New Activating Groups

One of the most interesting aspects of poly(aryl ether) research is the multitude of polymer structures which have been prepared with a wide variety of functionality and backbone constituents. Structural diversity can be introduced through either the bisphenol or dihalo monomer. One important means of introducing new functional and heterocyclic groups is based on their use as the activating group in the dihalo monomer. Activating groups can be either *ortho* or *para* relative to the leaving group, or fused to the aryl halide in the case of heterocycles. The important features of activating groups are that they display high electron affinity and possess a site of unsaturation which can stabilize negative charge by resonance to a hetero atom. The applicability of a potential activating group can often be assessed by the deshielding of the protons *ortho* to the group of interest in the ^{1}H-NMR, where chemical shifts of 7.5 and greater indicate probable fluoro-displacement *(41)*. Alternatively, correlation of aryl fluoride reactivity with fluorine chemical shift by ^{19}F-NMR is a useful predictive tool since relative reactivities of non-related activated aryl fluorides have been shown to map well with the fluorine chemical shift *(42)*. In addition to NMR techniques, aryl fluoride reactivity has been estimated using Hückel molecular orbital calculations to determine the net charge density at the C-F carbon atom *(34)*.

There has been a large number of reports over the past five years on poly(aryl ethers) prepared via nucleophilic aromatic substitution polymerization from new activating groups. The resurgence in this area stems from both the microelectronic and composite industries which require materials with a broad scope of thermal, mechanical and dielectric properties. The activating groups can be categorized into several categories including fused heterocycle, pendent heterocycle, perfluoroalkyl, carboxylic acid derivative and phosphine oxide. The heterocycle-based activating groups will be covered together in the following section with the heterocycle containing bisphenols.

Perfluoroalkyl-activated ether synthesis was pursued as a means of preparing highly fluorinated polymers which are readily soluble in common organic solvents *(43-45)*. The electron withdrawing effect of the perfluoroalkyl group was determined to be similar to that of a ketone by comparison of the Hammet σ value of the trifluoromethyl group to a ketone group (σ:CF_3=0.54, carbonyl=0.50) and the deshielding of aromatic protons ortho to perfluoroalkyl groups ($\delta = 7.6$). However, relative to other activating groups, a perfluoroalkyl group is structurally unique because it has no site of unsaturation. Therefore, activation by a perfluoroalkyl group requires the participation of hyperconjugation and/or stabilization through the carbon-carbon sigma bond to lower the activation energy of the transformation. High molecular weight polymers were obtained from polymerizations of 1,6-bis(fluorophenyl) perfluorohexane, pendent perfluoroalkyl difluorides and dinitro monomers with various bisphenolates.

Carboxylic acid derivatives such as amides and esters have been shown to be effective activating groups for nucleophilic aromatic substitution polymerizations. The monomer 1,4-bis(4-fluorobenzamido)benzene was polymerized with various bisphenols in the presence of base to high molecular weight polymer with no evidence of esterification or hydrolysis *(46-48)*. In the case of the ester-activation, high molecular weight was obtained only by using the silyl ether derivative of the bisphenol and a CsF catalyst *(49)*. Other activated carbonyl derivatives include bis(fluorophenyl)benzil *(50)* and azine *(51)* monomers. High molecular weight polymer was obtained in each case, and subsequent transformations were possible on the polymer backbone to form heterocycle-containing poly(aryl ethers).

McGrath and co-workers have demonstrated that the phosphine oxide moiety is an effective activating group for nucleophilic displacement reactions *(52-64)*. A wide variety of phosphorous containing poly(aryl ethers) have been prepared which show high thermal stability, flame resistance, oxygen plasma resistance and have been used as modifiers for thermosetting resins and as second order nonlinear optical materials. In addition, the phosphine oxide activated displacement was used to prepare new diamines, new polyimides and thermosetting resins.

Poly(Aryl Ethers) Containing Heterocyclic Units. During the last several years, a significant effort has been devoted to incorporating heterocyclic units into the backbone of poly(aryl ethers) *(65,66)*. The primary motivation for pursuing this chemistry has been to prepare more processable analogs of thermally stable polymers, e.g., poly(phenylquinoxalines), poly(benzazoles), etc., analogous to poly(ether imides). To this end, aryl ether analogs of thermally stable polymers have been prepared either by nucleophilic displacement of carbonyl- or sulfone-activated aryl halides with bisphenols containing heterocyclic units, or by using heterocyclic-activated bis(aryl halides) in conjunction with conventional bisphenols. In the case of the heterocyclic-activated displacement, the heterocyclic group can be either fused or pendent to the aryl fluoride. An advantage of using poly(aryl ether) synthesis as the route to these polymers is that it allows for the preparation of functional oligomers suitable for block copolymer synthesis *(67,68,22)* and for the thermoplastic toughening of thermosets *(69,70)*.

One of the first examples of heterocyclic-activated ether synthesis is the activation of nitro displacement by a fused imide ring in the synthesis of poly(ether imides) *(4b)*. More recently, the fused pyrazine in the quinoxaline ring system has been shown to activate fluoro-displacement from the 6- or 7-position, and was used to prepare poly(aryl ether phenylquinoxalines) (Scheme VI) *(41)*. The activation of fluoro-displacement by the relatively nonpolar pyrazine moiety is a result of the electron-poor pyrazine ring and the ability of the quinoxaline ring system to stabilize the negative charge formed in a stabilized intermediate (and/or transition state) through a Meisenheimer-like complex. Quinoxaline-activated poly(aryl ether) synthesis

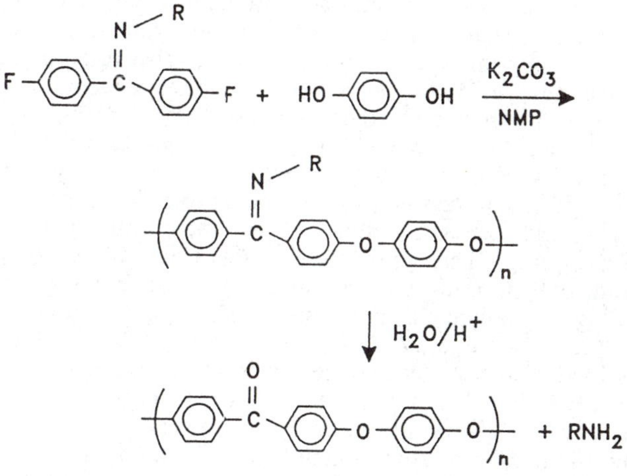

R
N
C
F
F
+
HO
OH
K_2CO_3
NMP
O
n
H_2O/H^+
+ RNH_2

Scheme IV.

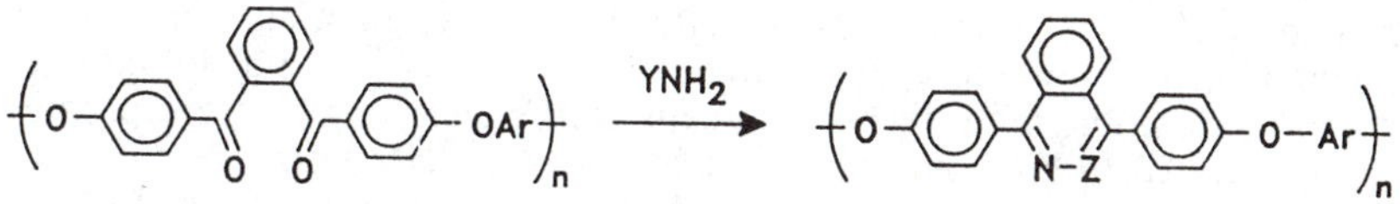

O
OAr
n
YNH_2
N-Z
O-Ar

Y = NH_2, Z = N

Y = $PhCH_2$, Z = PhC

Scheme V.

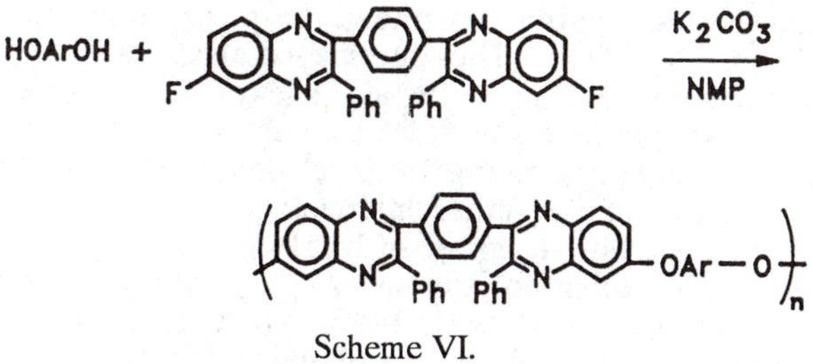

HOArOH +
N
F
Ph
Ph
K_2CO_3
NMP
OAr—O
n

Scheme VI.

has also been exploited in the self-polymerization of an A-B monomer, 2-(4-hydroxyphenyl)-3-phenyl-6-fluoroquinoxaline *(71,72)*, and an AB_2 monomer *(73)*.

Many pendent heterocyclic rings activate halo-displacement by analogy to ketone and sulfone groups. The quinoxaline ring system has been shown to activate fluoro-displacement from phenyl groups substituted at the 3-position of the quinoxaline *(74,75)*. The reactivity at the pendent phenyl ring was observed to be somewhat lower than the fused ring for fluoro-displacement. Benzazole-activation was pursued as a general route to aryl ether-based poly(benzothiazoles *(76)*, poly(benzoxazoles *(77)*, and poly(benzimidazoles) (Scheme VII) *(78)*. The *para*-substituted benzazole ring behaves as a masked carboxylic acid and exerts a strong electron withdrawing influence on the pendent phenyl group. The appropriately substituted dihalo compounds were prepared and polymerized with a variety of bisphenols yielding soluble, processable, high T_g polymers. In a similar fashion, poly(aryl ethers) based on phthalazine *(33)*, isoquinoline *(34)*, oxadiazole *(2,79)*, and triazole *(80)*, heterocycles have also been synthesized, where the ring system activates displacement of a para-substituted fluoro group by phenolates. Poly(heterocyclic aryl ethers) have also been prepared by chloro-displacement from carbonyl-linked 2-chlorothiophene *(81,82)* and furan *(83)* monomers. Higher reactivity was observed for these monomers relative to aryl chlorides.

A second synthetic route to poly(aryl ethers) containing heterocyclic units involved the reaction of an aromatic dihydroxy heterocycle with conventional activated difluoro compounds. Bisphenols containing quinoxaline, phenylimidazole, oxadizaole, pyrazole, triazole, phenolphthalein, phenolphthalimidine, and phenolphthalein anilide heterocycles were prepared and polymerized to high molecular weight *(84-95)*. However, the most noteworthy examples are the poly(aryl ether benzimidazoles), prepared from bis(4-hydroxyphenyl) bibenzimidazole, due to their unique combination of adhesive, thermal and mechanical properties (Scheme VIII) *(93-95)*.

Masked Bisphenol Methods. Poly(aryl ether) synthesis by displacement polymerization involves the use of basic/ionic conditions to generate the phenolate nucleophile. Kricheldorf has developed polymerization methodology based on the use of silylated bisphenols in the presence of cesium fluoride in lieu of the bisphenolate, allowing poly(aryl ethers) preparation under virtually neutral conditions without the presence of ionic reagents or products (Scheme IX) *(96)*. The fluoride catalyst leads to only a small amount of phenolate ion, or activated siloxy intermediate, during the polymerization and produces volatile trimethylsilyl fluoride rather than potassium fluoride, greatly limiting the nucleophilic species capable or aryl ether cleavage, hence limiting ether interchange. The polymerizations are carried out in the melt with a variety of bisphenols to prepare a broad scope of poly(aryl ethers) including polysulfones *(97)*, poly(aryl ether benzonitriles) *(98)*, poly(aryl ether pyridines) *(99)*, and poly(aryl ether ketones) *(100)*. In the last case, high molecular weight amorphous and crystalline poly(aryl ether ketones) were synthesized. A modified procedure allowing the use of less expensive activated aryl chlorides involves solution polymerization of silylated bisphenols in the presence of potassium carbonate *(101)*. This approach has also been extended to solution polymerizations to prepare microphase separated xylenyl ether-aryl ether sulfone copolymers *(102)*.

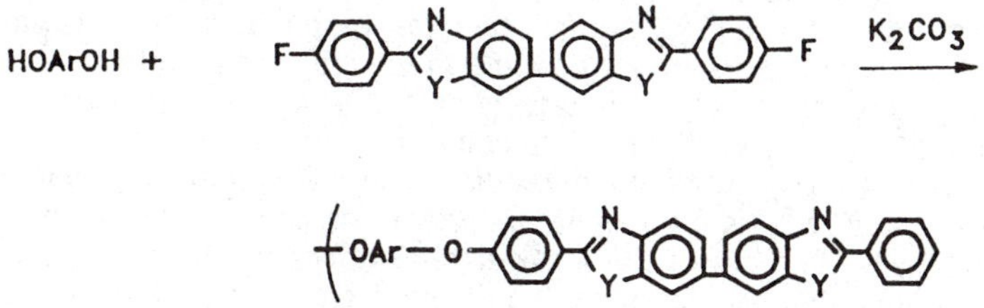

HOArOH + $\xrightarrow{K_2CO_3}$

Y = S,O,NH

Scheme VII.

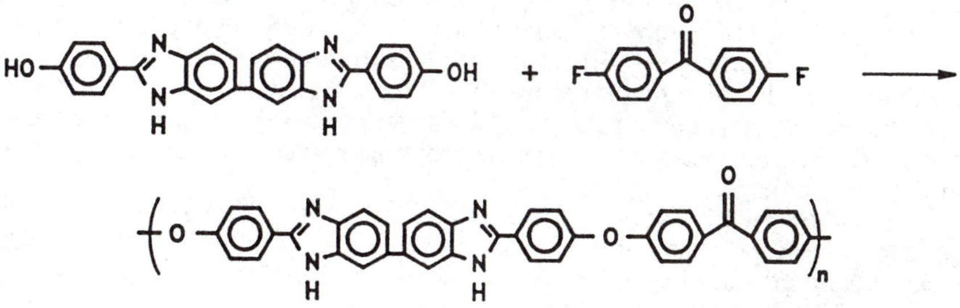

Scheme VIII.

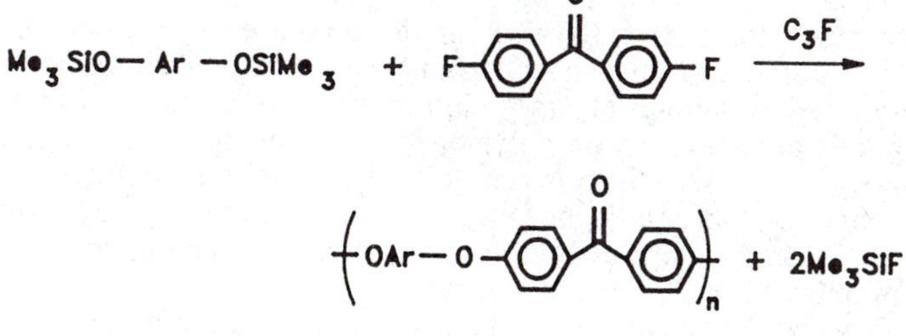

$Me_3SiO—Ar—OSiMe_3$ + $\xrightarrow{C_3F}$ + $2Me_3SiF$

Scheme IX.

Hay has reported the use of carbamates as masked bisphenols for poly(aryl ether) synthesis *(103)*. The carbamates are easily prepared by reaction of the bisphenol with n-propylisocyanate, and are purified by recrystallization. The carbamate group is rapidly cleaved in the presence of potassium carbonate at 155 – 165 °C to generate a potassium phenolate *in situ*, which participates in the displacement of an activated fluoride. High polymer was obtained in a relatively short period of time (1 – 2 h) using this procedure. Other masking groups such as carbonates and acetates were examined but were less successful. One interesting source of masked phenol used was biphenol-A polycarbonate (Lexan™), which degrades under the potassium carbonate/NMP reaction conditions to produce phenoate which polymerizes with the activated dihalide. The carbamate methodology was extended to masked bisthiophenols to provide a new approach to poly(aryl sulfides) *(104)*. The use of the masked bisthiophenol alleviates the difficulty in handling unstable bisthiophenol monomers which are easily oxidized in air.

Other Synthetic Routes to Poly(Aryl Ether)s

In addition to nucleophilic aromatic substitution, there are a number of other synthetic routes to poly(aryl ethers). Friedel-Crafts condensation of arylsulfonyl chlorides and aryl carboxylic acid derivatives with aryl ethers has been employed to prepare polysulfones *(2b)* and poly(ether ketones) *(105,106)*, respectively. Direct polycondensation of various benzoic acids containing a phenyl ether structure has been carried in 1:10 phosphorous pentoxide/methanesulfonic acid *(107)*. The success of this method is a consequence of the high selectivity of the electrophilic reagent for substitution para to the ether linkage.

Transition-metal catalyzed carbon-carbon bond formation has been reported as the polymer-forming reaction for poly(aryl ether) synthesis. The nickel catalyzed coupling of diaryl chlorides has been reported for the synthesis of high molecular weight polysulfones *(108)*. The method is particularly useful for the introduction of biphenyl moieties into the polymer backbone. The nickel-catalyzed coupling chemistry was used by Wang to produce a bis(p-fluorobenzoyl)biphenyl monomer for displacement polymerization with bisphenolates *(30)*. Polymerization of bis(aryloxy) monomers has been shown to occur in the presence of an iron (III) chloride catalyst via a cation radical mechanism (Scholl reaction) *(109)*. Like the nickel-catalyzed coupling of aryl chlorides, this method involves carbon-carbon bond formation in the polymer-forming reaction. The cation-radical methodology was used to prepare soluble poly(ether sulfones), poly(ether ether ketones) and aromatic polyethers. Palladium-catalyzed cross-coupling of aromatic diacid chlorides and bis(trimethylstannane) monomers was utilized to prepare poly(aryl ether ketones) *(110)*.

Poly(aryl ethers) have been prepared by Ullman coupling of bisphenols and dibromoarylenes in the presence of a copper catalyst *(111)*. This methodology does not require the presence of an activating group for halo-displacement; hence aromatic ether polymers can be prepared without polar functionality.

Mixed aromatic-aliphatic polyethers have received attention of late as a means of forming liquid-crystalline and dendritic polymer structures. Liquid-liquid phase-transfer conditions were used for the polymerization of bisphenols with α,ω-dibromoalkanes *(112)* and bismesylates of cyclohexanedimethanol *(113)* to afford main-chain liquid crystalline polyethers. Solid-liquid phase-transfer was employed by Hawker in the synthesis of dendritic polyethers from an AB_2 monomer, such as 5-bromo-methyl-1,3-dihydroxybenzene *(114)*. Recently a silicon-assisted modification of the polymerization of bromomethylphenol was reported by Fréchet, where polycondensation of a O-trimethylsilyl-substituted bromomethylphenol was shown to give high polymer in the presence of base *(115)*. The higher molecular weights relative to the classical Williamson ether synthesis were attributed to the trimethylsilyl group increasing the solubility of oligomers, avoiding premature precipitation.

Poly(Aryl Ether) From Cyclic Precursors

One area of active research has been the synthesis poly(aryl ethers) from cyclic precursors. This approach has been successfully applied to poly(carbonates) *(116)* and has the advantage of allowing polymer formation to occur in the melt from low viscosity cyclic oligomers, without the liberation of volatile by-products. The synthesis of the cyclic aryl ether oligomers has been carried out by reaction of a bisphenolate and a dihalophenyl sulfone or ketone under high dilution *(117,118)*. A high yield synthesis was recently reported by Hay using 1,2-bis(4-fluorobenzoyl)benzene as a comonomer with bisphenolates *(119)*. This monomer allows for more concentrated solutions in the cyclics synthesis, and produced oligomers which ranged from 2,000 to 10,000 g/mol. Cyclic ketone oligomers were reported using a nickel-catalyzed coupling of aryl chloride terminated oligomers. The polymerization of the cyclic oligomers are initiated by nucleophiles to afford linear polymers.

Literature Cited

1. Rose, J. B. In *High Performance Polymers: Their Origin and Development*; Seymour, R. B., Kirshenbaum, G. S., Eds.; Elsevier: New York, 1986; p. 187.
2. (a) Johnson, R. N.; Farnham, A. G.; Clendinning, R. A.; Hale, W. F.; Merriam, C. N. *J. Polym. Sci.: Polym. Chem. Ed.* **1967**, *5*, 2375. (b) Rose, J. B. *Polymer* **1974**, *15*, 456. (c) Attwood, T. E.; Dawson, P. C.; Freeman, J. L.; Hoy, L. R. J.; Rose, J. B.; Staniland, P. A. *Polymer* **1981**, *22*, 1096.
3. (a) Maiti, S.; Mandal, B. K. *Prog. Polym. Sci.* **1986**, *12*, 111. (b) Rose, J. B. In *Recent Advances in Mechanistic and Synthetic Aspects of Polymerization*; Fontanille, M., Guyot, A., Eds.; Pergamon: Oxford, 1989; p. 483.
4. (a) Radlmann, V. E.; Schmidt, W.; Nischk, G. E. *Die Makromol Chem.* **1969**, *130*, 45. (b) Takekoshi, T.; Wirth, J. G.; Heath, D. R.; Kochanowski, J. E.; Manello, J. S.; Webber, M. J. *J. Polym. Sci.: Polym. Chem. Ed.* **1980**, *18*, 3069. (b) White, D. M.; Takekoshi, T.; Williams, F. J.; Relles, H. M.; Donahue, P. E.; Klopfer, H. J.; Loucks, G. R.; Manello, J. S.; Mathews, R. O.; Schlvenz, R. W. *J. Polym. Sci.: Polym. Chem. Ed.* **1981**, *19*, 1635.
5. Labadie, J. W.; Carter, K. R.; Hedrick, J. L.; Jonsson, H.; Twieg, R.; Kim, S. Y. *Polym. Bull.* **1993**, *30*, 25.

6. Viswanathan, R.; Johnson, B. C.; McGrath, J. E. *Polymer* **1984**, *25*, 1827.
7. (a) Attwood, T. A.; Newton, A. B.; Rose, J. B. *Br. Polym. J.* **1972**, *4*, 391. (b) Miller, J. *Aromatic Nucleophilic Substitution*; Elseivier: Amsterdam, 1968.
8. Jonsson, H.; Hedrick, J. L.; Labadie, J. W. *Polymer Preprints* **1992**, *33(1)*, 394.
9. (a) Markezich, R. L.; Zamek, C. S.; Donahue, P. E.; Williams, F. J. *J. Org. Chem.* **1977**, *42*, 3435. (b) Markezich, R. L.; Zamek, C. S. *J. Org. Chem.* **1977**, *42*, 3431.
10. Hergenrother, P. M.; Jensen, B. J.; Havens, S. J. *Polymer* **1988**, *29*, 358.
11. Percec, V.; Clough, R. S.; Rinaldi, P. L.; Litman, V. E. *Macromolecules* **1991**, *24*, 5889.
12. Percec, V.; Clough, R. S.; Grigoras, M.; Rinaldi, P. L.; Litman, V. E. *Macromolecules* **1993**, *26*, 3650.
13. Percec, V.; Clough, R. S.; Rinaldi, P. L.; Litman, V. E. *Macromolecules* **1994**, *27*, 1535.
14. Mani, R. S.; Zimmerman, B.; Bhatnager, A.; Mohanty, D. K. *Polymer* **1993**, *34*, 171.
15. Burnett, J. F. *Acc. Chem. Res.* **1978**, *11*, 413.
16. Bhatnager, A.; Mani, R. S.; King, B.; Mohanty, D. K. *Polymer* **1994**, *35*, 1111.
17. Kricheldorf, H. R. In *Handbook of Polymer Synthesis, Part A*; Kricheldorf, H. R., Ed.; Dekker: New York, 1992; p. 574.
18. Staniland, P. A. *Comprehensive Polymer Science*; Allen, G., Bevington, J. C., Eds.; Pergamon: Oxford, 1989; Vol. 5, p. 483.
19. Mullins, M. J.; Woo, E. P. *J. Macromol. Sci., Macromol. Chem. Phys.* **1987**, *C27*, 313.
20. Mohanty, D. K.; Lowry, R. C.; Lyle, G. D.; McGrath, J. E. *Int. SAMPE Symp.* **1987**, *32*, 408.
21. Mohanty, D. K.; Senger, J. S.; Smith, C. D.; McGrath, J. E. *Int. SAMPE Symp.* **1988**, *33*, 970.
22. Hedrick, J. L.; Volksen, W.; Mohanty, D. K. *J. Polym. Sci., Part A: Polym. Chem. Ed.* **1992**, *30*, 2085.
23. Kelsey, D. R.; Robenson, L. M.; Clendinning, R. A.; Blackwell, C. S. *Macromolecules* **1987**, *20*, 1204.
24. Risse, W.; Sogah, D. Y. *Macromolecules* **1990**, *23*, 4029.
25. Mohanty, D. K.; Lin, T. S.; Ward, T. C.; McGrath, J. E. *Int. SAMPE Symp. Exp.* **1986**, *31*, 945.
26. Fukawa, I.; Tsuneaki, T.; Dozona, T. *Macromolecules* **1991**, *24*, 3838.
27. Hergenrother, P. M.; Jensen, B. J.; Havens, S. J. *Polymer* **1988**, *29*, 358.
28. Wang, Z. Y.; Hay, A. S. *J. Polym. Sci., Part A: Polym. Chem. Ed.* **1991**, *29*, 1045.
29. (a) Ritter, H.; Thorwirth, R.; Muller, G. *Makromol. Chem.* **1993**, *194*, 1409. (b) Ohno, M.; Taketa, T.; Endo, T. *Macromolecules* **1994**, *27*, 3447.
30. Zhang, C.; Wang, Z. Y. *Macromolecules* **1993**, *26*, 3324.

31. Maier, G.; Yang, D.; Nugken, O. *Makromol. Chem.* **1993**, *194*, 1101.
32. Singh, R.; Hay, A. S. *Macromolecules* **1992**, *25*, 1017.
33. Singh, R.; Hay, A. S. *Macromolecules* **1992**, *25*, 1025.
34. Singh, R.; Hay, A. S. *Macromolecules* **1992**, *25*, 1033.
35. Wang, Z. Y.; Zhang, C. *Macromolecules* **1992**, *25*, 5851.
36. Wang, Z. Y.; Zhang, C. *Macromolecules* **1993**, *26*, 3330.
37. Wang, Z. Y.; Zhang, C.; Arnoux, F. *Macromolecules* **1994**, *27*, 4415.
38. Fukawa, I.; Tanabe, T. *J. Polym. Sci., Polym. Chem. Ed.* **1993**, *31*, 535.
39. Colquhoun, H. M.; Dudman, C. C.; Blundell, D. J.; Bunn, A.; Mackenzie, P. D.; McGrail, P. T.; Nield, E.; Rose, J. B.; Williams, D. J. *Macromolecules* **1993**, *26*, 107.
40. Hoffman, U.; Metzmann, F. H.; Klapper, M.; Mullen, K. *Macromolecules* **1994**, *27*, 3575.
41. Hedrick, J. L.; Labadie, J. W. *Macromolecules* **1990**, *23*, 1561.
42. Carter, K. R. *Macromolecules* **1995**, *28*, 6462.
43. Labadie, J. W.; Hedrick, J. L. *Macromolecules* **1990**, *23*, 5371.
44. Kim, S. Y.; Labadie, J. W. *Polym. Prepr.: Am. Chem. Soc., Div. Polym. Chem.* **1991**, *32(1)*, 164.
45. Labadie, J. W.; Kim, S. Y.; Carter, K. R. *Polym. Prepr.* **1993**, *34(1)*, 415.
46. Hedrick, J. L. *Macromolecules* **1991**, *24*, 812.
47. Lucas, M.; Hedrick, J. *Polym. Bull.* **1992**, *28*, 129.
48. Lucas, M.; Brock, P.; Hedrick, J. L. *J. Polym. Sci.: Polym. Chem. Ed.* **1993**, *31*, 2179.
49. Hilborn, J. *Polym. Prepr.* **1995**, *34(1)*, 415.
50. Strakelj, M.; Hedrick, J. C.; Hedrick, J. L.; Twieg, R. J. *Macromolecules* **1994**, *27*, 6277.
51. Carter, K. R.; Hedrick, J. L. *Macromolecules* **1994**, *27*, 3426.
52. Wachamad, W.; Cooper, K. L.; McGrath, J. E. *Polym. Prepr.* **1989**, *30(2)*, 441.
53. Wan, I.; Priddy Jr., D. B.; Lyle, G. D.; McGrath, J. E. *Polym. Prepr.* **1993**, *34(1)*, 806.
54. Pak, S.; Lyle, G. D.; Mercier, R.; McGrath, J. E. *Polymer* **1993**, *34(4)*, 885.
55. Priddy Jr., D. B.; Franks, M.; Konas, M.; Vrana, M. A.; Yoon, T. H.; McGrath, J. E. *Polym. Prepr.* **1993**, *34(1)*, 370.
56. Holzberlein, R. L.; Mohanty, D. K.; Smith, C. D.; Wu, S. D.; McGrath, J. E. *Polym. Prepr.* **1989**, *30(1)*, 293.
57. Smith, C. D.; Mohanty, D. K.; Holzberlein, R. L.; Wu, S. D.; Chen, D.; McGrath, J. E. *Int. SAMPE Electron. Conf.* **1989**, *3*, 1.
58. Smith, C. D.; Mohanty, D. K.; McGrath, J. E. *Int. SAMPE Symp. Exhib.* **1990**, *35*, 108.
59. Smith, C. D.; Gungor, A.; Keister, K. M.; Marand, H. A.; McGrath, J. E. *Polym. Prepr.* **1991**, *32(1)*, 93.
60. Smith, C. D.; Grubbs, H. J.; Webster, H. F.; Wightman, J. P.; McGrath, J. E. *Polym. Mater. Sci. Eng.* **1991**, *65*, 108.
61. Smith, C. D.; Gungor, A.; Keister, K. M.; Marand, H. A.; McGrath, J. E. *Polym. Prepr.* **1991**, *32(1)*, 93.

62. Grubbs, H. J.; Smith, C. D.; McGrath, J. E. *Polym. Mater. Sci. Eng.* **1991**, *65*, 111.
63. Webster, H. F.; Smith, C. D.; McGrath, J. E.; Wightman, J. P. *Polym. Mater. Sci. Eng.* **1991**, *65*, 113; Smith, C. D.; Webster, H. F.; Gungor, A.; Wightman, J. P.; McGrath, J. E. *High Performance Polymers* **1991**, *3(4)*, 211.
64. Priddy, D. B.; Fu, C. Y. S.; Pickering, T. C.; Hilary, S.; McGrath, J. E. *Mater. Res. Soc. Symp. Proc.* **1994**, *328*, 589.
65. Hergenrother, P. M.; Connell, J. W.; Labadie, J.; Hedrick, J. L. *Adv. Polym. Sci.* **1994**, *117*, 67.
66. Labadie, J. W.; Hedrick, J. L. *Die Makromol. Chem. Macromol. Symp. Ser.* **1992**, *54/55*, 313.
67. Hedrick, J. L.; Labadie, J. W.; Russell, T. P. *Macromolecules* **1991**, *24*, 4559.
68. Hedrick, J. L.; Hilborn, J.; Palmer, T.; Labadie, J. W.; Volksen, W. *J. Polym. Sci.: Polym. Chem. Ed.* **1990**, *28*, 2255.
69. Stenzenburger, H. D.; Roemer, W.; Hergenrother, P. M.; Jensen, B.; Breitigam, W. *SAMPE J.* **1990**, *26*, 75.
70. Smith, J. G.; Connell, J. W.; Hergenrother, P. M. *Polym. Prepr.* **1993**, *34*, 431.
71. Harris, F. W.; Korleski, J. E. *Polym. Mater. Sci. Eng. Proc.* **1989**, *61*, 870.
72. Labadie, J. W.; Hedrick, J. L.; Boyer, S. *J. Polym. Sci.: Polym. Chem. Ed.* **1992**, *30*, 519.
73. Twieg, R.; Hedrick, J. L. *Macromolecules*, submitted 1995.
74. Hedrick, J. L.; Twieg, R. J.; Matray, T.; Carter, K. *Macromolecules* **1993**, *26*, 4833.
75. Strukelj, M.; Hamier, J.; Elce, E.; Hay, A. S. *J. Polym. Sci.: Polym. Chem. Ed.* **1994**, *32*, 193.
76. Hedrick, J. L. *Macromolecules* **1991**, *24(23)*, 6361.
77. Hilborn, J. G.; Labadie, J. W.; Hedrick, J. L. *Macromolecules* **1990**, *23*, 2854.
78. Carter, K. R.; Jonsson, H.; Twieg, R.; Miller, R. D.; Hedrick, J. L. *Polym. Prepr.* **1992**, *33(1)*, 388.
79. Hedrick, J. L.; Twieg, R. *Macromolecules* **1992**, *25*, 2021.
80. Carter, K.; Miller, R.; Hedrick, J. L. *Macromolecules* **1993**, *26*, 2209.
81. DeSimone, J. M.; Stompel, S.; Samulski, E. T. *Polym. Prepr.* **1991**, *32(1)*, 172.
82. DeSimone, J. M.; Sheares, V. V.; Samulski, E. T. *Polym. Prepr.* **1992**, *33(1)*, 418.
83. Fukawa, I.; Tanabe, R. *J. Polym. Sci., Part A: Polym. Chem. Ed.* **1992**, *30*, 1977.
84. Connell, J. W.; Hergenrother, P. M. *Polym. Prepr.* **1988**, *29(1)*, 172; *Polymer* (in press).
85. Connell, J. W.; Hergenrother, P. M. *Polym. Mater. Sci. Eng. Proc.* **1989**, *60*, 527.
86. Connell, J. W.; Hergenrother, P. M. *J. Polym. Sci., Part A: Polym. Chem. Ed.* **1991**, *29*, 1667.
87. Connell, J. W.; Hergenrother, P. M. *High Perf. Polymers* **1990**, *2(4)*, 211.

88. Connell, J. W.; Hergenrother, P. M.; Wolf, P. *Polym. Mater. Sci. Eng. Proc.* **1990**, *63*, 366.
89. Connell, J. W.; Hergenrother, P. M.; Wolf, P. *Polymer* (in press).
90. Bass, R. G.; Srinivasan, K. R. *Polym. Prepr.* **1991**, *32(1)*, 619.
91. Bass, R. G.; Srinivasan, K. R.; Smith Jr., J. G. *Polym. Prepr.* **1991**, *31(2)*, 160.
92. Priddy, D. B.; Franks, M.; Konas, M.; Urana, M. A.; Yoon, T. H.; McGrath, J. E. *Polym. Prepr.* **1993**, *34*, 310.
93. Smith Jr., J. G.; Connell, J. W.; Hergenrother, P. M. *Polym. Prepr.* **1991**, *32(3)*, 193.
94. Hergenrother, P. M.; Smith Jr., J. G.; Connell, J. W. *Polymer* (in press).
95. Hergenrother, P. M.; Smith Jr., J. G.; Connell, J. W. *Polym. Prepr.* **1992**, *33(1)*, 411.
96. Kricheldorf, H. R. *Makromol. Chem., Macromol. Symp.* **1992**, *54/55*, 365.
97. Kricheldorf, H. R.; Bier, G. *J. Polym. Sci., Polym. Chem. Ed.* **1983**, *21*, 2283.
98. Kricheldorf, H. R.; Meier, J.; Schwarz, G. *Makromol. Chem., Rapid Commun.* **1987**, *8*, 529.
99. Kricheldorf, H. R.; Schwarz, G.; Erxleben, J. *Makromol. Chem.* **1988**, *189*, 2255.
100. Kricheldorf, H. R.; Bier, G. *Polymer* **1984**, *25*, 1151.
101. Kricheldorf, H. R.; Jahnke, P. *Makromol. Chem.* **1990**, *191*, 2027.
102. Hedrick, J. L.; Brown, H. R.; Hofer, D. C.; Johnson, R. D. *Macromolecules* **1989**, *22*, 2048.
103. Wang, Z. Y.; de Cavalho, H. N.; Hay, A. S. *J. Chem. Soc., Chem. Commun.* **1991**, 1221.
104. Wang, Z. Y.; Hay, A. S. *Polymer* **1992**, *33*, 1778.
105. Ueda, M.; Sato, M. *Macromolecules* **1987**, *20*, 2675.
106. Colquhoun, H. M.; Lewis, D. F. *Polymer* **1988**, *29*, 1902.
107. Ueda, M.; Oda, M. *Polym. J.* **1989**, *21*, 673.
108. Colon, I.; Kwiatkowski, G. T. *J. Polym. Sci.: Polym. Chem. Ed.* **1990**, *28*, 367.
109. Percec, V.; Wang, J. H.; Oishi, Y. *J. Polym. Sci.: Polym. Chem. Ed.* **1991**, *29*, 949.
110. Deeter, G. A.; Moore, J. S. *Macromolecules* **1993**, *26*, 2535.
111. Jurek, M. J.; McGrath, J. E. *Polym. Prepr.* **1987**, *28(1)*, 180.
112. Percec, V.; Kawasumi, M. *Macromolecules* **1991**, *24*, 6318.
113. Percec, V.; Yourd, R. *Macromolecules* **1988**, *21*, 3379.
114. Hawker, C. J.; Frechet, J. M. J. *J. Am. Chem. Soc.* **1990**, *112*, 7638.
115. Uhrich, K. E.; Frechet, J. M. J. *Polymer* **1994**, *35*, 1739.
116. Brunelle, D. J.; Boden, E. P.; Shannon, T. G. *J. Am. Chem. Soc.* **1990**, *112*, 2399.
117. (a) Mullins, M. J.; Galvan, R.; Bishop, M. T.; Woo, E. P.; Gorman, D. B.; Chamberlin, T. A. *Polym. Prepr.* **1991**, *32(2)*, 174; (b) Mullins, M. J.; Woo, E. P.; Murry, D. J.; Bishop, M. T. *CHEMTECH* **1993** (August), 25.
118. Xie, D.; Gibson, H. W. *Polym. Prepr.* **1994**, *35(1)*, 401.
119. Chan, K. P.; Wang, Y.; Hay, A. S. *Macromolecules* **1995**, *28*, 653.

RECEIVED December 4, 1995

Chapter 13

Synthesis and Properties of a Novel Series of Poly(arylene ether ketone)s Containing Unsymmetric Benzonaphthone Units

J. E. Douglas and Z. Y. Wang[1]

Department of Chemistry, Ottawa-Carleton Chemistry Institute, Carleton University, 1125 Colonel By Drive, Ottawa K1S 5B6, Canada

The design and synthesis of four unsymmetric 4,4'-dihalobenzo-1'-naphthone monomers and their polymerizations with various bisphenols, 4,4'-isopropylidenediphenol (BPA), 4,4'-hexafluoroisopropylidenediphenol (6F-BPA), 9,9-bis(4-hydroxyphenyl)fluorene (FBP) and 1,4-hydroquinone (HQ), are reported. The enhanced reactivity of the naphthoyl group, as measured by its ability to activate chlorides towards displacement with phenoxides in S_NAr-type polycondensation is clearly demonstrated. Copolymerizations of the dichloride and difluoride monomers in varying ratios were done with BPA. Model reactions were carried out and showed that chlorine on the benzoyl unit of the monomer exchanges more quickly with fluoride ion than chlorine on the naphthoyl moiety. The rate of this halogen exchange was found to be of little significance on the polyetherification time scale. An end-capped polymer with BPA was made and had an absolute number average molecular weight (Mn) of 58,300, corresponding to a monomer repeat length (n) of 127. Proton NMR shows the presence of three sequential isomers of this polymer series. The effects of N-methyl-2-pyrrolidinone and tetramethylene sulfone as solvents were also studied. The glass transition temperatures (Tgs) increased by 20-45 °C, relative to the Tgs of poly(arylene ether ketone)s containing a benzophenone unit. Decomposition temperatures for 5% weight loss as assessed by thermogravimetric analysis for all polymers were above 497 °C in air. Young's moduli ranged from 1.99 to 3.25 GPa.

Poly(aryl ether ketone)s (PAEKs) represent a class of engineering thermoplastics which possess properties (e.g. mechanical strength and durability) that are superior

[1]Corresponding author

0097–6156/96/0624–0226$12.00/0

to those of most commodity plastics. They compete with metals, ceramics and glass in a variety of applications. PAEKs are principally made by nucleophilic displacement polycondensation of bisphenoxides and activated dihaloarylketones (*1,2*). For example, PEEK is made by polymerization of 1,4-hydroquinone (HQ) with 4,4'-difluorobenzophenone (*3*). It is generally believed that these polycondensations proceed through a S_NAr mechanism involving an intermediate Meisenheimer complex (*4*). There are many factors which govern the success of these reactions. Some of these include solvent, temperature, the reactivity and chemical specificity of the pair of monomers used. It is well known that the rates of S_NAr reactions are greater for difluoroarylketones than their chloride analogues (*4*). Consequently, it is generally accepted that difluoroarylketones are activated toward nucleophilic displacement and can form high molecular weight polymers with little difficulty (*5-10*). However, the displacement of chlorine from dichloroarylketones with bisphenoxides to form high molecular weight PAEKs is more difficult (*5-10*). The latter are believed to undergo a competitive reductive dehalogenation reaction via a $S_{RN}1$ SET mechanism which accounts for the formation of low weight PAEKs (*6,8,9*). Exceptions include the use of (a) a diketo monomer having a 1:1 ketone/chlorine ratio as opposed to one with a 1:2 ketone/chlorine ratio (*5,10*), (b) a small amount of potassium fluoride in the presence of a phase transfer catalyst (*11*), (c) 'reactive' bisphenols such as 9,9-bis(4-hydroxyphenyl)fluorene (HPF) (*10*), and (d) 'reactive' dichloroheteroarylketones such as bis(5-chlorothienyl-2)ketone (*12*). According to the hard-soft-acid-base theory, certain bisphenols (as bases) have stronger attractions to particular dihaloarylketones (as acids) (*13,14*). Thus, the pairing up of bisphenols with appropriate dihaloarylketones becomes an important factor in achieving high molecular weight PAEKs (*5*).

For the past decade, there has been much research done in an effort to improve the thermal stability and mechanical strength of high performance thermoplastics. For a polymer to be considered 'thermally stable' or 'heat resistant', it should not decompose below 400 °C and should retain its useful properties close to its decomposition temperature (Td) over a long period of time (*15*). A common strategy to improve the thermal stability has been to increase the number of aromatic units in the repeat unit of the polymer backbone. Unfortunately, much of the progress made in enhancing thermal stabilities has come at a cost of poor processibility and solubility of the polymer. It has become the focus of scientists to develop new materials with improved solubilities and processibility, as well as high thermal and mechanical strengths.

In the late 80's, Hergenrother developed a PAEK containing a 2,6-dibenzoylnaphthalene unit with a Tg of 185 °C (*10*). In an effort to improve solubility, Ritter reported the synthesis of 1,5-naphthylene-based PAEKs with long alkoxy chains which resulted in poor thermal stability (*16*). Endo *et. al.* recently introduced the methyl groups at the 2,6-positions of a 1,5-naphthylene unit in PAEK to improve solubility and the Tg (222 °C) (*17*). We have recently designed and synthesized four unsymmetric dihalobenzo-1'-naphthone monomers **1a-d** containing chlorine and fluorine (Figure 1) (*18,19*). The structures of these monomers offer a unique unsymmetry previously rarely seen in PAEK history. It became possible now to study the effect the unsymmetric benzonaphthone unit would have on sequential ordering as well as thermal and mechanical properties.

Preliminary results have demonstrated that all four monomers **1a-d** can be polymerized effectively with many bisphenols to form high molecular weight PAEKs **2-5** (Figure 1) (*18,19*). As part of an ongoing systematic investigation of the effects of structural symmetry and the substitution pattern of the arylene core in PAEKs on the polymer properties, the detailed synthesis and properties of a series of PAEKs derived from **1a-d** and bisphenols are described herein.

Polycondensation

In the process of studying this series of monomers, it was found that both **1a** and **1b** could make high molecular weight polymers with BPA, whereas **1c** and **1d** could not (Table I). Reverse precipitation was performed in most cases, except entries 3 and 4 in Table I and entry 4 in Table II. This technique will narrow the polydispersity index and increase the apparent molecular weight (by GPC). But the effect to the solution viscosity will be minimal. Both **1c** and **1d** contain the 4-chlorobenzoyl moiety which has been reportedly difficult to polymerize with BPA (10). Whereas **1a** and **1b** contain the undoubtedly reactive 4-fluorobenzoyl moiety and the chloronaphthoyl group. The naphthyl ring appears to promote chlorine displacement in polyetherifications with bisphenoxides relative to the phenyl ring (18,19) This can be attributed to the increased conjugation and stability imparted to the Meisenheimer complex by the naphthyl ring. Hückel molecular orbital (HMO) calculations also suggest that the chloronaphthyl ring be more reactive than the chlorophenyl unit towards nucleophilic displacement (20).

Table I. Polymerizations of Dihalides 1a-d with Bisphenols (*19*)

Entry	*Polymer*	*Dihalide*	*Bisphenol*	*Mw x 10^{-4}*	*Mn x 10^{-4}*	*Mw/Mn*	η_{inh} *(dL/g)*
1	**2**	**1a**	BPA	16.0	5.70	2.80	0.90
2	**2**	**1b**	BPA	10.9	4.12	2.64	0.58
3	**2**	**1c**	BPA	3.67	1.88	1.95	0.28
4	**2**	**1d**	BPA	3.98	2.23	1.76	0.27
5	**3**	**1a**	HQ	7.20	2.26	3.19	0.61
6	**3**	**1b**	HQ	6.05	2.30	2.63	0.60
7	**3**	**1c**	HQ	6.73	2.66	2.53	0.56
8	**3**	**1d**	HQ	6.03	2.25	2.68	0.55
9	**4**	**1a**	6F-BPA	16.9	9.10	1.86	0.84
10	**4**	**1b**	6F-BPA	11.6	4.95	2.35	0.73
11	**4**	**1c**	6F-BPA	6.00	1.96	3.05	0.43
12	**4**	**1d**	6F-BPA	12.4	6.26	1.98	0.65
13	**5**	**1a**	FBP	18.8	10.0	1.88	0.76
14	**5**	**1b**	FBP	20.2	9.89	2.05	0.86
15	**5**	**1c**	FBP	11.9	5.15	2.30	0.55
16	**5**	**1d**	FBP	13.4	3.72	3.59	0.66

Copolymerizations of **1a** and **1d** in molar ratios of 1:1, 1:2, 1:4 and 1:9 with BPA were then performed (Table II). Reasonably high molecular weights were achieved with ratios of **1a** to **1d** as low as 1 to 4 (Entry 3, Table II), in comparison with polymerization of **1d** and BPA (Entry 4, Table I). In light of these results, a halogen exchange reaction between the chlorine on the phenyl part with *in-situ* generated fluoride ion during polycondensation seemed possible. Therefore, two model reactions were performed to investigate this phenomenon.

Table II. Copolymerization of monomers 1a and 1d with BPA

Entry	*Molar Ratio* ***1a:1d***	*Mw* $x\ 10^{-4}$	*Mn* $x\ 10^{-4}$	*Mw/Mn*	η_{inh} *(dL/g)*
1	1:1	10.6	5.40	1.96	0.55
2	1:2	6.18	2.89	2.14	0.44
3	1:4	4.97	2.68	1.86	0.37
4	1:9	4.62	2.16	2.14	0.29

In the first model reaction, **1d** was subjected to the same conditions used in all previous polymerizations (210 °C, $TMSO_2$/PhCl, and K_2CO_3) without BPA but with KF to promote the halogen exchange of chlorine with fluorine (Figure 2). It was found that **1b** formed faster and in a greater yield than **1c**. This demonstrated that the Cl-F exchange occurs more rapidly on the phenyl ring than on the naphthyl ring. A halogen exchange of this kind would facilitate polymerization.

A second model reaction was done in which fluoride ion was generated *in situ* by reacting two equivalents of phenol with **1a** (Figure 3). Monomer **1d** was then added to the reaction mixture to see if any Cl-F exchange occurred. Using the *in-situ* generated fluoride ion was believed to be closer to the actual polymerization environment than the addition of KF salt. It was discovered that **1b** was formed in greater quantity than **1c** but at a much slower rate than in the first model reaction. Although being confirmed, this slow halogen exchange perhaps plays a minor role in the actual polymerization mechanism.

An end-capping experiment was carried out with a 1% excess of difluoride **1a** and 2% of 3,5-di-*tert* -butylphenol relative to BPA. End-capping controls the molecular weight and allows one to calculate an exact molecular weight as opposed to a relative one as given by GPC. From the high resolution ^{1}H NMR, the relative integration values of the peaks at 1.6 ppm (the methyl group of BPA) and 1.3 ppm (*tert* -butyl group) was calculated and the number of repeat unit (n) was found to be 127 (Figure 4). Accordingly, the absolute number-average molecular weight (Mn) was calculated to be 58,300, larger than that obtained by GPC (Mn = 47,300). Therefore, the exact molecular weights of these polymers should be greater than the apparent ones determined by GPC. Inherent viscosity for this end-capped polymer **6** was found to be 0.47 dL/g.

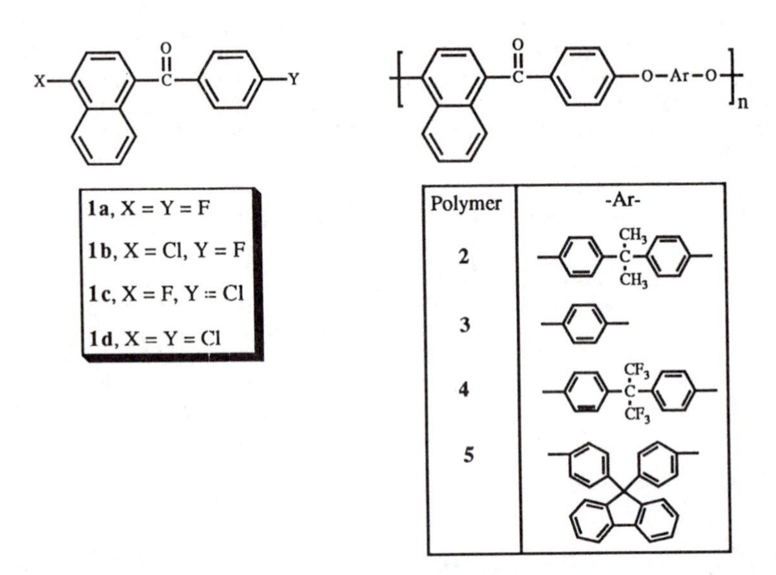

Figure 1. Poly(arylene ether ketone)s **2-5** derived from dihalobenzonaphthones **1a-d**.

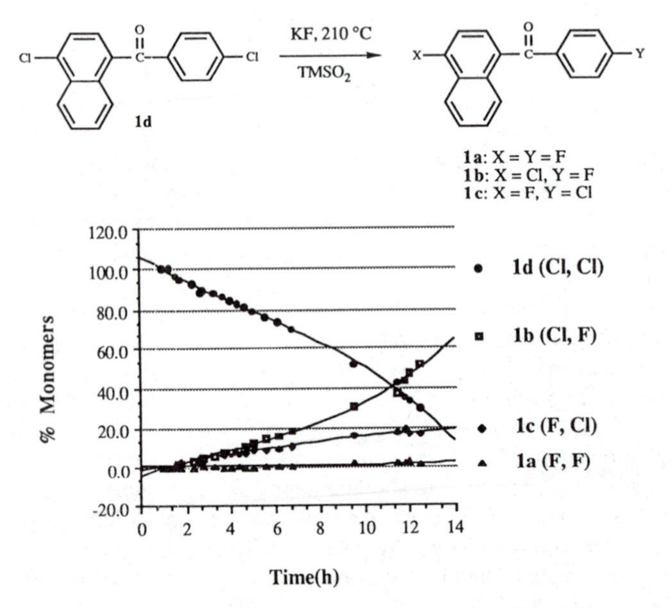

Figure 2. Model Cl-F exchange reaction using KF.

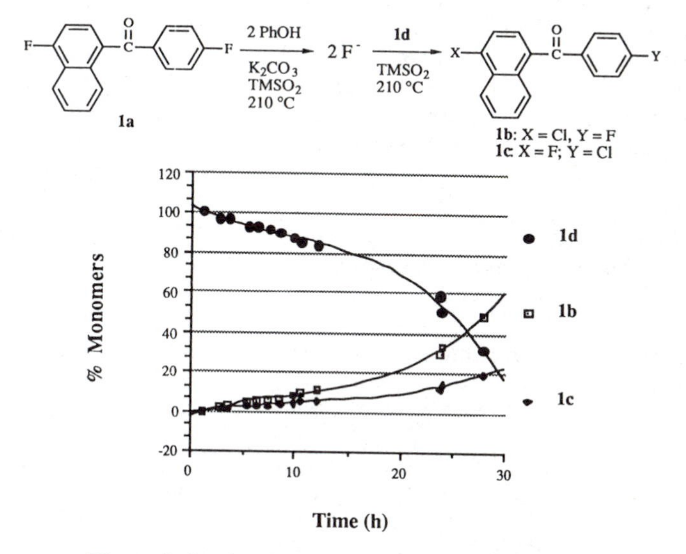

Figure 3. *In-situ* Cl-F exchange model reaction.

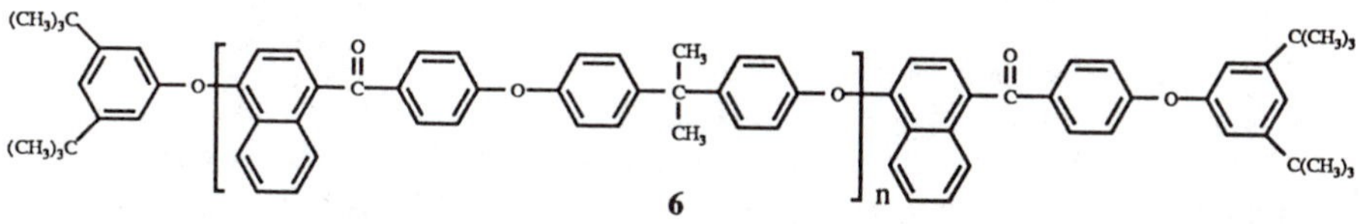

(CH3)3C
(CH3)3C
O
O
C
O
CH3
C
CH3
O
n
O
C
O
C(CH3)3
C(CH3)3
6

Figure 4. ^{1}H NMR of end-capped poly(arylene ether ketone) **6**.

Solvent Effects

Two solvent systems were investigated in this study. The first one involved NMP with toluene as an azeotroping solvent in a 20% solid content (Table III). It was possible to make high molecular weight polymers from **1a** with both HQ and BPA. But difficulties arose when trying to polymerize **1b** with BPA (Entry 3). A modest molecular weight of 3.64 $\times 10^4$ (η_{inh} = 0.36 dL/g) was achieved in NMP. A dark green color was also observed in the reaction mixture. Strukelj *et. al.* studied the solvent effects in the preparation of poly(aryl ether benzils)s and suggested that this green color stems from impurities in the commercially available NMP (21). Decomposition products of NMP or reactions of reactive species with NMP, also seem to have a dramatic effect on molecular weight and appear to degrade the polymer after molecular weight build-up. Percec *et. al.* also observed that reductive dehalogenation of diaryl halides can occur in NMP (6). In an effort to avoid these problems and increase molecular weight, a second solvent system, $TMSO_2$ and chlorobenzene, was used. A better result was obtained for polymerization of **1b** and BPA (Mw = 10.86 x 10^4 ; η_{inh} = 0.58 dL/g) at 210 °C in a 35% solid content when using this alternative solvent system.

Table III. Polymerization of 1a and 1b with BPA and HQ in NMP

Entry	*Monomer*	*Bisphenol*	*Temp.* (°C)	*Mw* x 10^{-4}	*Mn* x 10^{-4}	*Mw/Mn*	η_{inh} (dL/g)
1	**1a**	BPA	180	13.8	6.68	2.06	0.73
2	**1a**	HQ	180	7.63	3.89	1.96	0.57
3	**1b**	BPA	180	3.64	2.14	1.70	0.36
4	**1b**	HQ	180	8.09	3.79	2.13	0.54

The unsymmetry of the microstructure of this polymer series stems from the unsymmetric nature of the dihalobenzonaphthone monomers **1a-d**, since all four bisphenols used in this study are symmetrical. There are three possible diads in this series: (a) head to tail, (b) head to head, and (c) tail to tail (Figure 5). Evidence of these sequential isomers comes clearly from the ^{1}H NMR, as three singlets can be seen at 1.6 ppm for the end-capped polymer **6** (Figure 4). Although the relative abundance of three isomers can be quantitated (*ca.* 1:2:1), the peaks can not be assigned to any specific diads.

Thermal and Mechanical Properties

In all cases, insertion of the naphthyl ring into the PEEK-like polymer series increases the Tgs by 20-45 °C. Polymer **3** experiences the greatest increase in Tg (45 °C) over PEEK's value of 143 °C. But none of these PAEKs showed any sign of crystalline behavior according to differential scanning calorimetry (DSC).

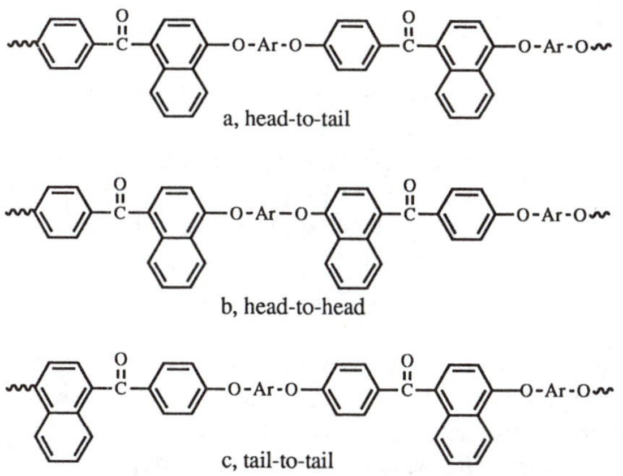

a, head-to-tail

b, head-to-head

c, tail-to-tail

Figure 5. Structures of three diads for PAEKs derived from unsymmetric dihalide **1**.

Although polymer **3** is very similar to PEEK in their primary structures, the bulky naphthyl ring and the random sequential order may prohibit the former from effective packing for crystallization. The decomposition temperatures (Td) ranged from 497 °C to 519 °C in static air for polymers **2-5** (Table IV), as assessed by thermogravimetry.

All values obtained from thermomechanical analysis (TMA) were done in the tensile stress-strain mode. Young's moduli for all polymers **2-5** fall within the same order of magnitude (1.99 - 3.25 GPa, Table IV), which are typical for engineering PAEKs. At elevated temperatures, films of these polymers maintain good mechanical properties in the gigapascal range up to 180 °C. Maximum tan δ values were in the range of 179 °C to 253 °C, which agree well with the Tg values obtained from DSC.

Table IV. Thermal and mechanical properties of polymers 2-5

PAEK	Td (° C)[a] (air)	Tg (°C) DSC	Tg (°C) TMA[b]	Young's Modulus (E', GPa)
2	497	189	179	1.99
3	506	188	183	2.39
4	517	199	185	2.59
5	519	272	253	3.25

a Decomposition temperature taken at 5% weight loss.
b Taken as the max. tan δ value in static air.

Experimental

Materials. Tetramethylene sulfone ($TMSO_2$) (99%), N-methyl-2-pyrrolidinone (NMP) (99%), N,N-dimethylacetamide (DMAc) (99%), toluene, chlorobenzene and 3,5-di-*tert*-butylphenol were purchased form Aldrich Chemical Co. and used without further purification. Anhydrous potassium carbonate was ground in a crucible before use. 1,4-Hydroquinone (HQ) was recrystallized from ethanol. 4,4'-Isopropylidenediphenol (BPA), 4,4'-hexafluoroisopropylidenediphenol (6F-BPA) and 9,9-bis(4-hydroxyphenyl)fluorene (FBP) were recrystallized from toluene.

Measurements. Melting points were measured on a Fisher-Johns Melting point apparatus. Mass spectra were taken from a VG7070E Mass Spectrometer. Apparent molecular weights were determined by gel permeation chromatography (GPC) using a PL-Gel column (5μ particle size) and chloroform as the eluting solvent. Molecular weights were based on calibrations with polystyrene standards and a UV detector was set at 254-nm wavelength. Inherent vicosities were measured using 0.5 g/dL chloroform solutions at 25.0 °C in a calibrated Ubbelohde dilution viscometer. ^{1}H and ^{13}C NMR spectra were obtained on a Bruker-400 instrument using deuterated chloroform. ^{19}F NMR was measured on a Varian XL-300 using deuterated DMSO as a solvent and CCl_3F as a reference. IR spectra were measured on a Perkin Elmer Series 1600 FTIR and Bomem-FTIR Michelson Series instrument. Thermal stabilities of the polymer samples were determined using a Seiko 220 TG/DTA analyzer run from ambient temperature to 1000 °C at 10 °C/min. Tests were done in nitrogen flushed at 200 mL/min. Decomposition temperatures (Td) were taken at 5% weight loss. The glass transition temperatures (Tg) were determined using a Seiko 220C DSC used in normal DSC mode, with a heating rate of 10 °C/min from ambient temperature to 35°C below Td. Three consecutive runs were performed for each polymer (with normal cooling in between scans), and Tg was taken from the third scan as the midpoint of the change in slope of the baseline. The tensile mechanical properties of the polymer films were obtained using a Seiko 120C TMA/SS analyzer operated in stress-strain mode at a heating rate of 3 °C/min to 40 °C above the Tg.

Monomer Synthesis. 1-Chloro and 1-fluoronaphthalene readily undergo the Friedel-Crafts reaction with the corresponding 4-halobenzoyl chlorides to form the four monomers **1a-d**. The following is a typical procedure for the synthesis of these monomers: To a 100 mL, three-necked, round-bottomed flask containing a magnetic stirring bar flame dried under nitrogen flow were added 4-fluorobenzoyl chloride (5.060 g, 31.83 mmol), 1-fluoronaphthalene (5.120 g, 31.83 mmol) and dry CH_2Cl_2 (30 mL). The reaction flask was then cooled in an ice bath (5 °C) before anhydrous $AlCl_3$ (35.10 g, 38.50 mmol) was added slowly. The solution turned yellow then a dark red/brown color. After an hour the solution was quenched with iced water and turned yellow. The organic layer was washed with aqueous HCl (5%), aqueous NaOH (5%), brine (23%) and water. The organic layers were combined and dried over anhydrous $MgSO_4$. The solvent was removed under reduced pressure to give **1a** as white crystals (crude yield 69%; 55 % yield after recrystallization from; mp 84-86 °C). IR (KBr) 1657.3 cm^{-1} (C=O); MS (EI, m/e, relative intensity %) 173 (M^{+} - ·PhF, 100), 268 (M^{+}, 87.6), 123 ($COPhF^{+}$, 68.9), 145 (M^{+} - ·COPhF, 63.3), 95 (PhF^{+}, 54.4); ^{1}H NMR (400 MHz) δ 8.16 (d, 1 H), 8.21 (d, 1 H), 7.85-7.91 (m, 2 H), 7.54-7.64 (m, 3 H), 7.12-7.21 (m, 3 H); ^{19}F NMR (282.2 MHz) δ -104.93 (Ph-F), -117.22 (Naph-F).

1b: white cubic crystals, 70% crude yield, 63% yield after recrystallization from hexane/cyclohexane; mp 117-118 °C; IR (KBr) 1656.4 cm^{-1} (C=O); MS (EI, m/e, relative intensity %) 123 ($COPhF^{+}$, 100), 284 (M^{+}, 92.0), 189 (M^{+} - ·PhF, 76.3), 95 (PhF^{+}, 66.6), 161 (M^{+} - ·COPhF, 41.7); ^{1}H NMR (400 Mhz) δ 8.39 (d, 1H), 8.05 (d, 1 H), 7.85-7.94 (m, 2 H), 7.63 (d, 1 H), 7.54-7.68 (m, 2 H), 7.47 (d, 1 H), 7.10-7.16 (m, 2 H)); ^{19}F NMR (282.2 MHz) δ -104.32.

1c: white fluffy needles (63% yield after recystallization from EtOH); mp 84 °C); IR (KBr) 1644.2 cm^{-1} (C=O); MS (EI, m/e, relative intensity %) 173 (M^{+} - ·PhCl, 100), 284 (M^{+}, 55.7), 145 (M^{+} - ·COPhCl, 55.3), 249 (M^{+} - ·Cl, 31.9), 139 ($COPhCl^{+}$, 36.9); ^{1}H NMR (400 MHz) δ 8.19 (m, 2 H), 7.79 (d, 2 H), 7.61(m, 2 H), 7.55 (d, 1 H), 7.45 (d, 2 H), 7.17 (d, 1 H); ^{19}F NMR (282.2 MHz) δ -116.79.

1d: white prismatic needles (45% crude yield, initially recrystallized in cyclohexane to remove orange impurities, then washed in methanol and finally recrystallized in EtOH with a yield of 14%); IR (KBr) 1644.6 cm^{-1} (C=O); MS (EI, m/e, relative intensity %) 189 (M^{+} - ·PhCl, 100), 139 ($COPhCl^{+}$, 73.2), 300 (M^{+}, 67.7), 265 (M^{+} - ·Cl, 54.3), 161 (M^{+} - ·COPhCl, 52.0); ^{1}H NMR (400 MHz) δ 8.39 (d, 1 H), 8.07 (d, 1 H), 7.79 (d, 2 H), 7.55-7.69 (m, 2 H) 7.63 (d, 1 H), 7.47 (d, 1 H), 7.44 (d, 2 H).

Polymerization. General procedures for both solvent systems used in this study are described below.

NMP/Toluene. A 50 mL, three necked round bottom flask equipped with a Dean-Stark trap connected with condenser and a nitrogen purge line was flame dried. Monomer **1a** (1.070 g, 4.000 mmol), BPA (0.910 g, 4.000 mmol) and K_2CO_3 (1.110 g, 8.000 mmol) were added to the flask with NMP (10 mL) and toluene (20 mL). The mixture was purged with nitrogen and stirred for 20 min, then heated slowly to 120-130 °C. The water and toluene were co-distilled while deoxygenated toluene was introduced. The solution turned yellow initially then darker orange. After 2-3 h, the temperature was increased to 180 °C. The solution turned green and became viscous after one hour at higher temperature. The reaction was monitored by GPC and stopped after 2 h. After letting it cool, it was diluted with NMP (9 mL), then added dropwise into methanol containing a few drops of concentrated HCl. All polymers were purified by dissolving in chloroform, filtering through Celite to remove any salts and precipitating into methanol again. The low molecular weight fractions were removed in all cases by reverse precipitation in this solvent system. This technique involved adding methanol dropwise into a solution made with chloroform (50 mL) and the polymer. This coagulated to form a gummy residue and a white cloudy solution consisting of the lower molecular weight fractions. The cloudy solution was then decanted and the gummy residue was dissolved in chlorofom (5-10 mL) and precipitated finally into methanol (150 mL).

$TMSO_2$/Chlorobenzene. A 50 mL, three-necked, round-bottomed flask equipped with a Dean-Stark trap connected with a condenser and a nitrogen purge line was flame dried. A solution of dihalide monomer (3.000 mmol), BPA (3.000 mmol), potassium carbonate (0.749 g, 5.400 mmol), $TMSO_2$ (4 mL) and chlorobenzene (4 mL) was heated to 210 °C in an oil bath in 30-40 min. Water formed during the reaction was removed as an azeotrope with chlorobenzene. The oil bath temperature was maintained at 210 °C for 2-4 h depending on the progress of the polymerization. The reaction was followed by GPC and the reaction was stopped when the molecular weight was seen to decrease after an initial increase. The polymer was then precipitated directly into methanol and collected by filtration. All polymers were purified by dissolving in $CHCl_3$, filtering through Celite to remove any salts and precipitating into methanol. Reverse precipitation, as described in the above procedure, was performed in all cases except entries 3 and 4, Table I. In these two cases, it was believed that the polymer's molecular weights (as measured by GPC) were too low and therefore too much of the polymer would be lost in reverse precipitation.

Copolymerization. Four copolymerizations were done in varying molar ratios of difluoro (**1a**) to dichloro (**1b**) monomer (ie 1:1, 1:2, 1:4, and 1:9) with BPA on a 3 mmol scale. A typical procedure for the copolymerization is as follows: A 50 mL, three-necked, round-bottomed flask equipped with a Dean-Stark trap connected with a condenser and nitrogen purge line was flame dried. A mixture of **1a** (0.402 g. 1.500 mmol), **1d** (0.450 g, 1.5 mmol), K_2CO_3 (0.746 g, 5.400 mmol), BPA (0.685 g, 3.000 mmol), $TMSO_2$ (4 mL) and chlorobenzene (4 mL) was heated to 210 °C in an oil bath in 40 min. Water formed during the reaction was removed as an azeotrope with chlorobenzene. The oil bath temperature was maintained at 210 °C for 2 h. Two mL of $TMSO_2$ were added to the system when stirring became difficult. The progress of the polymerization was followed by GPC and the reaction was stopped when the molecular weight was seen to decrease after an initial increase. The reaction mixture was then poured into methanol to precipitate the polymer which was collected by filtration, redissolved in solvent and filtered through Celite to remove any salts and precipitated into methanol. With the exception of entry 4, Table II, all copolymers were reverse precipitated like in the above procedures for the same reasons.

Model Reaction 1. A 50 mL, three-necked, round bottomed flask was set up with a Dean-Stark trap, condenser and nitrogen purge line as before. **1d** (0.990 g, 3.000 mmol), K_2CO_3 (0.749 g, 5.400 mmol), $TMSO_2$ (4 mL) and chlorobenzene (3 mL) were added to the system. This was then heated to 210 °C in 40 min and stirred. Standards of the four possible halogen exchange products (**1a**, **1b**, **1c** and **1d**) had been made previously. The progress of the reaction was monitored by HPLC and retention times were used to identify peaks. Relative abundance was determined using the peak area of the four monomers. This did not include the area of the unidentified compound at a lower retention time which emerged during the course of the reaction, small at first, larger at the end. This was believed to be either a decomposition product of $TMSO_2$ or a reductive dehalogenation product.

Model Reaction 2. A 50 mL, three-necked, round bottomed flask with Dean-Stark trap, condenser and nitrogen purge line was flame dried as before. **1a** (0.429 g, 1.500 mmol), phenol (0.282 g, 3.00 mmol), K_2CO_3 (0.749 g, 5.400 mmol), $TMSO_2$ (4 mL) and chlorobenzene (3 mL) were added to the system and heated to 210 °C. After 1h of heating, **1d** (0.450 g, 1.500 mmol) was added. The reaction was again monitored by HPLC.

End Capping. A Dean-Stark trap with a condenser and 50 mL, round-bottomed flask was set up as before and flame dried. **1a** (2.144 g, 8.000 mmol), BPA (1.806 g, 7.920 mmol), K_2CO_3 (2.208 g, 16.00 mmol) and 3,5-di-*tert*-butylphenol (0.055 g, 0.160 mmol), $TMSO_2$ (10 mL) and chlorobenzene (3 mL) were added to the reaction flask. The temperature was maintained at 170 °C for 1.5 h to azeotrope off water. Afterwards, the reaction was increased to 210 °C for 30 min. More $TMSO_2$ (6 mL) was added to allow stirring to continue. The reaction was stopped 30 min later and worked up as before with reverse precipitation. The polymer 6 was purified by reverse precipitation. η_{inh} = 0.47 dL/g; IR (film) 1653 cm^{-1} (C=O); ^{1}H NMR 8.38 (m, 1 H), 8.19 (m, 1 H), 7.82 (d, 2 H), 7.53 (m, 2 H), 7.47 (d,1 H), 7.25 (m, 4 H), 7.03 (d, 2 H), 6.96 (d, 4 H), 6.80 (d, 1 H), 1.69 (t, 6.91H), 1.30 (d, 0.33 H).

Conclusion

A series of poly(arylene ether ketone)s having different sequential diad structures derived from unsymmetric 4,4'-dihalobenzo-1'-naphthone and bisphenols are readily soluble in chloroform, 1,1,2,2-tetrachloroethane, and NMP. They appear to amorphous and show relative higher Tgs than the analogues derived from 4,4'-dihalobenzophenone and bisphenols, as assessed by DSC. The chloronaphthoyl group was found to be more reactive than the chlorophenyl group in the S_NAr polycondensation. Although model reactions show a Cl-F exchange reaction under the polymerization conditions, its contribution to the increase in molecular weight of the polymer is minimal.

Acknowledgments. We thank the Natural Sciences and Engineering Research Council of Canada for financial support.

Literature Cited

1. Parodi, F. in *Comprehensive Polymer Science*; Allen, G.; Bevington, J. C., Eds; Pergamon Press: Oxford, 1989; Vol. 5, p 561.
2. Saunders, K. J. *Organic Polymer Chemistry*; Chapman and Hall: New York, 1988; p 284.
3. ICI Americas, Inc. Wilmingtion, DE 19897, USA.
4. March, J. *Advanced Organic Chemistry*; 4th Ed., John Wiley and Sons, Inc.: Toronto, 1992; pp 641-653.
5. Percec, V.; Grigoras, M.; Clough, R. S.; Fanjul, J. *J. Poly. Sci.: Part A: Poly .Chem..* **1995**, *33*, 331.
6. Percec, V.; Clough, R. S.; Rinaldi, P. L.; Litman, V. E. *Macromolecules*, **1994**, *27*, 1535.
7. Percec, V.; Clough, R. S.; Fanjul, J.; Grigoras, M. *Polym. Prepr.*, **1993**, *34*(1), 162.
8. Percec, V.; Clough, R. S.; Rinaldi, P. L.; Litman, V. E.; *Macromolecules*, **1991**, *24*, 5889.
9. Percec. V.; Clough, R. S.; Rinaldi, P. L.; Litman, V.E. *Polym. Prepr.*, **1991**, *32*(1), 353.

10. Hergenrother, P. M.; Jensen, B. J.; Havens, S. J. *Polymer*, **1988**, *29*, 358.
11. Hoffman, U.; Helmer-Metzmann, F.; Klapper, M.; Müllen, K. *Macromolecules*, **1994**, *27*, 3574.
12. DeSimone, J. M.; Sheares, V. V. *Macromolecules*, **1992**, *25*, 4235.
13. Pearson, R. G.; Songstand, J. *J. Am. Chem. Soc.* **1967**, *89*, 1827.
14. Ho, T. L. *Chem. Rev.* **1975**, *75*, 1.
15. Stevens. M. P., *Polymer Chemistry- An Introduction*; 2nd Edition, Oxford University Press: NY, NY, 1990.
16. Ritter, H.; Thorwirth, R. *Makromol. Chem..* **1993**, *194*, 1469.
17. Endo, T.; Takata, T.; Ohno, M. *Macromolecules*, **1993**, *27*, 3447.
18. Douglas, J. E.; Wang, Z. Y. *Polym. Prepr.*, **1995**, *36*(1), 753.
19. Douglas, J. E.; Wang, Z. Y. *Macromolecules*, **1995**, *28*, 5970.
20. Hückel molecular orbital program is available from the Molecular Modeling System, Cambridge Scientific Computing Inc., 875 Massachusetts Ave., Suite 61, Cambridge, MA. The greater the net charge at the carbon where the halogen is bonded, the more reactive it is with a phenoxide in nucleophilic displacement polycondensation. 4,4'-Dichlorobenzophenone has a net charge of 0.025 at C-4 and generally cannot polymerize to high molecular weight, whereas 4,4'-difluorobenzophenone can (net charge = 0.050 at C-4). Calculations for monomers **1b** and **1d** showed that the carbon on the naphthyl ring to which chlorine is attached has a net charge of 0.044 and 0.045, respectively. These calculations, although not always accurate, indicate the increased aptitude the naphthyl ring has to displace chlorine over the phenyl ring.
21. Strukelj, M.; Hedrick, J.; Hedrick, J.; Tweig, R. *Macromolecules*, **1994**, *27*, 6277.

RECEIVED December 4, 1995

Chapter 14

Synthesis of Novel Poly(aryl ether)s

Allan S. Hay

**Department of Chemistry, McGill University,
801 Sherbrooke Street West, Montreal, Quebec H3A 2K6, Canada**

Several new classes of novel poly(aryl ether)s have been synthesized which have very high glass transition temperatures. Most of these polymers are soluble in solvents such as dichloromethane, chloroform or chlorobenzene at room temperature and they can be cast into flexible films. Poly(aryl ether)s containing phthalocyanine moietes have also beem synthesized. Methods have been developed to introduce reactive functionality, e.g diaryl acetylene and diaryl cyclopropane moieties, into the backbone of these polymers so that they can be thermally cross-linked after fabrication.

For the past several years we have been studying the synthesis of new polymers which are amorphous and have very high glass transition temperatures which allows them to be processed, at least from solution. Many of the polymers that we have synthesized are poly(aryl ether)s which have generally been prepared by standard methods by the reaction of novel activated dihalides with novel bisphenate salts in aprotic dipolar solvents in a nucleophilic displacement polymerization reaction. We have also developed new methods for the incorporation of functional groups into these polymers which allows them to be thermally cured to give insoluble materials. Furthermore, with some of these systems, we have been able to synthesize cyclic oligomeric precursors which can subsequently undergo ring-opening polymerization. The cyclic oligomers have very low melt viscosities which makes them much easier to fabricate.This paper will review some of these new developments.

I. Soluble Polymers with High Glass Transition Temperatures from Hindered Biphenols.

Highly hindered biphenols **1** (R= CH_3, Ph, Cl) were synthesized by oxidative coupling of the corresponding phenols. High molecular weight poly(aryl ether)s **3** were prepared from these biphenols by reaction with bis(4-fluorophenyl)sulfone or 4,4'-difluorobenzophenone in polar, aprotic solvents in the presence of potassium carbonate. The polymerization reactions were very slow because of the steric hindrance provided by the pendant phenyl groups in these biphenols. The polymers are amorphous and very thermooxidatively stable. They are very soluble in methylene chloride at room temperature and they can be cast into tough, flexible films.[i]

The polymers (**3**, R = CH_3 or Ph; **2**, or C=O) have Tg's in the range 265-285 °C. The poly(ether sulfone) from 3,3',5,5'-tetraphenyl-4,4'-biphenol (**3**; R=H; X = SO_2) could be prepared (Tg = 270 °C), however, the poly(ether ketone) **3** (R=H; X = C=O) could not be synthesized because it is highly crystalline and extremely insoluble and

0097–6156/96/0624–0239$12.00/0

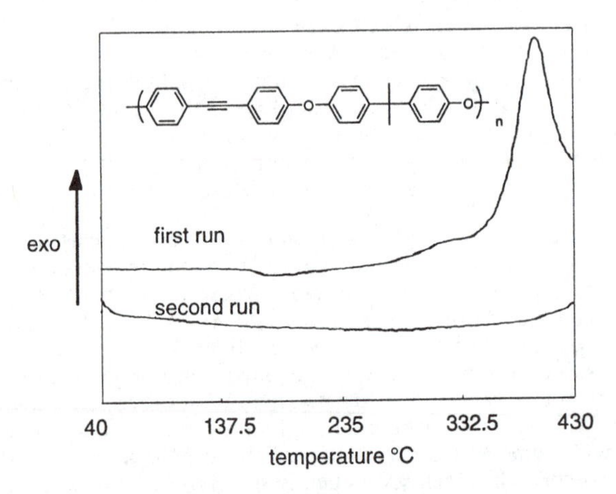

Figure 1. DSC scans of **10**.

comes out of solution during the polymerization reaction. Interestingly, when this biphenol has fluorine substituents in the meta- or para-positions of the pendant phenyl groups, the poly(ether ketone) is amorphous and high molecular weight polymers can be readily synthesized.[i]

*

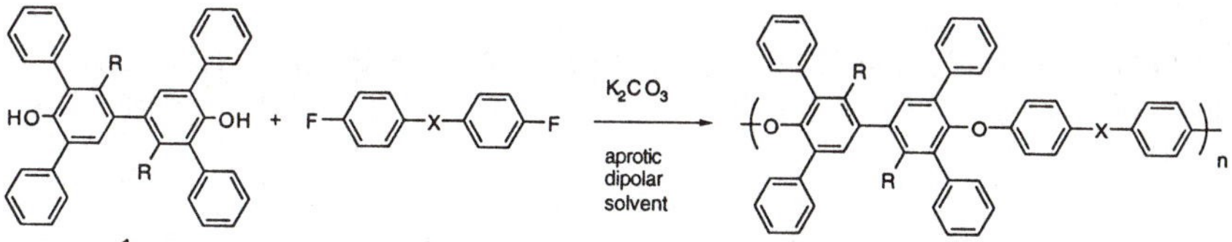

1 + 2 —(K_2CO_3, aprotic dipolar solvent)→ 3

II. Poly(aryl Ether)s Containing Cross-linkable Moieties.

A. Poly(aryl ether)s containing a diarylacetylene moiety. We have discovered that Schiff bases **4** react with N-benzyl derivatives **5** of benzotriazole or benzimidazole to give diarylacetylenes **7** in very high yield. These reactions, when benzotriazole is used, take place in less than one minute at 65 °C! The intermediate in the reaction is the enamine **6** and the benzotriazole and benzimidazole act as pseudohalogens under these conditions and are eliminated in the second step in the reaction.[ii]

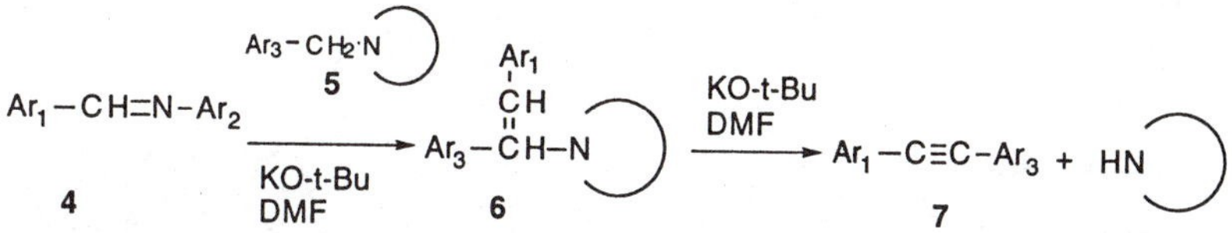

$Ar_1-CH{=}N-Ar_2$ (4) —(Ar_3-CH_2-N (5), KO-t-Bu, DMF)→ $Ar_3-C(=CH-Ar_1)-N$ (6) —(KO-t-Bu, DMF)→ $Ar_1-C{\equiv}C-Ar_3$ (7) + HN

HN = benzimidazole or benzotriazole

Fluoro-substituted diarylacetylenes can be readily prepared by this method and they undergo nucleophilic substitution with displacement of the fluoride, in the presence of excess potassium t-butoxide, to give the t-butyl ethers.[iv] The ethers can be readily converted to the phenols by removal of the t-butyl groups. On reaction of 4,4'-difluorodiphenyl acetylene with bisphenates, such as that from BPA **9**, high molecular weight linear polymers are formed.[v]

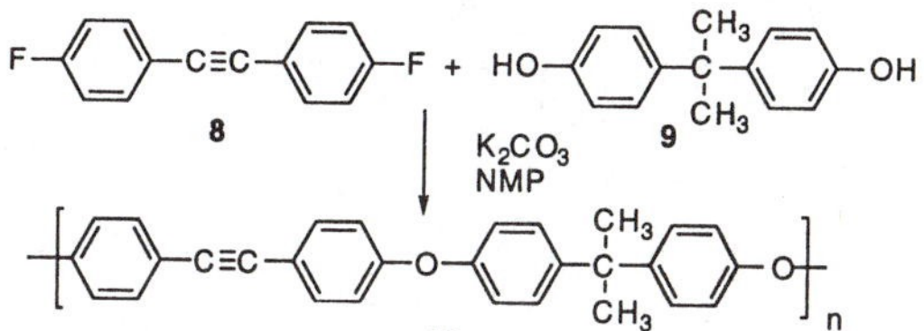

8 + 9 —(K_2CO_3, NMP)→ 10

At elevated temperatures these polymers readily cross-link.[vi] No Tg is apparent in the second scan in the DSC (Figure 1). The cross-linked polymers are now insoluble and maintain their moduli to much higher temperatures.

The curing of acetylene-terminated polyimides has been studied by ^{13}C NMR.[vi] Principally aromatic groups are shown to be formed during the curing reaction. This is apparently also the case with the present polymers since the 5% weight losses of these polymers are around 500 °C in air and in nitrogen which implies that very little unsaturation, which would be a site for air oxidation, is formed during the curing reaction.

B. Poly(aryl ether)s containing a 1,2-diphenylcyclopropane moiety. We have recently reported a very simple synthesis of the cyclopropane containing bisphenol **13** in two steps from the chalcone **11**.[vii]

HO–C6H4–CH=CH–C(=O)–C6H4–OH (**11**) $\xrightarrow[\text{CH}_3\text{CH}_2\text{OH, Reflux}]{\text{H}_2\text{NNH}_2}$ pyrazoline (**12**) $\xrightarrow[\text{NaOH, > 80\%}]{\text{210°C}}$ **13** (trans/cis = 9/1)

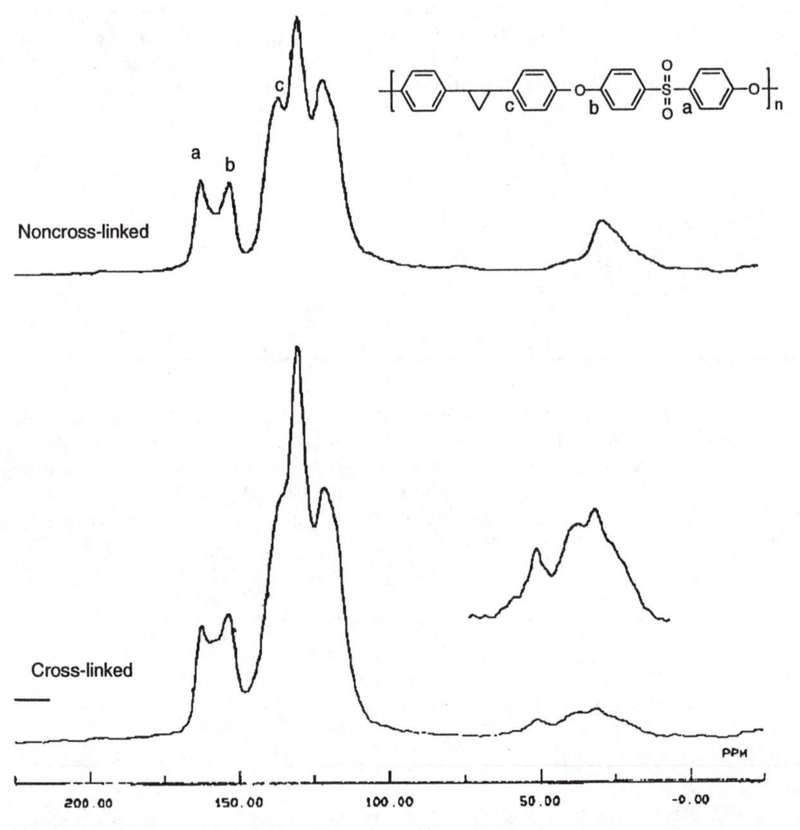

Figure 2. Solid state ^{13}C NMR spectra of polymer from **13**.

Poly(aryl ether)s prepared from **13**[ix] behave thermally very much like the acetylene containing poymers, undergoing an exothermic reaction above 300 °C and becoming insoluble due to cross-linking of the polymers. However, in this case, the resulting polymers are less stable in air than in nitrogen, as indicated by TGA, and the solid state ^{13}C NMR spectra (Figure 2) indicate that after curing there are numerous aliphatic carbons in the polymer which would be expected to be sites for oxidation at elevated temperatures.**C. Poly(aryl ether)s containing other cross-linkable moieties.** Polymers containing stilbene moieties,[x] the chalcone group[xi] and a dicyanoethylene group[xi] have been synthesized and all of these materials become cross-linked when heated. The dicyanostilbene monomer **14** can be converted, photochemically, to the corresponding phenanthrene **16** which, in turn, can also be used to synthesize novel poly(aryl ether)s **17**.

2 (4-fluorophenyl)acetonitrile → (40 % aq. NaOH; CCl_4, TMBA) → **14** → (hν) → **16** → (HOArOH, base) → **17**; **14** → (HOArOH, base) → **15**

III. Poly(aryl Ether)s Containing an o-Dibenzoylbenzene Moiety; Conversion to Phthalazines and Isoquinolines.

A. Linear polymers We have synthesized poly(ether ketone)s **19** by condensation of 1,2-dibenzoylbenzenes **18** with various biphenols. These o-substituted poly(ether ketone)s, as expected, are amorphous materials. We have synthesized polymers with 0, 2 and 4 pendant phenyl groups on the central ring.[xii]

a: $A_1 = A_2 = A_3 = A_4 = H$
b: $A_1 = A_4 = Ph$; $A_2 = A_3 = H$
c: $A_1 = A_2 = A_3 = A_4 = Ph$

18 → (H_2NNH_2) → **22**; **18** → (H-X-H, K_2CO_3, NMP, Y = F) → **19**; **18** → ($ArCH_2NH_2$) → **23**; **22** → (H-X-H, K_2CO_3, NMP) → **20**; **19** → (H_2NNH_2) → **20**; **19** → ($ArCH_2NH_2$) → **21**; **23** → (H-X-H, K_2CO_3, NMP) → **21**

An examination of a molecular model of 3,4,5,6-tetraphenyl-1,2-dibenzoylbenzene indicates that the phenyl substituents must be aligned almost perpendicular to the ring to which they are attached and that there is, therefore, a great deal of steric crowding which results in a relatively rigid structure. In these polymers the glass transition temperatures increase as the number of phenyl substituents on the central ring increases. The Tg's range from 182 °C to 313 °C. Despite the very high Tg's the polymers are very soluble in solvents such as chloroform at room temperature and can be cast into colorless, transparent films. They are very thermooxidatively stable as indicated by TGA measurements.

The poly(ether ketone)s can be converted to poly(ether phthalazine)s[xiv] **21** or poly(ether isoquinoline)s[xv] **22** by reaction with hydrazine and benzylamine, respectively. This conversion results in polymers in which there is considerable straightening of the polymer chains and this results in substantial increases in the glass transition temperatures (30 - 40 °C), in the solution viscosities, and in the apparent molecular weights as measured by GPC using polystyrene standards.

The difluorophthalazines **22** and difluoroisoquinolines **23** have also been synthesized and the fluoride groups are activated enough so that the poly(aryl ether)s can be prepared by direct reaction with bisphenates. The polymers obtained are identical with the polymers prepared by ring closure of the poly(ether ketone)s.

The poly(ether phthalazine)s and poly(ether isoquinoline)s synthesized to date are amorphous materials, soluble in solvents such as chloroform at room temperature and they can be cast into tough, transparent, pale yellow, flexible films.

B. Cyclic oligomers. We have now found that by performing the above reactions under high dilution conditions that cyclic oligomers can be formed in high yield.[xvi] The cyclic oligomers can then be polymerized at elevated temperatures in the presence of a catalyst, to give the high molecular weight linear polymers.

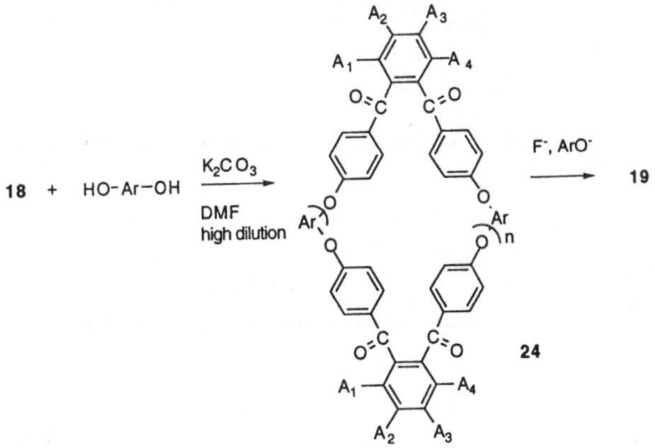

18 + HO-Ar-OH —(K_2CO_3, DMF, high dilution)→ 24 —(F^-, ArO^-)→ 19

The cyclic ether ketones **24** can also be converted to the corresponding phthalazines and isoquinolines which can also be polymerized at elevated tempertaures in the presence of a catalyst.

We have found that the phthalazine containing polymers cross-link and become insoluble at elevated temperatures and that the phthalazine group undergoes an unusual rearrangement to a quinazoline when heated. [xvii]

IV. New Monomers and Polymers Derived From Phenolphthaleins.

Phenolphthalein is currently prepared by the reaction of phthalic anhydride with phenol in the presence of a strong acid.[xviii] An ion exchange process, i.e. very similar to the present method for the manufacture of BPA from acetone and phenol, but at higher temperatures, has been reported.[xix]

A. Poly(aryl ether)s from imidoarylbiphenols. A series of biphenols **27** with different R substituents on the imide nitrogen from phenolphthalein, have recently been reported. [xx]

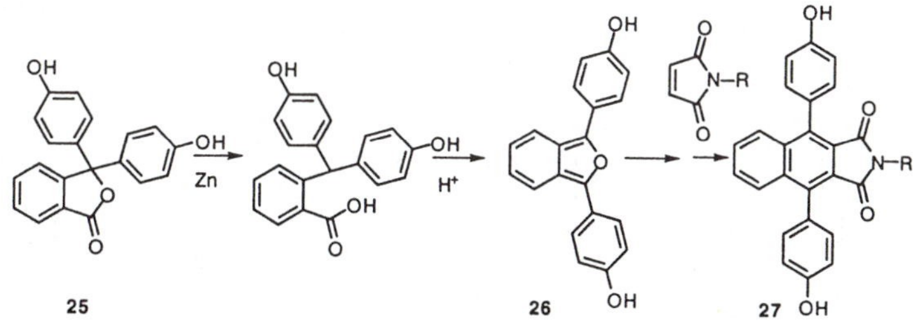

From these biphenols, polymers **28** with these different substituents on the imide nitrogen have been synthesized by reaction of the monomers with activated dihalides. The N-phenyl substituted biphenol, for example, gives a poly(ether sulfone) by reaction with bis(4-chlorophenyl sulfone) which has a Tg of 310 °C which has outstanding thermooxidatve stability

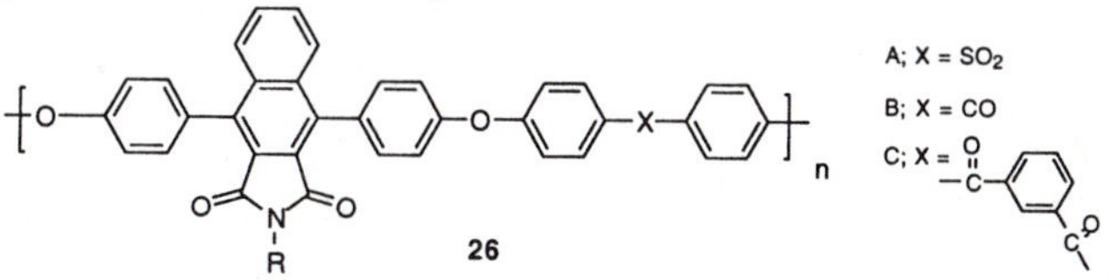

We have also synthesized the monomers **30** and polymers **31** using fumaronitrile as the dienophile. [xxi]

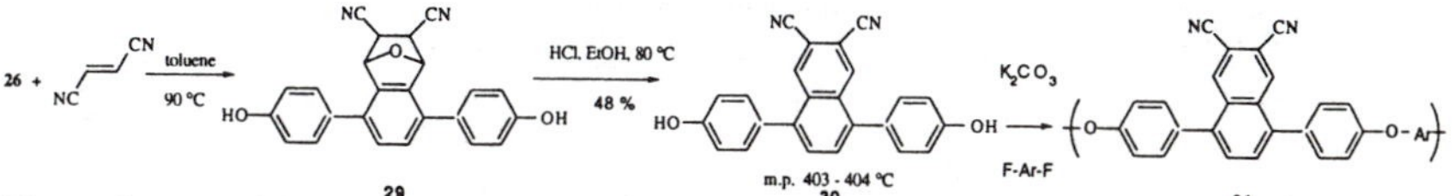

The polymers **31** have properties very similar to the imidoarylether polymers **28**. By reaction of these polymers and copolymers with excess phthalonitrile we were able to obtain polymers containing phthalocyanine groups in the chain. Copolymers containing the phthalocyanine moiety are soluble in solvents such as chloroform and can be cast into tough, flexible and intensely green colored films. The UV - visible spectra of a copolymer and of Zn-naphthalocyanine are shown in Figure 3.

B. Phthalazinones from phenolphthalein. Phenolphthalein reacts with hydroxylamine to give 4-aminophenol and 2(4-hydroxybenzoyl)benzoic acid **32** as products in very high yield.[xxi, xxii] We have recently demonstrated the synthesis of a novel monomer **33** from **32** by reaction with hydrazine.[xxiv] It exists in the phthalazinone form.

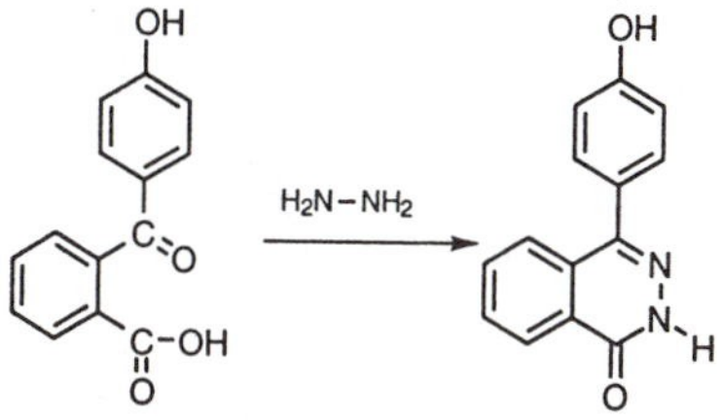

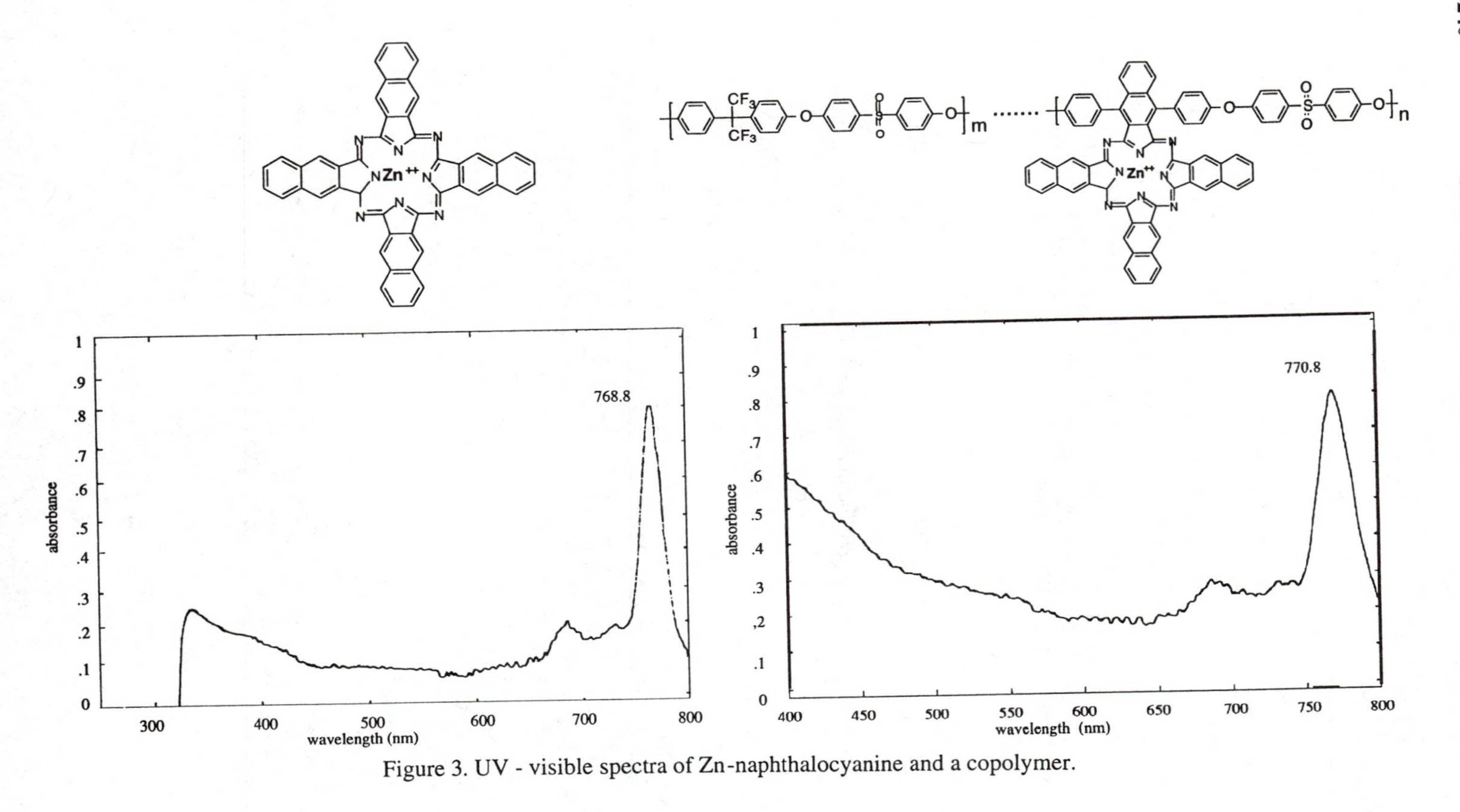

Figure 3. UV - visible spectra of Zn-naphthalocyanine and a copolymer.

In polymerization reactions the monomer **33** behaves just like a bisphenol. The poly(ether sulfone) and poly(ether ketone) can be readily prepared by reaction with bis(4-fluorophenyl sulfone) in N,N'-dimethylacetamide (DMAc) in the presence of potassium carbonate.[xxv] The polymer **34** has the structure shown with the NH group behaving just like a phenolic hydroxyl group in the polymerization reaction.

33 + F–⟨⟩–X–⟨⟩–F $\xrightarrow[NMP]{K_2CO_3}$ **34**

X = C=O, SO_2

The polymers have surprisingly high glass transition temperatures (X = C=O, Tg = 265 °C: X = SO_2, Tg = 290 °C). They can be cast into tough, colorless, flexible films. They are very thermooxidatively stable as indicated by the TGA analysis in air for the polysulfone (Figure 4)

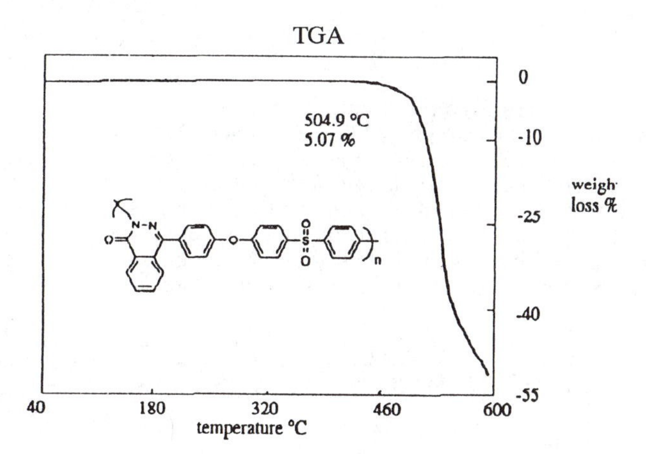

Figure 4. TGA analysis of a phthazinone sulfone polymer

V. Masked Biphenols in the Synthesis of Polyformals and Poly(Aryl Ether)s.

We have earlier prepared high molecular weight polyformals **35** by heating BPA **9** in an excess of methylene chloride with potassium hydroxide pellets in the presence of a phase-transfer catalyst or in an aprotic dipolar solvent.[xxvi]

9 + $ClCH_2Cl$ $\xrightarrow[NMP]{KOH}$ **35**

GPC analysis of the polymer indicated that it contained large amounts of low molecular weight cyclic oligomers which could not be eliminated completely by changing the reaction conditions. The BPA salt which initially forms is extremely insoluble, therefore, the reaction is in effect performed under high dilution conditions which favors the formation of cyclics.

A linear, high molecular weight polyformal **35** is readily synthesized by direct transformation of commercially available polycarbonate **36**.[xxvi]

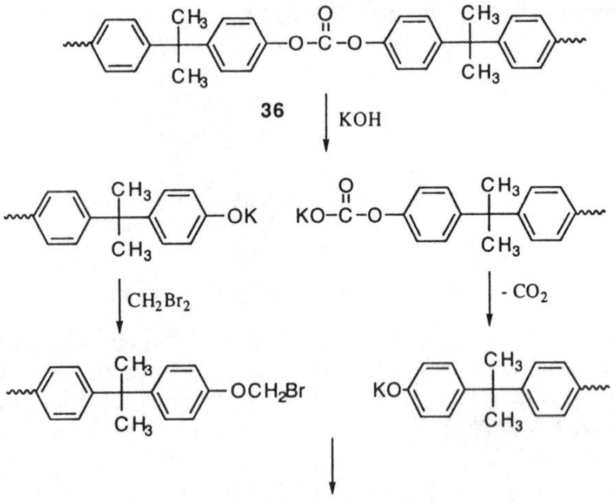

36 KOH

CH_2Br_2 $- CO_2$

35

The molecular weight of the resulting polymer is governed by that of the starting polycarbonate. No cyclics are formed in this reaction. We have also found that high molecular weight poly(arylene ether)s can be synthesized directly from polycarbonate and the corresponding activated dihalide.[xxvii] For example high molecular weight polysulfone **38** is prepared from polycarbonate **37** by the reaction with bis(4-fluorophenyl sulphone) in the presence of potassium bicarbonate. A dehydration step is not necessary in this reaction as is the case when bisphenols are used so that the overall reaction time is considerably shortened.

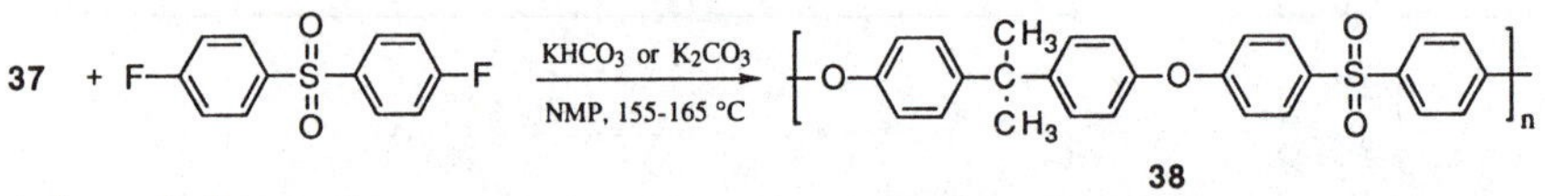

37 + F–C_6H_4–SO_2–C_6H_4–F $\xrightarrow[\text{NMP, 155-165 °C}]{KHCO_3 \text{ or } K_2CO_3}$ 38

Acknowledgements

This research was supported by the Natural Sciences and Engineering Research Council of Canada and the General Electric Company. Thanks are due also to the dedicated students noted in the references who carried out the research.

References

1. Kim, W. G.; Hay, A. S., *Makromol. Chem., Macromol. Symp.* **1992,** *54/55,* 331-336
2. Yang, H.; Hay, A. S., *Polym. Prepr. (Am. Chem. Soc., Div. Pol. Chem.),* **1993,** *34(1)*, 421-2
3. Paventi, M.; Elce, E.; Jackman, R. J.; Hay, A. S., *Tet. Letters* **1992,** *33*, 6405-6
4. Paventi, M.; Hay,A. S. *Tet. Letters*, **1993**, *34*, 999-1002)
5. Strukelj, M.; Paventi, M.; Hay, A. S., *Macromolecules*, **1993**, *26*, 1777-1778
6. Strukelj, M.; Paventi, M.; Hay, A. S. *Macromol. Symp.* **1994**, *77*, 369-378
7. Swanson, S. A.; Fleming, W. W. ; Hofer,D. C., *Macromolecules* **1992**, *25*, 582-588
8. Gao, C.; Hay, A. S., *Synthetic Communications*,**1995**, *25*, 1877-84
9. Gao, C.; Hay, A. S., *Macromolecules*, **1994**, *27* ,6708-13

10. Gao, C.; Hay, A. S., *J. Pol. Sci.*, submitted
11. Gao, C.; Hay, A.S., *J. Macromol. Sci*, in press 1995
12. Yeomans, K. A.; Hay, A. S., *Polym. Prepr.(Am. Chem. Soc., Div. Pol. Sci. & Eng.)*, **1993**, *69*, 240-41
13. Singh, R.; Hay,A. S., *Macromolecules*, **1992**, *25*, 1017-24,
14. Singh, R.; Hay,A. S., *Macromolecules*, **1992**, *25*, 1025-32,
15. Singh, R.; Hay,A. S., *Macromolecules*, **1992**, *25*, 1033-40
16. Chan, K. P.; Wang, Y.; Hay, A. S., *Macromolecules*, **1995**, *28*, 653-655
17. Chan, K. P.; Hay, A. S. *J. Org. Chem.* **1995**, *60*, 3131-3134)
18. Gamrath, H. G., *U.S. Patent* 2,522,940 (Sept. 19, 1950)
19. Prindle, H. B.; Ham, G. E., *U.S. Patent* 4,252,725 (Feb. 24, 1981)
20. Strukelj, M.; Hay, A. S., *Macromolecules*, **1992**, *25*, 4721-4729
21. Yang, H.; Hay, A. S., *J. Macromolecular Science*, **1995**, *32(5)*, 925-933
22. Friedlander, P., *Chem. Ber.* **1893**, *26*, 172
23. Lund,H., *Acta Chem. Scand.* **1960,** *14*, 359
24. Berard, N.; Hay, A. S., *Polym. Prepr. (Am. Chem. Soc., Div. Pol. Chem.)* **1993,** *34(1)*, 148-9
25. Berard, N.; Paventi, M.; Chan, K. P,; Hay, A. S.,*Die Makromolekulare Chemie, Macromol. Symp.* **1994**, *77*, 379-388
26. Hay, A. S.; Williams, F. J.; Relles, H. M.; Boulette, B. M., *J. Macromol. Sci. - Chem.*, **1984**, *A21 (8&9)*, 1065-1079
27. Wang, Z. Y.; Berard, N.; Hay, A. S., *J. Pol. Sci.*, **1992**, *30*, 299
28. Wang, Z. Y.; Nandin de Carvalho, H.; Hay, A. S.,*Chem. Comm.*, **1991**, 1221-2,

RECEIVED December 6, 1995

Chapter 15

Synthesis and Characterization of Photoresponsive and Thermosetting Aromatic Polyethers

Atul Bhatnagar[1] and Dillip K. Mohanty[2]

Department of Chemistry and Center for Applications in Polymer Science, Central Michigan University, Mount Pleasant, MI 48859

A wide variety of photo-responsive aromatic polyethers containing azo linkages in the main chain were prepared by nucleophilic aromatic displacement reactions. These polymers undergo cis-trans isomerism in solution when exposed to ultraviolet light. This photo induced isomerization was found to have a profound affect on solution viscosity of the polymers. The polymers were amorphous in nature and undergo crosslinking reactions at elevated temperatures. Tough films of the polymers could be obtained upon solution casting from chloroform.

An azo group is a photoisomerizable chromophore. Its incorporation into a variety of polymer structures has received considerable interest in the past. The photo induced isomerization of the azo group produces conformational changes in the polymer backbone. Various azo containing polymers including polyamides (*1*) and polyureas (*2*) have been studied extensively to observe such changes. Azoaromatic polyesters (*3*) have been investigated as potential polymeric non-linear optical substrates. On the other hand, incorporation of azo linkages in aromatic polyethers, (*4*)has not received significant attention in the past. Although Johnson et. al.(*5*) found the necessary activating influence of an azo group towards nucleophilic aromatic substitution reactions, detailed discussions on the synthesis and characterization of azoaromatic polyethers are lacking. In this paper we report the synthesis of a suitable monomer, a variety of azoaromatic polymers, model compound studies and the effects of photoisomerization on the solution viscosity behavior of the polymers. Also, the crosslinking reactions of these polymers at elevated temperatures has been investigated. The polymers have been characterized by spectroscopic, thermal and thermomechanical means.

EXPERIMENTAL

Materials. N-methylpyrrolidinone (NMP) and 4-fluoroaniline were dried over calcium hydride and then distilled at reduced pressure.

[1]Current address: Department of Chemistry, Virginia Polytechnic Institute and State University, Blacksburg, VA 24061

[2]Corresponding author

0097–6156/96/0624–0250$12.00/0

4,4'-Isopropylidenediphenol (bisphenol-A), kindly supplied by Dow Chemical, was purified by recrystallization from toluene. Tetramethyl bisphenol-A was prepared according to a reported procedure.(*6*) 4,4'-Sulfonyldiphenol (bisphenol-S) and *tert*-butylhydroquinone were recrystallized from acetone and ether/hexanes respectively. 9,9'-Bis 4-(hydroxyphenyl) fluorene, a gift from Ken Seika Corporation was used as received. Manganese dioxide was either prepared in the laboratory (*7*) or procured from commercial sources. All other reagents were used without further purification.

4,4'-Difluoroazobenzene. The reaction vessel consisted of a 2L three-necked round-bottomed flask equipped with a nitrogen inlet, an overhead stirrer and a Dean-Stark trap fitted with a condenser and a drying tube. Toluene (500 mL) and 4-fluoroaniline (48.64 g, 0.45 mole) were added to the flask. The reaction mixture was stirred vigorously and manganese dioxide (228.5 g, 2.6 moles) was added. The entire mixture was heated to reflux and water, the by-product of the reaction, was removed via the Dean-Stark trap by azeotropic distillation. The reaction was held at reflux for 24h then was allowed to cool. It was then filtered through celite and the filtrate was distilled to remove toluene. The residue was dissolved in small amounts of hexane and eluted on a silica gel column with hexane to afford **1** in 57% yield. The orange colored solid was purified further by recrystallization from hexane to obtain monomer grade material. **1:** mp 103 °C (DSC), (lit.[11] mp 101 °C); ^{1}H NMR ($CDCl_3$) δ 7.25 (m, 4H), 7.95 (m, 4H); ^{13}C NMR ($CDCl_3$) δ 164.39 (J_{CF}, 253 Hz), 161.5 (J_{CF}, 250 Hz), 149.03 (J_{CF}, 3 Hz), 124.31 (J_{CF},9 Hz), 122.60 (J_{CF}, 9 Hz), 116.22 (J_{CF}, 22 Hz), 116.34 (J_{CF}, 22 Hz); IR (KBr) 1498cm^{-1} (υ C-N); mass spectrum m/e 218 (M^+), 95. Anal. Calcd for $C_{12}H_8N_2F_2$: C, 66.06; H, 3.67; N, 12.87; F, 17.43. Found: C, 65.72; H, 3.77; N, 12.79; F, 17.12.

4,4'-Dichloroazobenzene. The reaction vessel consisted of a 2L three-necked round-bottomed flask equipped with a nitrogen inlet, an overhead stirrer and a Dean-Stark trap fitted with a condenser and a drying tube. Toluene (500mL) and 4-chloroaniline (51.02 g, 0.40 mole) were added to the flask. The reaction mixture was stirred vigorously and manganese dioxide (208.95 g, 2.4 moles) was added. The entire mixture was heated to reflux and water, the by-product of the reaction, was removed via the Dean-Stark trap by azeotropic distillation. The reaction was held at reflux for 24h then was allowed to cool. It was then filtered through celite and the filtrate was distilled to remove toluene. The residue was dissolved in small amounts of hexane and eluted on a silica gel column with hexane to afford **2** in 68% yield. The orange colored solid was purified further by recrystallization from hexane to obtain monomer grade material. **2:** mp 187 °C (DSC); ^{1}H NMR ($CDCl_3$) δ 7.5 (m, 4H), 7.95 (m, 4H); ^{13}C NMR ($CDCl_3$) δ 150.76, 137.21, 129.38, 124.18; IR (KBr) 1477 cm^{-1} (υ C-N); mass spectrum m/e 251 (M^+), 139, 111. Anal. Calcd for $C_{12}H_8N_2Cl_2$: C, 57.37; H, 3.19; N, 11.16; Cl, 28.29. Found: C, 57.55; H, 3.16; N, 11.18.

4,4'-(4-tert butylphenoxy)azobenzene. To a three-necked, 100 mL, round-bottomed flask fitted with a nitrogen inlet, a magnetic stirrer, a thermometer and a Dean-Stark trap fitted with a condenser and a drying tube, 4-*tert*-butylphenol (0.30 g, 0.002 mole), **1** (0.22 g, 0.001 mole) and anhydrous potassium carbonate (0.48 g, 0.0034 mole) were added. The weighing pans were washed with 25 mL of NMP and 15 mL of toluene with solvents flowing directly into the reaction vessel. The reaction mixture was heated to reflux and the reaction was allowed to continue at the reflux temperture (150 °C). The progress of the reaction was monitored by thin layer chromatography (TLC). After the completion of the reaction (18h), the reaction mixture was cooled and diluted with 50 mL of ethyl acetate. It was then filtered and the filtrate was distilled at reduced pressure to remove all solvents. The crude solid was recrystallized from hexane to afford **3** in 85% yield. **3**: mp 151 °C (DSC); ^{1}H NMR ($CDCl_3$) δ 1.85 (s, 18H), 7.05 (m, 8H), 7.40 (m, 4H), 7.90 (m, 4H); ^{13}C NMR ($CDCl_3$) δ 159.10, 152.80, 147.25, 146.02, 125.71, 123.39, 118.16, 117.18, 33.38, 30.47; IR (KBr) 1490 (υ C–N), 1240 (υ C–O-C) cm^{-1}; mass spectrum m/e 480 (M^+), 225, 210. Anal. Calcd for $C_{30}H_{34}N_2O_2$: C, 80.33; H, 7.11; N, 5.86. Found: C, 80.26; H, 7.32; N, 5.81.

General Procedure for the Preparation of Azoaromatic Polyethers. Polymerization reactions were carried out in a 100 mL four-necked, round-bottomed flask fitted with a nitrogen inlet, a thermometer, an over-head stirrer and a Dean-Stark apparatus in a heated oil-bath. The following representative procedure was used for the preparation of **6a**. Bisphenol-A (2.8532 g, 0.0125 mole) and **1** (3.7562 g, 0.0125 mole) were weighed and carefully transferred into the reaction vessel. Anhydrous potassium carbonate (3.75 g, excess) along with NMP (30 mL) and toluene (20 mL) were added to the reaction vessel. The reaction mixture was heated until toluene began to reflux (140 °C) which was periodically removed from the Dean-Stark trap along with water (by-product of the reaction). It took approximately 4-6h for the complete removal of water. The color of the reaction mixture turned deep orange and the viscosity increased gradually. The reaction was allowed to continue at 140 °C for 24h. The reaction mixture was then coagulated in approximately 10X volume of acetone. The polymer was collected by filtration and washed with acetone, water and then acetone again in a Soxhlet apparatus. The polymer was then dried in a vacuum oven at 80 °C for 24h.

Measurements. ^{1}H and ^{13}C NMR spectra were recorded using a General Electric QE-300 instrument. IR spectra were obtained with a Perkin Elmer Series 1600 FT-IR spectrophotometer. Absorption spectra were recorded with a Varian Cary 219 spectrophotometer. The thermal transition temperatures were measured either with a DuPont DSC 2100 or Perkin-Elmer DSC-7 at a heating rate of 10 °C min^{-1}unless otherwise indicated. The transition temperatures were reported as the maxima and minima of their endothermic and exothermic peaks. Glass-transition temperatures, T_g, were taken as the mid-point of the change in slope of the baseline. Thermogravimetric analysis (TGA) of the polymer samples were conducted at a heating rate of 10 °C in nitrogen. Intrinsic viscosity measurements were made on NMP solutions at 25 °C using a Cannon-Ubbelohde dilution viscometer. Polymer solutions in NMP were irradiated by a GE mercury lamp, 400-W, 120-V. During irradiaton the chamber was cooled by a fan and the solution were cooled using a cold finger.

RESULTS AND DISCUSSIONS

Monomer Synthesis. Both 4,4'-difluoroazobenzene **1**, and 4,4'-dichloroazobenzene **2**, were synthesized by the oxidative coupling reactions of haloanilines (Scheme 1) by manganese dioxide according to a reported general procedure (*6*). The yield of the dichloro compound was significantly lower (30%) as compared to the fluorinated analog (50%). The purity of these monomers was assessed both by the elemental analysis and DSC. The bisfluorinated azobenzene contained both cis and trans isomers. On the other hand, only the trans isomer of compound **2** could be isolated. These observations were confirmed by the ^{13}C NMR analysis. We are unable to offer a reasonable explanation for this phenomenon since arguments based on steric interactions alone can not be completely justified. However, this is consistent with our earlier observations with bishalides containing ketimine groups (*8, 9*)

Model Compound Studies. Both **1** and **2** have been used for the synthesis of aromatic polyethers in dimethyl sulfoxide (DMSO) (*5*), however high molecular weights were not obtained when **1** was used. On the other hand, significantly longer reactions times were required with **2** to yield oligomeric products only. We have undertaken extensive model compound studies in order to gain a better understanding as to why the dichloride affords only oligomeric products and to establish appropriate reaction conditions for the synthesis of high molecular weight polyethers from **1** and a variety of bisphenols.

It was possible to prepare the desired bisether **3**, in quantitative yield when **1** was allowed to react with two equivalents of 4-tert-butylphenol in the presence of anhydrous potassium carbonate in NMP at 140 °C (Scheme 2). However, a similar reaction of **2** afforded a mixture of products (Scheme 2). These compounds were separated on a silica gel column and identified to be the desired bisether **3**, the mono substituted compound **4**, and the dehalogenated derivative of the mono substituted product, **5**. This is consistent with earlier observation that both S_NAR and $S_{RN}1$ reactions compete during aromatic polyether formation (*10 - 13*). These reactions are illustrated in Scheme 3. Initiation (step 1) is provided by the electron transfer, presumably from the phenoxide anion, to form the radical anion of the halogenated azobenzene. In the second step, the radical anion expels the chloride anion to form a radical, which combines in the third step with a phenoxide to form a radical anion again. In the fourth step this newly formed radical anion transfers an electron to the halogen bearing nucleus of azobenzene to form the ether linkage and a new radical anion. Steps 2, 3 and 4, which can repeat themselves, are the propagation steps. It is highly likely that the radical **A**, formed in the second step can abstract a hydrogen atom from the solvent system, forming the dehalogenated product and terminating the reaction. The $S_{RN}1$ reactions compete effectively when the combination of a weak nucleophile and a substrate with poor leaving groups (e.g., chloride anion in the present case) is used. Furthermore, the choice of solvent plays an important role (*12*) Our observation suggests that the nucleophilicity of 4-tertbutyl phenoxide anion in NMP is somewhat similar to the poor nucleophilicity exhibited by a phenoxide anion (a weaker base) in DMAc. Dehalogenation reactions are one of the consequences of $S_{RN}1$ processes (Scheme 3). In this instance, the presence of dehalogenated compounds **5** indicates that $S_{RN}1$ reactions must be taking place when the bischloride is used. Therefore, during polymerization reactions with **2**, dehalogenation generates stoichiometric imbalance and only oligomeric products are formed. It should be pointed out that we were unable to detect the presence of azobenzene in the product mixture. On the other hand, it was possible to isolate the same when phenol was allowed to react with **2** under similar reaction conditions. On the basis of these findings, **1** was chosen as the monomer for polymerization reactions. Unidentified side products were observed when these reactions were conducted in DMAc, irrespective of the nature of the halide. Therefore, NMP was chosen as the solvent for conducting the polymerization reactions. In order to shorten the reaction time between **1** and 4-tert-butylphenol, reactions were carried out at elevated temperatures (>150 °C). Unfortunately, the color of the reaction mixture turned brown above a reaction temperature of 150 °C and the desired bisether, **3**, could not be detected. TLC analysis indicated the presence of highly polar base-line products possibly due to side reactions involving the carbonate anion attack on the carbon bearing the fluoride atom. This was

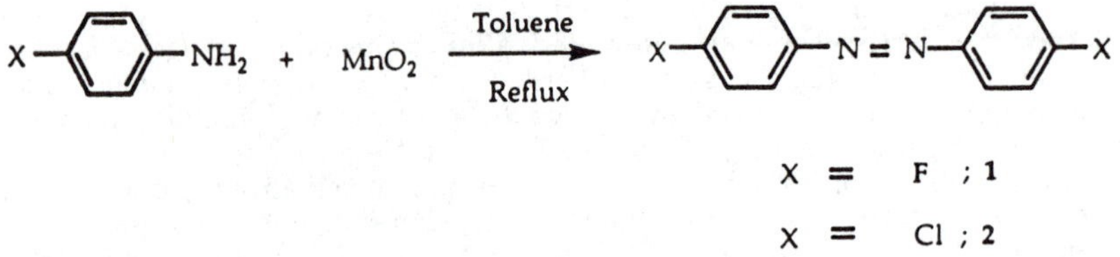

X—C₆H₄—NH₂ + MnO_2 —(Toluene, Reflux)→ X—C₆H₄—N = N—C₆H₄—X

X = F ; **1**

X = Cl ; **2**

Scheme 1.

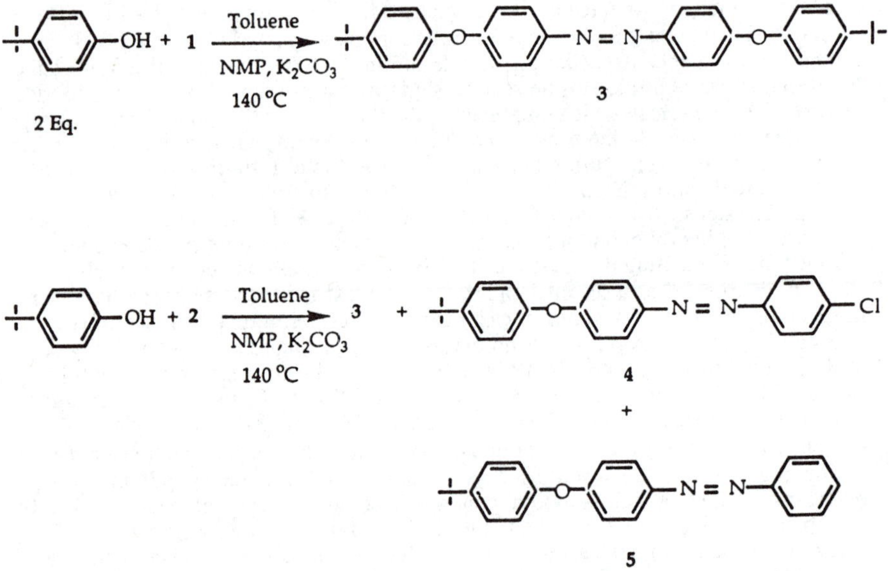

OH + **1** —(Toluene, NMP, K_2CO_3, 140 °C)→ **3**

2 Eq.

OH + **2** —(Toluene, NMP, K_2CO_3, 140 °C)→ **3** + **4** (—N = N—C₆H₄—Cl) + **5** (—N = N—C₆H₅)

Scheme 2.

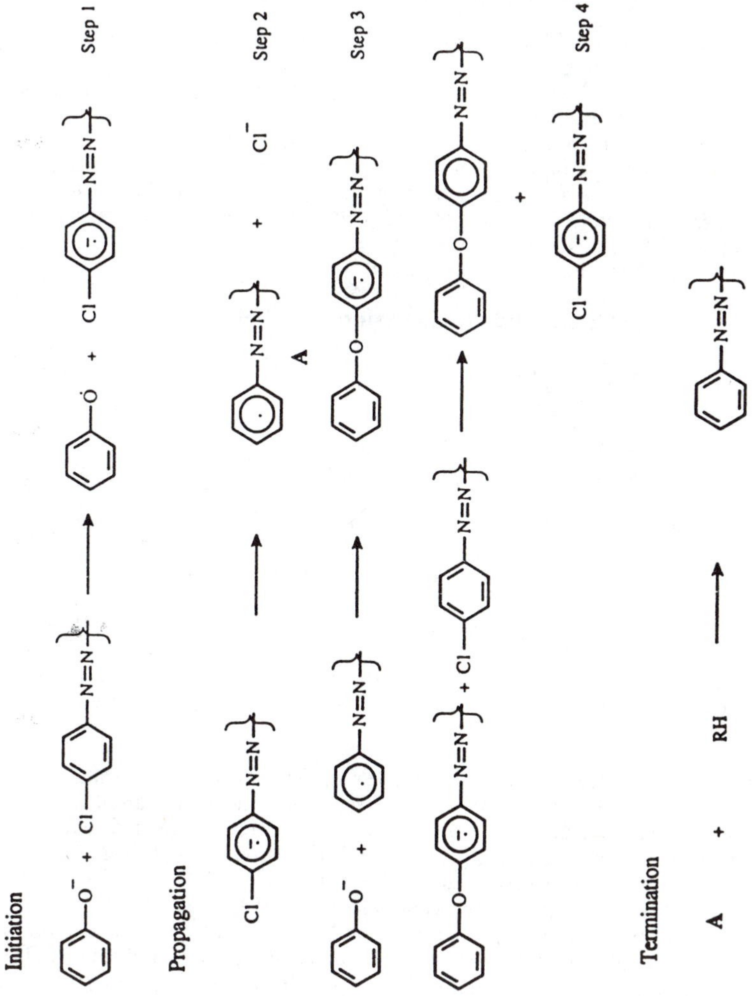

Initiation
Step 1
Propagation
Step 2
Step 3
Step 4
Termination
A
RH

Scheme 3.

confirmed by conducting the same reaction in the absence of 4-tert-butylphenol. A TLC analysis of the reaction mixture indicated the loss of the bisfluoride and the presence of similar products as before, within 2h. No attempt was made to isolate and identify the side products. To ascertain that the carbonate anion and the azo moiety do not react, a reaction of azobenzene with potassium carbonate in NMP at 150 °C was conducted. Only starting material could be recovered even after prolonged reaction periods. This led us to conclude that the actual polymerization reactions should be conducted preferably at 140 °C albeit the longer reaction time.

Polymer Synthesis and Characterization. High molecular weight polymers **6 a-e** were prepared by allowing difluoride, **1**, to react with appropriate bisphenols in NMP/toluene (2:1 v/v) solvent mixture in the presence of anhydrous potassium carbonate at 140 °C (Scheme 4). Reactions with more rigid bisphenols such as hydroquinone or biphenol failed to produce high molecular weight polymers. Oligomeric products precipitated out of the reaction mixture at early stages of the reaction. In order to maintain a homogeneous reaction mixture, the reaction temperature was increased to 150 °C. This resulted in discoloration and decomposition of the products.

The solubility behaviors of the polymers **6 a-e** are shown in Table I.

Table I: Solubility Behaviors of Polymers 6 a-e

Polymer	CH_2Cl_2	$CHCl_3$	THF	DMAc	DNSO	NMP
6a	sp*	s	s	sp*	sp*	s
6b	s	s	s	sp*	i	s
6c	sp*	s	s	i	i	s
6d	sp*	s*	s	s*	sp*	s
6e	sp*	sp*	i	s	i	s

s: soluble at room temperature, s*: soluble (hot), sp*: sparingly soluble (hot), i: insoluble (hot). (Reproduced with permission. Copyright 1995 Elsevier.)

The azoaromatic polyether **6e** exhibited exceptional solvent resistance. It had limited solubility in a series of common solvents including dichloromethane, chloroform, tetrahydrofuran and dimethylsulfoxide even at elevated temperatures. However, it was soluble in NMP, DMAc and trifluoromethanesulfonic acid. The presence of ether and sulfone groups in **6e** was established by IR (KBr) spectroscopy. The exceptional solvent resistance of **6e** can be attributed to the presence of the polar sulfone groups which results in strong inter and intra chain forces. Surprisingly, the polymers exhibited a low degree of solubility in DMSO. This may explain why earlier attempts [5] to prepare the polymer in DMSO were unsuccessful.

Polymers **6 a-d** produced clear finger nail creasable films upon solution casting from chloroform. The intrinsic viscosity values for polymers **6 a-e** in NMP at 25 °C are displayed in Table II. The values range from 0.50 to 1.02 dL/g, indicative of the

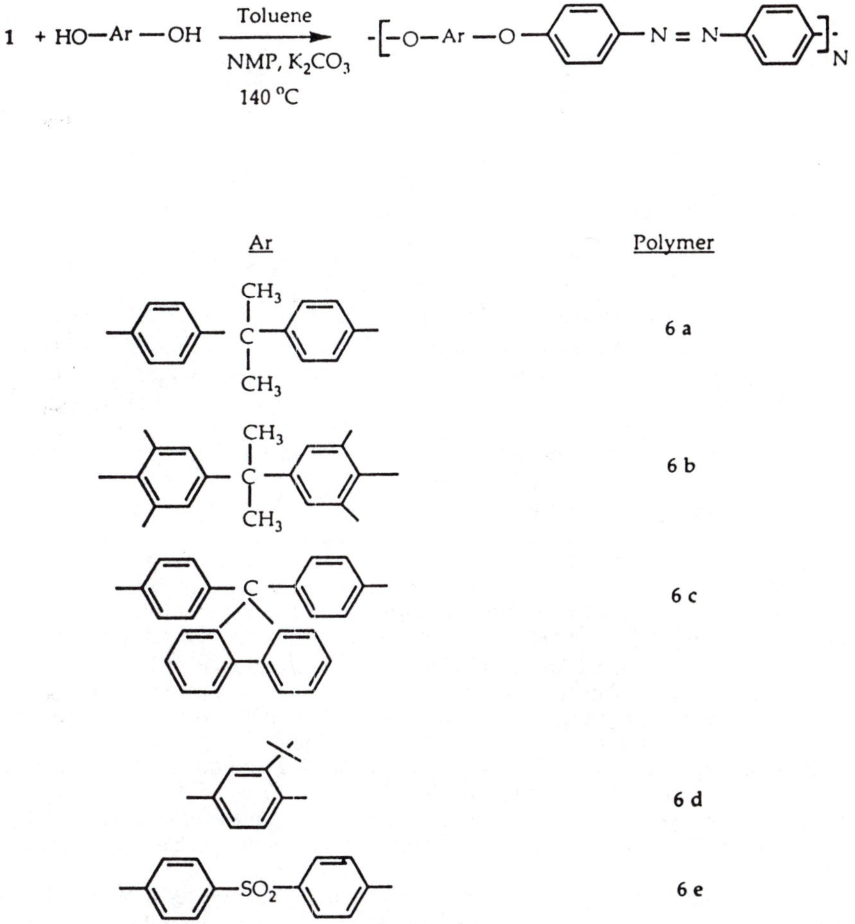

1 + HO—Ar—OH —Toluene, NMP, K_2CO_3, 140 °C→ -[—O—Ar—O—C₆H₄—N = N—C₆H₄—]$_N$

Ar	Polymer
—C₆H₄—C(CH_3)₂—C₆H₄—	6 a
—C₆H₂(CH_3)₂—C(CH_3)₂—C₆H₂(CH_3)₂—	6 b
—C₆H₄—C(fluorenylidene)—C₆H₄—	6 c
—C₆H₃(t-Bu)—	6 d
—C₆H₄—SO_2—C₆H₄—	6 e

Scheme 4.

Table II: Intrinsic Viscosity and Glass Transition Temperatures for Polymers 6 a-e

Polymers	η^a (dL/g)	Tg^b (°C)
6a	0.80	175
6b	0.91	235
6c	1.02	229
6d	0.50	187
6e	0.52	222

a: Solvent NMP at 25 °C
b: Second Heating; Heating Rate 10 °C min^{-1}

moderate to high molecular weight nature of the polymers. The repeat unit structures of the polymers **6 a-d** were characterized by ^{13}C NMR ($CDCl_3$), IR (film) and Resonance Raman (RR) spectroscopy. The ^{13}C NMR spectrum of **6a** contained ten absorbances which could be assigned to various carbon atoms in the repeat unit from the reported chemical shift constants (CSC) (*14* , *15*). This suggests that only the more stable trans isomer is present. This is in contrast to the monomer **1**, which consisted of a mixture of both cis and trans isomers. A solution of **6e** in trifluoromethanesulfonic acid was used to obtain ^{13}C NMR spectrum. Deuterium dioxide, in a sealed capillary tube was used as an internal standard. It was not possible to assign the absorbances to different carbon atoms in the repeat unit due to a lack of readily available CSC. The presence of ether linkages in the polymer backbone was confirmed by IR (film) spectroscopy. The RR spectrum of polymer **6a** is shown in Figure 1. An examination of the figure indicates absorbances at 1405 cm^{-1}due to the azo groups. Detailed RR studies will be presented elsewhere.

It is known that azobenzenes undergo cis-trans isomerization when exposed to UV light. The stable trans form is converted to the less stable cis form. Polymer **6a** was used to study the effect of irradiation on the solution viscosity. Intrinsic viscosity measurements were done after exposing a NMP solution of the polymer to UV light. The intrinsic viscosity value decreased to 57 % of the original at the photostationary state immediately after irradiation. A decrease of as much as 60 % with polyamides and 40 % for polyureas under similar conditions has been observed (*16*).

This cis-trans isomerization can be conveniently followed by monitoring the intense absorption at 320 nm due to π-π^* transition. The intensity of this absorbance decreases as more cis isomer is produced. The photogenerated cis form undergoes thermal transformation in the dark to trans form. This transformation was followed by UV spectroscopy. Similar results have been obtained with azoaromatic polyureas (*16*) and polyamides also. The increase in the peak intensity at 320 nm, in the dark, was monitored at various time intervals at six different temperatures (26, 29, 30, 33, 35 °C). It was found that the cis-trans isomerization follows first order kinetics. The rate of this isomerization was calculated from the slope of the plot of $\log(A_\infty - A_t)$ vs. time, t , where A_∞ and A_t are the absorbances at 360 nm before irradiation and at time t. It was possible to calculate the activation energies from the Arrhenius plots (e.g., for **6a** Fig. 2). The activation energies (E_a) for polymers **6 a-e** are presented in Table III. An examination of these values indicates that they are essentially identical

Table III: Activation Energy Values; E_a' (thermal recovery from cis to trans isomer) of Model compound 3 and Polymers 6 a-e

Model compd./Polymers	E_a (Kcal mol^{-1})
3	22.9
6a	23.6
6b	26.4
6c	26.7
6d	26.0
6e	24.6

Mean: 25.5 Kcal mol^{-1}; SD: 1.3

irrespective of the nature of the bisphenols. These values are similar to those obtained for a variety of polyureas (*16*) and polyamides. In addition, the calculated activation energy (22.9 KCal $mole^{-1}$) for low molecular weight model bisether **3** was similar to high molecular weight polymers. This is also consistent with earlier observations by Morawetz et. al. for low molecular weight model amides and high molecular weight polyamides containing a smaller number of azobenzene residues (*17*).

The azoaromatic polyethers exhibited well defined glass transition temperatures which are listed in Table II. The polymers where amorphous in nature except for **6e** which displayed a melt endotherm at 248 °C (10.8 J/g) in the first heating. It did not recrystallize upon slow cooling. The T_g values for the other polymers ranged from 175 °C to 235 °C depending on the nature of bisphenol used. The polymers exhibited a large exotherm when heated above their respective T_gs. These peak temperatures are presented in Table III. This can be attributed to the crosslinking reactions in which nitrogen gas is eliminated. A possible pathway of such a crosslinking reaction is presented in Scheme 6. The elimination of nitrogen gas was confirmed by the variable temperature thermogravimetric analysis. The TGA thermogram of polymer **6a** is shown in Fig. 3. The weight loss due to decomposition around 373 °C corresponds to the weight of a nitrogen molecule per repeat unit.

Polymers which had been heated to temperatures above the crosslinking temperatures were insoluble in the solvents in which the virgin polymers were soluble. Also, the polymers acquired a dark color. Decomposition of azo aromatic polymers at elevated temperatures with the elimination of nitrogen gas has been observed previously (*18*). The activation energies (E_a') for these crosslinking reactions were determined by conducting the DSC scans at five different heating rates (2.5, 5, 10, 15, 20 °C $min.^{-1}$) and measuring the exotherm peak temperatures (T_ps) (*19*). The E_a' was calculated from the slope of a plot of ln Ø vs. 1/T (Fig. 4), where Ø is the heating rate and T is the peak temperature (T_p) in °K. The activation energy values for crosslinking reactions (Table III) are essentially identical

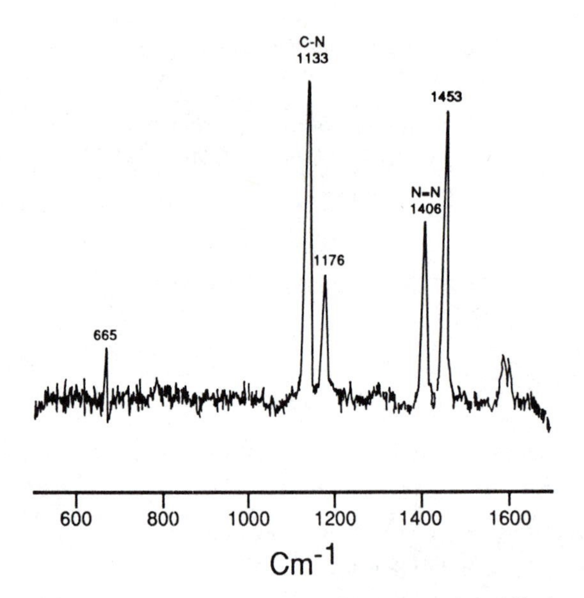

Figure 1: Resonance Raman Spectrum of **6a** (film)

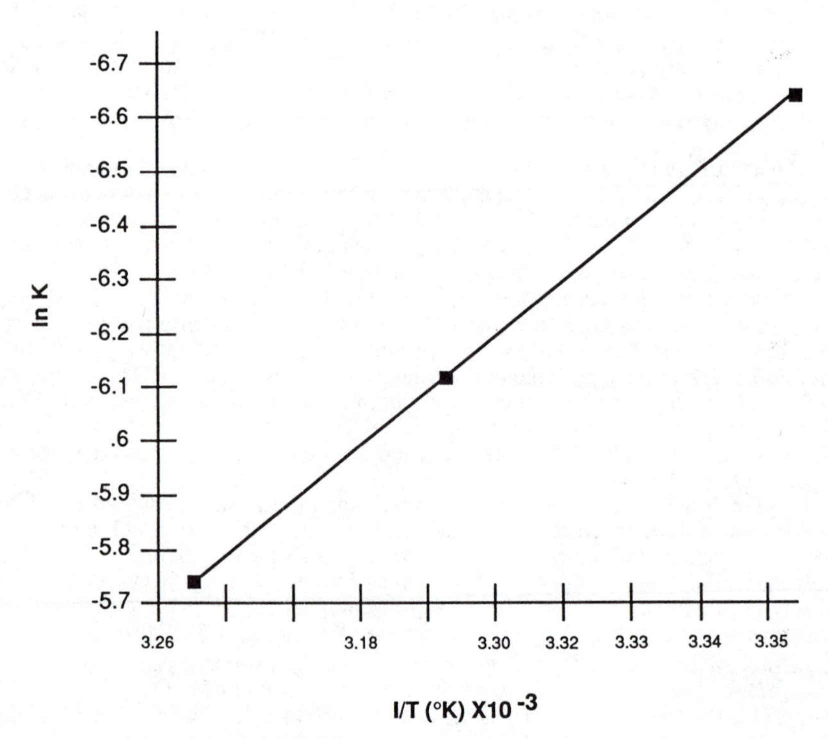

Figure 2: Plot of ln K vs. 1/T (°K)

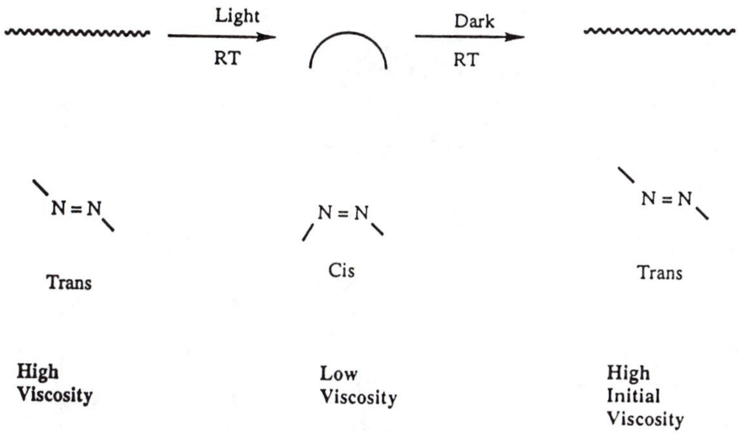

Light RT → Dark RT

N = N Trans — N = N Cis — N = N Trans

High Viscosity — Low Viscosity — High Initial Viscosity

Scheme 5.

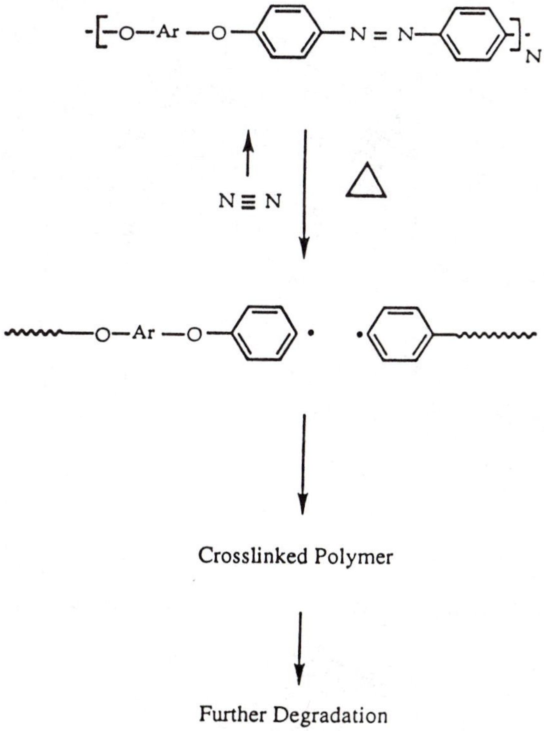

-[-O—Ar—O—C₆H₄—N = N—C₆H₄-]-$_N$

$N \equiv N$ ↑ ↓ △

~~~O—Ar—O—C₆H₄• •C₆H₄~~~

↓

Crosslinked Polymer

↓

Further Degradation

Scheme 6.
~~~

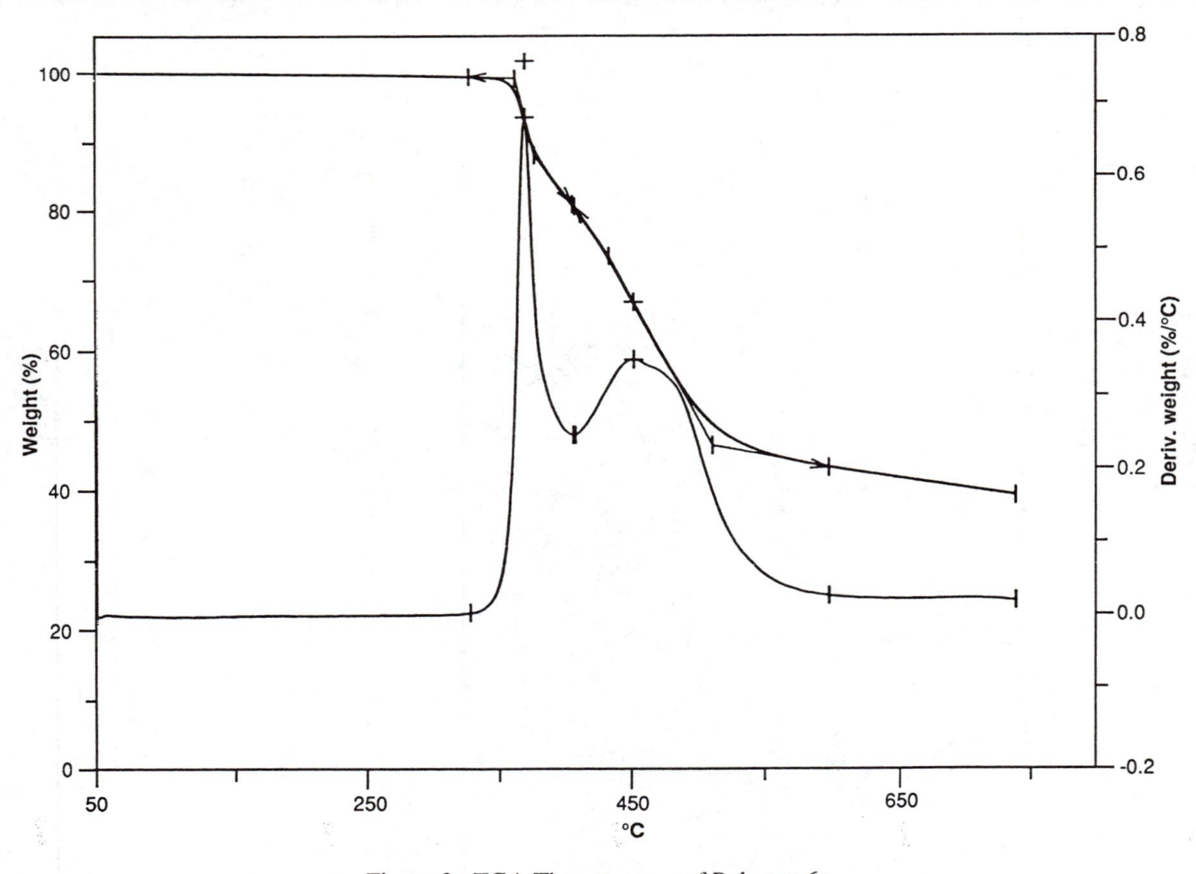

Figure 3: TGA Thermograms of Polymer **6a**

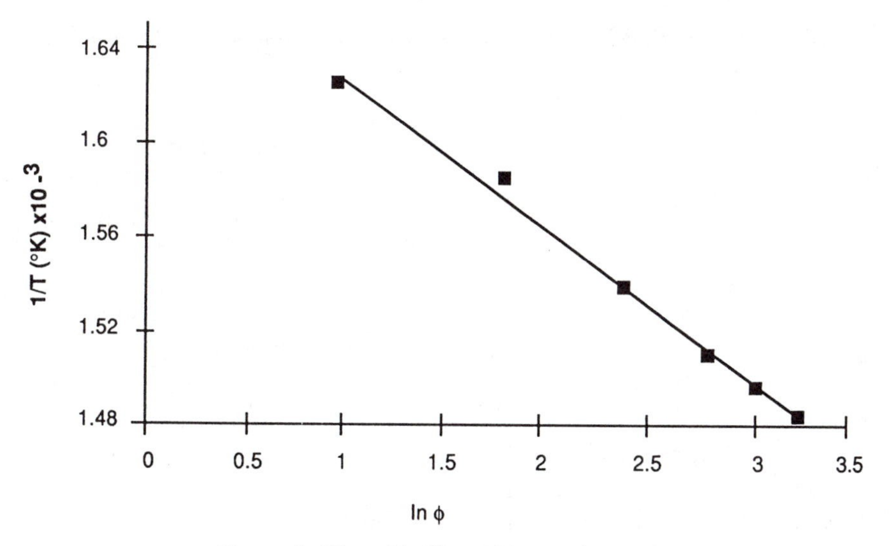

Figure 4: Plot of ln Ø vs 1/T for Polymer **6a**

irrespective of the nature of the polymer backbone. The calculated mean value of 34.8 KCal mole^{-1} is similar to the activation energy value of 30.8 KCal mole^{-1} reported for the decomposition of azoisobutyronitrile (*20*).

Both isothermal and variable temperature thermogravimetric analysis (TGA) was used to assess the thermal stability of these polymers. The onset of decomposition for these polymers ranged from 327 to 408 °C (Table IV). This is significantly lower than most conventional poly(aryl ether-ketones) and poly(aryl ether-sulfone) which have higher temperature of decomposition around 450 °C. The inferior thermal stability can once again be attributed to the presence of thermally labile azo linkages. However, the polymers showed excellent char yields at 750 °C in 30 to 65 % range. Polymers which decompose with high char yields have attracted attention for their improved flame-retardant characteristics (*21*). A weight loss of less than 10 % was observed by isothermal TGA at 300 °C after only one hour (Table V).

It was possible to obtain tough, transparent compression molded film of polymer **6a** only. Since it was necessary to increase the temperature of the plates to at least 100 °C above the respective glass transition temperature of the polymers, polymers **6 b-e** underwent crosslinking reactions at elevated temperatures.

CONCLUSIONS

A series of high molecular weight photoresponsive azoaromatic polyethers have been prepared. These polymers undergo trans-cis isomerization in solution. The viscosity of the polymer solution decreases upon irradiation and reverts to original value in the dark. All polymers exhibited well-defined glass transition temperatures and undergo crosslinking at elevated temperatures. Most of the polymers afford tough, transparent and highly colored films upon solution casting.

Table IV: DSC Peak Exotherm Temperature and Activation Energies, E_a (crosslinking)

Polymers	Tp^a (°C)	E_a'[b] (Kcal mole^{-1})
6a	377	31.2
6b	351	34.3
6c	393	34.8
6d	359	34.8
6e	364	36.1

a: Heating Rate 10 °C min^{-1}
b: Mean - 34.2 Kcal mol^{-1}; SD - 1.8

Table V: Thermogravimetric Analysis (in nitrogen) of Polymers 6 a-3

Polymers	Wt loss in isothermal aging at 300 °C in wt. % h^{-1}	Onset of decomposition temperature in °C	% Residue at 750 °C
6a	5.8	366	39
6b	4.6	327	35
6c	1.4	408	63
6d	9.3	340	40
6e	4.0	341	54

ACKNOWLEDGMENTS

The authors wish to acknowledge partial support of this project by Grant #62811 made available to DKM by the Department of Chemistry and the Michigan Polymer Consortium, Grant #48450, from the FRCE committee, Central Michigan University, and Grant #62845 from the Research Excellence Fund, State of Michigan. A University Research Professorship award to DKM by Central Michigan University was extremely helpful. Partial support for the purchase of the GE QE - 300 NMR Spectrometer used in this work was provided by NSF/ILI grant #USE-8852049.

REFERENCES

1. a) Jayaprakash, D., Balasubramanian, M., Nanjan, M. *J. Polym. Sci., Polym.. Chem. Ed.* **1985**, *23*, 2319. b) Irie, M., Hayashi, K. *J. Macromol., Sci., Chem.* **1979**, A *13*, 511. c) Irie, M., Hirano, K., Hashimoto, S., Hayashi, K. *Macromolecules* **1981**, *14*, 262. d) Blair, H. S., Pogue, H. I., Riordan, J. E. *Polymer* **1980**, *21*, 1195.
2. Sudesh Kumar, G., DePra P., Necker, D.C. *Macromolecules* **1984**, *17*, 1912.
3. Hall, H. K., Kuo, T., Leslie, T. M. *Macromolecules* **1989**, *22*, 3525.
4. a) Johnson, R. N., Harris, J.E., *Encyclopedia of Polymer Science and Engineering*, 2nd Ed.; Mark, H. F., Bikales, N. M., Overberger, C. G., Mengers, G., Eds.; John Wiley & Sons. New York, 1988, Vol. 13, p 196. b) Parodi, F. In *Comprehensive Polymer Science*; Annen, G., Bevington, J. C., Eds.; Pergamon: Oxford, 1989, Vol. 5, p 561. c) S. Maiti, B.K. Mandal, *Prog. Polym. Sci.* **1986**, *12*, 111.
5. Johnson, R. N., Farnham, A. G., Callendinning, R. A., Hale, W. F. and Merian, C. N. J., *J. Polym. Sci., Polym. Chem. Ed.*, **1967**, *5*, 2375.
6. Mohanty, D. K., McGrath, J. E. in *Advances in Polymer Synthesis* **1985**, *31*, 113, Culbertson, B. M., J. E., Eds; Penum Press, New York and London, 1985.
7. Wheeler, O.H., Gonzalez, D. *Tetrahedron Lett.* **1964**, *20*, 189. b) Sandler, S. R., Karo, W. "Organic Functional Group Preparations," Academic Press, Vol. 2, 1971, p. 321.
8. a) Yeakel, C., Gower, K., Mani, R. S., Allen, R. D., Mohanty, D. K. *Macromol. Chem.* **1993**, *194*, 2779. b) Carter, K. K., Hedrick, J. L., *Macromolecules*, 1994, **27**, 3426.
9. DeSimone, J. M., Stompel, S., Samulski, E. T., Wang, Y. Q., Brennan, A. B., *Macromolecules* **1992**, *25*, 2546.
10. a) Rossi, R. H. and deRossi, R. H. "Aromatic Substitution by the $S_{RN}1$ Mechanisms," American Chemical Society., Washington D. C., 1983. b) Percec, V., Clough, R.S., Rinaldi, P. L. and Litman, V. E. *Macromolecules* **1991**, *24*, 5889. b) Percec, V., Clough, R. S., Fanjul, J. and Griogoras, M. *Polym. Prep. Am. Chem. Soc.* **1993**, *34 (1)*, 162. c) Percec, V., Clough, R. S., Fanjul, J. and Griogoras, M. *Macromolecules* **1993**, *26*, 3650 d) Bhatnagar, A., Mani, R. S., King, B., Mohanty, D. K. *Polymer* **1994**, 35 (5), 1111. e) Pross, A., in *Nucleophilicity* '; Harris, J. H., McManus, S. P., Eds., American Chemical Society, Washington D. C., 1987, p. 332 and references therein.
11. Levy, G. C., Lichter, R. L., Nelson, G. L. "Carbon- 13 Nuclear Magnetic Resonance Spectroscopy," John Wiley and Sons, New York, 2nd Ed., p 111. b) Ewing, D. F. *Org. Mag. Reson.* **1979**, *12*, 499.
12. Sudesh Kumar, G., DePra P., Necker, D. C. *Macromolecules* **1984**, *17*, 2463.
13. Paik, C. S., Morawetz, H., *Macromolecules* **1970**, *5*, 171.
14. Srinivasan, P. R., Srinivasan, M., Mahadeva, V. *J. Polym. Sci., Polym. Chem. Ed.* **1982**, *20*, 1145.
15. Prime R. B., in *Thermal Characterization of Polymeric Materials*, Turi E., Ed., 1981, p. 543.
16. Swarin, S. J., Wims, A. M. *Anal. Calorim.* **1976**, *4*, 155.
17. Krevelen van D. W. *Polymer* **1975**, *16*, 516.

RECEIVED January 25, 1996

Chapter 16

Poly(aryl ether benzazole)s

Self-Polymerization of AB Monomers via Benzimidazole-Activated Ether Synthesis

T. J. Matray, R. J. Twieg, and James L. Hedrick[1]

Research Division, IBM Almaden Research Center, 650 Harry Road, San Jose, CA 95120–6099

The aromatic poly(benzimidazoles) (PBI) comprise a class of heterocyclic polymers that show excellent thermal stability (*1*). The high degree of molecular rigidity in the backbone together with intramolecular interactions produces a high modulus polymer with a variety of applications including fibers and in circuit boards. In spite of these attractive properties, the use of PBIs has been somewhat limited since many are only soluble in strong acids and cannot be processed from organic solvents. Improved solubility and processability has been accomplished by the incorporation of meta substituted linkages, aryl ether linkages and both pendent and main-chain alkyl substituents which can be built into the monomers prior to polymerization. The poly(benzimidazoles) are a subset of the group of poly(benzazoles) which also includes the benzoxazole and benzothiazole heterocycle containing analogs.

The PBIs are generally synthesized by the step growth polymerization of an aromatic bis(o-phenylene diamine) with an aromatic diacid (or diacid derivative such as a diacid chloride). These polymerizations are often run in poly(phosphoric acid) (PPA) as this medium solvates the monomers and the subsequent polymer formed, activates both functional groups toward condensation and reacts with the water formed by the polycondensation to effectively dehydrate the system (*2*). An alternative synthesis method for the poly(benzazoles) is available when diarylether linkages are present in the polymer backbone. It has been generally recognized that incorporation of aryl ether linkages in a polymer backbone provides improved solubility and sustains many of the desired thermal and mechanical properties. As such, these polyethers are then attractive from both a synthetic and functional viewpoint. Using this polyether approach, the preformed heterocyclic unit is already fully elaborated in one (or both) of the comonomers which are combined in the final poly(ether) synthesis.

Examples of high temperature heterocycle containing polymers which have been modified by the incorporation of aryl ether linkages include imides (*3*), phenylquinoxalines (*4,5*), triazoles (*6,7*), benzoxazole (*8,9*), oxadiazoles *10,11*), benzothiazoles (*12*), isoquinolines *13*) and phthalazines (*14*). Recently, the synthesis of many of these polymers was accomplished by

[1]Corresponding author

0097–6156/96/0624–0266$12.00/0

heterocycle ring activated aromatic nucleophilic substitution chemistry involving displacement of an aryl halide with a phenoxide in a polar solvent. Conventional activating groups for aromatic nucleophilic substitution are simple electron withdrawing groups such as ketone or sulfone, but it is now widely recognized that appropriately substituted heterocycles also can activate halides towards displacement. The characteristics common to all these activating groups is that they are electron withdrawing, usually have a site of unsaturation and can stabilize the negative charge developed in the displacement through resonance to a heteroatom involving the formation of a Meisenheimer complex. Connell and co-workers have prepared poly(aryl ether benzimidazoles), where the preformed benzimidazole heterocycle was introduced in the bisphenol and polymerized with conventional activated bis(aryl halides) in the presence of base (AABB condensation) (*15* – *17*). The resulting polymers could be prepared in high molecular weight in common organic solvents and could be processed from solution or the melt. These polymers were shown to have exceptional thermal and mechanical properties.

We have recently focused our efforts on the preparation and polymerization of benzimidazole, benzoxazole and benzothiazole heterocycles which have 4-fluorophenyl groups pendent on the 2-position of the benzazole. Here the heterocycle acts as a conjugating and electron withdrawing group activating the aryl fluoride towards aromatic nucleophilic substitution polymerization. Polymer formation by aromatic nucleophilic substitution reaction is usually accomplished in an AABB fashion, e.g., between a bisphenol and a counterpart containing two 2-(4-fluorophenyl)benzazole units. It has previously been demonstrated by the preparation and self-polymerization of 6-fluoro-2-(4-hydroxyphenyl)-3-phenylquinoxaline that AB synthesis is an alternative route to new poly(aryl ether phenylquinoxalines) (*18*). By utilizing this A-B monomer, rigorous control of the stoichiometry, as required for conventional poly(aryl ethers), was no longer necessary. We have now extended this AB strategy to 2-phenyl(benzazoles) in which the fluorine leaving group is on the para position of a benzene ring appended to the 2-position of the activating benzazole heterocycle and the phenolic functionality is also on the same monomer (directly on the annulated benzene ring of the benzazole or attached via some intervening structure). As a representative case, the synthetic details for an benzimidazole AB monomer and its elaboration into a polymer is provided.

Experimental

Synthesis of 1-phenyl-2-(4-fluorophenyl)-6-(4-hydroxyphenoxy)benzimidazole

2-nitro-5-fluorodiphenylamine (1). In a 1000 ml round bottom flask equipped with stir bar, reflux condenser and nitrogen inlet was placed 2,4-difluoronitrobenzene. (95.45g, 600 mmol), methylsulfoxide (300 ml), aniline (69.8g, 750 mmol) and anhydrous potassium carbonate (82.8g, 600 mmol). The resulting slurry was gradually warmed to 90 °C and maintained at that temperature for three hours. After this time, only a trace of the difluoronitrobenzene remained, so additional aniline (5g, 53.7 mmol) was added and the reaction maintained an additional hour at 90 °C. The slurry was cooled and transferred to a 2 liter erlenmeyer flask with the aid of some water and then 10% aqueous HCl was added dropwise with stirring until the total volume was 1800 ml. The orange precipitate was isolated by suction filtration, washed well with water and recrystallized from a mixture of acetic acid and water to give 122.2g (88%) of orange crystals in two crops: mp 90.0−92.2 C; ^{1}H NMR ($CDCL_3$) δ 9.66 (br s, 1H), 8.32−8.22 (dd, J=6.0, 6.1 Hz, 1H), 7.49−7.39 (m, 2H), 7.35−7.20 (m, 3H), 6.85−6.77 (dd, J=2.4, 2.7 Hz, 1H), 6.53−6.43 (m, 1H). Anal. Calcd for $C_{12}H_9N_2O_2F$: C, 62.07; H, 3.91; N, 12.06; F, 8.18. Found: C, 61.99; H, 3.83; N, 11.99; F, 7.93.

2-nitro-5-(4-methoxyphenoxy)diphenylamine (2). In a 1000 mL round bottom flask was placed 2-nitro-5-fluorodiphenylamine (46.4g, 200 mmol), p-methoxyphenol (27.28g, 220 mmol), anhydrous potassium carbonate (30.36g 220 mmol) and 350 mL N-methylpyrrolidinone. The mixture was heated to 130 °C and stirred for three hours. After cooling, 500 mL of 10% HCl was added dropwise resulting in an oily residue. The mixture was extracted with ethyl acetate and washed with water and 10% HCl. The organic phase was dried and evaporated giving a orange solid which was recrystallized from methanol to give 56.86 g (84.9%) of the product as orange crystals: m.p. 83−85 C; ^{1}H NMR ($CDCL_3$) δ 9.71 (s, 1H), 8.17 (d, J=9.5 Hz, 1H), 7.43−7.13 (m, 5H), 6.98 (d, J=9.2 Hz, 2H), 6.89 (d, J=9.1 Hz, 2H), 6.68 (s, 1H), 6.27 (d, J=9.2 Hz, 2H), 3.80 (s, 3H). Anal. Calcd for $C_{19}H_{16}N_2O_4$: C, 67.85; H, 4.79; N, 8.33. Found: C, 68.16; H, 4.55; N, 8.33.

2-amino-5-(4-methoxyphenoxy)diphenylamine (3). In a 500 mL round bottom two-neck flask equipped with stir bar, gas inlet, reflux condenser (fitted with a balloon), was placed 2-nitro-5-(4-methoxyphenoxy)-diphenylamine (10.0g, 30 mmol), 10% palladium hydroxide on activated carbon (1.0g) and 90 mL of ethylene glycol dimethylether. The system was flushed with nitrogen two times by filling the balloon and allowing it to discharge through the flask. The same was done with hydrogen after which the second aperture was sealed with a septum. The balloon was refilled with hydrogen and the yellow solution was warmed to 40 °C and allowed to stir for five hours. After this time only a small amount of starting material was found to remain by TLC analysis. The reaction was allowed to stand overnight at room temperature under hydrogen, resulting in a clear solution. The amine was not isolated but converted to its 4-fluorobenzamide derivative in situ.

N-(4-fluorobenzoyl)-2-anilino-4(4-methoxyphenoxy)aniline (4). The crude reduction mixture containing the 2-amino-5-(4-methoxyphenoxy)diphenylamine was swept with a stream of nitrogen to remove any remaining hydrogen and then pyridine (2.37g, 30 mmol) was added followed by 4-fluorobenzoyl chloride (4.76g, 30 mmol). The system was kept under nitrogen and allowed to stir for two hours at room temperature. The mixture was then filtered (with the aid of some ethyl acetate) through a pad of silica gel, and the resulting solution was stripped of solvent leaving 13.23 g of an off white solid. This material was used without further purification: mp 125.6−126.6 C; ^{1}H NMR (DMSO) δ 9.71 (br, s, 1H), 8.06−7.94 (m, 2H), 7.59 (br, s, 1H), 7.39 (d, J=8.5 Hz, 2H), 7.32 (d, J=8.2 Hz, 2H), 7.25−7.14

(m, 2H), 7.09 – 6.92 (m, 5H), 6.87 – 6.76 (m, 2H), 6.52 (d, J=8.8 Hz, 1H), 3.74 (s, 3H).

1-phenyl-2-(4-fluorophenyl)-5-(4-methoxyphenoxy)benzimidazole (5). Into a 1000 mL round bottom flask fitted with a stir bar, reflux condenser, and nitrogen inlet was put N-(4-fluorobenzoyl)-2-anilino-4-(4-methoxyphenoxy)aniline (13.25g, 30 mmol) and 250 mL acetic acid. The solution was heated to reflux (150 °C) for three hours. After cooling, 750 mL of water was added dropwise giving a cloudy mixture which was extracted with 300 mL of ethyl acetate and neutralized with sodium bicarbonate. The organic extract was dried, filtered, and evaporated leaving a brown crystalline solid. Purification by low-pressure liquid chromatography (5%-10% gradient of EtOAc/Toluene) gave a white solid which was recrystallized from EtOH and water leaving 6.94g (56.1%) of white fluffy crystals: mp 134.2 – 136.1 C; ^1H NMR ($CDCL_3$) δ 7.78 (d, J=10.1 Hz, 1H), 7.61 – 7.46 (m, 5H), 7.41 (d, J=6.7 Hz, 2H), 7.29 – 7.16 (m, 2H), 7.03 – 6.86 (m, 5H), 6.68 (s, 1H), 3.73 (s, 3H). Anal. Calcd for $C_{26}H_{19}N_2O_2F$: C, 76.08; H, 4.66; N, 6.82; F, 4.63. Found: C, 76.34; H, 4.56; N, 6.85; F, 4.42.

1-phenyl-2-(4-fluorophenyl)-5-(4-hydroxyphenoxy)benzimidazole (6). Into a 500 mL round bottom flask equipped with a stir bar, reflux condenser, and nitrogen inlet, was placed 2-(4-fluorophenyl)-3-phenyl-6-(4-methoxyphenoxy)benzimidazole (6.5g, 15.8 mmol) and pyridine hydrochloride (45.5g, 40 mmol). The solid mixture was heated to 190 °C, and after 4.5 hrs. the solution was allowed to cool to a solid to which 500 mL of water was added. The resulting white precipitate was filtered and washed well with water. Recrystallization from a mixture of isopropanol and water gave 4.86 g (77.6%) of white crystals. Anal. Calcd for $C_{25}H_{17}N_2O_2F$: 75.75; H, 4.32; N, 7.07; F, 4.79. Found C, 75.38; H, 4.28; N, 7.20; F, 4.30.

Polymer Synthesis. A typical synthesis of a poly(aryl ether benzimidazole) was conducted in a three-neck flask equipped with a nitrogen inlet, mechanical stirrer, Dean-Stark trap, and a condenser. A detailed synthetic procedure designed to prepare a poly(aryl ether benzimidazole) based on **6** is provided. The flask was charged with **6** (1.250 g, 3.15 mmol) and carefully washed into the flask with 25 mL of DMPU. Toluene (20 mL) and K_2CO_3 (0.87 g, 6.5 mmol) were added. Note that the K_2CO_3 was used in 40 – 50% excess. The reaction mixture was then heated until the toluene began to reflux. An optimum reflux temperature range was achieved when the oil bath was maintained between 150 and 170 °C. Toluene was periodically removed from the Dean-Stark trap and replaced with deoxygenated dry toluene to ensure dehydration. The reaction mixture was maintained at 160 °C until the presence of water was no longer observed in the Dean-Stark trap. This usually took between 4 and 8 h, and, during this stage of the reaction, the solvent underwent several color changes. For example, during the initial formation of the phenoxide, a yellow-brown color was observed and as the refluxing proceeded, the color changed to brown. Upon dehydration, the temperature was slowly increased to 180 °C and the toluene was removed through the Dean-Stark trap. The polymerization was heated at

180 °C for approximately 20 h, and completion or near completion was qualitatively estimated by the point where the viscosity increased dramatically. The high molecular weight product was diluted with 50 mL of NMP and filtered hot to remove the inorganic salts. The filtered solution was cooled, and several drops of weak acid (e.g., acetic acid) were added to neutralize the phenoxide end groups. The polymer solution was then coagulated in approximately 10× volume of methanol and then boiled in water to remove trapped salts. The polymer was then dried in a vacuum oven (80 °C) to a constant weight. In each case, the yield was essentially quantitative.

Glass transition temperatures were measured on a Du Pont DSC 1090 instrument with a heating rate of 10 °C/ min. Thermal gravimetric analysis (TGA) on the polymer films was conducted with a heating rate of 5 °C/ min for the variable scans. Intrinsic viscosity measurements were determined by using a Cannon Ubbelohde dilution viscometer in NMP.

Results and Discussion

The rationale for the benzazole-activated aromatic nucleophilic displacement is similar to that of the conventional systems in which the site of attack is polarized (made electron deficient) by both the leaving group and the activating group and the transition state is stabilized by having an efficient means for charge delocalization. The effectiveness of a benzazole as an activating group may be estimated by comparison of its group Hammett coefficients relative to other common activating groups. In fact, it is fortunate that Hammett coefficients for some of these benzazoles has already been determined (*19,20*). These heterocycles have the following σ_p values when appended at their 2-positions (on the carbon between the two electronegative heteroatoms): benzoxazole, 0.34; benzthiazole, 0.34; N-phenylbenzimidazole, 0.24. These sigma values are not particularly large but are in the range of those for typical carbonyl activating groups: PhCO, 0.46 and PhNHCO, 0.35. For comparison, a very potent activating group is the phenylsulfone system found in diphenylsulfone polymers: $PhSO_2$, 0.70. The feasibility of the benzazole-activated fluoro-displacement was first evaluated in model reactions of 2-(4-fluorophenyl) benzoxazole, 2-(4-fluorophenyl) benzothiazole and 2-(4-fluorophenyl) benzimidazole with m-cresol in NMP containing base, affording the desired aryl ethers in each case in approximately 95% yields (*21*). The reactions occurred with high conversion and yield, demonstrating the relevance of benzazole-activation for poly(aryl ether) syntheses.

The synthesis of the requisite AB monomers is interesting in its own right in that multiple aromatic nucleophilic substitution reactions are employed. For the benzimidazole series, 2,4-difluoronitrobenzene is employed as a starting material. Reaction with an amine with 2,4-difluoronitrobenzene can be easily controlled to produce only the ortho monoadduct with high regiospecificity and yield. In the case of reaction of 2,4-difluoronitrobenzene with aniline, the monoadduct arising from selective reaction at the ortho position is isolated in excellent yield. The remaining fluorine at the para position is next substituted by a monoprotected hydroquinone (in this case 4-methoxyphenol was utilized). With the benzene ring now fully functionalized, the 2-substituted imidazole ring is constructed next; in a one-pot reaction, the nitro group is first reduced by catalytic hydrogenation, then immediately converted to its 4-fluorobenzamide derivative. The resulting orthoaminophenylbenzamide is isolated and then closed to the imidazole ring by simply boiling in acetic acid. With the fully substituted benzimidazole prepared, the only remaining task is to deprotect the methyl ether and this is readily accomplished by simply heating the methyl ether precursor in pyridine hydrochloride to afford the final AB monomer. The overall yield for this five step process is about 20%. See Scheme 1.

The polymerizations require the use of dipolar aprotic solvents such as N-methylpyrrolidone (NMP), dimethyl acetamide (DMAC), dimethyl sulfoxide (DMSO) or N,N′-dimethylpropylene urea (DMPU). Nucleophilic aromatic substitution polymerizations are typically performed in a high boiling aprotic polar solvent with the monomer(s) reacted in the presence of a base, potassium carbonate, at elevated temperatures (ca. 180 °C). Potassium carbonate is used to convert the phenol into the potassium phenolate and since K_2CO_3 is a weak base, no hydrolytic side reactions are observed. Dipolar aprotic solvents are used in these poly(aryl ether) syntheses, since they effectively dissolve the monomers and solvate the polar intermediates and the final polymer. DMPU has been shown to be an excellent solvent for poly(ether) syntheses, particularly for those polymers which are only marginally soluble in other dipolar aprotic solvents (*22*). Furthermore, DMPU allows higher reaction temperatures (260 °C). We have observed that DMPU, when used in conjunction with toluene as a dehydrating agent, accelerates many nucleophilic substitution reactions.

The self-polymerization of the A-B monomers were carried out in a DMPU/toluene (2/1) solvent mixture containing K_2CO_3. Stringent monomer delivery criteria, typical of most step-growth polymerizations to maintain the correct stoichiometry, was not required in the self-polymerization of the A-B monomers, since 1:1 stoichiometry is inherent to the monomers. The solids composition was maintained between 20 and 25 wt% analogous to conventional poly(aryl ether) synthesis. The water generated by phenoxide formation in the initial stages of the polymerization was removed as the toluene azeotrope and the reaction mixture was observed to reflux at the desired rate when the temperature of the oil bath was maintained between 150 and 160 °C. Toluene was periodically removed through the Dean-Stark trap and replaced with fresh deoxygenated toluene. Upon dehydration, the polymerization temperature was increased to 180 – 190 °C to effect the nucleophilic aromatic displacement reaction. High molecular weight polymer was obtained as judged by the dramatic increase in viscosity (24 h). The resulting polymer was filtered (to remove inorganic salts), coagulated in excess methanol, washed with boiling water (to remove remaining salts) and dried in a vacuum oven (80 °C) to a constant weight. See Scheme 2.

This general procedure was useful for each of the A-B benzazole monomers, but in the case of the N-phenylbenzimidazole containing monomers longer reaction times were required to achieve high viscosity (i.e., 48 h at 180 °C). Attempts to reduce the polymerization time by increasing the temperature (240 °C), often resulted in the formation of gel. Conversely, the imidazole-containing A-B monomer yielded high polymer in less than 24 h at 190 °C in DMPU (*15 – 17*). Interestingly, the shorter reaction times of benzimidazole containing monomers in nucleophilic aromatic substitution polymerizations was also observed by Connell, et al. In each of the above polymerizations, the subsequent polymers remained soluble during the course

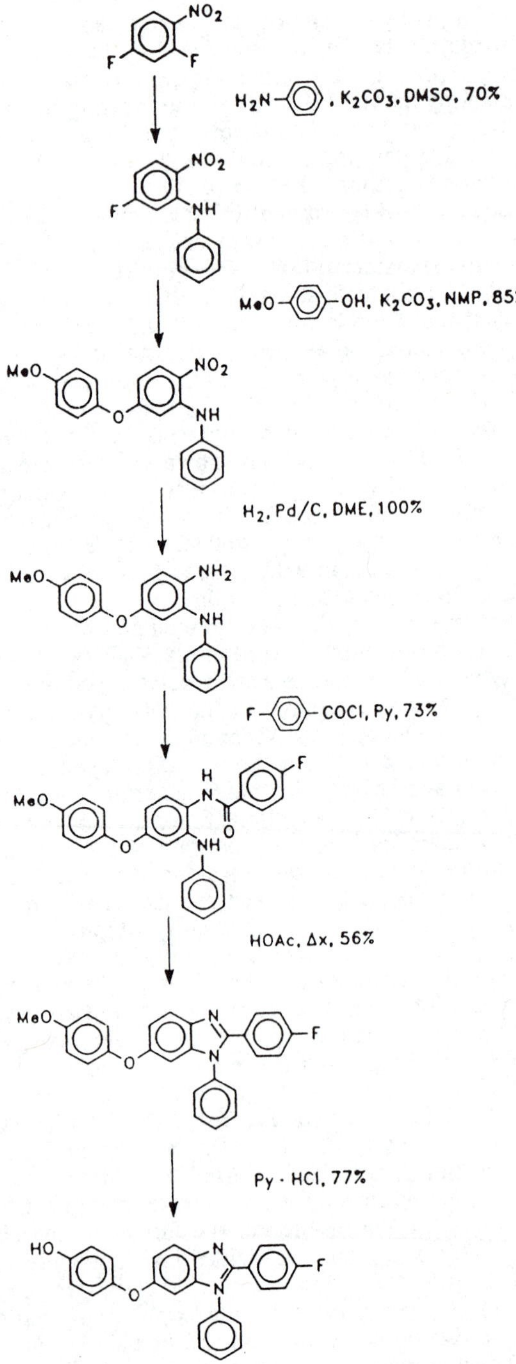

H_2N–C6H5, K_2CO_3, DMSO, 70%

MeO–C6H4–OH, K_2CO_3, NMP, 85%

H_2, Pd/C, DME, 100%

F–C6H4–COCl, Py, 73%

HOAc, Δx, 56%

Py · HCl, 77%

Scheme 1.

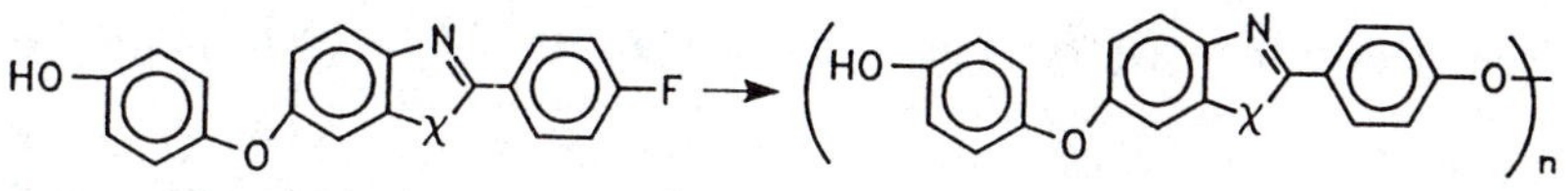

χ	$[\eta]_{NMP}^{25°C}$ dL/g
NH	0.60
NPh	0.45

Scheme 2.

of the reaction, in spite of their rigid structures, allowing the formation of high molecular weight.

The possibility that phenoxide formation might inhibit self polymerization of the AB monomer was of some concern. It was anticipated that the electron rich phenoxide might diminish the activation capacity of the heterocycle which is required for aromatic nucleophilic substitution (or, vis-a-vis, the heterocycle may diminish the nucleophilicity of the phenol). This "bridging" affect would be of particular concern if the phenoxide were formed quantitatively and was situated immediately on the benzimidazole heterocycle (rather than appended as in the specific system described here). However, as proposed by Labadie et al., it is possible that an equilibrium exists between K_2CO_3 and $KHCO_3$ as well as the phenoxide and free phenol. If this is the case, then it is likely that the initial fluoro-displacement will occur at the protonated phenolic monomers. However, it is noteworthy to point out that these A-B monomers took substantially longer to polymerize than their benzoxazole and benzothiazole analogs which may have resulted, to some extent, from possible bridging affects.

Summary

A new class of poly(aryl ether benzimidazoles) have been prepared by heterocyclic-activated displacement polymerization. In these reactions, the aryl ether linkages are generated in the polymer forming reaction. We have demonstrated that the benzimidazole heterocyclic unit is sufficiently electron withdrawing to activate aryl-fluorides towards nucleophilic displacement by a variety of nucleophiles. High molecular weight polymer was readily achieved and had T_g's in the 218 – 250 °C range, depending on the monomers used in the synthesis. This represents another example of the synthesis of poly(aryl ethers) based on a heterocyclic activated halo displacement, and this synthesis can be considered the benzimidazole analogue of the poly(ether imide) synthesis. Moreover, the heterocyclic activated nucleophilic displacement chemistry provides a general methodology to high-temperature, high-T_g poly(aryl ethers). See Table I.

Table I. Characteristics of Poly(aryl ether benzozoles)

Polymer	$[\eta]_{NMP}^{25°C}$ dL/g	Tg °C	Polymer Decomposition Temperature, °C	Isothermal Aging wt. loss/n 300 °C	350 °C
imidazole	0.58	215.	470.	0.120	0.100
phenylimidazole-1	0.45	240	480	0.006	0.084

Literature Cited

1. Vogel, H., Marvel, C. S., *J. Polym.Sci.*, **1961**, *2*, 511.
2. Wolfe, J. E. *Encyl. Polym. Sci. Technol.* **1988**, *11*, 601.
3. White, D. M.; Takehoshi, T.; Williams, F. J., II; Relles, M.; Donahue, P. E., II; Klopfer, I.; Loucks, G. R.; Manello, J. S.; Mathews, R. O.; Schluenz, R. W. *J. Polym. Sci., Polym. Chem. Ed.* **1981**, *19*, 1635.

4. Hedrick, J. L.; Labadie, J. W. *Macromolecules* **1988**, *21*, 1883.
5. Connell, J. W.; Hergenrother, P. M. *Polym. Prepr.* **1987**, *29(1)*, 172.
6. Connell, J. W.; Hergenrother, P. M.; Wolf, P. *Polymer* **1992**, *33*, 3507.
7. Carter, K. R.; Twieg, R.; Hedrick, J. L.; Jonsson, H.; Miller, R. D. *Polym. Prep.* **1992**, *33(1)*, 388.
8. Connell, J. W.; Hergenrother, P. M. *Polym. Mat. Sci. Eng. Proc.* **1989**, *60*, 527.
9. Hilborn, J.; Labadie, J.; Hedrick, J. L. *Macromolecules* **1990**, *23*, 2854.
10. Connell, J. W.; Hergenrother, P. M. *Polym. Prepr. (Am. Chem. Soc., Div. Polym. Chem.)* **1987**, *29(1)*, 172.
11. Hedrick, J.; Twieg, R. *Macromolecules* **1992**, *25*, 2021.
12. Hedrick, J. L. *Macromolecules* **1991**, *24*, 6361.
13. Singh, R.; Hay, A. S. *Macromolecules* **1992**, *25*, 1025.
14. Singh, R.; Hay, A. S. *Macromolecules* **1992**, *25*, 1025.
15. Smith, J. G.; Connell, J. W.; Hergenrother, P. M. *Polym. Prepr.* **1991**, *32(3)*, 193.
16. Hergenrother, P. M.; Smith, J. G.; Connell, J. W. *Polymer.* (in press).
17. Hergenrother, P. M.; Smith, J. G.; Connell, J. W. *Polym. Prepr.* **1992**, *33(1)*, 411.
18. Labadie, J. W.; Hedrick, J. L.; Boyer, S. *J. Polym. Sci.: Part A: Polym. Chem.* **1992**, *30*, 519.
19. Bystrov, V. F.; Belaya, Zh. N.; Gruz, B. E.; Syrova, G. P.; Tolmachev, A. I.; Shulezhko, L. M.; Yagupol'skii, L. M. *Zh. Obshchei Khim.* **1968**, *38*, 1001.
20. Exner, O. In "A Critical Compilation of Substituent Constants," Chapter 10 in *Correlation Analysis in Chemistry—Recent Advances,* Chapman, N. B., Shorter, J., Eds.; Plenum Press: New York, 1978.
21. Labadie, J.; Hedrick, J. L. *J. Makromol. Chem., Macromol. Symp.* **1992** *54/55*, 313.
22. Labadie, J. W.; Carter, K. R.;, Hedrick, J. L.; Jonsson, H.; Twieg, R. J.; Kim, S. Y. *J. Polym. Bull* **1993** *30*, 25.

RECEIVED December 6, 1995

Chapter 17

Investigation of Monomer Reactivity in Poly(aryl ether) Synthesis Utilizing ^{19}F NMR Spectroscopy

Kenneth R. Carter

Research Division, IBM Almaden Research Center, 650 Harry Road, San Jose, CA 95120–6099

NMR is a valuable tool for evaluating the electron withdrawing effect of substituents present on phenyl rings. When an electron withdrawing group is present on a phenyl ring, a partial positive charge develops at the ortho and para positions through resonance interactions. The reactivity of a number of aryl fluoride monomers used in nucleophilic aromatic substitution polymerization was explored utilizing ^{19}F NMR experiments. The ^{19}F shifts reflect the reactivity of the individual monomers examined. Taft inductive and resonance parameters were calculated for a series of monomers from ^{19}F data and used to identify activating forces for the monomers. NMR data were compared with calculated net atomic charges. Relative reactivity studies were also performed in order to verify the utility of this fast and convenient NMR probe of monomer reactivity.

The development of high-temperature polymers for use in thermally stable microelectronic devices has led to the examination of a number of polymer systems, including poly(aryl ethers). One common route towards obtaining these polymers is through nucleophilic aromatic substitution polymerization. This route, first reported by Johnson and co-workers (1) in 1967, involves the nucleophilic displacement of activated dihalo aryl derivatives by bisphenol salts to yield high molecular weight poly(aryl ethers) where the generation of an aryl ether linkage is the polymer-forming reaction. The commercially available poly(aryl ethers) [e.g. polysulfones and polyetherketones] are prepared in this manner and the reaction mechanisms and conditions used in those reactions are rather well understood (2,3).

The nucleophilic displacement of a halogen from an activated aryl halide system occurs in a two step addition-elimination reaction (S_NAr). The nucleophile adds to the electron-deficient aryl halide, forming a negatively charged Meisenheimer complex from which the

0097–6156/96/0624–0276$12.00/0

halide is eliminated leading to the formation of an aryl-ether linkage. The activating group present in the aryl halide serves two purposes. The group must be an electron withdrawing moiety, which decreases the electron density at the site of the reaction, and secondly, its presence must lower the energy of the transition state for the reaction by stabilizing the anionic intermediate formed. These S_NAr reactions proceed at a significant rate only if the electron withdrawing substituent is located either the ortho or para position relative to the halide.

The most commonly employed activating groups in these reactions have been ketones, sulfones and more recently phosphine oxides, which are all strongly electron withdrawing substituents. Aryl fluorides have been observed to be the most effective substrates of all the aryl halides for a number of reasons (4). Poly(aryl ethers), made by nucleophilic aromatic substitution polymerization, contain the activating group as a part of the main-chain of the polymer. Consequently, the resulting poly(aryl ethers) properties are often influenced by the activating group, as well as any other functionalities, present in the monomers (5).

Other functional groups, some weakly electron withdrawing, can also activate aryl fluorides towards nucleophilic aromatic substitution (1,6). A variety of heterocycles [for examples, see references 1,7,8,9,10,11,12] as well as other functional groups [e.g. perfluoroalkyl groups (13), azines (14), acetylenes (15), etc.] can effectively activate aryl fluorides toward S_NAr reactions and many of these groups have been successfully used in the preparation of the corresponding poly(aryl ethers).

Though a number of different activating groups have been employed in poly(aryl ether) syntheses, little work has been done to determine the relative reactivities of various monomers. No compilation of monomer reactivity exists, though certain trends have been noted. In earlier investigations of heterocyclic-activation of aryl fluorides, the predictive use of ^1H NMR chemical shift data, involving the protons ortho to the electron withdrawing group, was reported (16). In those examples, the ortho proton resonances of potential monomers were compared with those found present in aryl fluoride systems containing conventional activating groups [e.g. ketones or sulfones]. Hay and co-workers have reported Hückel molecular orbital calculations for the determination of the net charge densities at the C-F carbon atoms of a number aryl fluoride monomers (15,17,18). Credible predictions of monomer reactivity containing several types of activating groups were made on the basis of HMO calculations.

We felt it would be informative to find a spectroscopic probe that would allow the evaluation of a large number of potential monomers in regards to their ability to undergo these S_NAr-type polymerization reactions. Work performed in the preparation of poly(aryl ether triazoles) (8) and poly(aryl ether quinoxalines) (9) indicated that ^{19}F NMR could be used as a sensitive and convenient probe of a monomers ability to undergo transformation under standard S_NAr conditions. Early results of the use of ^{19}F NMR in the evaluation of monomer reactivity has been reported (19,20).

For this current study, a wide range of aryl fluoride monomers were studied by ^{19}F NMR spectroscopy and it was shown that there is indeed a correspondence between the observed chemical shifts and monomer reactivity. The structures of monomers studied are shown in Figure 1. Competitive relative reactivity studies were performed in order to corroborate the observed ^{19}F NMR chemical shifts correlations. The MNDO-PM3 semiempirical method was also used to calculate charge densities of the C-F carbons of selected aryl fluorides. These values were compared to experimentally measured ^{19}F NMR chemical shifts.

Experimental

Characterization. NMR spectra were recorded on either a IBM WP 250 spectrometer operating at 250.1 MHz (^{1}H), 62.9 MHz (^{13}C) or a IBM WP 300 spectrometer operating at 282.3 MHz (^{19}F). Tetramethylsilane (Me_4Si) was used as a reference for ^{1}H and ^{13}C NMR measurements while $CFCl_3$ was used as an internal standard for the ^{19}F NMR measurements with the reference peaks being assigned at 0.0 ppm. Chemical shifts up field of the reference are assigned a negative sign and are reported in ppm. All samples for ^{19}F NMR were prepared as dilute solutions (5% w/w) in d_6 - DMSO. The ^{13}C NMR measurements were performed using $CDCl_3$ as the solvent. Molecular geometries and net charge densities were calculated using the semiempirical PM3 procedure (21,22) within the commercially available MOPAC package. Initial molecular geometries were obtained by MMX molecular mechanics techniques contained within the ALCHEMY molecular graphics package.

Materials. All aryl fluorides were either commercially available or were synthesized according to references found herein.

Model Compounds. Model compounds were synthesized by reacting the desired aryl fluoride with *m*-cresol in the presence of K_2CO_3 in N-methyl-2- pyrrolidinone (NMP) while azotropically removing water formed with toluene. The model compounds were used as standards for relative reactivity studies.

Competitive Relative Reaction Rates. A series of experiments was performed in which equimolar amounts of two aryl difluorides were reacted with a limited amount of *m*-cresol under standard S_NAr conditions [NMP, K_2CO_3, 170 °C, azeotropic removal of water by toluene] (Scheme 1). A known amount of biphenyl was added in each experiment as an internal standard. Aliquots of the reaction mixtures were taken every 15 minutes and analyzed by HPLC for product formation, so that plots of product formation vs. reaction time could be made. These reactions were performed with pairs of monomers of varying reactivity; **{1+2}**; **{2+3}**; **{3+4}**; **{4+13a}**.

Results and Discussion

NMR is a valuable tool for evaluating the electron withdrawing effect of aryl substituents in potential monomers. When a resonance electron withdrawing group is present on a

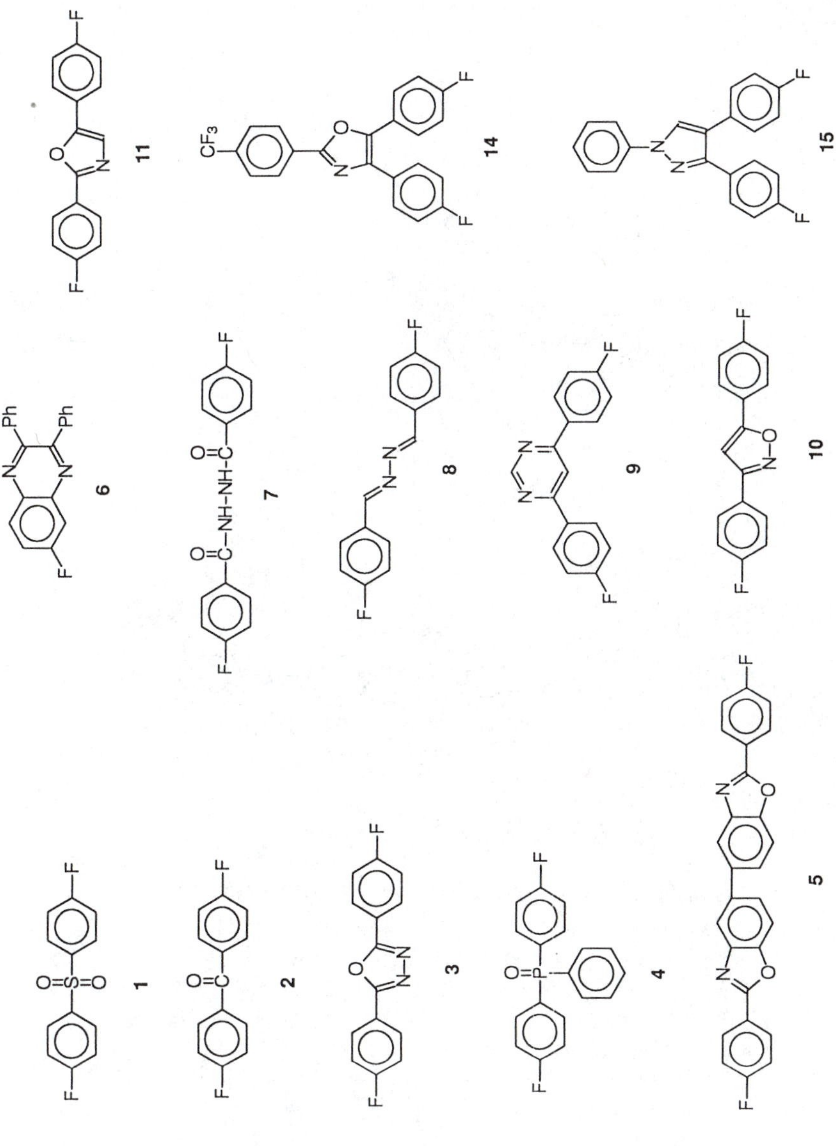

Figure 1. Structures of Various Aryl-Fluorides

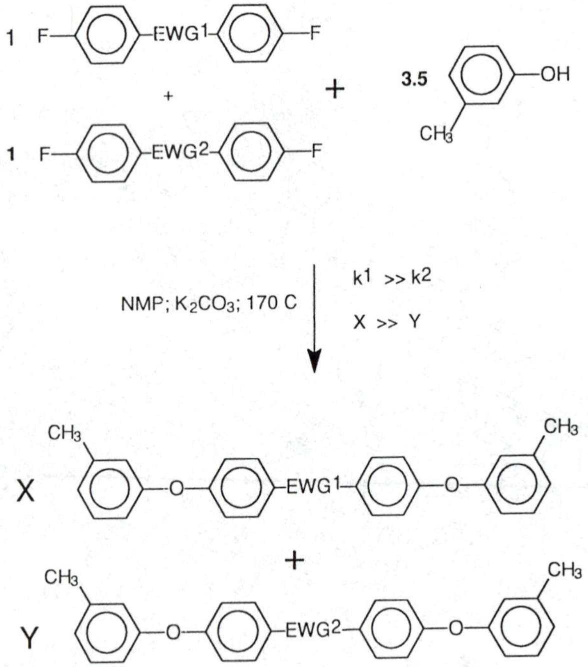

1 F–C$_6$H$_4$–EWG1–C$_6$H$_4$–F + 1 F–C$_6$H$_4$–EWG2–C$_6$H$_4$–F + 3.5 m-cresol (CH$_3$, OH)

NMP; K_2CO_3; 170 C

$k1 >> k2$

$X >> Y$

X: CH$_3$–C$_6$H$_4$–O–C$_6$H$_4$–EWG1–C$_6$H$_4$–O–C$_6$H$_4$–CH$_3$

+

Y: CH$_3$–C$_6$H$_4$–O–C$_6$H$_4$–EWG2–C$_6$H$_4$–O–C$_6$H$_4$–CH$_3$

Scheme 1. Competitive Relative Reactivity Experiment

phenyl ring, a partial positive charge develops at the ortho and para positions through resonance interactions. NMR chemical shifts are very sensitive to electron density at the particular nuclei of interest and in the case of the aryl fluoride monomers, there are three NMR probes that can be utilized; ^{1}H, ^{13}C, and ^{19}F.

In previous investigations of heterocycle activation, the predictive use of ^{1}H NMR chemical shift data for the protons ortho to the electron withdrawing group was demonstrated (23). In those examples, the ortho protons resonances of potential monomers were compared with those of conventional activating groups [e.g. ketones or sulfones]. While some valid predictions of monomer reactivity based on ^{1}H NMR data have been made, the technique suffers from several disadvantages. There is a small chemical shift range of this nucleus, amounting to only about $\Delta\delta = 1$ ppm observed between non-activated compounds and highly activated ones. Additionally, ^{1}H NMR shielding can vary significantly due to local intramolecular anisotropic effects, and the technique only surveys the site ortho to the electron withdrawing group rather than the actual site of the substitution reaction. For these reasons other NMR probes were sought.

Both ^{13}C and ^{19}F NMR can be used to probe the electron density at the actual site of nucleophilic reaction, i.e. the C-F bond of aryl fluorides. A large number of studies involving the use of ^{13}C NMR spectroscopy in the study of substituent effects have been reported and reviewed (21,24,25).

For the aryl fluorides of interest in this study, the ^{13}C NMR chemical shifts of the carbon atoms para to electron withdrawing groups in monomers which successfully undergo polymerization were found to range from δ 164.5 to 166.2 ppm while the ^{13}C resonance of the unreactive molecule, fluorobenzene is observed at δ 162.8 ppm. This corresponds to an usable frequency range of about 213 Hz when measured on a 250 MHz instrument, with a resolution of ±2 Hz. ^{13}C chemical shift data correspond well with the strength of various electron withdrawing groups as shown in Table 1. Unfortunately, when performing ^{13}C NMR experiments on dilute samples, many scans are required due to the low sensitivity and low natural abundance of ^{13}C nuclei. Quite often, the C-F carbon can be difficult to detect since it is a quaternary carbon atom and due to strong C-F coupling ($J_{CF} = 250$ Hz). For these reasons the use ^{19}F NMR was explored.

Since we are concerned with the level of activation of the aryl fluoride bond, it is possible to use the fluorine atoms as probes of the strength of various electron withdrawing groups. The use of ^{19}F NMR chemical shifts proved to be the most sensitive probe of reactivity with a chemical shift range spanning 9 ppm (2500 Hz) between the most activated monomer, 4,4′-difluorophenyl sulfone (-104.08 ppm) and non-activated fluorobenzene (-112.77 ppm). Due to the high sensitivity of ^{19}F NMR and the ease of detection, this technique was used to study a series of aryl fluoride monomers.

The ^{19}F chemical shifts of the series of monomers studied are listed in Table 1, and where available, the ^{13}C NMR chemical shift of the ipso carbon is also given. The monomers

Table 1. NMR Chemical Shifts Of Various Aryl Fluorides[a]

Compound	^{13}C Chemical Shift[b]	^{19}F Chemical Shift[c]
4,4′-difluorodiphenyl sulfone **(1)**	165.31	-104.08
bis(4-fluorophenyl)methanone **(2)**	165.27	-106.01
2,5-bis(4-fluorophenyl)-1,3,4-oxadiazole **(3)**	165.55	-106.71
bis(4-fluorophenyl)phenylphosphine oxide **(4)**	165.05	-106.71
2,2′-bis(4-fluorophenyl)bibenzoxazole **(5)**	-	-106.83
2,3-diphenyl-6-fluoroquinoxaline **(6)**	164.8	-107.81
bis[(4-fluorophenyl)carbonyl]hydrazine **(7)**	-	-107.81
1-4-bis(4-fluorophenyl)-2,3-diaza-1,3-butadiene **(8)**	164.50	-108.00
4,6-bis(4-fluorophenyl)pyrimidine[10] **(9)**	-	-109.42
3,5-bis(4-fluorophenyl)isoxazole[10] **(10)**	-	-109.81 -110.73
2,5-bis(4-fluorophenyl)oxazole[11] **(11)**	164.07 162.70	-109.91 -112.76
2,5-bis(4-fluorophenyl)-1-aryl-1,3,4-triazole **(12a)**	163.38	-110.49
2,3-bis(4-fluorophenyl)quinoxaline **(13a)**	163.10	-111.99
4,3-bis(4-fluorophenyl)-2-[4′-(trifluoromethyl)-phenyl]oxazole **(14)**	-	-110.54 -112.27
Fluorobenzene	162.82	-112.77
3,4-bis(4-fluorophenyl)-1-phenylpyrazole **(15)**	-	-113.29 -114.88

a) ^{13}C Chemical shifts are reported in ppm relative to TMS = 0.0 ppm. ^{19}F Chemical shifts are reported in ppm relative to CCl_3F = 0.0 ppm; b) Performed on dilute solutions in $CDCl_3$.; c) Performed on dilute solutions in DMSO d_6.

are listed in order of decreasing chemical shift and Figure 2 shows a plot of ^{13}C and ^{19}F data. The plot shows that there is a good correlation between both the ^{13}C and ^{19}F NMR shifts (r = 0.98). The relative rates of reaction, as experimentally estimated by the time needed to form high molecular weight polymer, is related to the magnitude of the ^{19}F NMR chemical shifts; i.e. the lower the chemical shift, the lower the reactivity of that monomer. A correlation of these observations to actual monomer reactivity was performed and will be discussed below.

^{19}F NMR Data vs. Calculated Net Atomic Charges. The electronic effects of various aryl fluoride monomers was also studied by calculating net charge densities using the semiempirical PM3 procedure within the MOPAC package. The electron densities of the ispo carbons (C-F carbons) of **1,2,3,8,12a,** and fluorobenzene were calculated (Table 2) and the values obtained were plotted against the measured ^{19}F shift values (Figure 3).

Table 2. Calculated Net Atomic Charges

Compound	Net Atomic Charge[a]
4,4′-difluorodiphenyl sulfone **(1)**	+0.1175
bis(4-fluorophenyl)methanone **(2)**	+0.0876
2,5-bis(4-fluorophenyl)-1,3,4-oxadiazole **(3)**	+0.0892
1-4-bis(4-fluorophenyl)-2,3-diaza-1,3-butadiene **(8)**	+0.083
2,5-bis(4-fluorophenyl)-1-aryl-1,3,4-triazole **(12a)**	+0.0748
Fluorobenzene	+0.0647

a) Calculated using the semiempirical PM3 procedure within the MOPAC package.

There is a fairly good agreement of the ^{19}F data and the calculated net atomic charge. One notable exception is the lack of correlation observed for the sulfone monomer, **1**. The sulfone monomer, which has been observed to be the most reactive monomer examined in this study, does not correlate to the slope of net atomic charge vs. chemical shift magnitude. The C-F carbon of the sulfone monomer, **1**, clearly has a high net atomic charge, but the observed ^{19}F shift value is not as high as one would predict based upon the slope of the line in Figure 3 Perhaps the PM3 calculations are not accurate for this particular monomer, not properly estimating the effect of the sulfone group. Though this one monomer does not correlate well with the others, in whole the ^{19}F NMR data follow the calculated net atomic charges order fairly well [^{19}F, r = 0.98 when excluding the sulfone monomer].

^{19}F NMR Shifts as a Measure of Reactivity. The ^{19}F NMR technique allowed the screening of a large number of potential fluoro-monomers in order to assess the feasibility of usage in nucleophilic substitution polymerization. Aryl fluorides with ^{19}F NMR chemical shifts equal to or less than -112.8 ppm (fluorobenzene) were deemed non-polymerizable, and indeed, attempts to obtain high molecular weight polymers with these monomers were always unsuccessful regardless of the reaction conditions employed.

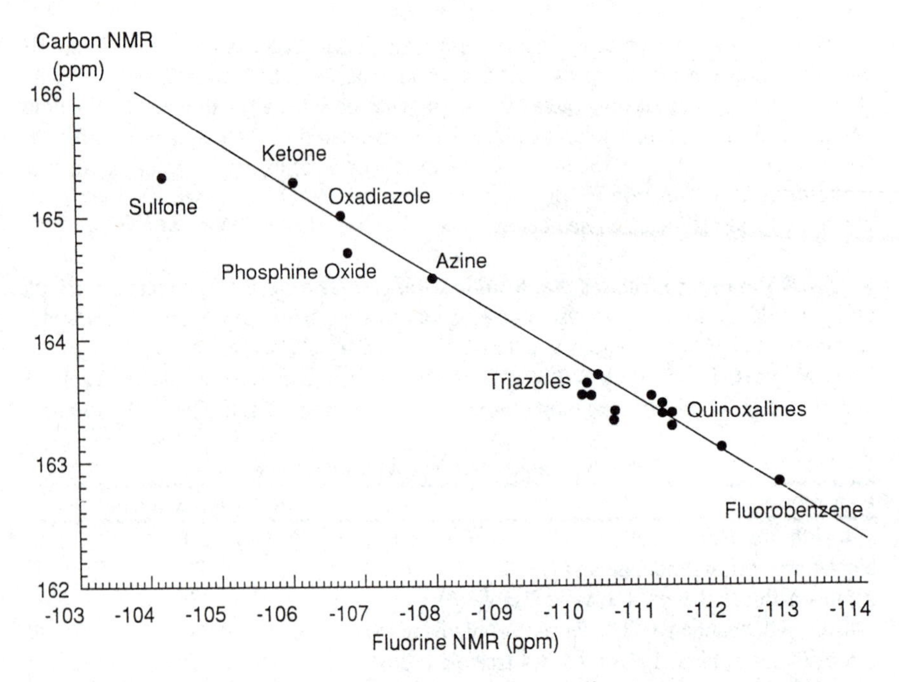

Figure 2. A Plot of ^{13}C NMR Chemical Shifts vs ^{19}F NMR Chemical Shifts

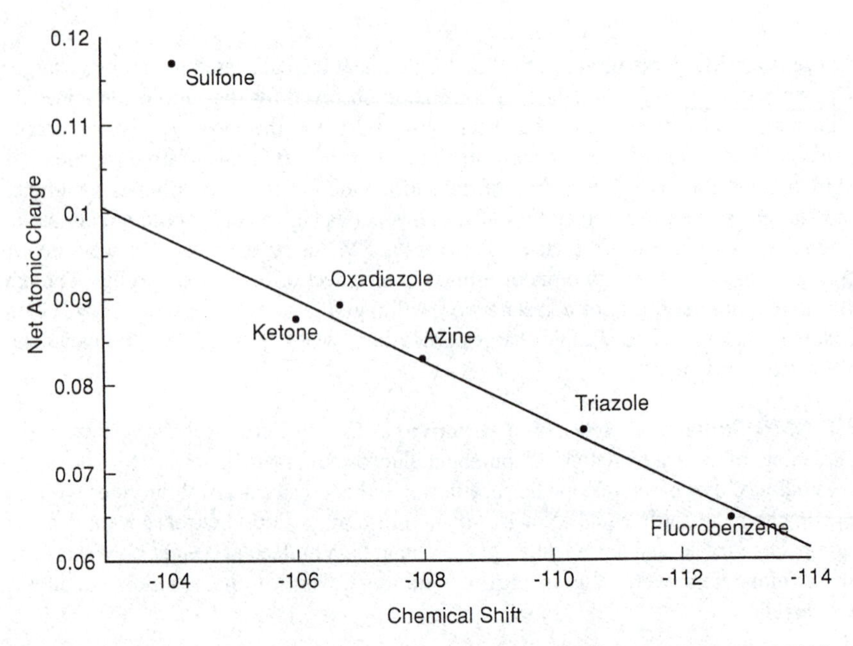

Figure 3. A Plot of Calculated Net Atomic Charges vs ^{19}F NMR Chemical Shifts

In cases where the chemical shifts ranged between -110 and -112 ppm, interesting results were found. Two monomer classes studied by our research group fall within this range, the 2,5-bis(4′-fluorophenyl)-1-aryl-1,3,4-triazoles (8), δ = -110 to -110.5, **(12a-h)** and the 2,3-bis(4′-fluorophenyl)-6-quinoxalines (9) , δ = -111 to -111.9, **(13a-e)**. Initial polymerization attempts were carried out utilizing polymerization conditions commonly used with strongly activated monomers; NMP solvent with toluene reflux, K_2CO_3, with reaction temperatures around 180 °C for 24 h. Under these conditions, no high molecular weight polymers were obtained in these systems, due to the weak activating character of the monomers. When the reactions were performed in the solvent DMPU (dimethylpropyleneurea) at higher reaction temperatures (190 °C), high molecular weight polymers were obtained. DMPU has been shown to facilitate nucleophilic substitution reactions in cases where conventional polar aprotic solvents are ineffective (26). The successful incorporation of these monomers into poly(aryl ether) systems shows that even polymers derived from these weakly activated monomers, as identified by NMR, can be attained through modification of reaction conditions. The conditions and reaction times required for polymerization of the quinoxalines were more stringent than for the triazoles, which corresponds to a lower reactivity of the quinoxalines as predicted by NMR. Even reactivity differences within the monomer series were noted and they also corresponded to the ^{19}F NMR data (8,9).

Bass *et al.* have studied some other weakly activated aryl fluoride monomers as well, examining pyrimidine and isoxazole heterocycles as activating groups (10). The ^{19}F shifts of the isoxazole monomer **[10]** were reported at δ = -109.81 and -110.73 (two resonances are observed due to the non-symmetrical nature of the monomer) and that of the pyrimidine monomer **[9]** at δ = -109.42. These ranges suggest that both should be able to yield high molecular weight poly(aryl ethers) if reaction conditions were altered to account for the low monomer reactivity. Though no information is given about attempts to polymerize the isoxazole monomer, the pyrimidine monomer was successfully incorporated into a poly(aryl ether). The authors do report that the polymer obtained from **9** is of low molecular weight, but the polymerization conditions described in their study were different than those that have been employed by our group. Similarly, Maier *et al.* have reported the successful polymerization of a 2,5-bis(4-fluorophenyl)oxazole **[11]** (11). In that case, the monomer had ^{19}F shifts at δ = -109.91 and -112.76. The low value of -112.76 suggests that the aryl fluoride should have very low reactivity. That is, in fact, the case with the monomer giving high molecular weight only when the reaction is performed at high temperature (200 °C), in DMPU, for 20 h. The ability of this monomer to undergo polymerization under these conditions is an indication of its high thermal stability and lack of side-reactions that one can experience under these conditions.

Attempts at polymerizations and model reactions of 4,5-bis(4′-fluorophenyl)-2-[4′-trifluormethyl)phenyl]oxazole, **14**, which has ^{19}F δ = -110.54 and -112.27 were repeatedly performed. Reaction conditions similar to those employed in the synthesis of the 1,3,4-triazoles and 2,3-bis(4-fluorophenyl)quinoxalines were employed (NMP or DMPU; >180 °C). No desired products were obtained. Model reaction studies revealed that in addition to a variety of degradation products, only one of the fluorophenyl groups of **14** undergoes substitution by the nucleophile, the aryl fluoride resonance at $\delta = -112.27$ is still observed in the product of the reaction of **14** and *m*-cresol. The failure of this monomer to polymerize shows that as the activation of the aryl fluoride decreases, a point is reached where the S_NAr reaction will either no longer take place or that other high temperature degradation reactions preclude the formation of high molecular weight polymer. We examined various other aryl fluorides [e.g. 3,4-bis(4′-fluorophenyl)-1-phenylpyrazole, **15**] that were shown to be electron donating as compared to fluorobenzene and found that none of these would undergo nucleophilic aromatic substitution.

In addition to providing relative reactivity information about aryl fluorides, ^{19}F NMR can be used to calculate the Taft substituent parameters σ_I and σ_R^0 (27-29) (see Table 3).

Table 3. Relative Reactivities of Selected Aryl-Fluorides

Electron Withdrawing Group	σ_I	σ_R
NO_2-	+0.56	+0.20
COPh-	+0.19	+0.18
SO_2Ph-	+0.52	+0.14
azine (C=N-N=C)-	+0.18	+0.14
CF_3-	+0.38	+0.10
2,5-triazole rings	+0.17	+0.05
H-	+0.08	0.00
1-triazole ring	+0.49	-0.02
CH_3-	-0.08	-0.15
$N(CH_3)_2$-	+0.10	-0.54

These parameters are a relative measure of inductive and resonance effects in substituted aryl rings, where $\int H^{m-x}$ and $\int H^{p-x}$ are the differences in chemical shift of meta- and para-

$$\sigma_I = 0.1409\,(0.6 - \int H^{m-x})$$

$$\sigma_{R^0} = 0.0339\,(\int H^{m-x} - \int H^{p-x})$$

substituted aryl fluorides as compared to fluorobenzene (28,29). The value of σ_R^0 is a better measure of conjugative interactions between the phenyl ring and the substituent than σ_I which is mainly influenced by inductive effects. Many of the more weakly activated monomers rely on their ability to stabilize the anionic intermediate through resonance interactions. These same monomers have little contribution due to electron withdrawal through inductive effects. Conversely, the highly activated monomer, difluorophenyl sulfone, has a very large σ_I, +0.52, which accounts for its high reactivity even though its σ_R^0 at 0.14 is not significantly different than values found for other activating groups. The trifluoromethyl group also has considerable contribution from inductive effects. Indeed, we have shown that trifluoromethyl groups can activate ortho-nitro groups towards S_NAr reactions to yield new poly(aryl ethers) (30). Labadie *et. al.* have also shown that other perfluoroalkyl groups can serve as good activating groups for S_NAr reactions (31,32).

A comparison of the Taft parameters calculated for a number of aryl fluorides is shown in Table 3. It can be seen that the conjugative electron withdrawal (σ_R^0) by the triazole ring from the 2,5-diphenyl substituents is small, ranging somewhere between that of CF_3 and H. The inductive parameter, σ_I, also shows that the triazole ring is mildly activating. Both values are smaller than those found in conventional monomers such as ketone or sulfone activated systems.

In the case when an azino group is the activating group **[8]**, one sees a larger value for σ_R^0 than observed with the triazole ring, indicating a greater amount of electron withdrawal through resonance effects with the azino functionality. Experimental results show that aryl fluorides activated by the azino group are much more reactive than the triazole activated monomers. The calculation of the Taft reactivity parameters σ_I and σ_R^0 from NMR data can give a good deal of information on the degree of activation of various substituent groups.

In order to corroborate the reactivity observations with the NMR data, competitive relative reactivity studies were conducted. The competitive reactions were designed to assess activated fluoro-monomers reactivity in relation to each other. The reaction conditions were kept as close to those employed in the synthesis of poly(aryl ethers) as possible. Equimolar amounts of two aryl fluorides were allowed to react with a limiting amount of *m*-cresol. As the nucleophilic aromatic substitution reactions proceeded, small aliquots of the reaction mixture were quantitatively analyzed by HPLC. The individual relative reactivity reactions were performed between pairs of monomers of decreasing reactivity; **{1+2}**, **{2+3}**, **{3+4}**, **{4+13a}**. Though all experiments were carried out under similar reaction conditions,(NMP solutions of the monomers, *m*-cresol, biphenyl heated to 170°C in the presence of K_2CO_3 with azeotropical removal of water with toluene), small temperature variations during heating and slight concentration differences between different reactions prevented the calculation of any absolute reactivity rates. The data obtained were a good measure of relative reactivity between the monomer pairs studied and are shown in Table 4.

In all cases, the observed relative reactivities followed the order predicted based upon the ^{19}F NMR data. The reaction between **{3+4}** was of interest because the ^{19}F NMR

Table 4. Relative Reactivites of Selected Aryl-Fluorides

Compound	Relative Reactivity[a]	^{13}C Shift	^{19}F Shift
4,4′-difluorodiphenyl sulfone **(1)**	57140	165.31	-104.08
bis(4-fluorophenyl)methanone **(2)**	9080	165.27	-106.01
2,5-bis(4-fluorophenyl)-1,3,4 oxadiazole **(3)**	510	165.55	-106.71
bis(4-fluorophenyl)phenylphosphine oxide **(4)**	230	165.05	-106.71
2,3-bis(4-fluorophenyl)quinoxaline **(13a)**	1	163.10	-111.99

a) Reactivites relative to 13a.

chemical shifts of those monomers were both observed at -106.71, though the ^{13}C NMR data suggested that the oxadiazole monomer, **3**, should be slightly more activated. Indeed that was the case with **3** reacting slightly faster than the phosphine oxide monomer, **4**. The rate of the reaction was only about 2 times greater in this case, which was the smallest observed difference between any two monomer pairs studied.

These competitive reactivity experiments clearly show that there is a relationship between monomer reactivity and observed ^{19}F NMR chemical shifts. Assuming that the magnitude of the ^{13}C or ^{19}F shift is proportional to the activation energy of the substitution reaction, E_a, and using the relative rates obtained experimentally, one can see if the experimental data obtained follows the Arrhenious equation, $k = Ae^{-Ea/RT}$ Therefore $k \sim$ relative reaction rate $\sim e^{-Ea/RT}$, so *ln(*relative reaction rate) $\sim E_a$. A comparison of ^{13}C and ^{19}F chemical shifts vs relative reactive rates was performed and shown in Figures 4 and 5. The ^{19}F NMR data fits well (r = 0.97) while the ^{13}C NMR data gives a much less satisfactory fit (r = 0.85). While the series of monomers studied clearly does not represent all of the many classes of activating groups available, it is a good data set to represent monomers with a wide range of relative reactivities. Using this series as a guide, one can measure the ^{19}F NMR shift of a monomer and be able to predicts with reasonable certainty whether it is strongly, moderately or weakly activated. More work needs to be done to determine the precise empirical relationship between ^{19}F NMR shifts and E_a.

Conclusions

The use of ^{19}F NMR spectroscopy to evaluate potential aryl fluoride monomers as candidates for nucleophilic aromatic substitution polymerization has shown to be an accurate and time saving technique. Since the ^{19}F shift is controlled by the electron density of the carbon to which it is attached, the magnitude of the ^{19}F chemical shift can

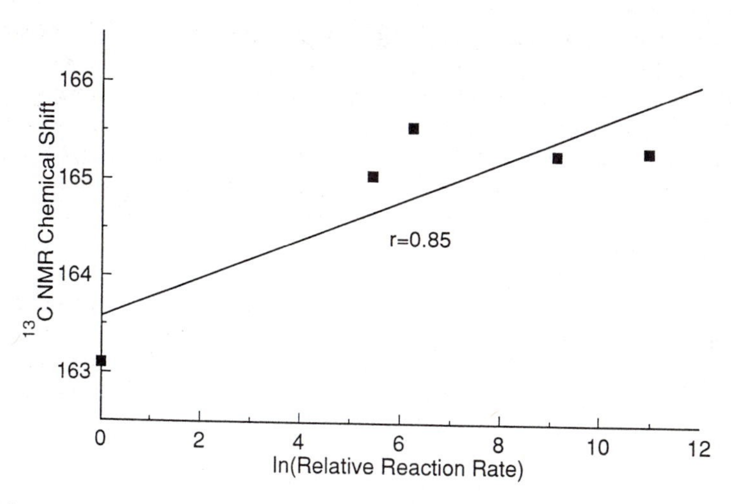

Figure 4. A Plot of ^{13}C NMR Chemical Shift vs ln(Relative Reaction Rate)

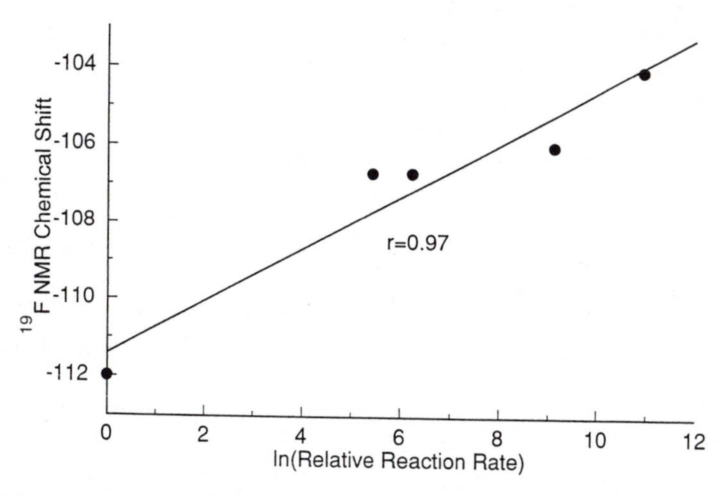

Figure 5. A Plot of ^{19}F NMR Chemical Shift vs ln(Relative Reaction Rate)

be related to the degree of activation of aryl fluorides by various electron withdrawing groups. The larger the magnitude of the shift, the more activated the compound, with reactivity dropping off as the observed chemical shift approaches that of fluorobenzene.

In order to take advantage of the use of ^{19}F NMR, care should be taken during sample preparation and when interpreting the NMR data. All samples must be prepared at low dilutions and comparisons can only be made when all spectra are obtained in the same solvent. It is important to note that even though the use of ^{19}F shifts can give a good deal of information about the electron withdrawing ability of various substituents, an activated aryl fluoride still may not be a suitable monomer if side reactions involving other functional groups present in the molecule preclude successful polymerization.

Acknowledgments

The author wishes to thank Gregory May and Mark Sherwood for their assistance in performing the ^{19}F NMR measurements, and Jeff Labadie for advice on relative reactivity experiments. Discussions with Gerhard Maier proved very valuable and also thanks to James Hedrick, Robert Miller and Robert Twieg for the donation of several of the aryl fluorides used in this study.

Literature Cited

1. Johnson, R. N.; Farnham, A. G.; Clendinning, R. A.; Hale, W. F.; Merriam, C. N. *J. Polym. Sci., A-1*, **1967**, *5*, 2375.
2. Maiti, S.; Mandai, B. K. *Prog. Polym. Sci.*, **1986**, *12*, 111.
3. Rose, J. B. *Polymer*, **1974**, *15*, 456.
4. Bunnett, J. F.; Zahler, R. E. *Chem. Rev.*, **1951**, *49*, 273.
5. Maiti, S.; Mandal, B. K. *Prog. Polym. Sci.*, **1986**, *12*, 111.
6. Labadie, J. W.; Hedrick, J. L. *Makromol. Chem., Macromol. Symp.*, **1992**, *54/55*, 313.
7. Carter, K. R.; Twieg, R.; Hedrick, J. L.; Jonsson, H.; Miller, R. D. *Polym. Prep.*, **1992**, *33(1)*, 388.
8. Carter, K. R.; Miller, R. D.; Hedrick, J. L. *Macromolecules*, **1993**, *26*, 2155.
9. Hedrick, J. L.; Twieg, R. J.; Matray, T.; Carter, K. R. *Macromolecules*, **1993**, *26*, 4833.
10. Herbert, C. G.; Bass, R. G.; Watson, K. A.; Connell, J. W. *Polym. Prep.*, **1994**, *35*(2), 703.
11. Schneider, J. M.; Maier, G.; Nuyken, O. *Makromol. Reps.*, **1994**, *A3*, 179.
12. Hay, A. S.; Singh, R. *Macromolecules*, **1992**, *25*, 1033.
13. Carter, K. R.; Kim, S. Y.; Labadie, J. W. *Polym. Prep.*, **1993**, *34*, 415.
14. Carter, K. R.; Hedrick, J. L. *Macromolecules*, **1994**, *27*, 3426.
15. Strukelj, M; Paventi, M.; Hay, A. S. *Macromolecules*, **1993**, *26*, 1777.
16. Hedrick, J. L.; Labadie, J. W. *Macromolecules*, **1990**, *23*, 1561.

17. Yeomans, K. A.; Hay, A. S. *ACS Polym. Mater. Sci. Eng.* **1993**, *69*, 241.
18. Strukelj, M.; Hamier, J.; Elce, E.; Hay, A. S. *J. Polym. Sci., Polym. Chem.*, **1994**, *32*, 193.
19. Carter, K. R. *Proc. PMSE,* **1993**, *69*, 432.
20. Carter, K. R. *Macromolecules,* **1995**, *28*, 6462.
21. Stewart, J. *Comput.-Aided Mol. Des.*, **1990**, *4*, 1.
22. Stewart, J. *J. Comput. Chem.*, **1989**, *10*, 209.
23. Hedrick, J. L.; Labadie, J. W. *Macromolecules,* **1990**, *23*, 1561.
24. Craik, D. J.; Brownlee, R. T. C *Prog. Phys. Org. Chem.*, **1983**, *14*, 1.
25. Nelson, G. L.; Williams, E. A. *Prog. Phys. Org. Chem.*, **1976**, *12*, 229.
26. Labadie, J. W.; Carter, K. R.; Hedrick, J. L.; Jonsson, H.; Twieg, R. J.; Kim, S. Y. *Polymer Bulletin,* **1993**, *30*, 25.
27. Hehre, W. J.; Taft, R. W.; Topsom, R. D. *Prog. Phys. Org. Chem.*, Volume 12; John Wiley & Sons: New York, 1976.
28. Taft, R. W.; Price, E.; Fow, I. R; Lewis, I. C.; Andersen, K. K.; Davis, G. T. *J. Am. Chem. Soc.*, **1963**, *85*, 3146.
29. Kaplan, L. J.; Martin, J. C. *J. Am. Chem. Soc.*, **1973**, *85*, 793.
30. Carter, K. R.; Kim, S. Y.; Labadie, J. W. *Polym. Prep.*, **1993**, *34(1)*, 415.
31. Kim, S. Y.; Labadie, J. W. *Polym .Prep*, **1991**, 32(1), 164.
32. Labadie, J. W.; Hedrick, J. L *Macromolecules,* **1991**, *24*, 812.

RECEIVED December 6, 1995

General Step-Growth Polymerization Topics

Chapter 18

A Survey of Some Recent Advances in Step-Growth Polymerization

Jeff W. Labadie[1], James L. Hedrick[2], and Mitsuru Ueda[3]

[1]Argonaut Technologies, Inc., 887 Industrial Road, San Carlos, CA 94070
[2]Research Division, IBM Almaden Research Center, 650 Harry Road, San Jose, CA 95120–6099
[3]Department of Materials Science and Engineering, Yamegata University, 4–3–16 Jonan, Yonezawa, Yamagata 992, Japan

The field of step-growth polymers encompasses many polymer structures and polymerization reaction types. This chapter attempts to cover topics in step-growth polymerization outside of the areas reviewed in the other introductory chapters in this book, i.e., poly(aryl ethers), dendritic polymers, high-temperature polymers and transition-metal catalyzed polymerizations. Polyamides, polyesters, polycarbonates, poly(phenylene sulfides) and other important polymer systems are addressed. The chapter is not a comprehensive review but rather an overview of some of the more interesting recent research results reported for these step-growth polymers, including new polymerization chemistries and mechanistic studies.

Aromatic Polyamides

Aromatic polyamides (aramids) are important in the commercial fibers industry because of their high tensile strength and good flame resistance. The demand for these materials is increasing as new applications are found. One example is their use in composites where high-use temperatures, light weight, chemical resistance, and dimensional stability are crucial.

The most widely employed synthetic route to aramids is based on the polycondensation of dicarboxylic acids with diamines in the presence of condensing agents. Good reviews on the synthesis of aramids have recently appeared (*1 – 3*). Recently, promising alternative synthetic routes to aramids have been reported and are described herein. These include the polycondensation of N-silylated diamines with diacid chlorides, the addition-elimination reaction of dicarboxylic acids with diisocyanates, and the palladium-catalyzed carbonylation polymerization of aromatic dibromides, aromatic diamines and carbon monoxide.

The synthesis of aramids from N-silylated amines has been employed because N-silylated aromatic amines show higher reactivity relative to the parent diamines and the resulting trialkylsilyl halide does not lower the reactivity of unreacted amine functionality as is the case with amine protonation

0097–6156/96/0624–0294$12.00/0

when HCl is released (*4*). This method has been successfully applied in the preparation of fluorine-containing aramids which were unobtainable by a conventional method (*5*). Polycondensation of tetrafluoro-m-phenylene diamine with aromatic diacid chlorides gave low molecular weight polymers because of the poor nucleophilicity of the fluoro-diamine. However, polycondensations carried out with N-silylated tetrafluoro-m-phenylene diamine with aromatic diacid chlorides in NMP at 0 °C afforded polymers with inherent viscosities up to 0.47 dL/g. This method has also been applied to the synthesis of N-phenylated aramids of high molecular weights from N,N′-bis(trimethylsilyl)-p-dianilinobenzene and aromatic diacid chlorides (*6*).

Isocyanates have been shown to undergo a high yield reaction with carboxylic acids to afford amides. Polymerization of aromatic diisocyanates with aromatic dicarboxylic acids was carried out in the presence of 3-methyl-1-phenyl-2-phospholene 2-oxide to afford aramids (*7*). The polymerization was carried out in sulfolane at 200 °C. Polymers with inherent viscosities of up to 1.8 dL/g were obtained.

Ring-opening polymerization of aromatic cyclic oligomers has been an area of great interest since they can offer unique advantages in the manufacture of important products such as molding and composite resins. The synthesis of aramids by ring-opening polymerization of macrocyclic oligoamides has been reported employing this approach (*8*). Polymerization occurred at 300 °C in the presence of 1-methyl-3-n-butylimidazole-2-thione and phenylphophinic acid as the condensing agent. Polymers with inherent viscosities as high as 0.82 dL/g were obtained.

Transition-metal catalyzed polymerizations are an attractive method for the synthesis of condensation polymers, many of which are inaccessible by other methods. The palladium-catalyzed reaction of aromatic bromides, amines and carbon monoxide yields aromatic amides in high yields (*9*). Imai extended this chemistry to polymer synthesis through the use of bifunctional monomer pairs (*10*). Carbonylative polymerization of aromatic dibromides and aromatic diamines is carried out under an atmosphere of carbon monoxide and in the presence of a palladium catalyst, 1,8-diazabicyclo[5,4,0]-7-undecene (DBU), and phosphine ligands. A dipolar aprotic solvent was used at a polymerization temperature of 115 °C. The polymers produced had inherent viscosities ranging from 0.2 to 0.8 dL/g. Higher molecular weight aramids were prepared using aromatic diiodides as substrates in place of aromatic dibromides under higher CO pressure (*11*).

Transition metals have also been employed as complexing agents to increase the solubility of aramids. The low solubility of aramids often necessitates extreme synthesis and processing conditions. Recently, a new method involving chromium carbonyl-arene complexation has been reported which affords improved solubility and processability (*12*). Polycondensation of (p-phenylenediamine)$Cr(CO)_3$ with terephthaloyl chloride in N,N-dimethylacetamide proceeded as a homogeneous solution to yield high molecular weight poly(p-phenylene terephthalamide)$Cr(CO)_3$ complex. Solution casting afforded a strong, air-stable polymer film. Decomplexation of the chromium from the aramid was effected by heating or oxidation with iodine.

Polyesters

Aromatic polyesters have been prepared by condensation of aromatic diacid chlorides and silylated bisphenols in the presence of chloride ions as a catalyst (*13*). Lower temperatures were employed than those required for polymerization of acetylated diphenols and aromatic carboxylic acids, affording less side products and higher molecular weights in some cases (*14,15*). Similarly, trimethylsilyl 4-acetoxybenzoate can be polymerized at temperatures between 350 – 400 °C in Marlotherm-S to give poly(4-hydroxybenzoate) (Scheme II) (*16*). In this case, the silyl monomer is lower in reactivity than 4-acetoxybenzoic acid, which is normally used to prepare poly(4-hydroxybenzoate). Polymerization of the trimethylsilyl ester requires higher temperatures and gives lower polymer yields due to lower reactivity relative to the free acid monomer. However, the trimethylsilyl ester route affords highly crystalline poly(4-hydroxybenzoate) with a whisker-like morphology. This was attributed to the formation of fewer side-products than the polymerization of 4-acetoxybenzoic acid, which allows the generation of more perfect crystals than the poly(4-hydroxybenzoate) whiskers.

Polycarbonates

Polycarbonates derived from bisphenol-A are important commercial thermoplastics. In order to expand the potential applications of aromatic polycarbonates, significant research has been devoted towards improving the heat resistance while maintaining the other desirable properties of bisphenol-A polycarbonate. One approach to increasing the dimensional stability of bisphenol-A polycarbonate (T_g = 150 °C) is to introduce pendent methyl groups to the polymer backbone. Polycarbonates of tetramethyl bisphenol-A have been reported and display an increase in T_g to 203 °C, however, with a concomitant reduction in ductility and impact strength (*17*). Recently, Freitag reported the use of a new bisphenol derived from hydrogenated isophorone, 4,4′-trimethylcyclohexylidene)diphenol, TMC (*18*) (Scheme III). The TMC polycarbonate has a T_g of 239 °C, while ductility and melt processability are maintained. The TMC polycarbonate and copolymers with bisphenol-A are commercial polymers marketed by Bayer as Apec HT resins.

An important advancement in the synthesis of polycarbonates in recent years is the polymerization of cyclic oligocarbonates (*19*). The cyclic precursors are synthesized by a trialkylmine-catalyzed, *pseudo*-high dilution hydrolysis/condensation reaction of bisphenol-A chloroformate. The oligomers are formed in 85 – 90%, with high polymer as the side product, and are comprised of between 2 and 20 bisphenol-A carbonate repeat units. The oligomers are polymerized to high polymer in the presence of an appropriate catalyst. This methodology has the advantage of allowing the processing of the low viscosity cyclic oligomers, which can be polymerized in a subsequent step without outgassing of volatile products of polymerization.

Aliphatic polycarbonates have been investigated as potential ceramic binders, and in high resolution photoresist schemes. These applications rely on the clean thermal degradation of aliphatic polycarbonates, which is accelerated in the presence of photogenerated acids. The use of solid-liquid phase transfer catalysis in conjunction with bis(carbonylimidazolides) (*20*) or bis(p-nitrophenylcarbonates) (*21*) was developed by Frechet for the synthesis of novel tertiary copolycarbonates (Scheme IV). The instability of tertiary chloroformates renders tertiary polycarbonates inaccessible through conventional chloroformate monomers or intermediates. The bis(carbonylimidazolide) monomer was shown to polymerize with both tertiary and secondary alcohols, demonstrating the utility of the method in forming polycarbonates from less reactive sterically hindered monomers.

Poly(formal)s

Polyformals of bisphenols and dichloromethane were prepared by polyetherification in dimethylsulfoxide at 80 °C (*22*). The poly(formal)s of bisphenol-A, tetramethylbisphenol-A (TMBA), and their copolymers were generated by this method. High molecular weight polymers were obtained, except in the case of TMBA homopolymers where crystallization led to premature precipitation. Incorporation of 70% TMBA afforded an increase in T_g from 88 °C to 113 °C. The 1,4-dihydroxy-2-cyclohexenols were prepared under phase-transfer conditions with dibromomethane (*23*). These polymers were evaluated as self-developable resists.

Poly(phenylene Sulfide)s

Poly(phenylene sulfide) (PPS) is an important engineering thermoplastic which is produced by the polymerization of sodium sulfide-hydrate and p-dichlorobenzene at 200 – 280 °C in NMP (*24*) (Scheme V). The classical synthesis, characterization and properties of PPS has been reviewed (*25*). In recent years, research on the mechanism of PPS formation, and several new routes to PPS have been reported and will be surveyed here.

Proposals for the mechanism of PPS formation include nucleophilic aromatic substitution (S_NAr) (*26*), radical-cation (*27*), and radical-anion processes (*28,29*). Some of the interesting features of the polymerization are that the initial reaction of the sodium sulfide-hydrate with NMP affords a soluble NaSH-sodium 4-(N-methylamino)butanoate mixture, and that polymers of higher molecular weight than predicted by the Caruthers equation are produced at low conversions. Mechanistic elucidation has been hampered by the harsh polymerization conditions and poor solubility of PPS in common organic solvents. A detailed mechanistic study of model compounds by Fahey provided strong evidence that the ionic S_NAr mechanism predominates (*30*). Some of the evidence supporting the S_NAr mechanism was the selective formation of phenylthiobenzenes, absence of disulfide production, kinetics behavior, the lack of influence of radical initiators and inhibitors, relative rate Hammet values, and activation parameters consistent with nucleophilic aromatic substitution. The radical-anion process was not completely discounted and may be a minor competing mechanism.

An alternative route to PPS involves polymerization of p-halothiophenols as A-B monomers (*26*). Copper 4-bromothiophenoxide polymerizes to PPS in quinoline or quinoline/pyridine mixtures at temperatures of 200 – 230 °C and atmospheric pressures (*31*). Mechanistic studies support that polymerization of the copper salt proceeds by an $S_{RN}1$ radical-anion mechanism at the early stages of the reaction and may contribute in the later stages of polymerization as well (*32*). Debromination has been observed as a molecular weight lim-

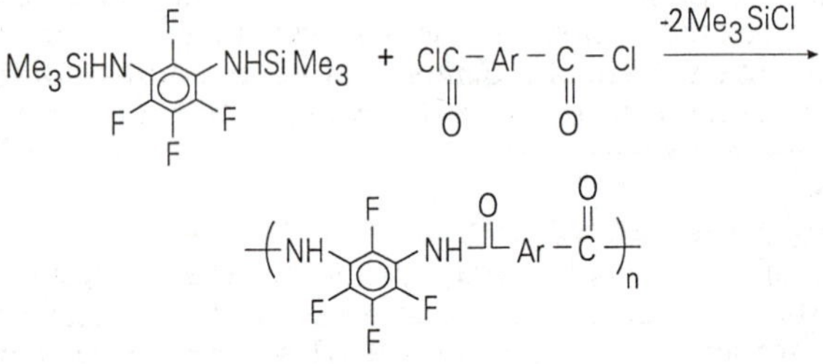

$Me_3SiHN-C_6F_4-NHSiMe_3 + ClC(=O)-Ar-C(=O)-Cl \xrightarrow{-2Me_3SiCl} -\!\!\left(NH-C_6F_4-NH-C(=O)-Ar-C(=O)\right)_n\!\!-$

Scheme I.

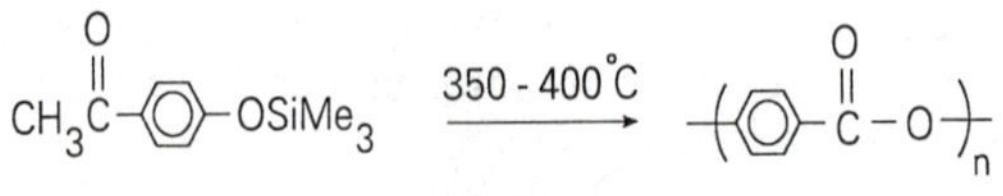

$CH_3C(=O)-C_6H_4-OSiMe_3 \xrightarrow{350 - 400^{\circ}C} -\!\!\left(C_6H_4-C(=O)-O\right)_n\!\!-$

Scheme II.

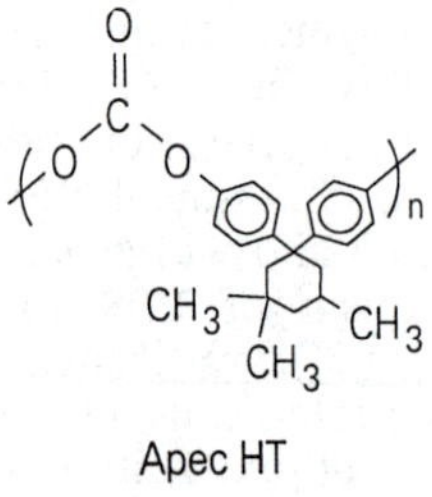

O, C, O, O, n, CH_3, CH_3, CH_3

Apec HT

Scheme III.

$X-C(=O)-O-C(CH_3)_2-R-C(CH_3)_2-OC(=O)-X$

$+ HO-R'-OH \xrightarrow{PTC} \left[O-C(=O)-O-C(CH_3)_2-R-C(CH_3)_2-O-C(=O)-O-R\right]$

$R = CH_2CH_2-$, $-C_6H_4-$

$X = -N$ (imidazolyl) , $-O-C_6H_4-NO_2$

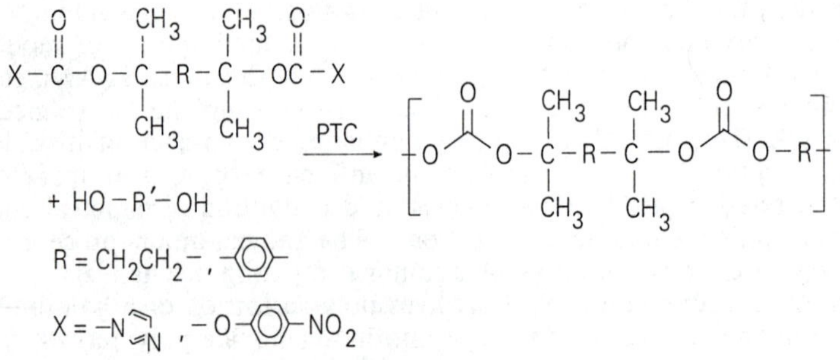

Scheme IV.

iting side reaction which can be eliminated by carrying out the polymerization between 180 – 200 °C (*33*). A similar dechlorination process has been observed in poly(aryl ether) synthesis and has been attributed to an S_{NR}I-type mechanism (*34*).

Melt polymerization of diiodobenzene with elemental sulfur in the presence of air, followed by further heating in the solid state affords high molecular weight PPS with the extrusion of iodine (*35*) (Scheme VI). The resulting PPS is nearly equivalent to commercial PPS, with the exception that a low level of disulfide linkages are retained, and much lower levels of ionic impurities are present in the isolated polymer. The polymerization mechanism appears radical in nature, which was substantiated by an electron paramagnetic resonance spectroscopic study of the polymerization (*36*). A related route to high molecular weight PPS involves thermolysis of bis(4-iodophenyl disulfide) in diphenyl ether at 230 – 270 °C (*37*). The method was also applied to the synthesis of poly(naphtylene sulfides). This procedure was extended to bromide analogs through the addition of KI (*37b,38*).

Diphenyl disulfides have been converted to oligo(phenylene sulfides) using either cationic (*39*) or oxidative polymerization conditions (*40*). The oxidative polymerizations have been reported in the presence of either hydroquinones (*40a*) or a oxovanadium/oxygen catalyst system (*40b*). The polymerizations are carried out in chlorinated solvents at room temperature, which limits the molecular weight obtainable to 1,000 daltons due to oligomer precipitation. The oligomers have phenyl disulfide end groups and can be functionalized with an iodophenoxyisophthalic acid end group and converted to polyamide-graft-oligophenylene sulfide copolymers (*41*). High molecular weight poly(phenylene sulfide) has been prepared by oxidative polymerization of methyl-4-(phenylthio)phenyl sulfoxide in trifluoromethane sulfonic acid to afford poly(sufonium cation) as a soluble precursor polymer (*42*) (Scheme VII). This material was subsequently demethylated in refluxing pyridine to afford PPS which precipitated as deprotection neared completion. This methodology was extended to methyl phenyl sulfide, where the sulfide was oxidized to the sulfoxide *in situ* (*43*).

Polyenaminonitriles

Polyenaminonitriles are a unique class of step-growth polymers that rely on the condensation of bis(chlorovinylidene cyanide) monomers with aromatic (*44*) or aliphatic diamines (*45*) (Scheme VIII). The strong electron-withdrawing nature of the nitrile groups impart reactivity similar to a carbonyl group to the vinylidene cyanide moiety. The polyenaminotriles are soluble, processable materials which can be subsequently cyclized thermally to a poly(aminoquinoline). This situation is similar to that of polyimides, where poly(amic acid) precursor polymers are used, however, no volatile side products are released in the enaminonitrile to aminoquinoline transformation. The chemistry of the chlorovinylidene cyanide group has been extended to reactions with arylhydrazines and utilized in the synthesis of polypyrazoles (*46*).

Poly(p-phenyl Vinylene). With the discovery that poly(p-phenylvinylene) (PPV) can be used in polymeric light emitting diodes (*47*), these materials

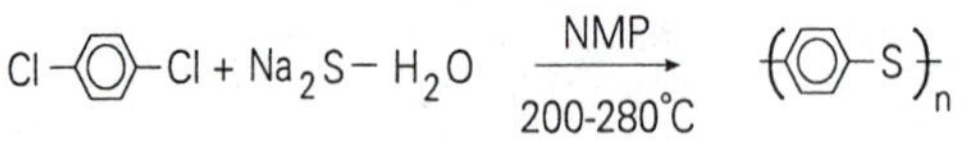

$$\mathrm{Cl{-}C_6H_4{-}Cl} + \mathrm{Na_2S{-}H_2O} \xrightarrow[200\text{-}280^\circ C]{NMP} \mathrm{{+}C_6H_4{-}S{+}_n}$$

Scheme V.

$$8\,\mathrm{I{-}C_6H_4{-}I} + \mathrm{S_8} \longrightarrow \mathrm{{+}C_6H_4{-}S{+}} + 8\,\mathrm{I_2}$$

Scheme VI.

$$\mathrm{C_6H_5{-}S{-}C_6H_4{-}S(=O){-}Me} \longrightarrow \mathrm{{+}C_6H_4{-}S{-}C_6H_4{-}S^{\oplus}(Me){+}_n} \xrightarrow{\text{Pyridine, } \Delta} \mathrm{{+}C_6H_4{-}S{+}_{2n}}$$

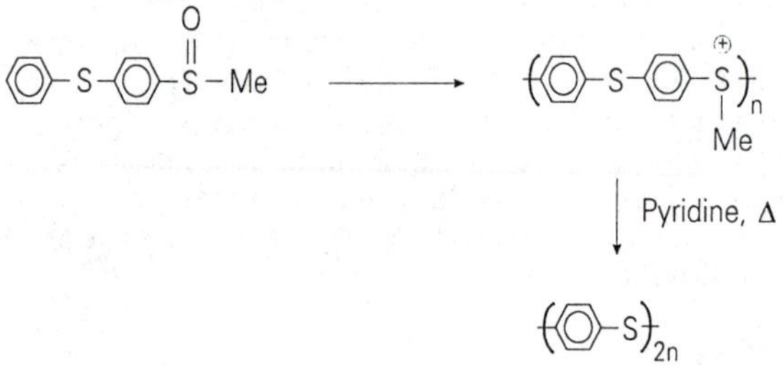

Scheme VII.

have recently attracted a great deal of interest. Classical step-growth polymerization methods for PPV synthesis, e.g., Wittig and Knoevenagel condensation, have been reviewed (*48*). A problem inherent with these synthetic methods is the insoluble nature of PPV, which leads to premature precipitation. One approach used to overcome this problem is to incorporate alkoxy groups on the aromatic ring of PPV (*49*). Alternatively, benzyl-substituted poly(xylylene) precursor polymers have been shown to afford PPV after a thermal or chemical step. Xylylene chloride (*50*) and Xylylene sulfonium salts (*51*) have been used as monomers and are polymerized in the presence of base (Scheme IX). Transition-metal catalyzed coupling has also been utilized as a route to PPV systems and is reviewed in the chapter describing this polymerization method.

Benzocyclobutene Polymers

Polymers derived from monomers containing the benzocyclobutene moiety were reported independently by Kirchoff and Hahn at Dow (*52*) and Arnold (*53*). The Dow workers have focused on bisbenzocyclobutene monomers (BCBs) containing α,β-alkenyl and bis(alkenylsiloxane) linkages between the benzocyclobutene moieties (Scheme X), whereas Arnold reported maleimide- and phenylethynyl-benzocyclobutene copolymers. Dow has commercialized several BCB systems for microelectronics applications. The Dow BCBs are prepolymerized to processable oligomers, which are cured to the final network after film deposition without outgassing of volatiles. The polymerization chemistry is based on thermal opening of the cyclobutane ring to afford an o-quinodimethane reactive intermediate. The o-quinodimethane can undergo either Diels-Alder reactions with alkenes, alkynes or self-reaction to give cyclic dimers or linear polymer (Scheme XI). Mechanistic studies on arylvinylbenzocyclobutene (*54*) and BCB (*55*) demonstrated that Diels-Alder addition of the o-quindimethane to the alkene is the predominate reaction pathway. In cases where one double bond is present per two benzocyclobutene units, quinodimethane self-reaction ensues after consumption of the double bonds by the Diels-Alder process. A high degree of Diels-Alder reaction was also observed in the copolymerization of bismaleimide and BCBs (*53c*). Self-polymerization of an A-B benzocyclobutene-phenylethynyl phtthalimide monomer occurred with a mixture of cross- and self-condensation of the reactive endgroups (*53a*).

Perfluorocyclobutane Aromatic Polyethers

Novel polymers containing alternating perfluorocyclobutane and aromatic ether subunits have been prepared by polymerization of aryl trifluorovinyl ether monomers via the thermal $[2\pi + 2\pi]$ cyclodimerization of the trifluorovinyl ether functionality (Scheme XII) (*56*). The dimerization of the trifluorovinyl ether moiety is not a concerted pericyclic reaction, rather proceeds through a diradical intermediate. Dimerization affords predominately 1,2-substituted perfluorocyclobutane rings with a mixture of cis and trans isomers. Both linear and crosslinked polymer systems were synthesized using bis- and tris(trifluorovinyoxy) monomers, respectively. These polymers display both a low dielectric constant (≤ 2.5) and very low moisture absorption.

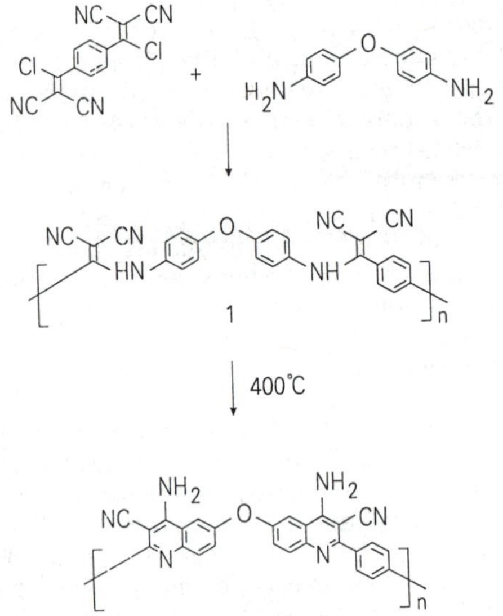

Scheme VIII.

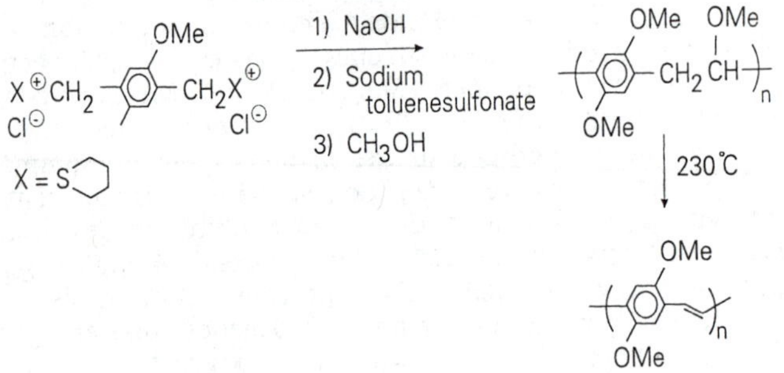

Scheme IX.

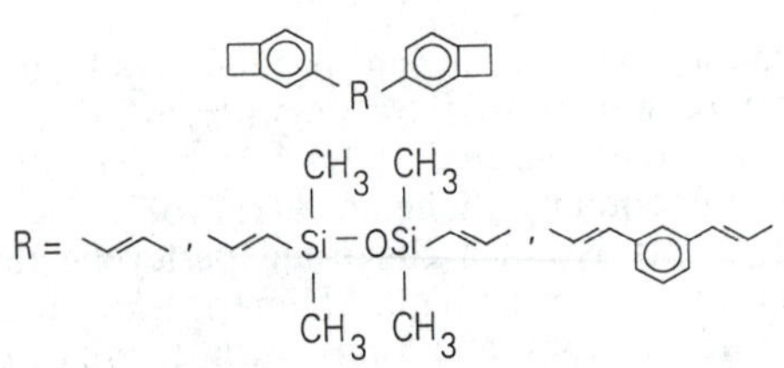

Scheme X.

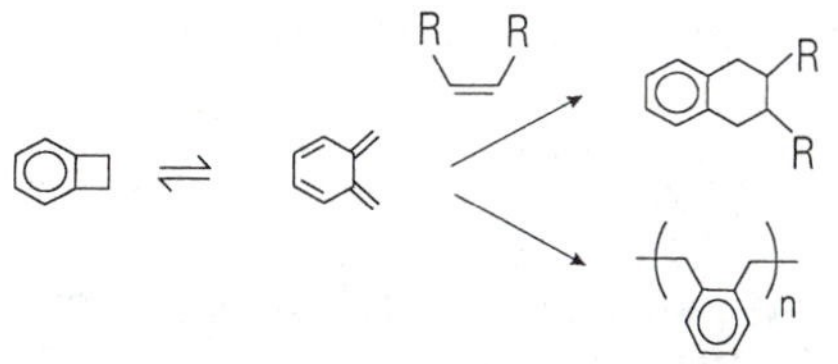

Scheme XI.

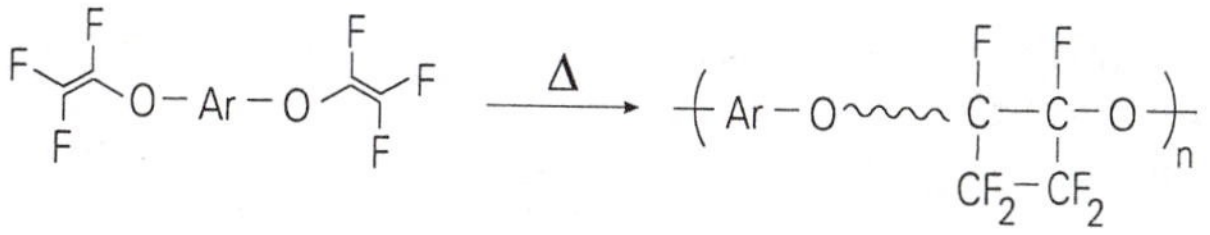

Scheme XII.

Literature Cited

1. Yang, H. H. *Aromatic High Strength Fibers*; Wiley: New York, 1989.
2. Sekiguchi, H.; Coutin, B. In *Handbook of Polymer Synthesis, Part A*; Kricheldorf, H. R., Ed.; Dekker: New York, 1992; p. 807.
3. Volbracht, L. *Comprehensive Polymer Science*; Allen, G., Bevington, J. C., Eds.; Pergamon: Oxford, 1989; Vol. 5, p. 375.
4. Imai, Y. *Makromol. Chem., Macromol. Symp.* **1992**, *54/55*, 389.
5. Oishi, Y.; Harada, S.; Kakimoto, M.; Imai, Y. *J. Polym. Sci., Part A, Polym. Chem.* **1992**, *30*, 1203.
6. Oishi, Y.; Kakimoto, M.; Imai, Y. *J. Polym. Sci., Part A, Polym. Chem.* **1987**, *25*, 2493.
7. Otsuki, T.; Kakimoto, M.; Imai, Y. *J. Polym. Sci., Part A, Polym. Chem.* **1989**, *27*, 1775.
8. Memeger, W., Jr.; Lazar, J.; Ovenall, D.; Leach, R. A. *Macromolecules* **1993**, *26*, 3476.
9. Heck, R. F. *Palladium Reagents in Organic Syntheses*; Academic: New York, 1985; p. 18.
10. Yoneyama, M.; Kakimoto, M.; Imai, Y. *Macromolecules* **1988**, *21*, 1908.
11. Perry, R.; Turner, S. R.; Blevins, B. W. *Macromolecules* **1993**, *26*, 1509.
12. Dembek, A. A.; Burch, R. R.; Feiring, A. E. *J. Am. Chem. Soc.* **1993**, *115*, 2087.
13. Kricheldorf, H. R. *Makromol. Chem., Macromol. Symp.* **1992**, *54/55*, 365.
14. Kricheldorf, H. R.; Englehardt, J. *J. Polym. Sci., Part A* **1988**, *26*, 1621.
15. Kricheldorf, H. R.; Weegen-Schulz, B.; Englehardt, J. *Makromol. Chem.* **1991**, *192*, 631.
16. Kricheldorf, H. R.; Schwarz, G.; Ruhser, F. *Macromolecules* **1991**, *24*, 3485.
17. Morbitzer, L.; Grigo, U. *Angew. Makromol. Chem.* **1988**, *162*, 87.

18. Freitag, D.; Westeppe, U. *Makromol. Chem., Rapid Commun.* **1991**, *12*, 95.
19. (a) Brunelle, D. J.; Boden, E. P.; Shannon, T. G. *J. Am. Chem. Soc.* **1990**, *112*, 2399; (b) Brunelle, D. J.; Boden, E. P. *Makromol Chem., Makromol Symp.* **1992**, *54/55*, 397.
20. Houlihan, F. M.; Bouchard, F.; Frechet, J. M. J.; Willson, C. G. *Macromolecules* **1986**, *19*, 13.
21. Frechet, J. M. J.; Bouchard, F.; Houlihan, F. M.; Eichler, B.; Kryczka, B.; Willson, C. G. *Macromol. Chem. Rapid. Commun.* **1986**, *7*, 121.
22. Andolino Brandt, P.J.; Senger, E.; Patel, N.; York, G.; McGrath, J. E. *Polymer* **1990**, *31*, 180.
23. Frechet, J.M. J.; Willson, C. G.; Iizawa, T.; Nishikubo, T.; Igarashi, K.; Fahey, J. *ACS Symposium Series 412, Polymers in Microelectronics, Materials and Processes*; Reichmanis, E., MacDonald, S. A., Iwayagami, T., Eds.; American Chemical Society: Washington, D.C., 1989; Chap. 7, p. 100.
24. Edmonds, J. T.; Hill, H. W. U.S. Patent 3,354,129.
25. Lopez, L. C.; Wilkes, G. L. *J.M.S.–Rev. Macromol. Chem. Phys.* **1989**, *C29*, 83.
26. Lenz, R. W.; Handlovits, C. E.; Smith, H. A. *J. Polym. Sci.* **1962**, *58*, 351.
27. (a) Koch, W.; Heitz, W. *Makromol. Chem.* **1983**, *184*, 779; (b) Koch, W.; risse, W.; Heitz, W. *Makromol. Chem. Suppl.* **1985**, *12*, 105.
28. (a) Annenkova, V. Z.; Antonik, L. M.; Vakulshaya, T. I.; Voronkov, M. G. *Dohl. Akad. Navk. SSR* **1986**, *286*, 1400; (b) Annenkova, V. Z.; Antonik, L. M.; Shafeeva, I. V.; Vakulskaya, T. I.; Vitkovskii, V. Y.; Voronkov, M. G. *Vysokomol. Soed., Ser. B* **1986**, *28*, 137.
29. Burnett, J. F. *ACC Chem. Res.* **1978**, *11*, 413.
30. Fahey, D. R.; Ash, C. E. *Macromolecules* **1991**, *24*, 4242.
31. (a) Port, A. B.; Still, R. H. *J. Appl. Polym. Sci.* **1979**, *24*, 1145; (b) Lovell, P. A.; Still, R. M. *Makromol. Chem.* **1987**, *188*, 1561.
32. Lovell, P. A.; Archer, A. C. *Makromol. Chem., Macromol. Symp.* **1992**, *54/55*, 257.
33. Archer, A. C.; Lovell, P. A. *Polymer International* **1994**, *33*, 19.
34. (a) Percec, V.; Clough, R. S.; Rinaldi, P. L.; Litman, V. E. *Macromolecules* **1991**, *24*, 5889; (b) Percec, V.; Clough, R. S.; Grigoras, M.; Rinaldi, P. L.; Litman, V. E. *Macromolecules* **1993**, *26*, 3650; (c) Mani, R. S.; Zimmerman, B.; Bhatnager, A.; Mohanty, D. K. *Polymer* **1993**, *34*, 171.
35. (a) Rule, M.; Fagerburg, D. R.; Watkins, J. J.; Lawrence, P. B. *Makromol. Chem., Rapid Commun.* **1991**, *12*, 221; (b) Rule, M.; Fagerburg, D. R.; Watkins, J. J.; Lawrence, P. B.; Zimmerman, R. L.; Cloyd, J. D. *Makromol. Chem., Macromol. Symp.* **1992**, *54/55*, 233.
36. Lowman, D. W.; Fagerburg, D. R. *Macromolecules* **1993**, *26*, 4606.
37. (a) Wang, Z. Y.; Hay, A. S. *Macromolecules* **1991**, *24*, 333; (b) Wang, Z. Y.; Hay, A. S. *Makromol. Chem., Macromol. Symp.* **1992**, *54*, 247.
38. Tsuchida, E.; Yamamoto, K.; Jikei, M.; Miyatake, K. *Macromolecules* **1993**, *26*, 4113.

39. (a) Tsuchida, E.; Yamamoto, K.; Nishide, H.; Yoshida, S. *Macromolecules* **1987**, *20*, 2031; (b) Tsuchida, E.; Yamamoto, K.; Nishide, H.; Yoshida, S.; Jikei, M. *Macromolecules* **1990**, *23*, 2101.
40. (a) Tsuchida, E.; Jikei, M. *Macromolecules* **1990**, *23*, 930; (b) Yamamoto, K.; Tsuchida, E.; NIshide, H.; Jikei, M.; Oyaizu, K. *Macromolecules* **1993**, *26*, 3432.
41. Tsudida, E.; Yamamoto, K.; Oyaizu, K.; Suzuki, F.; Hay, A. S.; Wang, Z. Y. *Macromolecules* **1995**, *28*, 409.
42. Tsuchida, E.; Shouji, E.; Yamamoto, K. *Macromolecules* **1993**, *26*, 7144.
43. Tsuchida, E.; Suzuki, F.; Shouji, E.; Yamamoto, K. *Macromolecules* **1994**, *27*, 1057.
44. (a) Moore, J. A.; Robello, D. R. *Macromolecules* **1989**, *22*, 1084; (b) Moore, J. A.; Kim, S. Y. *Makromol. Chem., Macromol. Symp.* **1992**, *54/55*, 423; (c) Moore, J. A.; Mehta, P. G. *Macromolecules* **1995**, *28*, 854.
45. Moore, J. A.; Mehta, P. G.; Kim, S. Y. *Macromolecules* **1993**, *26*, 3504.
46. Moore, J. A.; Mehta, P. G. *Macromolecules* **1995**, *28*, 444.
47. Burroughs, J. M.; Bradley, D. D. C.; Brown, A. R.; Marks, R. N.; Mackay, K.; Friend, R. H.; Burns, P. L.; Holmes, A. B. *Nature* **1990**, *347*, 539.
48. Hörhold, H.-H.; Helbig, M. *Makromol. Chem., Macromol. Symp.* **1987**, *12*, 229.
49. (a) Lenz, R. W.; Han, C. C.; Lux, M. *Polymer* **1989**, *30*, 1041; (b) Askari, S. H.; Rughooputh, S. D.; Wadl, F.; Megen, A. J. *Polymer Preprints* **1989**, *30*, 157.
50. (a) Gilch, H. G.; Wheelwright, W. L. *J. Polym. Sci., Part A-1* **1966**, *4*, 1337; (b) Swatos, W. J.; Gordon, B. *Polymer Preprints* **1990**, *31(1)*, 505; (c) Sarnecki, G. J.; Burn, P. L.; Kraft, A.; Friend, R. H.; Holmes, A. B. *Synthetic Metals* **1993**, *55*, 914.
51. (a) Lahti, P. M.; Modarelli, D. A.; Denton, F. R.; Lenz, R. W.; Karasz, F. E. *J. Am. Chem. Soc.* **1988**, *110*, 7258; (b) Tokito, S.; Smith, P.; Heeger, A. J. *Polymer* **1991**, *32*, 464.
52. (a) Kirchoff, R. A.; Carriere, C. J.; Bruza, K. J.; Rondon, N. G.; Sammler, R. L. *J. Macromol. Sci.-Chem.* **1991**, *A28*, 1079; (b) Hahn, S. F.; Townsend, P. H.; Burdeaux, D. C.; Gilpin, J. A. *Polymeric Materials for Electronics and Interconnections*; Lupinski, J. H., Moore, R. S., Eds.; ACS Symposium Series 407; American Chemical Society: Washington, D. C., 1989.
53. (a) Tan, L. S.; Arnold, F. E. *J. Polym. Sci., Polym. Chem. Ed.* **1987**, *25*, 3159; (b) Tan, L. S.; Arnold, F. E. *J. Polym. Sci., Polym. Chem. Ed.* **1988**, *26*, 1819; (c) Tan, L. S.; Arnold, F. E. *J. Polym. Sci., Polym. Chem. Ed.* **1988**, *26*, 3103.
54. Hahn, S. F.; Martin, S. J.; McKelvy, M. L. *Macromolecules* **1992**, *25*, 1539.
55. Hahn, S. F.; Martin, S. J.; McKelvy, M. L.; Patrick, D. W. *Macromolecules* **1993**, *26*, 3870.
56. Babb, D. A.; Ezzell, B. R.; Clement, K. S.; Richey, W. F.; Kennedy, A. P. *J. Polym. Sci., Polym. Chem. Ed.* **1993**, *31*, 3465.

RECEIVED January 25, 1996

Chapter 19

Mechanism of Step-Growth Thermal Polymerization of Arylacetylene

S. Gandon, P. Mison, and B. Sillion

Unité Mixte de Recherche, UMR 102, B.P. 3, 69390 Vernaison, France

The reaction mechanism of arylacetylene thermal polymerization has been studied on 4-(1-hexyloxy)-phenylacetylene 1, used as a monofunctional model compound. Its linear dimers 2-5 [a diyne and 3 enyne isomers] were also synthetized and their thermal behavior investigated. Reaction products were analyzed by chromatography (HPLC, SEC), spectroscopy (^{1}H and ^{13}C NMR) and spectrometry (SIMS) techniques.
The lowest molecular weight components were isolated and their structure established. There are naphthalenic dimers (2 isomers) and benzenic trimers (3 isomers) and they represented 30% in weight of the overall thermal reaction product of 1. We have shown that linear dimers 2-5 are not reaction intermediates of the thermal polymerization of 1. Moreover, under prolonged curing, a depolymerization process was pointed out, indicating that the highest polymerization steps could give termination by intramolecular reactions inducing chain cleavage leading to lower molecular weight entities.
A bimolecular reaction mechanism generating diradical intermediate species is proposed and its implications are discussed.

Many acetylene terminated resins have been described *(1,2)*. Although some α,ω-diethynylimide oligomers were commercialized in the seventies, the reaction mechanism of the thermal polymerization of the arylacetylene group is not yet well established. The polymerization takes place above 130°C, without catalyst *(1,2)*. The properties of the resulting networks are dependent on the length and the nature of the aromatic or heterocyclic moieties, as well as on polymerization and post-cure conditions. Different assumptions concerning the polymerization mechanism were based on kinetic data, electron spin resonance (ESR) analyses, structures determination of low molecular weight species and nuclear magnetic resonance (NMR) studies of the oligomeric products.

0097–6156/96/0624–0306$12.00/0

The radical character of the polymerization was early evidenced *(3)* and later confirmed by the use of radical inhibitors *(4)* or initiators *(5)*, the study of effect of air atmosphere *(6)* and quantitative ESR determinations *(7)*. Concentration of radical species was plotted as a function of acetylene conversion degree and it was shown that the acetylene group disappears faster than radical species appear *(7,8)*.

During the polymerization of ethynylpyridine, dimers with quinoline and isoquinoline structure were isolated together with benzenic trimers *(9)*. The authors pointed out the trimers were formed through an elimination reaction from the growing chains. However, Pickard *et al. (10)* mentioned that the molecular weight of the polymeric fraction were not dependent on the reaction temperature and they suggested that the first step of the polymerization was the formation of a diradical intermediate. From ^{1}H and ^{13}C NMR data, the same research group *(11)* proposed that the polymeric products were polyenic systems with a trans-cisoid structure. It was also pointed out *(3, 12)* that the reactivity decreases as the average molecular weight of products increases.

Solid state ^{13}C NMR cross polarization magic angle spinning data, obtained from ^{13}C isotope enriched samples, allowed Swanson *et al. (13)* to suggest that the polymerization can occur according to different reaction pathways: cyclotrimerization, biradical mechanism, Glaser and Straus coupling, Friedel-Craft addition and Diels-Alder cycloaddition. Diyne and enyne linear dimers formed from Glaser and Straus coupling reactions might be produced during the first steps of the thermal reaction *(14,15)*. They were implied as reaction intermediates to explain the formation of aromatic dimers and trimers *(14,15)*. Nevertheless, they were not explicitly identified from reaction products. Two naphthalenic dimers and three benzenic trimers have been isolated using preparative high performance liquid chromatography (HPLC) and their structures established on the basis of ^{1}H and ^{13}C NMR criterions *(16)*.

Others authors *(17)* suggested from kinetic data, that the initiation step of the thermal polymerization arylacetylenes is a bimolecular process.

Despite the large number of available literature data, the complete understanding of this unusual radical polymerization is still not fully understood. In this paper, we report the results of our studies on the thermal polymerization of 4-(1-hexyloxy)-phenylacetylene 1 and its four linear dimers to try to answer two questions:

1) Are the linear dimers intermediates during the monomer polymerization?

2) Is there an effect of the polymerization temperature on the average molecular weight and the molecular distribution of the reaction products?

EXPERIMENTAL

Preparation of model compounds 1-5 (scheme 1)

Monomer 1 was prepared, with an overall yield of 71%, through a three step synthetic sequence, using known experimental procedures *(18-20)*. Diyne dimer 2 was synthetized (94%) according to a modified Glaser oxidative catalytic coupling of 1 *(21)*. Straus enyne dimer 3 (Z configuration) was obtained (83%),

Ar—C≡C—H

1 (71%)

Ar—C≡C—CH=CH—Ar

3 (Z) (83%) 4 (E) (50%)

Ar—C≡C—C≡C—Ar

2 (94%)

Ar—C≡C—C(Ar)=CH_2

5 (42%)

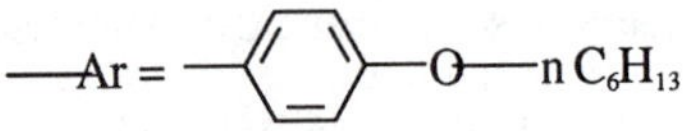

—Ar = —C_6H_4—O—n C_6H_{13}

Scheme 1.

with a complete diastereoselectivity, by reducing 2 with activated zinc (22). The E configuration isomer 4 was prepared (50%) by photochemical isomerization (iodine catalyzed) of 3 (*23, 24*). The most difficult to obtain was the Sabourin type dimer 5 ; in fact the main problem is the thermal instability of 5. 5 was prepared by synthetizing first a diaryl ynone (*25*) (70%), which was transformed to 5 (60%), at low temperature, through a Wittig reaction (*26*).

Thermolysis conditions and analysis techniques
All reactions implying 1 and 5 were performed with freshly prepared samples. Thermal polymerizations were conducted in glass sealed tubes (under argon atmosphere unless otherwise specified) without solvent addition, in the dark and in a programmed oven. Three curing procedures were used: two isothermal treatments at 180°C and 280°C for 64 and 7 hours, respectively, or a stepwise temperature ramp as indicated in figure 1.

The reaction products were analyzed chromatographically [high pressure liquid chromatography (HPLC) and size exclusion chromatography (SEC)] and spectroscopically [proton nuclear magnetic resonance (^{1}H NMR) and secondary ion mass spectrometry (SIMS)].

RESULTS

A) Monomer 1

Thermal polymerization using the cure ramp a. The HPLC analysis crude reaction mixture is shown in figure 2. The lowest retention time signals correspond to compounds 6 to 10 (scheme 2) which were isolated by preparative HPLC and identified from their ^{1}H and ^{13}C NMR characteristics. Moreover these assignments are in agreement with literature data *(15, 16)*.

By preparative column chromatography on silica gel we determined that the aromatic species 6-10 represent 30% in weight of the overall crude reaction mixture. Moreover, ^{1}H NMR analyses of these separated aromatic product

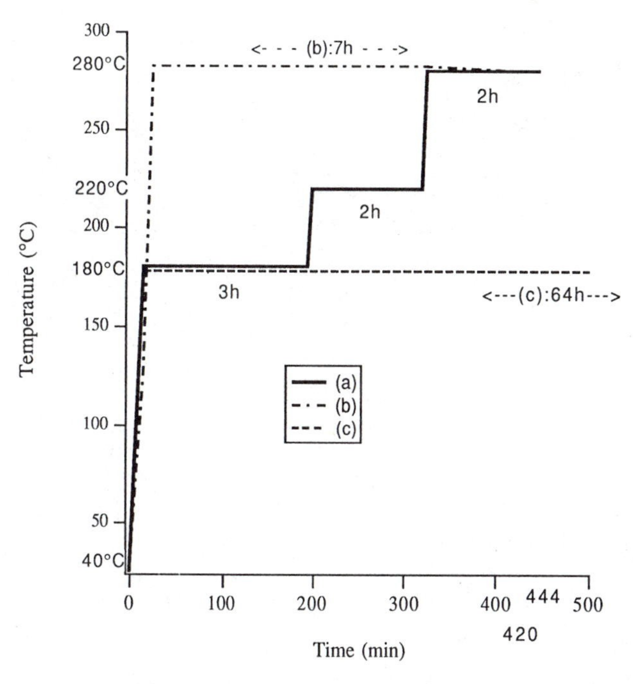

Figure 1: Graphical representation of the three kind of curing procedures: (a): stepwise ramp cure ; (b): isothermal cure (280°C) ; (c): isothermal cure (180°C).

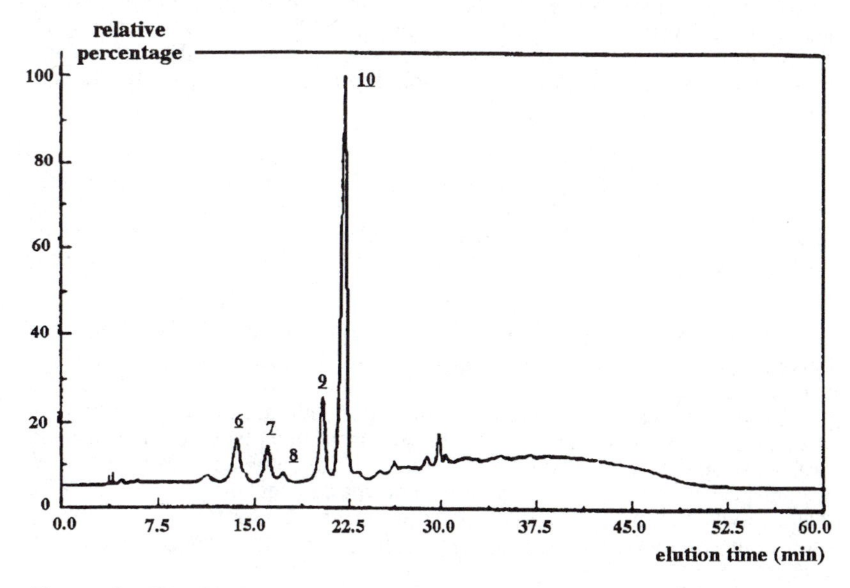

Figure 2: HPLC chromatogram of the thermal reaction (cure ramp a) of 1. Structures of compounds 6-10 are given in scheme 2. [Relative intensities must be considered with care due to ultraviolet detection (see text)].

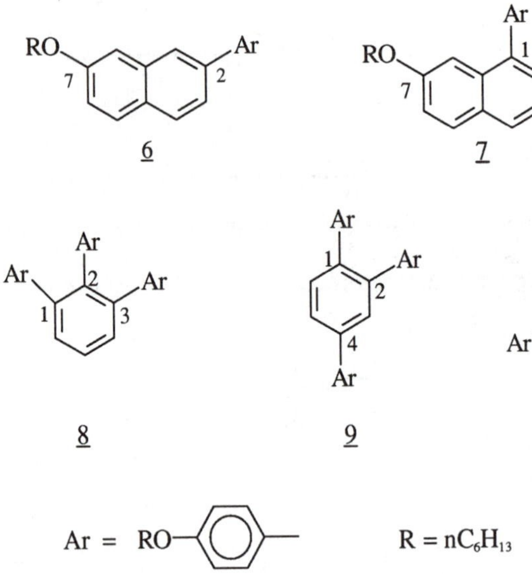

Ar = RO–C$_6$H$_4$–   R = nC_6H_{13}

Scheme 2.

fractions indicate that the relative amounts of the aromatic entities are: 40% of 6, 40% of 10, 15% of 9 and < 5% of 7 ; trimer 8 cannot be quantified.

A SIMS analysis (figure 3) confirms the large abundance of dimers (m/z = 404) and trimers (m/z = 606) species and allowed to detect oligomer entities until a polymerization degree (PD) of 14 (m/z = 2828). The relative intensities are decreasing along an exponential curve. Such an analysis fully confirms the polystyrene standards SEC analyses (Table I) which gave an average PD of 7 (cure ramp a).

A kinetic study was performed by analyzing the relative amounts of monomer 1 disappearance and products 6-10 formation, as well as the polymeric ones. Results are presented in figure 4. It must be pointed out that the appearance of the first products (6 and 10) takes place only after 1.5 hours of reaction and at that reaction time 20% of the monomer has already disappeared.

Isothermal curings at 180°C/64 hours (c) and 280°C/7 hours (b). The general shapes of the HPLC analyses were similar, except a small slide of the centre of polymer fractions towards smaller retention time. The SEC analyses are given in figure 5. It is worth to note that the elution curves of polymeric species C and D (curves A and B correspond to naphthalenic and benzenic products respectively) are narrowing down and shifting towards lower molecular weight species as the reaction temperature increases (from diagram # 1 to # 3). The overall intensities measurements are summarized in Table I. Thereby an increase in reaction temperature and/or reaction time induces a depolymerization process.

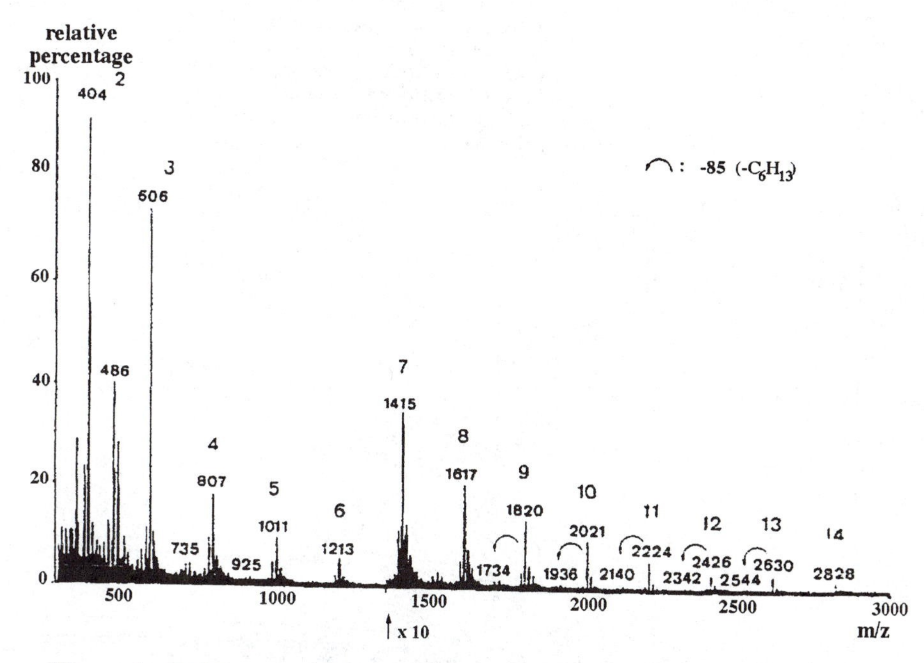

Figure 3: SIMS spectrum of the thermal reaction of 1 (cure ramp a). Numbers above the molecular ions indicate the polymerization degrees [from m/z = 1400 (mass unity) the ordinate scale is expanded by a factor of 10].

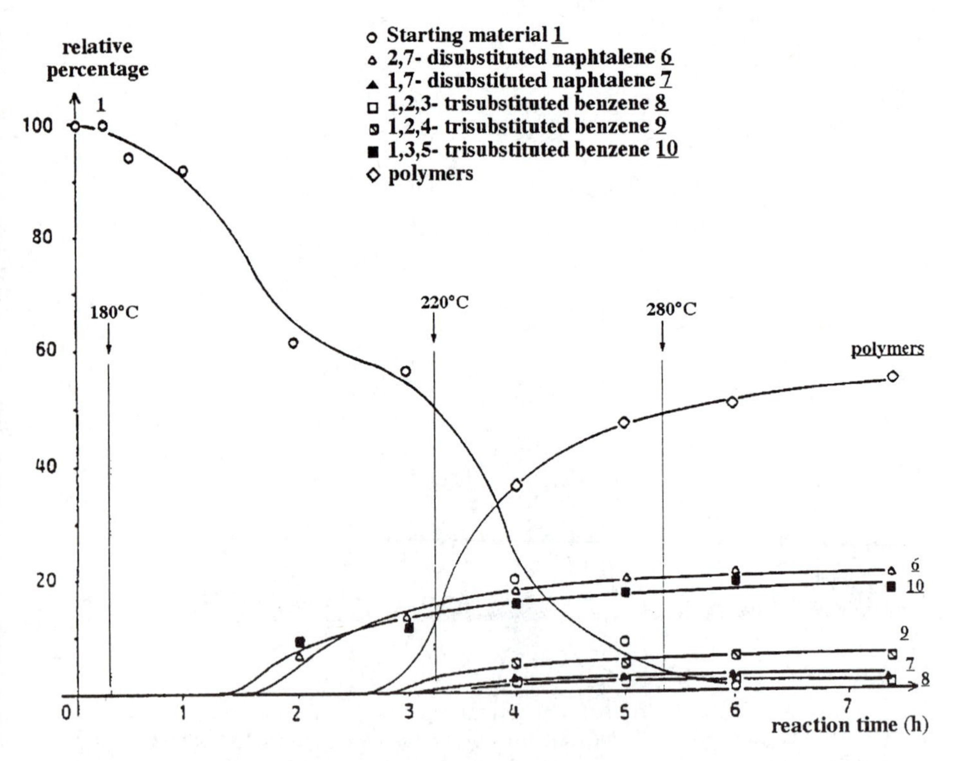

Figure 4: HPLC kinetic profile of thermal polymerization of 1: Relative amounts of formed products as a function of reaction time.

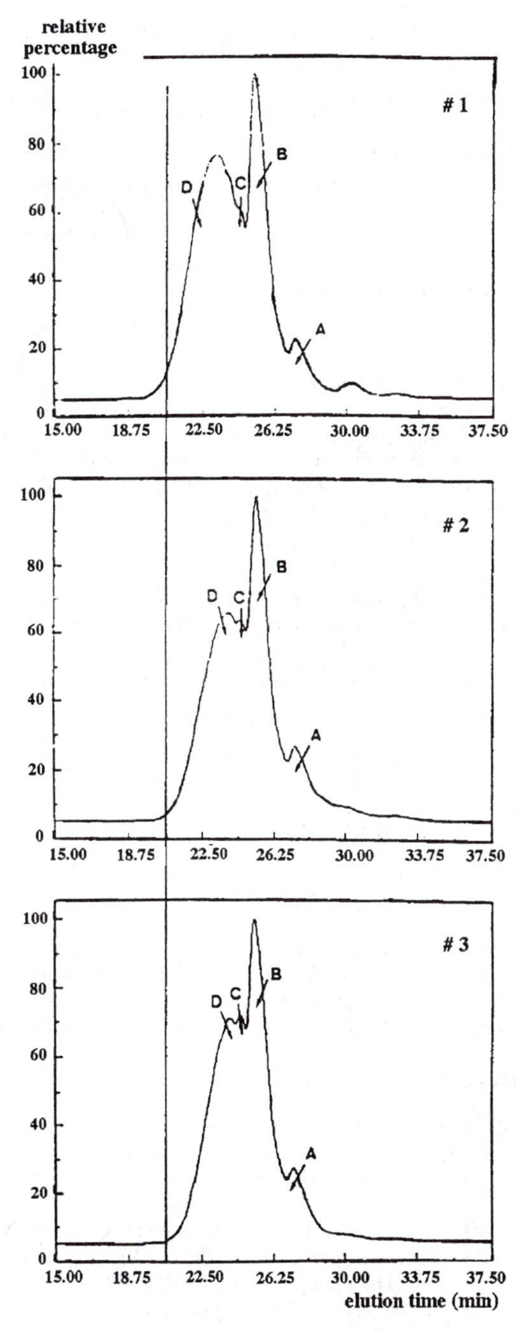

Figure 5: SEC chromatograms of crude products of the thermal polymerization of 1: # 1, isothermal curing at 180°C (c) ; # 2, ramp curing (a) ; # 3 isothermal curing at 280°C (b)

Table I. Average molecular weight ($\overline{Mw}$) and polymolecularity ratio (I) as a function of thermal treatment of 1

Experimental conditions	180°C/64 h (c)	Cure ramp (a)	280°C/7 h (b)
$\overline{Mw}$	1706	1439	1270
I	1.62	1.51	1.47

An experiment conducted at 280°C with a prolonged reaction time (Table II) confirmed these previous observations. The average molecular weight of the reaction products reach a maximum after 2 hours of reaction and then decrease, until a constant composition was obtained after 7 hours of thermal treatment. The same kind of statements can be made on the polymolecularity ratios: a narrowing down in molecular weight distribution is observed until 9 hours of reaction. Such results are favouring the assumption that some aromatic species were formed from the growing chain *(9)*.

Table II. Average molecular weight ($\overline{Mw}$) and polymolecularity ratio (I) as a function of reaction time for the curing of 1 at 280°C

Reaction time	1 h	2 h(*)	3 h	5 h	7 h	9 h	11 h	24 h	34 h
$\overline{Mw}$	1362	1377	1334	1305	1273	1267	1241	1203	1230
I	1.55	1.53	1.51	1.50	1.51	1.48	1.48	1.49	1.47

(*)No remaining monomer

The HPLC and SEC analyses (figure 6) of the crude products obtained by curing 1 in air atmosphere, showed that the naphthalenic dimers 6 and 7 (signal A on the SEC diagram) are formed in larger amounts if compared to benzenic and oligomer species. The formation of new products is not observed.

B) Linear dimers 2-5

Enynes linear dimers (such as 3-5) have been proposed as intermediates to explain the formation of naphthalenic and benzoic products *(15,16)* during the thermal polymerization of arylacetylenes.

A comparative differential scanning calorimetry (DSC) study was conducted between 50°C and 450°C, using a heating rate of 10°C per minute. The results are summarized in Table III. They show that with the exception of dimer 5, linear dimers 2-4 polymerized at higher temperatures than monomer, but similar values of polymerization enthalpies were measured.

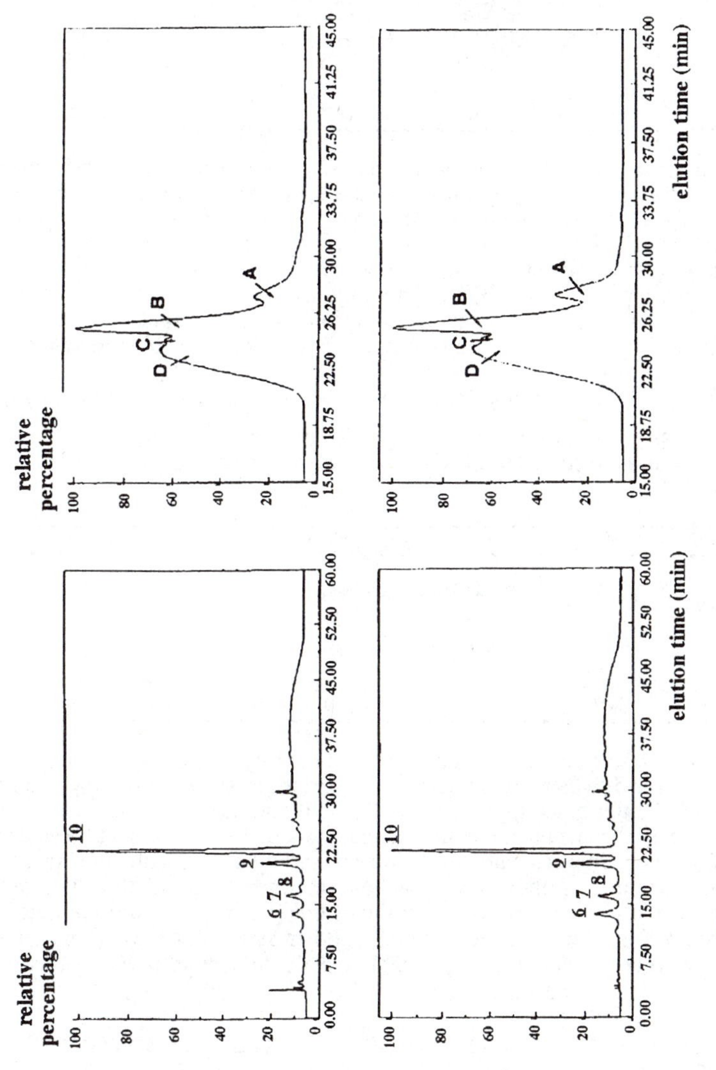

Figure 6: HPLC and SEC analyses of the thermal polymerization of 1 (ramp curing a): under argon atmosphere (upper traces) and air atmosphere (lower traces).

Table III. Enthalpies (ΔH) beginning (T) and maximum (Tmax) temperature of polymerisation of monomer 1 and linear dimers 2-5

Compound	2	3	4	5	1
T (°C)	220	200	220	80-90	130-140
T_{max} (°C)	290	320	320	130	250
ΔH (kJ/mol)	198	183	142	70	165

Analytical determination were conducted on the thermal polymerization products of linear dimer 2-5 obtained according to ramp curing a (figure 1). Results of SEC analyses are given in Table IV. A slightly higher average molecular weights of curing products were observed for the linear enyne dimers 3-5 that it was for monomer 1 product. However, this correspond with a slightly lower polymerization degree (PD). The polymerization product of the diyne dimer 2 indicated both higher average molecular weight and higher polymerization degree. These results were confirmed by SIMS measurements.

Table IV. SEC analyses of curing products (ramp a) of linear dimers 2-5 and monomer 1: average molecular weights ($\overline{Mw}$), polymolecularity ratios (I) and polymerization degrees (PD)

Compound	2	3	4	5	1
$\overline{Mw}$	3835	1417	1426	1628	1144
I	1,51	1,53	1,61	1,31	1,58
PD	9-10	3-4	3-4	4	5-6

Kinetic studies of the polymerization of linear dimers by way of HPLC analyses. Except for enyne 5, which is completely polymerized as soon as the temperature reached 180°C, the other linear dimers 2-4 polymerized at lower rate than monomer 1 one. The fact that the thermal reactivity of 1 is higher than the linear dimers 2-4 ones, implies that these dimers should accumulate during the polymerization of 1 if they were reaction intermediates and this was not experimentally observed. Moreover, from all linear dimers, we did not detect the formation of naphthalenic derivative 7 and benzene product 8-10, only the formation of a very small amount of naphthalenic compound 6 was observed from E and Z Straus dimers 3 and 4.

These two kinds of results demonstrate that dimers 2, 3 and 4 are not reaction intermediates of the thermal polymerization of 1. The same conclusion can be applied to dimer 5 considering the HPLC and SEC analyses of its curing product, which is basically different of the monomer 1.

DISCUSSION

The radical character of the arylacetylene curing reaction was already well established *(7,8)* and our results confirm it. The initiation step could be seen as a monomolecular process *(27,28)* corresponding to the homolytical cleavage of the terminal acetylene C-H bond. However such a mechanism could not explain the formation of the 1,2,3-trisubstituted aromatic trimer 8, even by considering the possibility for the monoradicals to rearrange. It could not either explain the delay between the first 20% of 1 consumption and the appearance of the first formed product 6 and 10 (Figure 4).

Bimolecular *(15-17)* and diradical *(9-10)* mechanisms have been invoked to explain results of some acetylene thermal polymerization. Though a thermal [2+2] reaction *(15,16)* is not favoured according to Woodward-Hoffmann rules, a stepwise pathway involving diradical species can be considered *(29)*. Such a reaction for triple bonds would give cyclobutadiene derivatives, which cannot be isolated, except where these rearrange or react to give stable compounds *(30)*. Thereby, we propose a bimolecular mechanism in which cyclobutadiene derivatives should be considered as transient species and would allow us to explain some rearrangements (scheme 3).

The initiation of the bimolecular mechanism is a reaction between two monomer molecules. This will account for the monomer consumption without formation of reaction products during the first hour of reaction. The two molecules can combine, as usual, following three possible ways (heat-to-head, head-to-tail and tail-to-tail) to form diradical species d_1, d_2, d_3, d_4.

From this stage, each diradical species d_1-d_4 can follow two reactions pathways. Firstly a termination process, by intramolecular cyclization, followed by an irreversible aromatization giving naphthalenic species ; secondly a propagation process by reacting either with carbon-1 or carbon-2 of a new molecule of monomer to form trimer diradicals t_1, t_2, t_3, etc... These two possibilities are explicited on diradical d_4 in scheme 3. Such a mechanism explains the formation of naphthalenic products 6 and 7 (from d_1 and d_3 respectively) but also the 2,6-disubstituted isomer one, which was not found experimentally.

At the trimer diradical (t_n) step, two reaction pathways are again possible: termination by cyclization-aromatization giving trisubstituted benzene species or propagation by addition to a monomer molecule to give tetramer diradicals T_n (scheme 3). The cyclization process will give directly the three aromatic trimers 8-10 observed experimentally. This is an important point, because monoradical mechanism *(27, 28)* could not afford the formation of the 1,2,3-trisubstituted isomer type, such as compound 8.

The propagation by diradical coupling could occur. However, it should be limited to low molecular moieties because we do not observe high molecular weight polymers (scheme 3, *e.g.*: $d_2 + d_3$).

The most favoured termination step is probably the intramolecular cyclization of the diradical species. Although we do not have a direct experimental proof, it may be suggested that one end-chain radical could attack a neighbouring benzenic ring, without cleavage of the chain (scheme 4). Abstraction of an hydrogen radical gives a naphthalenic end-group.

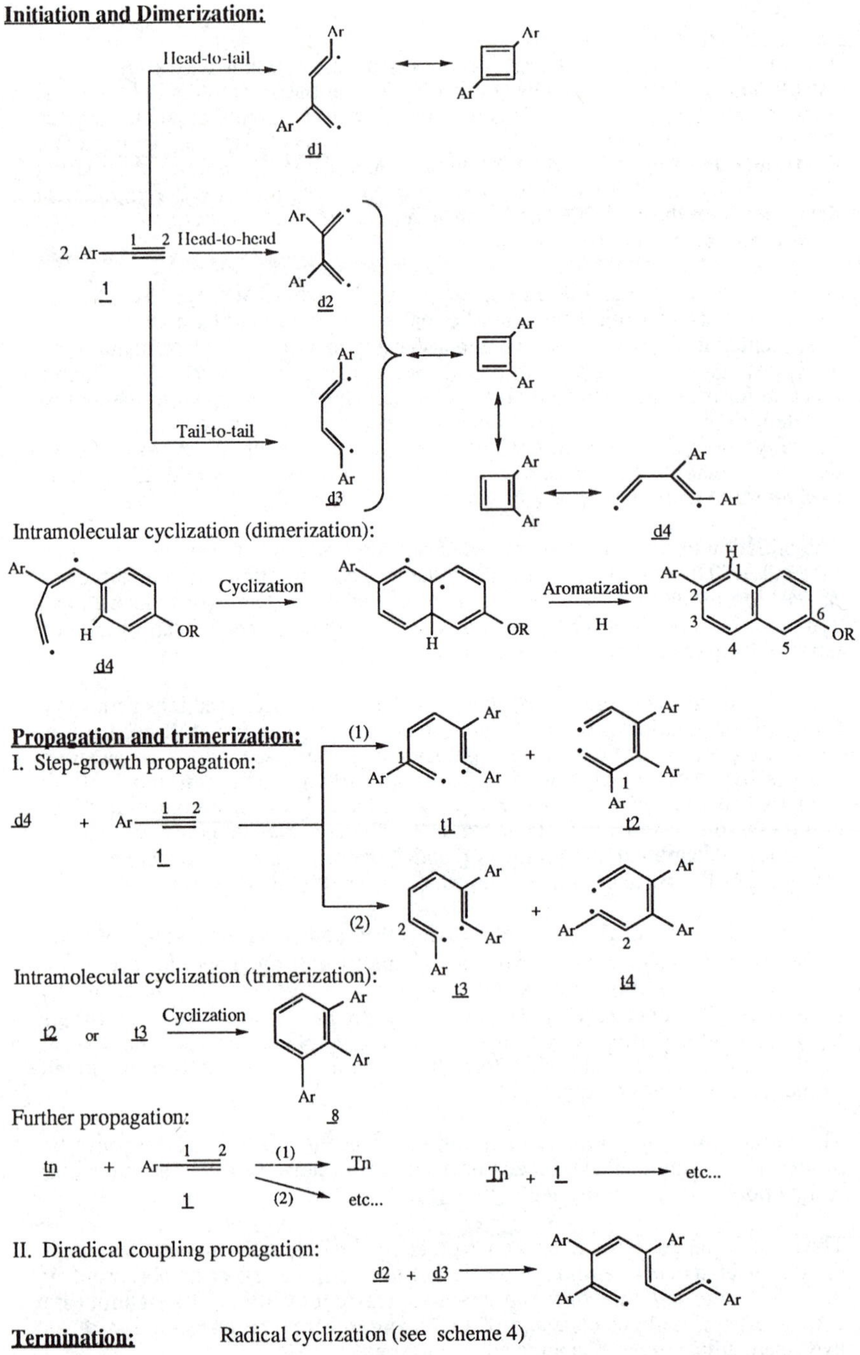

Initiation and Dimerization:

Head-to-tail

Ar

Ar

Ar

Ar

d1

Ar

2 Ar— 1 ≡ 2 Head-to-head

1

Ar

d2

Ar

Ar

Ar

Tail-to-tail

Ar

d3

Ar

Ar

Ar

Ar

d4

Intramolecular cyclization (dimerization):

Ar

H

OR

d4

Cyclization

Ar

H

OR

Aromatization

H

H

Ar

1

2

3

4

5

6

OR

Propagation and trimerization:

I. Step-growth propagation:

d4 + Ar— 1 ≡ 2

1

(1)

Ar

1

Ar

Ar

t1

+

Ar

Ar

1

Ar

t2

(2)

Ar

2

Ar

Ar

t3

+

Ar

Ar

Ar

2

t4

Intramolecular cyclization (trimerization):

t2 or t3 Cyclization

Ar

Ar

Ar

8

Further propagation:

tn + Ar— 1 ≡ 2 (1) → Tn

1 (2) → etc...

Tn + 1 ——→ etc...

II. Diradical coupling propagation:

d2 + d3 ——→

Ar

Ar

Ar

Ar

Termination: Radical cyclization (see scheme 4)

Scheme 3.

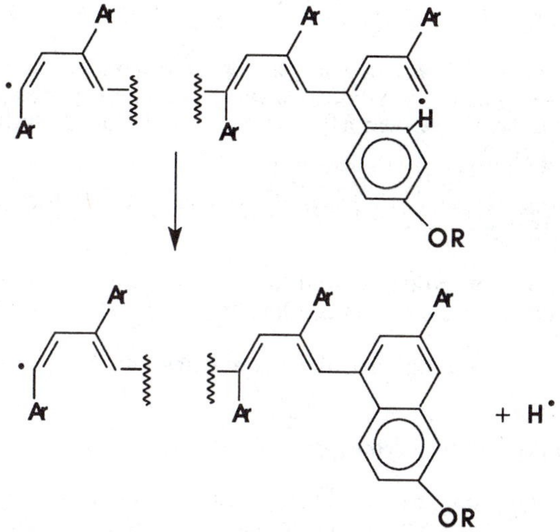

Scheme 4.

The depolymerization process can be seen as the formation of aromatic species (naphthalenic or benzenic) by chain cleavage (scheme 5) to generate shorter dienic chain diradical.

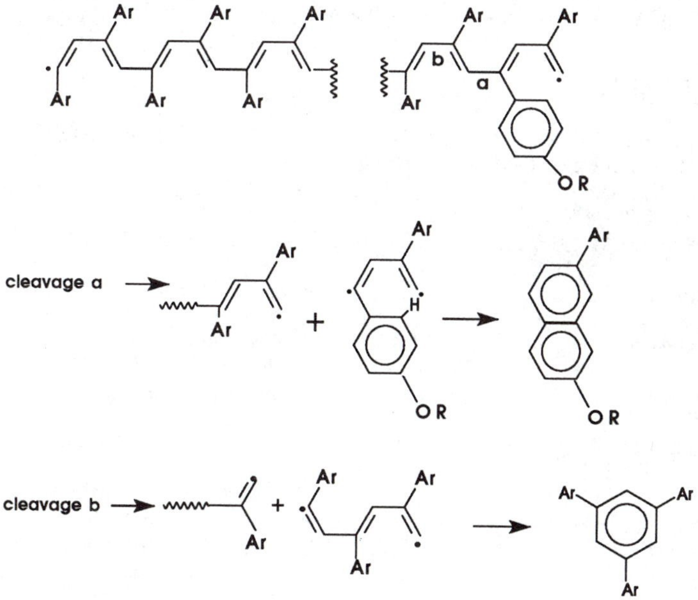

cleavage a

cleavage b

Scheme 5.

CONCLUSION

The 4-(1-hexyloxy)-phenylacetylene 1 and its corresponding four linear dimers (E and Z enynes 3 and 4, vinylidene enyne 5 and diyne 2) exhibited different behaviors when there were thermally polymerized. For the first time, important results were observed:

- the terminal acetylene 1 polymerized faster than its linear dimers (with the exception of vinylidene dimer 5) ;

- their polymerization products are different ; compound 1 gave naphthalenic and benzenic products, the dimers 2-5 did not ;

- the linear dimers are not reaction intermediates of the polymerization of monomer 1 ;

- under prolonged curing, a depolymerization process took place.

Henceforth, we proposed a "non-classical" mechanism for the thermal polymerization of 1: a bimolecular initiation giving rise to dimer diradicals, followed by a step-growth propagation through diradical species.

At each reaction step, each new radical can behave according to two reaction pathways: addition with monomer (propagation) and/or cyclization (termination).

The depolymerization process was explained by the expelling of aromatic moieties to form shorter chain diradical species.

REFERENCES

1. Hergenrother , P.M. *Encyclopedia of Polymer Science and Engineering*, John Wiley, Chichester, **1985**, Vol 1, pp 61-86.
2. Alam, S. ; Kandpal, L.D. ; Varma I.K. *J. Macromol. Sci., Rev. Macromol. Chem.*, **1993**, C 33, 291.
3. Barkalov, I.M. ; Berlin, A.A. ; Gol'danskii, V.I. ; Min-Gao, G. *Polym. Sci.* SSSR, **1963**, 5, 1025.
4. Ruhbander, R.J. ; Aponyi, T.J. *Proceedings of the 11th National Sampe Technical Conference*, Boston, Nov. 13-15, **1979**, pp 295-308.
5. Picklesimer, L.G. ; Lucarelli, M.A. ; Jones, W.B. ; Helminiak, T.E. ; Kang, C.C. *Polym. Prepr., Am. Chem. Soc., Div. Polym. Chem.*, **1981**, 22, 97.
6. Kuo, C.C. ; Lee, C. Y-C. *Am. Chem. Soc., Div. Org. Coast. Plast. Prepr.*, **1982**, 47, 114.
7. Sandreczki, T.C. ; Lee, C. Y-C. *Polym. Prepr., Am. Chem. Soc., Div. Polym. Chem.*, **1982**, 23, 185.
8. Levy, R.L. ; Lind A.C. ; Sandreczki, T.C. *Proceedings of the 15th National Sampe Technical Conference,* Cincinnati, oct. 4-6 **1983**, pp 21-30.
9. Chauser, M.G. ; Rudionov, Yu. M. ; Cherkashin, M.I. *J. Marcomol. Sci. Chem.*, **1977**, A11, 1113.
10. Pickard, J.M. ; Jones, E.G. ; Goldfarb, I.J. *Macromolecules*, **1979**, 12, 895.
11. Pickard, J.M. ; Chattoraj, S.C. ; Loughran, G.A. ; Ryan, M.T. *Macromolecules*, **1980**, 13, 1289.

12. Amdur, S. ; Cheng, A.T.Y. ; Wong, C.J. ; Ehrlich, P. ; Allendoerfer, R.D. *J. Polym. Sci., Polym. Chem. Ed.*, **1978**, 16, 407.
13. Swanson, S.A. ; Fleming, W.W. ; Hofer, D.C. *Macromolecules*, **1992**, 25, 582.
14. Kovar, R.F. ; Ehlers, G.F.L. ; Arnold, F.E. *J. Polym. Sci., Polym. Chem. Ed.*, **1977,** 15, 1081.
15. Hergenrother, P.M. ; Sykes, G.F. ; Young, P.R. *Polym. Prepr., Am. Chem. Soc., Div. Petr. Chem.*, **1979**, 24, 243.
16. Grenier-Loustalot, M-F. *High Perform. Polym.*, **1994**, 6, 347.
17. Sekiguchi, H. ; Kang, H-C. ; Tersac, G. ; Sillion, B. *Makromol. Chem., Macromole. Symp.*, **1991**, 47, 317.
18. Bruce, D.W.; Dunmur, D.A. ; Lalinde, E. ; Maitlis, P.M. ; Styring, P. *Liq. Cryst.*, 1988, 3, 385.
19. Takahashi, S. ; Kuroyama, Y. ; Sonogashira, K. ; Hagihara, N. *Synthesis*, **1980**, 627.
20. Austin, W.B. ; Bilow, N. ; Kelleghan, W.J. ; Lau, K.S.Y. *J. Org. Chem.*, **1981**, 46, 2280.
21. Fritzsche, U. ; Hunig, S. *Tetrahedron Lett.*, **1972**, 4831.
22. Brandsma, L. *"Preparative Acetylenic Chemistry", Studies in Organic Chemistry,* 2nd Edition, Elsevier, Amsterdam, **1988**, Vol. 34, pp 284-288.
23. Sonnet, P.E. *Tetrahedron*, **1980**, 36, 557.
24. Muller, E. ; Staub, H. ; Rao, J.M. *Tetrahedron Lett.*, **1970**, 773.
25. Birkoffer, L. ; Ritter, A. ; Uhlenbrauck, H. *Chem. Ber.*, **1963**, 96, 3280.
26. Wittig, G. ; Schoellkopf U. *Chem. Ber.*, **1954**, 87, 1318.
27. Ratto, J.J. ; Dynes, P.J. ; Hamermesh, C.L. *J. Polym. Sci., Polym. Chem. Ed.*, **1980**, 18, 1035.
28. Baklouti, M. ; Chaabouni, R. ; Fontanille, M. ; Villenave, J-J. *Eur. Polym. J.,* **1995**, 31, pp 215.
29. Carruthers, B. *Tetrahedron Organic Chemistry Series*, Pergamon Press, Oxford, **1990**, Vol 8, pp 332-367.
30. March, *J. Advanced Organic Chemistry*, John Wiley and Sons, New York, 3rd Edition, **1985,** pp 763-775.

RECEIVED January 25, 1996

Chapter 20

Synthesis of Aromatic Polyethers Containing Nonlinear Optical Chromophores

Joseph J. Kulig[1], Collin G. Moore, and William J. Brittain[2]

Maurice Morton Institute of Polymer Science, University of Akron, Akron, OH 44325–3909

Aromatic polyethers based on copolymers of bisphenol-A (BPA) and two nonlinear optical chromophores (**1** and **2**) have been prepared. The polyethers with the highest glass transition temperatures were polyformals which were prepared by a phase transfer catalyzed process in which the organic solvent (CH_2Cl_2 or CH_2Br_2) served as the co-reactant. The homopolymer of **1** displayed a glass transition temperature of 194 °C compared to a value of 92 °C for the homopolymer of BPA. The glass transition temperatures for copolymers displayed intermediate values for materials made from either **1** or **2**.

As part of our continuing program on the synthesis of nonlinear optical (NLO) polymers, this report describes the synthesis of aromatic polyethers based on bisphenol monomers **1** and **2**. Polyethers of **1** and **2** are expected to display NLO behavior because of the donor-acceptor nature of these monomer units. The ether linkage in the polyformals will be a good electron donating group. Moylan, et al. (*1*) and Cheng, et al. (*2*) have demonstrated that the dimethoxy derivative of **1** is a good guest molecule in the preparation of thermoplastic guest-host NLO systems. Therefore, we reasoned that it would be desirable to prepare polymers based

[1]Current address: Research Division, IBM Almaden Research Center, 650 Harry Road, San Jose, CA 95120–6099
[2]Corresponding author

0097–6156/96/0624–0322$12.00/0

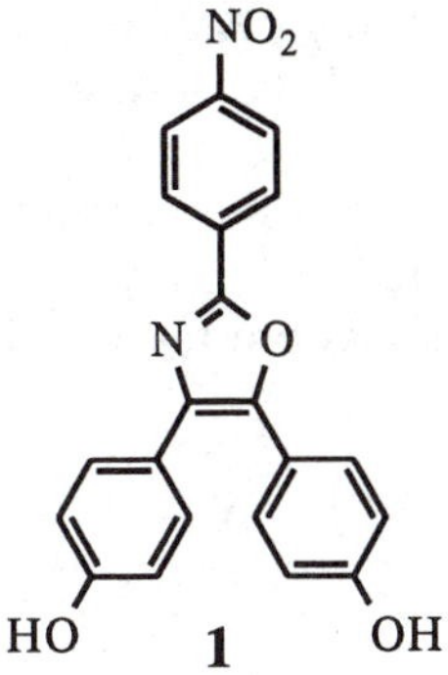

NO_2 NO_2

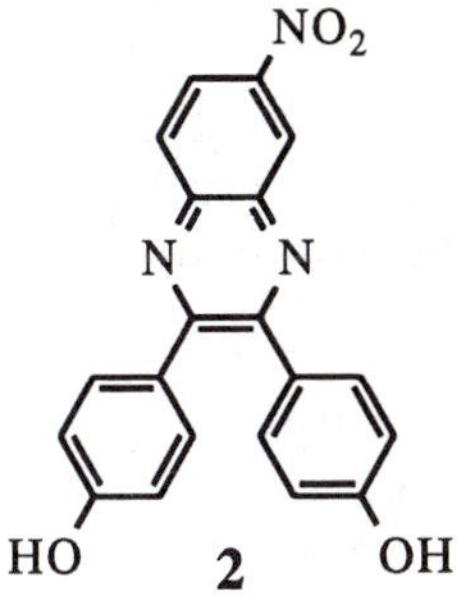

N O N N

HO **1** OH HO **2** OH

on **1** and **2** and examine their utility as polymeric NLO materials.

We have previously described the preparation of polycarbonates from oxazole-**1** (*3,4*). The polycarbonates were prepared by the ring-opening polymerization of oligomeric macrocycles. This approach to NLO polycarbonates has a reactive processing potential for fabricating devices. Incorporation of **1** into polycarbonate copolymers with bisphenol A (BPA) increased the glass transition temperature (T_g) by 64 degrees for the composition containing 31% **1**. The electro-optic coefficient of the copolymer containing 13 mole % **1** was 0.6 pm/V (*3*). More recent work has further defined the processing window for the ring-opening polymerization of these macrocyclic carbonates.

Polyether Synthesis

Initially, polyethers based chromophore **1** were synthesized to explore the chemistry of the dye and to develop discotic liquid crystalline materials. These materials were prepared by phase transfer catalyzed Williamson etherification (Scheme 1). The first prepared polyether was based on **1** and had a spacer length of four methylene units (n=4, x=0). This demonstrated the chemistry, but the material possessed limited solubility in common organic solvents. Consequently, longer spacers (n=10, x=0) were incorporated to prepare soluble homopolymers of **1**. The ^{1}H-NMR spectrum confirmed the structure but the polymer displayed a glass transition temperature of 100 °C. Furthermore, the absence of a melting point and a WAXD crystalline structure indicated the materials were amorphous.

Polyformals. Since the polyethers with long methylene spacers did not form liquid crystals, electric-field poling would be needed to orient the NLO chromophore. The low glass transition temperatures of polyethers with n=10 prompted the synthesis of polyformals (n=1).

Hay and co-workers (*5,6*) described the preparation of polyformals from the reaction of bisphenols and excess methylene chloride with sodium hydroxide or potassium hydroxide in the

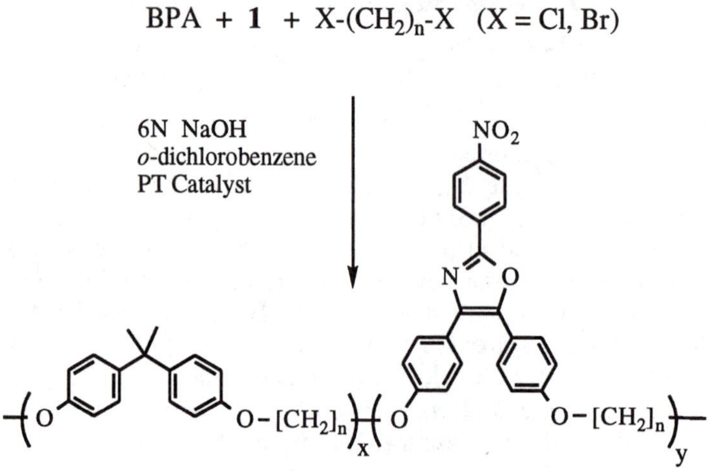

$$\text{BPA} + \mathbf{1} + \text{X-(CH}_2)_n\text{-X} \quad (\text{X} = \text{Cl, Br})$$

6N NaOH
o-dichlorobenzene
PT Catalyst

NO_2

N O

O O-$[CH_2]_n$ O O-$[CH_2]_n$ x y

Scheme 1. Synthesis of Polyethers Based on **1**

presence of a phase transfer catalyst. In a study of the polyformals based on BPA, large amounts of cyclic oligomers were obtained (4-50%) depending on the reactions conditions. The initial formation of an insoluble disodium or dipotassium salt apparently creates conditions for a high dilution reaction which favors cyclic formation. Higher temperature, higher concentration of phase transfer catalyst or the use of a more polar solvent minimizes, but does not eliminate, the formation of cyclic oligomers. The polyformal of BPA displays a glass transition temperature at 94°C, more than 50° below the corresponding polycarbonate. The T_g of polyformal samples containing cyclic oligomers is suppressed; for a sample with 21% cyclics, the T_g drops to 85 °C. Hay and co-workers prepared a series of polyformals from different bisphenols using similar reaction conditions. Of particular note is the polyformal of 9,9-bis(4-hydroxyphenyl)fluorene which has a T_g of 214°C.

The polyformal obtained from triphenyloxazole monomer **1**

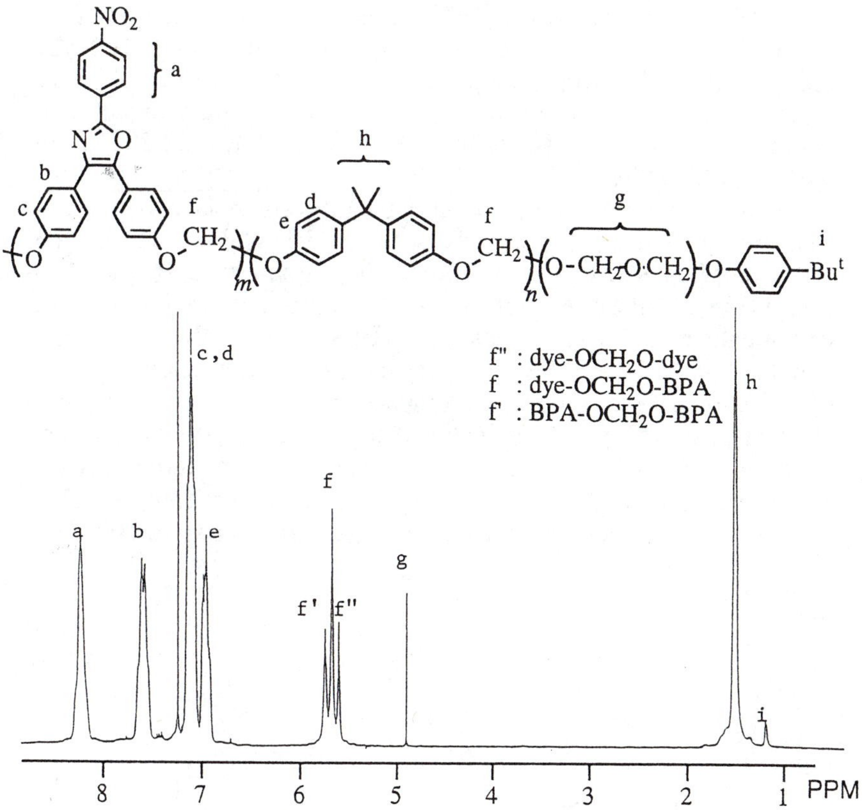

Figure 1. 200 MHz ^1H NMR of 2:1, BPA:**1** polyformal in $CDCl_3$.

and CH_2Cl_2 has a glass transition temperature of 194°C and is insoluble in most solvents except for hot DMSO. Therefore, it was difficult to determine the molecular weight. Based on this result, copolymers with BPA were prepared to obtain polymers with better solubility. The preparation of 1:1, **1**:BPA polyformal gave a polymer whose ^1H NMR reflected the feed ratio and was soluble in CH_2Cl_2 and THF. The ^1H NMR resonance for the methylene hydrogens (Figure 1) is composed of three peaks which can be attributed to the three possible chemical shift environments: BPA-OCH_2O-BPA, BPA-OCH_2O-**1**, and **1**-OCH_2O-**1**. The peak molecular weight was 48,000 g/mol with a broad molecular distribution and ≈ 26% cyclics based on SEC. The T_g was lowered to 148°C by incorporation of 50% BPA; however, the system displayed good thermal stability with 2% wt. loss observed at 356°C.

We also prepared the BPA copolymer of phenylquinoxaline monomer **2** using a 1:1 feed ratio (Scheme 2). Based on ^{1}H NMR analysis, the polyformal contained a 0.9:1 ratio of **2**:BPA. Similar to the 1:1, BPA:**1** polyformal, the 1:1, BPA:**2** polyformal displayed a T_g of 149°C. Based on SEC analysis, M_n = 22,200 g/mol with ≈ 20% cyclics.

We have found that methylene dibromide is a better coreactant due to its higher reactivity. However, the polymerization was complicated by side reactions; the haloether is attacked by hydroxide. This does not terminate chain growth but creates oligomethoxy linkages (Scheme 3). These oligomethoxy linkages (OCH_2-O-CH_2O) were easily measured by a characteristic proton NMR resonance at 4.9 ppm. Comparison of integrated intensities for this peak versus the intensity of the peaks for the polyformal linkage (O-CH_2-O) revealed an increase in oligomethoxy defects for copolymers with increasing amounts of **1**, see Table I. This may be a reflection of the lower reactivity of **1** relative to BPA.

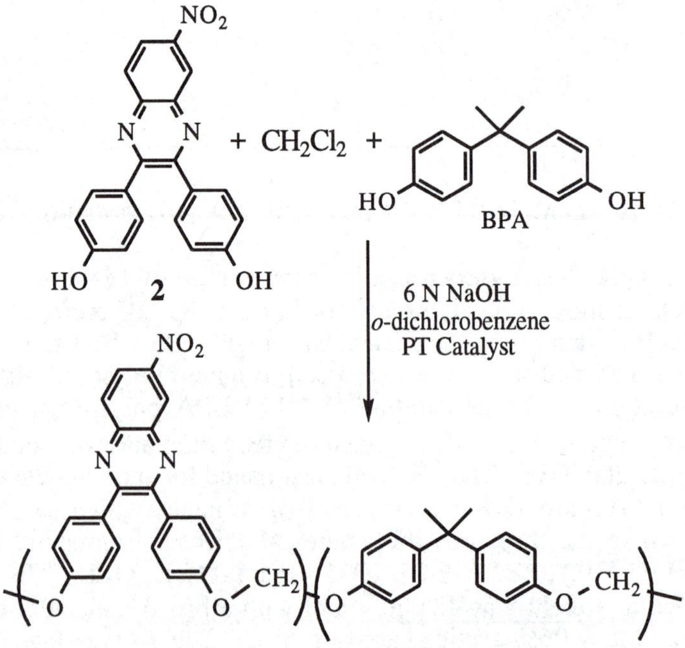

Scheme 2. Synthesis of **2**/BPA Polyformal

Table I. Co-polyformals of BPA and **1**

Composition, x[a]	T_g, °C	M_n, g/mol[b]	% Defects[c]
1.0	92		0
0.75	121	24,700	5
0.67	138	14,000	-
0.50	157	29,000	11
0.25	158	-	-
0.0	194	-	-

[a] x = mole fraction of BPA, see Scheme 1. [b] Determined by ^1H NMR. [c] Defects are oligomethoxy linkages: $-O-CH_2OCH_2-O-$

Another potential side-reaction is branching through the nucleophilic aromatic substitution of phenoxide at the nitro group of **1**.

During the phase transfer etherification the bisphenols were slowly transferred into the reaction. This slow addition of monomer, the highly reactive haloether chain ends and the flexibility of the formal linkage lead to the formation of cyclic polyformals; typical yields = 20 wt%. The cyclics eluted as the low molecular weight tail in the size exclusion chromatograph. Reverse precipitation by the addition of non-solvent to a polymer solution removed the cyclic component (see Figure 2).

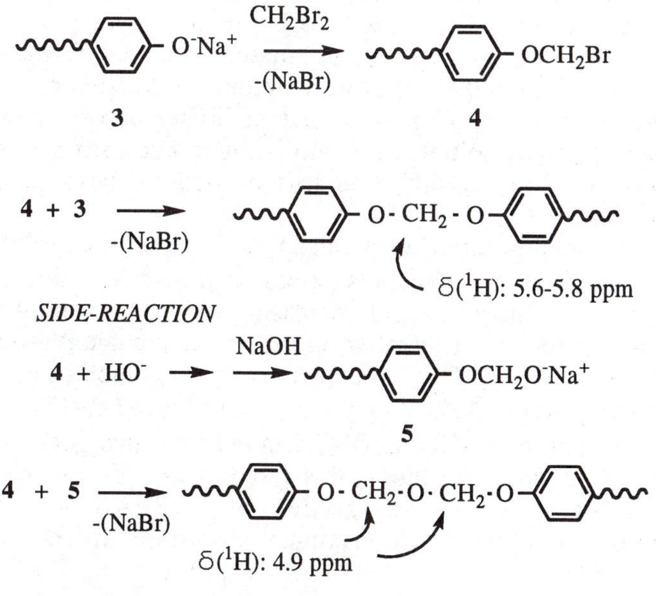

O^-Na^+ — CH_2Br_2, -(NaBr) → OCH_2Br

3 **4**

4 + **3** —-(NaBr)→ $O-CH_2-O$

δ(^{1}H): 5.6-5.8 ppm

SIDE-REACTION

4 + HO^- → NaOH → $OCH_2O^-Na^+$

5

4 + **5** —-(NaBr)→ $O-CH_2-O-CH_2-O$

δ(^{1}H): 4.9 ppm

Scheme 3. Reaction Steps in Polyformal Synthesis

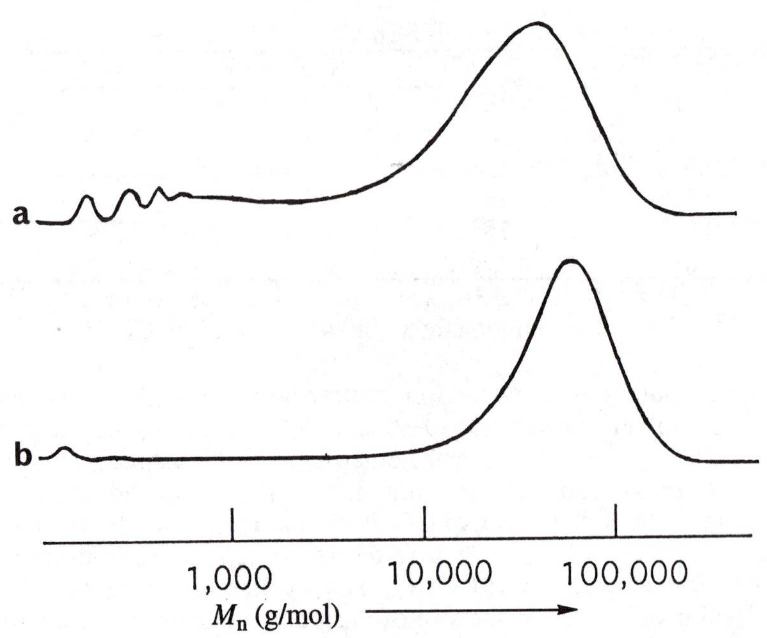

Figure 2. GPC analysis of 3:1, BPA:**1** polyformal before (**a**) and after (**b**) reverse precipitation.

The solubility of the copolymer decreased markedly as the chromophore content increased. BPA rich copolymers were soluble in methylene chloride/acetone mixtures. Copolyformals of 25 to 50 mol% **1** (x = 0.75, 0.5) were soluble in tetrahydrofuran and methylene chloride. Copolyformals of higher chromophore content were partially soluble in tetrahydrofuran but a stronger solvent such as hot dimethylsulfoxide was needed to dissolve the homopolymer of **1**.

The incorporation of **1** raised the T_g of the copolyformals relative to the BPA homopolyformal as shown in Table I. The glass transition increased with increasing weight fraction of the dye until nearly 50% **1**. Thereafter, increasing dye content produced a modest increase in T_g. The increasing oligomethoxy units in copolymers above 50% **1** may have also reduced the T_g. Thus a comparison of the NMR and DSC data indicates that good synthetic control of polymer structure in this system was achieved up to 50% dye content. The UV-vis spectrum of the BPA/**1**-50% copolyformal in chloroform had a maximum absorption at 392 nm and a cutoff at 575 nm.

Currently, we are examining the NLO properties of these polyformals.

Summary

We have prepared polyformal copolymers using BPA and one of two different NLO chromophores. A phase transfer catalyzed process with either methylene dichloride or methylene dibromide provided high yields of the corresponding polymers. Reverse precipitation was required to remove cyclic byproducts. As the amount of either chromophore **1** or **2** was increased in the copolymers, the glass transition temperature increased. In the case of triphenyloxazole-**1**, the homopolymer has a glass transition temperature of 194 °C. For 50:50 copolymers, the glass transition temperatures for the copolyformals were 157 °C for **1**:BPA and 149 °C for **2**:BPA. These glass transition temperatures are sufficiently high to make these polyformals potential candidates for optical applications.

Experimental Section

General Procedures and Materials. ^{1}H NMR spectra were recorded with a Varian Gemini 200 spectrometer. IR spectra were recorded with a Nicolet System 730 spectrometer. Gel permeation chromatography was performed with a Waters system using two PLgel 5 mm MIXED-D columns with THF eluent at a flow rate of 1 mL/min. Thermal gravimetric analysis was performed using a DuPont Model 951 and thermal analysis was performed using a DuPont Model 910 Differential Scanning Calorimeter. Unless noted otherwise, all other reagents were used as received. The preparation of 6-nitro-2,3-bis(4-hydroxyphenyl)phenylquinoxaline (**2**) has been described previously (*7*).

Polyether prepared from **1** and 1,10-dibromodecane. A 50-mL flask was charged with **1** (1.602 g, 4.28 mmol), 7.8 mL of 10 N sodium hydroxide, 5 mL water, 8 mL of 1,2-dichlorobenzene, 1,10-dibromodecane (1.17 g, 3.90 mmol), and tetrabutylammonium hydrogen sulfate (524 mg, 1.66 mmol). The mixture was heated at 80 °C under argon for 5 h. The product was isolated by diluting with methylene chloride, washing with dilute acid and water and precipitation into methanol. The yield was 1.57 g (78%); ^{1}H-NMR (200 MHz, $CDCl_3$) δ:8.25 (m, 4H, aromatic), 7.57 (m, 4H, aromatic), 6.9 (m, 4H, aromatic), 3.95 (t, 4H, methylene), 1.75

(m, 4H, methylene), 1.4 (b m, 12H, methylene); $Tg = 100$ °C; $Mn = 21{,}000$ g mol^{-1}; $Mw = 43{,}000$ g mol^{-1}; $Mw/Mn = 2.03$; $\eta_{inh} = 0.28$ dL g^{-1}.

Polyformal of BPA/**1**, 75/25. BPA (1.697 g, 7.43 mmol), **1** (927 mg, 2.48 mmol), and 4-*tert*-butylphenol (15 mg, 0.10 mmol) were combined in a 50-mL flask. Separately, sodium hydroxide (4.04 g, 101 mmol), 16 mL of water and tetrabutylammonium hydrogen sulfate (3.396 g 10.0 mmol) were added. Lastly, 10 mL of 1,2-dichlorobenzene and 4.9 mL methylene bromide were added. The system was purged with argon and the mixture was stirred at 80 °C. The reaction was complete within 45 min as indicated by the clearing of water phase. The reaction mixture was washed with water (2x), dilute hydrochloric acid (1x) and water (3x). The organic phase was diluted with methylene chloride, dried over anhydrous magnesium sulfate, filtered and reverse precipitated with acetone. The yield of the linear polymer was 2.74 g (39%); ^1H-NMR (200 MHz, $CDCl_3$) δ: 8.27 (m, aromatic), 7.59 (m, aromatic), 6.98 (m, aromatic), 5.78, 5.70, 5.63 (s, methylene, ratio 0.08, 0.36, 0.56) 1.60 (s, isopropylidiene), 1.26 (s, *t*-butyl);$Tg = 121$ °C; SEC (PS equiv.) $Mn = 20{,}500$ g mol^{-1}; $Mw = 35{,}600$ g mol^{-1}; $Mw/Mn = 1.74$; $\eta_{inh} = 0.51$ dL g^{-1}. Other copolymer compositions of polyformals were based on the procedure given above.

Polyformal of BPA/**2**, 50/50. BPA (0.318 g, 1.4 mmol) and **2** (0.5 g, 1.34 mmol) were combined in a 50-mL flask. Separately, 4.63 mL 50 % sodium hydroxide, 3 mL of CH_2Cl_2, 6.2 mL *o*-dichlorobenzene, 4-*sec*-butylphenol (4.2 mg, 0.028 mmol) and tetrabutylammonium hydrogen sulfate (0.94 g 2.8 mmol) were added. The reaction was stirred rapidly at 80°C for 5 h and precipitated in methanol to afford 0.816 g (99% yield) of polymer; Tg = 149°C; ^1H NMR (200 MHz, $CDCl_3$): δ 8.95, 8.43, 8.16 (m, 3H, phenylquinoxaline), 7.62, 7.32 (m, 8H, aromatic), 7.18, 6.9 (m, 8H, BPA aromatic), 5.65 (m, 2H, OCH_2O), 1.63 (s, 6H, isopropylidene). The ratio of **2**/BPA incorporated into the polyformal was 0.89 based on NMR integration. M_n (SEC, polystyrene standards) = 22,200 g/mol. Approximately 20% of the product was cyclic oligomers based on SEC analysis.

Acknowledgments

Acknowledgment is made to the donors of The Petroleum Research Fund, administered by the American Chemical Society for support of this research. The authors acknowledge helpful discussions with R. J. Twieg (IBM).

Literature Cited

1. Moylan, C. R.; Miller, R. D.; Twieg, R. J.; Betterton, K. M.; Lee, V. Y.; Matray, T. J.; Nguyen, C. *Chem. Mater.* **1993**, *5*, 1499.
2. Li, F.; Kim, K.-H.; Kulig, J. J.; Savitski, E. P.; Brittain, W. J.; Harris, F. W.; Cheng, S. Z. D.; Hubbard, S. F.; Singer, K. D. *J. Mater. Chem.* **1995**, *5*, 253.
3. Kulig, J. J.; Brittain, W. J.; Gilmour, S.; Perry, J. W. *Macromolecules* **1994**, *27*, 4838.
4. Brittain, W. J.; Kulig, J. J.; Moore, C. G. U.S. 5 405 926, 1995.
5. Hay, A. S.; Williams, F. J.; Relles, H. M.; Boulette, B. M.; Donahue, P. E.; Johnson, D. S. *J. Polym. Sci., Polym. Lett. Ed.* **1983**, *21*, 449.
6. Hay, A. S.; Williams, F. J.; Relles, H. M.; Boulette, B. M. *J. Macromol. Sci., Chem.* **1984**, *A21*, 1065.
7. Kulig, J. J.; Moore, C. G.; Brittain, W. J. *Polym. Prepr., Am. Chem. Soc., Div. Polym. Chem.* **1994**, *35(1)*, 492.

RECEIVED December 4, 1995

Chapter 21

Polycondensation of Carboxylic Acids and Carbinols in Heterogeneous Media

Lei Jong[1] and John C. Saam

Michigan Molecular Institute, Midland, MI 48640–2696

Glycols and dicarboxylic acids condense in acidic water-in-oil (w/o) emulsion, microemulsion or solid-in-liquid (s/l) dispersion to form moderate molecular weight polyesters at temperatures substantially below those normally employed in direct polyesterifications. Removal of by-produced water is unnecessary and effective catalysts are the strongly acidic oil-soluble surfactants or cation exchange resins. The formation of linear oligomeric polyesters having temperature sensitive reactive end groups demonstrate the absence of side reactions such as alcoholysis, acidolysis and transesterification. Controlling factors are elucidated with kinetic data based on a system intended to simulate alkyd resins.

Industrial polyesterifications are typically melt processes where a viscous polymerizing mass of carboxylic acids and glycols is held at temperatures as high as 250°C to drive the polymerization and remove the by-produced water. Among the consequences are troublesome side reactions that can include etherification of hydroxyl groups, ene or Diels-Alder additions, decarboxylation, chain cleavage by ester pyrolysis, acidolysis, alcoholysis and ester redistributions*(1)*, *(2)*, *(3)*. This leads to broadened molecular weight distributions*(4)*, sporadic gelation and loss of functionality. Thus, while adequate for many materials, the process is unsuitable for preparing polyesters with ordered or temperature sensitive structures due to the extreme processing conditions.

An unexplored alterative is to conduct esterifications in heterogeneous media so that the water produced forms in a hydrophobic phase. Transfer of water to a hydrophilic phase can then reduce the net free energy to favorably shift the equilibrium:

Hydrophobic Phase | Hydrophilic Phase

$$\sim\!\sim CO_2H + \sim\!\sim OH \underset{}{\overset{K_1}{\rightleftharpoons}} \sim\!\sim CO_2 \sim\!\sim + H_2O_o \overset{K_2}{\rightleftharpoons} H_2O_w \qquad (1)$$

[1]Current address: Teepack, Inc., Danville, IL 61832

0097–6156/96/0624–0332$12.00/0

Where H_2O_o and H_2O_w represent water in the hydrophobic and hydrophilic phases respectively, K_1 is the equilibrium constant for homogeneous esterification and K_2 a constant for the transfer of water between phases. Neglecting activity coefficients, letting bracketed terms represent concentrations and $K_2 = [H_2O_w]/[H_2O_o]$, the overall apparent equilibrium constant, K_{app}, can be defined as:

$$K_{app} = \frac{K_1K_2}{[H_2O_w]} = \frac{[\sim CO_2 \sim]}{[CO_2H]\,[OH]} \tag{2}$$

and the rate of ester bond formation in this scheme will be given by:

$$rate = k_f[\sim CO_2H]\,[\sim OH] - k_r[\sim CO_2 \sim] \tag{3}$$

Where k_f and k_r are the overall forward and reverse rate constants containing identical time-invariant terms for catalyst concentration and where k_r contains an additional time-invariant term, $[H_2O_o]$ so that $k_f/k_r = K_{app}$.

The term $[H_2O_w]$ by convention is unity when concentrations of solute are insignificant or when H_2O_w is present as a separate phase, but with solutes in the hydrophilic phase $[H_2O_w]$ will be less than unity. Factors influencing K_2 can be the inherent insolubility of water in the hydrophobic phase as well as osmotic pressure differences resulting from solutes present in the hydrophilic phase. Further, when the hydrophilic phase comprises the cores of inverse micelles or microemulsion particles, a tendency to expand particle radii through adsorption of H_2O_o will pervade in order to relieve appreciable internal pressures due to their minute sizes *(5)*. These forces can combine to give a K_2 much greater than unity and consequently values of K_{app} substantially larger than K_1 reported for esterification in homogeneous media*(6,7)*. This could eliminate, at least in some circumstances, the need for continuous removal of water by distillation.

Literature on catalysis in inverse w/o micelles or in inverse microemulsions appears to be limited so far to ester hydrolysis*(5,8)*. Early unrecognized examples of such esterifications, however, might have been the low temperature conversion of carboxylic acids to their methyl esters in methylene or ethylene chloride *(9)*. The process was catalyzed by sulfuric or sulfonic acids, products were obtained in high yield and the usual requirement for continuous removal of water by azeotropic distillation was eliminated. The produced water formed the hydrophilic phase in-situ and progress of esterification was evident from the appearance of cloudiness and the separation of a second aqueous layer. Ion exchange resins, which could serve as a solid hydrophilic phase, have been long known to catalyze esterifications but water was almost always removed by azeotropic distillation *(10)*. An interesting exception, however, was the formation of methyl or ethyl esters of acrylic acid with excess alcohol in the presence of a sulfonated polystyrene-divinylbenzene cation exchange resin. Yields approached 90% and removal of water by distillation was not required *(11)*. Despite this, water was invariably removed by continuous distillation in the few reported examples where cation exchange resins catalyzed polyester synthesis *(12, 13)*.

More recently we have shown that oligomeric polyesters form in aqueous oil-in-water (o/w) emulsion from molten mixtures dicarboxylic acids and diols below 100°C *(14, 15)*. In this case the emulsion particles served as a hydrophobic phase and catalysts were surface-active sulfonic acids. The method was restricted, however, to water-insoluble monomers that were liquid at the polymerization temperature.

We now report polyesterification of carboxylic acids and carbinols in water-in-oil (w/o) emulsion or microemulsion where the interior of acidic emulsion particles or the cores of acidic inverse surfactant micelles can serve as the hydrophilic phase. The interface between phases can serve as a reaction site as well as the continuous oil phase so the process, depending on the surfactant's solubility, can assume aspects of both homogeneous and heterogeneous catalysis. A strongly acidic cation exchange resin, where the hydrophilic phase consists of hydrated sulfonic acid groups present on nearly every repeat unit of the resin, is also evaluated in solid-in-liquid (s/l) dispersions. The inherent mobility surfactants in this case will be lost and catalysis is expected to be restricted to the interface with the hydrophobic phase.

The approach is demonstrated in the synthesis of oligomeric polyesters, including some with temperature sensitive structures that would never survive a thermally driven process with the accompanying acidolysis, alcoholysis and ester redistribution. Factors controlling the kinetics and equilibria are also elucidated using a model system where linoleic acid is grafted to poly(styrene-co-allyl alcohol) (PSAA). The system is intended to simulate the kind of steric crowding encountered in alkyd coating resins, but without the complicating multiple reactivities of the several reactants typically present in alkyd formulations.

Experimental

Materials. Carboxylic acids,carbinols, sodium dodecylbenzenesulfonate (NaDBSA), N,O-bis(trimethylsilyl)acetamide (TMSA)and Amberlyst-15 (4.83 mequiv. of H^+/g) were purchased from Aldrich Chemical Co. Samples of poly(styrene-co-allyl alcohol), 5.5% OH, M_n = 1500 (PSAA1), and 7.7% OH, M_n = 1150 (PSAA2) were from Polysciences Incorporated. Compositions were verified by determination of hydroxyl equivalent weight and ^{13}C NMR. Dodecylbezenesulfonic acid (DBSA), originally sourced from Stepan Company, was donated by Dow Corning Corporation.

Instrumentation. Number (M_n) and weight (M_w) average molecular weights were determined by gel phase chromatography (GPC) using tetrahydrofuran at a flow rate of 1.0 ml/min on a Waters system calibrated with polystyrene standards and equipped with a refractive index detector. ^{13}C and ^1H NMR were determined on a Brucker WM 360 MHz NMR spectrometer and infrared spectra on a Nicolet FTIR-20 spectrometer.

Polycondensation of Azelaic Acid and 1,10-Decanediol. Catalysis by DBSA in w/o Microemulsion. Mixtures consisting each of 3.764 g. (20.00 mmol) of azelaic acid, 3.486 g. of 1,10-decanediol (20.00 mmol), 10.0 ml of the solvents and 3 mmol of the catalysts specified in Table I were warmed with stirring to about 50°C for 5-10 min. to dissolve the monomers, cooled to room temperature and stirred in closed containers for 48 hours. When run in methylene chloride a suspension of dicarboxylic acid crystals gradually dissolved as the reaction progressed. Other runs were clear from the beginning with the exception of the run where 10 ml. of water was added to form an emulsion. After 48 hours all but one of the visually clear dispersions showed a Tyndall effect when placed in a beam of columnated white light, Table I. The polymers were isolated by evaporation of the solvent at room temperature, washing the residue in hot methanol, filtering the solids and drying at room temperature. A typical run in toluene produced 5.0 g. of poly(decamethylene azelaate), 77% yield. GPC: M_n = 4320, M_w = 6460, M_w/M_n = 1.5. FTIR gave a strong characteristic absorption for ester carbonyl at 1734 cm^{-1} and ^{13}C NMR (360 MHz, $CDCl_3$, TMS) was consistent with poly(decamethylene azelaate). δ(ppm): 24.7-25.7 (2, 2C, $\underline{C}H_2$); 28.7-29.3 (4, 4.5C, $\underline{C}H_2$); 34.1 (1, 1C, $\underline{C}H_2CO_2R$); 64.2 (1, 1C, $\underline{C}H_2O$); 174.0 (1, $R\underline{C}O_2R$); 177.6 (1, v. weak, $R\underline{C}O_2H$).

Low Molecular Weight Oligomer from Azelaic Acid and 1,4-Cyclohexanedimethanol. Catalysis by an Acidic Cation Exchange Resin. A solution of 5.987 g (31.81 mmol) of azelaic acid and 9.194 g (63.76 mmol) of 1,4-cyclohexanedimethanol in 15 ml. of toluene was stirred with 5.0 g of Amberlyst-15 (24.1 mequiv. of H^+) and heated to 80°C in a partial vacuum adjusted so that the solution would just reflux. Refluxing and stirring were continued to maintain the resin beads in suspension while any water collected in a trap or exited the system. After 6.3 hours the mixture was cooled to room temperature, diluted with toluene to reduce viscosity and the resin beads filtered. The solvent was immediately evaporated from a sample for analysis. FTIR indicated essentially complete conversion to ester carbonyl. ^1H NMR (300 MHz, $CDCl_3$,TMS) was consistent with poly(1,4-cyclohexanedimethylene azalaate). δ(ppm): 1.01 and 1.82 [m, CH_2, cyclohexyl (8x+8)H], 1.32 [m, CH_2 (acyclic), (6x)H], 1.60 [m, CH_2+CH (acyclic, cyclohexyl), (8x)H], 2.29 [t, CH_2-CO_2R, (4x)H], 3.47 [2d, CH_2OH, 4H)], 3.89 [2d, RCO_2CH_2, (4x)H], x=CH_2-CO_2R/CH_2OH =1.8. Data from other runs are summarized in Table II.

2-Methacryloxyethyl Terminated Oligomer. A suspension of 8.0 g of Amberlyst 15 (38.6 mequiv of H^+) in a solution of 8.703 g (46.24 mmol) of azelaic acid, 5.004 g (34.70 mmol) of 1,4-cyclohexanedimethanol and 30 ml. of toluene was reacted for 16 hours in a manner similar to the previous example. Then the suspension was cooled, 3.10 g (23.8 mmol) of 2-hydroxyethyl methacrylate was introduced with 0.15 g. of hydroquinone and a partial vacuum adjusted so that the mixture would reflux at 50-60°C. After 9.3 hours the mixture was cooled to room temperature, diluted, the resin beads filtered and the solution was washed with water. The resulting solution showed a tendency to gel on storage at room temperature The polymer was therefore immediately separated from a sample by precipitation with excess methanol. ^1H NMR (300 MHz, $CDCl_3$,TMS) δ (ppm): 1.01 and 1.82 [m, CH_2, cyclohexyl (8x)H], 1.32 [m, CH_2 (acyclic), (6x)H], 1.60 [m, CH_2+CH (acyclic, cyclohexyl), (8x)H], 1.95 (weak, d, =CCH_3), 2.29 [t, CH_2-CO_2R, (4x+4)H], 3.66 [s, CH_3O, 1.3 H], 3.89 [2d, RCO_2CH_2, (4x)H], 4.33 (t, OCH_2 CH_2O), 5.57 (d, HC=C, 1.0 H), 6.10 (d, HC=C, 1 1.0 H), x = 6.3.

Measuring Conversion to Ester. Conversions in esterifications of PSAA1 with linoleic acid were monitored by following the appearance of ester carbonyl in FTIR spectra of thin polymer films. It was necessary first to eliminate hydrogen bonding with carbonyl by trimethylsilylation of any unreacted hydroxyl groups in the PSAA to obtain a symmetrical absorption band. Samples, immediately after removal from the reaction medium, were neutralized with 1 N sodium hydroxide to terminate the reaction and to convert unreacted carboxyl groups to carboxylate. Solvent and water were removed in a vacuum and about 0.1-0.2 g of residue were dissolved in 2 ml of chloroform and 1.0 ml of TMSA. After two hours at room temperature the trimethylsilylated polymer was precipitated with excess methanol and dried in a vacuum at room temperature. FTIR from 4000 to 400 cm^{-1} were obtained on films cast on KBr plates from 3% chloroform solutions. Absence absorptions in the 3200-3600 cm^{-1} region indicated complete removal of OH and the ester carbonyl band at 1735 cm^{-1} was symmetrical and free of overlapping absorptions from hydrogen bonding. An aromatic peak at 700 cm^{-1} was taken as an internal reference and the relative absorbance of carbonyl, $A_{C=O}/A_{Ar}$, correlated to conversion, P , according to:

$$P = 0.5435(A_{C=O}/A_{Ar}) \tag{4}$$

The ratio $A_{C=O}/A_{Ar}$ followed Beer's law when measured on thin films of known mixtures of trimethylsilylated PSAA1 and ethyl linoleate.

Rates of Esterification of PSAA with Linoleic Acid. Kinetic runs were conducted in a closed vessel with a stirrer and ports for sample removal. In a typical run 23 g. of PSAA1 (87 mequiv.), 2.7 g. of NaDBSA (7.8 mmol) were dissolved at room temperature in 42.7 g. of toluene containing 0.19 g of 2,6-di(t-butyl)-4-methylphenol as an antioxidant. After a homogeneous solution formed, 24.3 g. of linoleic acid (87 mmol) was added with stirring and the temperature was maintained within ±0.5°C that specified. Then 13.7 ml of 30% (7.4N) sulfuric acid (100 mequiv) was introduced which emulsified spontaneously with stirring. If the surfactant was DBSA instead of NaDBSA, it was introduced at this point. Temperature was adjusted and 6 ml samples were periodically removed over the course of the reaction and immediately terminated and analyzed by FTIR as described. In separate determinations of the apparent equilibrium constant, mixtures were stirred in sealed reactors for at least 100-200 h depending on temperature and analyzed by the described FTIR procedure.

The product was isolated for purposes of identification in a separate run. A solution of 4.430 g. of PSAA1 (14.5 meq.), 4.4049 g of linoleic acid (14.5 mmol) and 0.57 g. of NaDBSA (1.63 mmol) in 9.0 ml of methylene chloride was vigorously stirred with 4.0 ml of 10% sulfuric acid (~2N) at room temperature to give an emulsion. Stirring was continued in a closed container at room at room temperature for 72 hours. The emulsion was separated by centrifugation at 10,000 rpm in a refrigerated centrifuge and the nonaqueous phase was washed repeatedly with water and separated by centrifugation until the wash water was neutral. Evaporation of the solvent at room temperature in a stream of air and then finally in a vacuum gave 6.0 g of a tacky, viscous resin. The original PSAA and linoleic acid were soluble in methanol but the resulting resin was insoluble.

The ^{13}C NMR of the methanol insoluble resin was complex due to the overlapping signals from aromatic with unsaturated structures as well as the multiple methylene structures but gave a distinctive chemical shift for ester carbonyl while the carboxyl carbonyl from the original linoleic acid was absent. The positions and relative intensities of $\underline{C}H_2(CH{=}CH)_2$ and $\underline{C}H{=}\underline{C}H$ were that of the original linoleic acid while those of the aromatic $\underline{C}$, aromatic-$\underline{C}H$<, $\underline{C}H_2OC{=}O$ and $\underline{C}H_2C{=}O$ were in agreement with the anticipated product. Weaker extraneous signals in the 125-135, 20-30 and 14 ppm regions were due to the approximately 25% oleic and traces of linolenic acids present in the original linoleic acid. ^{13}C NMR (360 MHz, $DCCl_3$, TMS), δ (ppm): 173.8 ($CH_2O\underline{C}{=}O$), 145.1, 128.0, 127.8, 125.2 (aromatic $\underline{C}$) 130.1, 128.9, 128.1, 127.9 ($\underline{C}H{=}\underline{C}H$), 50.5 ($\underline{C}H_2OC{=}O$), 40.3 (aromatic-$\underline{C}H$<), 34.1 ($\underline{C}H_2C{=}O$), 31.5 (C16 $\underline{C}H_2$), 29.5-29.0 (C4-7 and C15 $\underline{C}H_2$), 27.1 ($\underline{C}H_2CH{=}CH$), 25.6 ($\underline{C}H_2(CH{=}CH)_2$), 24.6 (C3 $\underline{C}H_2$), 22.5 (C17 $\underline{C}H_2$) and 14.0 (CH_3).

Results

Polymer Synthesis. The examples in Table I illustrate the room temperature polyesterification of azelaic acid with 1,10-decanediol in w/o emulsion, in w/o microemulsion and in solution. The by-produced water remained in all the examples. The isolated yields represent the methanol insoluble portion of the polydecamethlyene azelaate and reflect the losses incurred during work-up. Polydispersities of the polymers were between 1.5 and 1.7 and somewhat lower than expected for typical statistical step growth polymerizations. This might be due to partial removal of lower molecular weight species during the polymer isolation by precipitation from methanol.

Molecular weights in the first three examples in Table I exceed those of our earlier work in o/w emulsion by approximately a factor of two (*15*).

Reaction mixtures containing the DBSA catalyst were visually clear in the absence of added water, but in a columnated beam of white light distinctive Tyndall effects indicated the presence of a second phase in the form of either microemulsion or micelle particles. Essentially the same yields and molecular weights resulted as long as the process was conducted in hydrophobic solvents as o/w emulsions or microemulsions. Yields and molecular weights decreased noticeably, however, when excessive water was added or when tetrahydrofuran replaced the more hydrophobic toluene (last two entries, Table I). No polymer formed when the polymerization in tetrahydrofuran was catalyzed by the non-surface active methanesulfonic acid. The absence of a Tyndall effect in this case indicated a solution process and emphasized the importance of heterogeneous media.

Table I. Room Temperature Polycondensation of Equimolar Azelaic Acid and 1,10-Decanediol in w/o Emulsion and Microemulsion[(a)]

Solvent	*Surfactant-Catalyst*	*Type of Dispersion*	*Polymer Yield (%)*	$10^{-3}X$ M_n	M_w/M_n
toluene	DBSA	w/o emulsion[b]	77	3.47	1.40
toluene	DBSA	w/o microemulsion[c]	81	4.44	1.69
methylene chloride	DBSA	w/o microemulsion[c,d]	74	4.10	1.75
tetrahydro-furan	DBSA	w/o microemulsion[c]	66	2.92	1.55
tetrahydro-furan	$MeSO_3H$	solution[e]	nil	---	---

[a]Procedure given in the experimental section. [b]38 wt. % water present as an aqueous phase. [c]Aqueous phase forms *in situ*, strong Tyndall effect. [d]Reagents initially partly soluble, but dissolve as reaction progresses. [e]No Tyndall effect.

Catalysis by the acidic sulfonated cation exchange resin in s/l dispersion is illustrated with the polycondensation of 1,4-cyclohexanedimethanol and azelaic acid, equation 5.

$$HOH_2C-C_6H_{10}-CH_2OH + HO_2C(CH_2)_7CO_2H \xrightarrow{\text{Amberlyst-15}} H_2O + H[OCH_2-C_6H_{10}-CH_2OC(=O)(CH_2)_7C(=O)]_nOH \quad (5)$$

The results in Table II illustrate polymerizations conducted at 90°C where molar ratios of diacid to diol were adjusted to give either carbinol or carboxyl-ended oligomers and molecular weights low enough so that chain ends could be characterized by ^1H NMR.

Conversions were nearly complete, again without the necessity of removing by-produced water. The process appeared more rapid, however, if the by-produced water were partly removed in a partial vacuum. In this case reactions were essentially complete within 6-18 hours but little difference was seen in yields or molecular weights from the run where by-produced water remained in the reaction.

The ^{1}H NMR of the products indicated the anticipated carbinol or carboxyl terminated oligomers free of ether structures, but that somewhat less carbinol was present than originally used in the polymerization. This was attributed to the absorption of carbinol into the resin noticed previously by Karpov et. al. in polyesterifications catalyzed by similar cation exchange resins but where water was continuously removed*(12)*. Nevertheless, the ratio of repeat units incorporated, $\underline{CH_2}CO_2R$, to chain ends, $\underline{CH_2}OH$ or $\underline{CH_2}CO_2H$, seen in the ^{1}H NMR gave estimated molecular weights close to those expected from the molar ratios of the diacid to diol incorporated in the polymer assuming essentially complete conversion of the reactive groups. Molecular weights determined by gel phase chromatography also agreed with the anticipated values.

Table II. Condensation of Azelaic Acid and 1,4-Cyclohexanedimethanol Catalyzed by Cation Exchange Resins, Methacrylate Ended Oligomers

Mol Ratio Diacid/Diol[a]	*Type of End Group*	10^{-3} x M_n *Predicted*	10^{-3} x M_n *By* ^{1}*H NMR*	10^{-3} x M_n *By GPC*	M_w/M_n
0.900[b]	carbinol	2.7	3.3[f]	2.9	2.0
0.909[c]	carbinol	3.0	3.1[f]	3.4	2.5
0.840[c,d]	methacrylate	1.8	2.1[g]	1.7	2.1
1.16[c,e]	2-hydroxyethyl methacrylate	2.0	1.8[g]	1.8	2.2

[a]Based ^{1}H NMR of the product. [b]Run at 90°C, no removal of water. [c]Run at 90°C, refluxed in a partial vacuum, water partly removed. [d]Post reaction at 60°C of the OH ended polymer with methacrylic acid , 1 equiv./chain end. [e]Post reaction at 60°C of the CO_2H ended polymer with 2-hydroxyethyl methacrylate, 1 equiv./chain end. [f]Based on $\underline{CH_2}OH$, 3.47 ppm (TMS). [g]Based on $\underline{CH_2}$=C (doublet), 5.57, 6.10 ppm.

The last two entries in Table II illustrate oligomers terminated with methacrylate or 2-methacryloxyethylene units. These were introduced in a second stage of the polymerizations by adding an equivalent/chain end of either methacrylic acid (MA) to the carbinol ended oligomer or 2-hydroxyethyl methacrylate (HEMA) to the carboxyl terminated oligomer, equation 6. The solvent and catalyst were retained from the first stage polyesterification and the second stage continued for another 9-10 hours at 60°C.

$$\sim CH_2OH + HO\text{-}C(=O)\text{-}C(CH_3)=CH_2 \longrightarrow CH_2=C(CH_3)\text{-}C(=O)\text{-}O\sim\sim O\text{-}C(=O)\text{-}C(CH_3)=CH_2$$

$$\sim CH_2CO_2H + HO\text{-}CH_2CH_2\text{-}O\text{-}C(=O)\text{-}C(CH_3)=CH_2 \longrightarrow \tag{6}$$

$$CH_2=C(CH_3)\text{-}C(=O)\text{-}O\text{-}CH_2CH_2\text{-}O\text{-}C(=O)\sim\sim C(=O)\text{-}O\text{-}CH_2CH_2\text{-}O\text{-}C(=O)\text{-}C(CH_3)=CH_2$$

The products, which undergo free radical polymerization and cross-linking, clearly show signals for terminal methacryloxy units in the ^1H NMR, Figure 1. About 85% of the chain ends of the carbinol ended oligomer convert to methacrylate and 65% of the ends in the carboxyl ended oligomer convert to 2-methacryloxyethylene units. The M_n, based on GPC and area intensities of ^1H NMR signals from total chain ends including methacryloxy, are close to that expected of a polycondensation where no additional chain ends form by acidolysis, alcoholysis or transesterification. Were these significant M_n would be substantially lower than expected and signals from carboxyl would be apparent at about 0.1 ppm down-field from $C\underline{H_2}CO_2R$ in Figure 1A showing the carbinol-ended oligomer terminated with methacrylic acid. When HEMA is used to terminate the carboxyl-ended oligomer, the accompanying ethylene bridges with appropriate area intensities are apparent, Figure 1B. The integrity of the 2-methacryloxyethylene structure is retained since distinguishable signals that would result from its cleavage are absent. These would include $C\underline{H_2}OH$ from either the CHDM ends (doublet, 3.5 ppm) or the ethylene glycol ends (triplet, est. 3.8-3.9 ppm), Their absence or negligible response indicates suppressed acidolysis and alcoholysis and suggests the method may be applied to synthesis of other structured polyesters via sequential polycondensations of carboxylic acids and carbinols.

Rates, Equilibria. Figure 2 illustrates results from the three methods of heterophase esterification where linoleic acid is grafted to equal equivalents of the pendant carbinol groups in poly(styrene-co-allyl alcohol), PSAA-1. The methods include esterifications conducted in w/o microemulsion catalyzed by DBSA, in s/l dispersion catalyzed by an acidic cation exchange resin and, finally, in a w/o emulsion catalyzed either by DBSA with water as an aqueous phase or by NaDBSA with 7.4N H_2SO_4 as an aqueous phase. In all runs about 46 wt % toluene was present as a diluent and no attempt was made to remove the by-produced water. A ^{13}C NMR of the isolated product (See experimental section) indicated a straightforward esterification of the PSAA and preservation of the linoleate structure in the grafted side chains. The esterifications were monitored by the appearance of ester carbonyl in FTIR using the procedure outlined in the experimental section.

Virtually superimposable conversions vs. time result from esterifications conducted at 25°C in w/o microemulsion catalyzed by DBSA or in a w/o emulsion catalyzed by NaDBSA/7.4N H_2SO_4, Figure 2. Apparently ion exchange can rapidly occur between the aqueous 7.4N H_2SO_4 phase and the NaDBSA to give conditions at the reaction site similar to those in the o/w microemulsion with just the DBSA catalyst. Neither NaDBSA or 7.4N H_2SO_4 alone are effective catalysts under these conditions.

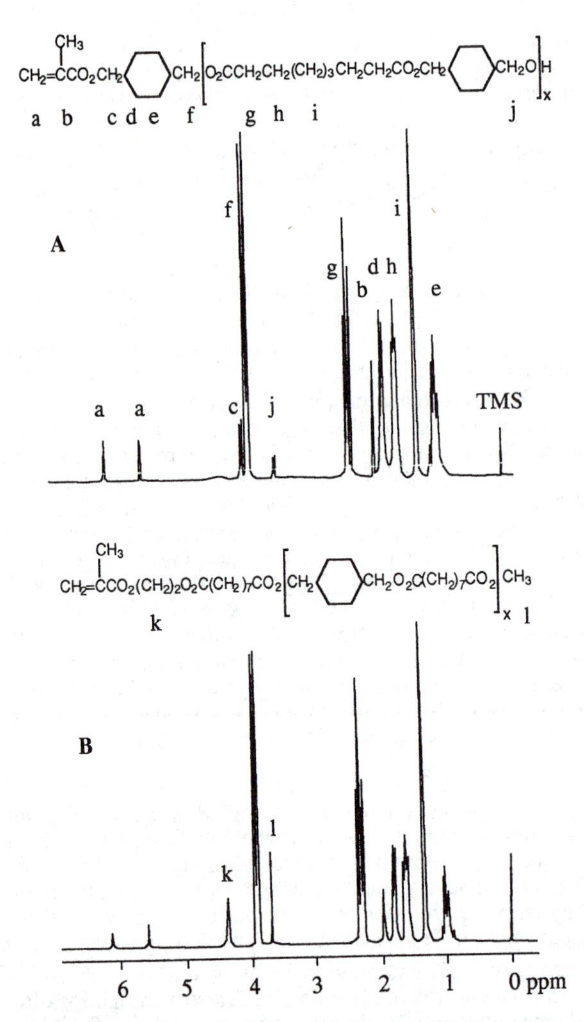

Figure 1. ¹H NMR of oligomers of poly (1,4-dimethylenecyclohexane azalaate) terminated with methacryloxy (A) or 2-methacryloxyethyl (B) ends by reacting the carbinol ended oligomer with methacrylic acid (A) or the carboxyl ended oligmer with HEMA (B) in a second stage of the esterification. Residual carboxyl ends in B were methoxylated.

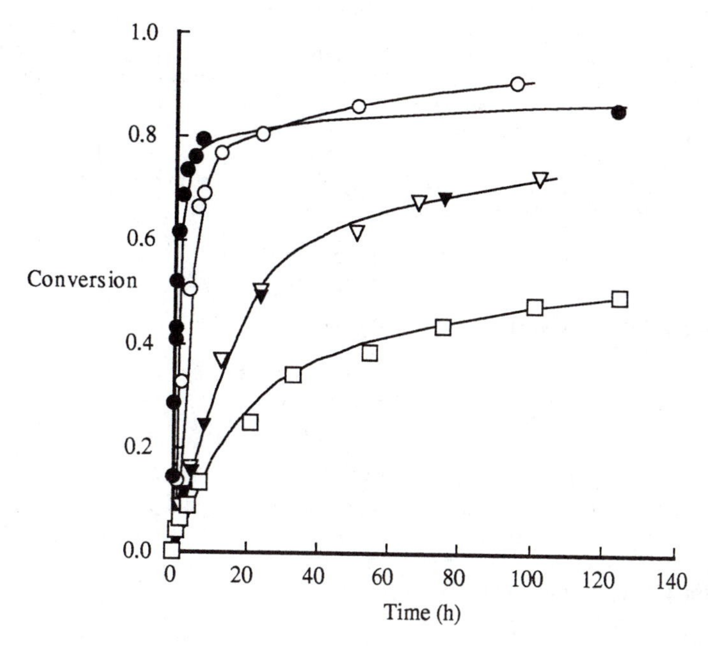

Figure 2. Time-conversion plots for the grafting of linoleic acid to PSAA-1 in o/w emulsion, o/w microemulsion and in a s/l dispersion of an acidic cation exchange resin. O: Amberlyst s/l dispersion, 80°C, 830 meq H^+/ kg. ●: w/o emulsion, NaDBSA/7.4 N H_2SO_4, 80°C, 83 meq/kg NaDBSA. ∇: w/o emulsion, NaDBSA/H_2SO_4, 25°C, 83 meq/kg NaDBSA. ▼: w/o microemulsion, 25°C, DBSA 87 meq/kg, no added aqueous phase. □: DBSA, 87 meq/kg, 25°C, in w/o aqueous emulsion.

[In an earlier communication we reported NaDBSA catalyzes polyesterification in o/w emulsions*(15)*. We now suspect the NaDBSA contained acidic impurities.] Replacing the 7.4 N H_2SO_4 with the same volume of water and equimolar DBSA for NaDBSA in the w/o emulsion, gave markedly reduced rates, presumably due to dissolution of DBSA in the aqueous phase and its depletion at the reaction sites. Rates in the NaDBSA/7.4N H_2SO_4 w/o emulsions accelerate with increased temperature and are more rapid at 80°C than the s/l dispersion catalyzed by the cation exchange resin at the same temperature, despite the nearly ten-fold greater amount of acid present in the resin. Apparently only a fraction of the SO_3H units in the resin are available for catalysis. The reaction in this case is most likely restricted to the s/l interface and excluded from the resin interior due to steric constraints.

Conversions in w/o emulsion or microemulsion asymptotically approach equilibrium values between 60 and 80% while those from s/l dispersions catalyzed by the cation exchange resin reach or exceed 95% despite the slower rates. Although not apparent in Figure 2, it will be shown later that equilibrium conversions show a weak inverse dependence on temperature for runs conducted in the NaDBSA/7.4N H_2SO_4 w/o emulsion.

Equation 3 best describes the observed time-conversion data which, in terms of conversion,P, to ester carbonyl at any time, t, becomes

$$dP/dt = A_0 k_f / r(1-P)(1-rP) - k_r P \tag{7}$$

and from equation 2 the apparent equilibrium constant is given by

$$K_{app} = \frac{rP_e}{A_0(1-P_e)(1-rP_e)} \tag{8}$$

Where P_e is the equilibrium conversion, A_0 is the smaller of the initial functional group concentrations, r is the ratio of A_0 to the larger initial functional group concentration and $[H_2O]_o$ or $[H_2O]_w$, defined by K_2, are assumed constant with respect to time at constant temperature. In all runs conducted here r = 1.

Integration of Equation 7 between the limits $P = 0$ and P_e gives

$$at + b = \mathrm{Ln}\left[\frac{(1/P_e - rP)}{r(P_e - P)}\right], \quad a = \frac{A_0 k_f (1 - rP_e^2)}{rP_e}, \quad b = -\mathrm{Ln}\left[rP_e^2\right] \tag{9}$$

Plots of the right side of Equation 9 against time are linear up to at least 70% of P_e and k_f is determined from the slope, a, A_0 and P_e. P_e is obtained from the maximum conversion and then confirmed by comparing the observed intercept, b, with that from substituting P_e in the right hand expression for b in Equation 9. Then, with experimentally determined P_e, K_{app} is calculated from Equation 8 and k_r from k_f/K_{app}. The examples in Figure 3 illustrate the fit of Equation 9 for esterifications conducted in w/o microemulsion at varied concentrations of DBSA, [DBSA] . Other rate laws, including zero, first order, first order reversible and simple second order fail to fit the data nearly as well, especially beyond the initial stages of the process. In runs catalyzed by the cation exchange resin where P_e reaches 0.95 or better, k_r is too small for meaningful estimation and the data fit the simple second order rate law in addition

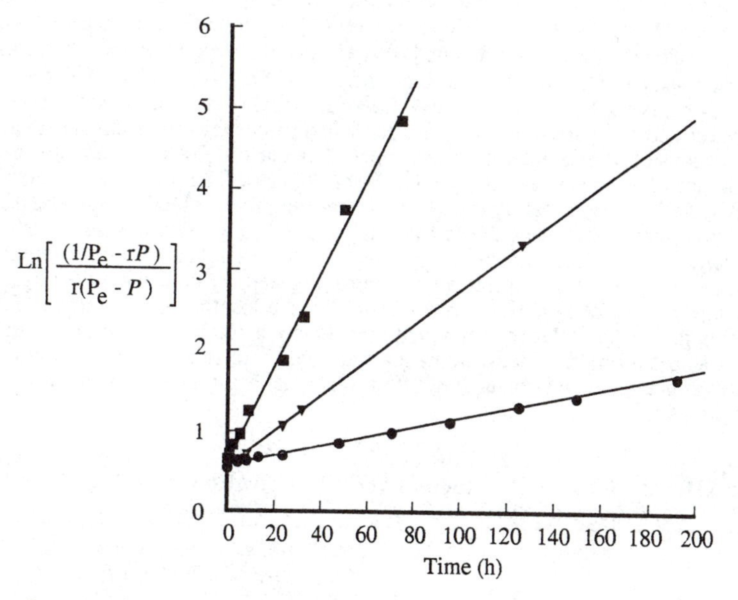

Figure 3. Time conversion data fit to the integrated rate equation 9. Examples are for esterifications at 25° C in w/o microemulsions at varied concentrations of DBSA. ●: 17.9 mmol/kg, ▼: 83.5 mmol/kg, ■: 238 mmol/kg

to Equation 9. In this case the system is essentially irreversible and both rate laws give the same value for k_f.

The constants, k_f and k_r, increase linearly with [DBSA] in reactions conducted in w/o microemulsions as long as [DBSA] is below 175 mmol/kg, Figure 4. However, at higher concentrations k_f shows a sharp discontinuity becoming essentially independent of [DBSA]. The behavior is reminiscent of enzyme or surface catalysis where reaction sites can become saturated so that further adsorption of a reactant or catalyst is impeded*(16)*. Similar phenomena were observed in DBSA catalyzed condensations of siloxanol terminated polydimethylsiloxane in aqueous o/w emulsions where the o/w interface was the reaction site *(see Table IV in (17))*. The values of k_r, on the other hand, exhibit a continuous linear dependence on [DBSA] suggesting that ester hydrolysis is occurring in a different environment.

Despite indications of a heterogeneous mechanism, temperature dependencies for k_f and k_r adhere well to the transition-state format for esterifications conducted in NaDBSA/7.4N H_2SO_4 w/o emulsions, Figure 5. Data are analyzed in terms of a first order dependence of both k_f and k_r on [DBSA] since the maximum concentration in these runs that can form by ion exchange with NaDBSA, 83 mmol/kg, falls on the linear segment of the curve in Figure 4. Table III shows the resulting activation parameters and compares them with the published values for homogeneous melt esterifications of octadecanoic acid with octadecanol at 160-200°C both in the presence and absence of a toluenesulfonic acid catalyst *(18), (19)*. The present activation enthalpy, ΔH^*, is significantly greater, while the negative activation entropy, ΔS^*, is nearly half that reported for either the uncatalyzed or catalyzed processes in the melt. Consideration of molecular models suggests that the larger ΔH^* observed here might be due to steric constraints during a rate controlling step, particularly if the oligomeric carbinol were subject to conformational restrictions in locating at an interface. The smaller negative ΔS^* relative to the melt process, on the other hand, implies reduced requirements for ordering in the transition state and is consistent with a process where reactants concentrate and organize by adsorption at an interface prior to a rate controlling step.

Table III. Activation Parameters for Esterification of PSAA-1 and Linoleic acid in w/o NaDBSA/7.4N H_2SO_4 Emulsion

	For k_f	*For k_r*	*Literature, k_f[a] (No Catalyst)*	*Literature, k_f[a] ($MeC_6H_4SO_3H$ Catalyst)*
ΔH^* (kJ/mol)	63.1	76.5	51.0	18.5
ΔS^* (kJ/mol-K)	-109.0	-88.2	-210	-210

[a] From the estrification of octadecanoic acid with octadecanol in the melt *(18,19)*.

Differences between values of ΔH^* observed for k_f and k_r and values of ΔS^* for the same constants allow estimates of the thermodynamic parameters, Table IV. Thus ΔG and K_{app} calculated in this fashion are in agreement with those of K_{app} based on P_e measured at 25°C. The negative values of ΔG and ΔH indicate a moderate inverse dependence of K_{app} on temperature. It is interesting to note that the value of K_{app} at 25°C exceeds by a factor of 260 the equilibrium constant reported for ethyl acetate in equilibrium in solution where the activity of water was taken to be unity *(6)*.

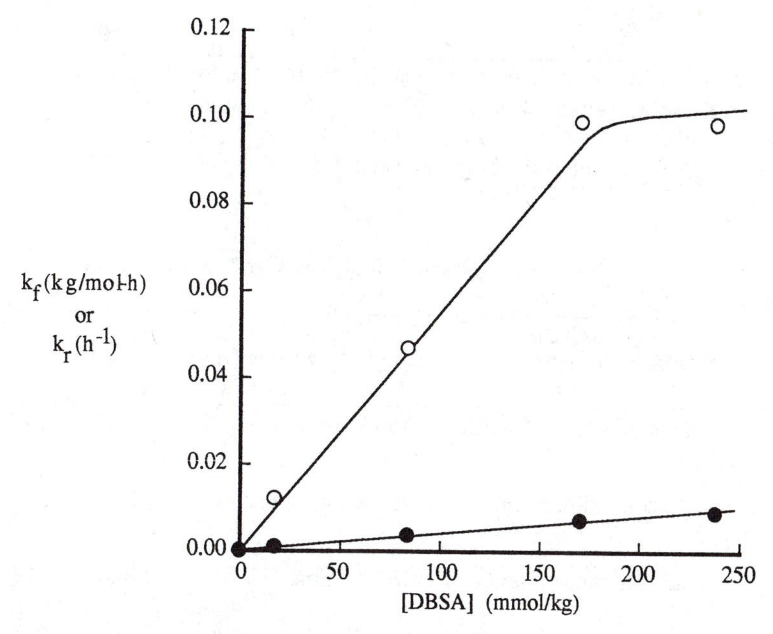

Figure 4. Effect of DBSA concentration on forward, k_f, and reverse, k_r, rate constants in the esterification of PSAA-1 with linoleic acid in o/w microemulsion at 25°C. O: k_f, ●: k_r.

Table IV. Thermodynamic Parameters for Esterification of PSAA and Linoleic acid in w/o NaDBSA/7.4N H_2SO_4 Emulsion at 25°C

	ΔH (kJ/mol)	*ΔS (kJ/mol-K)*	*ΔG (kJ/mol)*	K_{app}
reported here	-13.4[a]	-22.1[a]	-6.9[a]	16[a]
reported here	---	---	-6.5[b]	14.0[b]
reported here	---	---	-6.8[c]	15.6[c]
literature	---	---	6.7[d]	0.061[d]

[a]From the differences in observed activation parameters for k_f and k_r in table 3. [b]Based on K_{app} observed in w/o microemulsion at 25° C. [c]Based on K_{app} observed in NaDBSA/7.4N H_2SO_4 w/o emulsion at 25° C. [d]From K_1 (reference 6) for ethyl acetate at equilibrium with water in solution at 25° C, activity of water = 1.0.

Table V. Factors Influencing k_f and K_{app} in Grafting Carboxylic Acids to PSSA at 25°C [a]

Type of Dispersion	*Solvent /Carboxylic Acid*	*Catalyst/Aqueous Phase*	$10^3 x\ k_f$ *kg/mol/h*	K_{app} *kg/mol*
w/o microemulsion	toluene/linoleic	DBSA/forms *in•situ*	46.6	14.0
w/o microemulsion	dioxane/linoleic	DBSA/forms *in•situ*	26.0	5.5
w/o microemulsion	toluene/acetic	DBSA/forms *in•situ*	216.0	18.9
w/o emulsion	toluene/linoleic	NaDBSA/7.4N H_2SO_4	47.5	15.6
w/o emulsion	toluene/linoleic	DBSA/water	9.9	---
w/o dispersion	toluene/linoleic	None/7.4N H_2SO_4	2.4[b]	16.6[b]
w/o dispersion	toluene/linoleic	None/7.4N HCl	13.3[b]	15.6[b]
s/l dispersion	toluene/linoleic	Amberlyst/forms *in•situ*	6.6[c]	624.0[c]

[a]All runs with 83 meq./Kg of DBSA or NaDBSA except 870 meq./Kg H^+ in the s/l dispersion. [b]Based on the measured 49°C value and the parameters in Tables III and IV. [c]Based on the measured 80°C value and the parameters in Tables III and IV.

Other factors influencing k_f and K_{app} include the use of a hydrophobic solvent, steric effects, electrolytes in the aqueous phase and the use of surface active catalysts. These effects are illustrated in Table V. Thus, with the less hydrophobic dioxane instead of toluene as a solvent, both k_f and K_{app} are suppressed in runs conducted in w/o microemulsions. The sensitivity of the system to steric bulk is apparent when linoleic acid is replaced with acetic acid to give a five-fold increase in k_f and an enhanced K_{app}. The beneficial effect of electrolytes in the aqueous phase, already illustrated in Figure 3 where NaDBSA/7.4N H_2SO_4 is replaced with identical amounts of DBSA/ water, is shown here in terms of k_f. Rates are reduced in the absence of a surfactant with only 7.4N H_2SO_4 or HCl present in the aqueous phase, although equilibrium is eventually reached. The cation exchange resin, while giving slower rates per equivalent of H^+ than the NaDBSA/7.4N H_2SO_4 system, produces extraordinarily high values of K_{app}. The exposed SO_3H groups on resin surface are considered to catalyze the reaction while the SO_3H groups in the interior can hydrate and efficiently remove by-produced water from the reaction sites to give a higher K_{app}.

Conclusions

Two concurrent reaction pathways can account for the observations. One is proposed to occur in the hydrophobic phase and the other at the interfacial boundary between the hydrophobic and hydrophilic phases, Figure 6. Esterification is considered to occur at both sites so the forward rate constant will be composed of the rate constants from both reactions ($k_f = k_f' + k_f''$). Hydrolysis (k_r), on the other hand, is considered to occur primarily in solution in the hydrophobic phase where it is limited by the solubility of water in the hydrophobic media, as determined by the constant K_2. The interface consists of a boundary between the hydrophobic and the polar regions of w/o emulsion particles, or the micelles in w/o microemulsions. In the s/l dispersions the interface is between the surface of the cation exchange resin particles and the hydrophobic medium. In either case catalysis at the interface is considered essentially irreversible while the process in the hydrophobic phase is a typical reversible esterification in solution defined by K_1 in equation 1.

The process is considered to be a version of the classical AAC2 mechanism for ester formation and hydrolysis *(20)* but modified here so that esterification can occur at the interface. After its formation at the interface the ester bond is considered to diffuse (k_d) and remain preferentially in the hydrophobic phase due to its reduced polarity and relative inability to hydrogen bond compared with the reactants. The more polar carbinol and carboxylate groups along with surfactant, on the other hand, are more predisposed to locate at the interface (K_r) where the known associations between carbinols and anionic surfactants *(21)* can provide further impetus for accumulation there. The locally high concentrations of reactants and catalyst can then enhance forward reaction rates at the interface, k_f'. Surfactant, while primarily locating at the interface, distributes between the hydrophobic and hydrophilic phases (K_s, K'_s) depending on the relative dominance of its polar and nonpolar structural units. Thus catalysis with surfactant can be both at the interface where the process is considered essentially irreversible or in the hydrophobic phase where the process will be a typical reversible solution esterification. The catalytic centers of the cation exchange resin, on the other hand, are restricted in their mobility and cannot migrate. In this case the process will be limited to the resin interface, providing little or no opportunity for the catalysis of reversible esterification in the hydrophobic phase.

Over-all rates (k_f) and reversibility (K_{app}) in such a scheme will be determined by factors influencing the solubilities of ester, water, and surfactant in the hydrophobic phase and the ability of the reactants and catalyst to locate at the interface. These factors

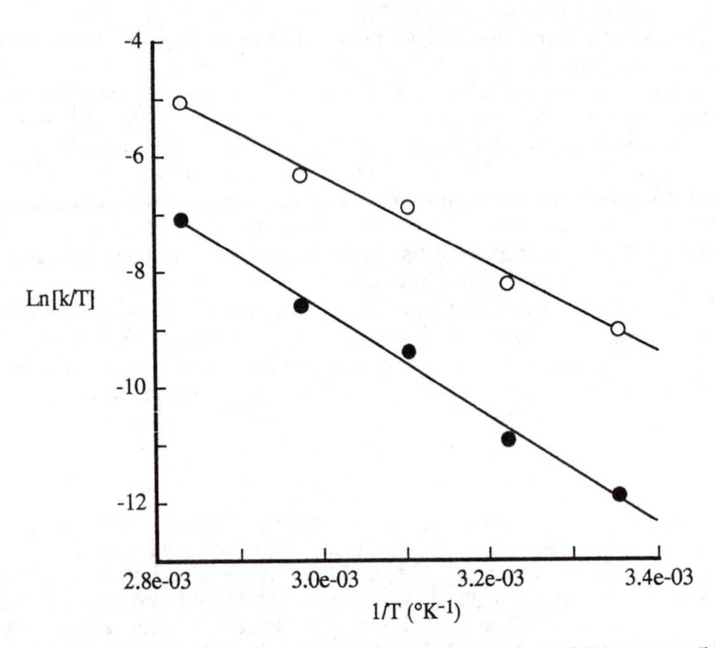

Figure 5. Temperature dependence for k_f/[DBSA] (upper) and k_r/[DBSA] (lower) plotted in the format of the transition-state theory. Runs are conducted in NaDBSA/7.4N H_2SO_4 w/o emulsion. Results are in Table III.

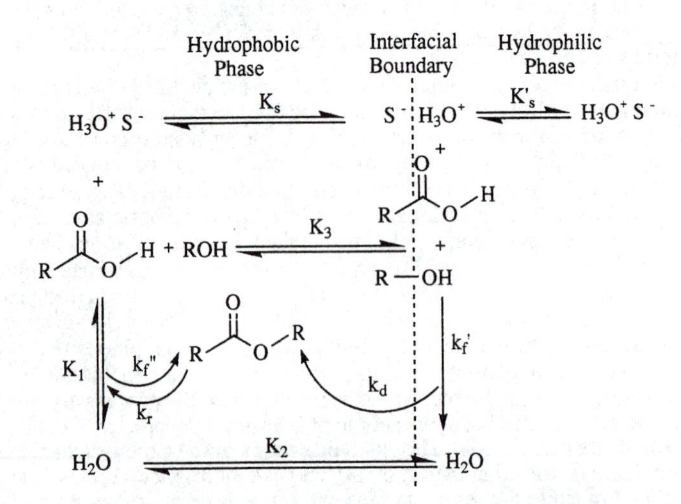

Figure 6. Proposed reaction sequence for esterification in heterogenious media. R represents groups preferentialy soluble in the hydrophobic phase and S $^-$ the surfactant anion.

will include temperature, hydrophobicity of the solvent, the polarities of reactants and product, electrolytes in the aqueous phase and the nature of the surfactant. Adjustment of these parameters can lead to a significant suppression of ester hydrolysis and, combined with the absence randomizing side reactions, provide opportunities for synthesis of oligomeric polyesters with organized or temperature-sensitive structures.

Acknowledgements

The authors acknowledge the National Science Foundation Industry/University Cooperative Research Center in Coatings for financial support. Vicki Sible and Robin Hickerson of the Dow Corning Corporation are acknowledged for contributing the GPC analysis.

Literature Cited

(1) Holmberg, K. and Johansson, J. *In Organic Coatings Science and Technology;* Parfitt, G.and Patsis, A., Ed., Marcell Dekker: New York, N.Y. 1984; Vol. 6, pp 23.
(2) Walz, G. *J. Oil Color Chem. Assoc.* **1977,** *60* , 11.
(3) Saunders, J. H.; Dobson, F. In *Comprehensive Chemical Kinetics;* Bamford, C. H.; Tipper, C. F. H. Ed.; Elsevier: New York, N.Y. 1976, Vol. 15; pp 500.
(4) Kangas, S. L.; Jones, F. N. *J. Coatings Technol.* **1987,** *59* , 89.
(5) Fendler, J. H. *Membrane Mimetic Chemistry*; John Wiley and Sons: New York, N. Y., 1982; pp 11.
(6) Jencks, W.P. and Gilchrist, M. *J. Am. Chem Soc.* **1964,** *86* ,4651.
(7) Chegolya, A.S., Shevchenko, V.V., Mikhailov, G. D. *J. Polym. Sci., Polym. Chem. Ed.* **1979,** *17*, 889.
(8) Fendler, E. J.; Chang, S. A.; Fendler, J. H.; Medary, R. T.; El Seoud, O.A.; Woods, V. A. in *Reaction Kinetics in Micelles*; Cordes, E. Ed., Plenum Press: New York, N.Y. 1973; pp127.
(9) Clinton, R. O.; Laskowski, S. C. *J. Am. Chem. Soc.* **1948,** *70*, 3135.
(10) Helffrich, F. *Angew. Chem.* **1954,** *66*, 241.
(11) Malinowski, M.; Tokarzewska, M. *Przem. Chem.* **1973,** *52*, 800. *Chem. Abs. 80:* 108896n.
(12) Karpov, O. N.; Kazragis, A.; Fedosyuk, L. G. *Leit. TSR Aukst. Mokyklu Mokslo Darb., Khim. Khim. Tekhnol.* **1970,** *12*, 193. *Chem. Abs.76*: 113631u.
(13) Krasnobaeva, V. S.; Sukov, V. D.; Mogilevich, M. M. *Izv. vyssh. Ucheben. Zaved. Khim. Khim. Tekhnol.* **1988** *31*, 80. *Chem. Abs. 110:* 39407c
(14) Saam, J. C.; Chou, Y. J. (to the Dow Corning Corporation) *U. S. Patent 4,355,154* **1982.**
(15) Baile, M.; Chou, Y. N.; Saam, J. C. *Polym. Bull.* **1990,** *23*, 251.
(16) Benson, S. W. *The Foundations of Chemical Kinetics*; McGraw-Hill: New York, N.Y., 1960, pp 655.
(17) Saam J. C.; Huebner, D. J. *J. Polym. Sci., Polym.Chem.Ed.* **1982,** *20*, 3351.
(18) Fradet, A.; Marechal, E. *J. Polym. Sci., Polym. Chem. Ed.* **1981,** *19*, 2905.
(19) Fradet, A.; Marechal, E. *J. Macromol. Sci. Chem.* **1982** *A17,* 859.
(20) Ingold, C. K. *Structure and Mechanism in Organic Chemistry;* Cornell University Press: Ithica, N. Y., 1969, pp 1129-31.
(21) Kung, H. C.; Goddard, E. D. *J. Phys. Chem.* **1964,** *68*, 3465.

RECEIVED December 4, 1995

Chapter 22

Polycarbonates Derived from *trans*-1,4-Cyclohexane Units

Jeff W. Labadie[1], E. Hadziiouannou[1], and B. Haidar[2]

[1]**Research Division, IBM Almaden Research Center, 650 Harry Road, San Jose, CA 950–6099**
[2]**Centre National de la Recherche Scientifique, Mulhouse, France**

The polycondensation of trans-1,4-bis(chlorooxycarbonyl)cyclohexane (**1**) with diols afforded high molecular weight polycarbonates (**2**). Monomer **1** was prepared from a cis:trans mixture of 1,4-cyclohexanediol and phosgene, where the pure trans isomer was isolated by crystallization from the concentrated reaction mixture. Polymerization of **1** with primary α,ω–diols in the presence of pyridine afforded high molecular weight polycarbonates. The resulting polymers were highly crystalline with melting points up to 205 °C for the polycarbonate derived from 1,4-butanediol (**2a**). Secondary diols gave lower molecular weight materials. Block copolymers of **2a** and poly(propylene oxide) were prepared with 25, 50 and 75 wt.% polypropylene oxide. The copolymers displayed a heterogeneous morphology with discrete PO domains, and the crystallinity of the polycarbonate was maintained for the 25 and 50 wt.% poly(propylene oxide) compositions. The expected clean thermal degradation was observed for all homopolymers and copolymers.

Aliphatic polycarbonates have been much less studied relative to aromatic polycarbonates for reasons, which include lower Tg, mechanical toughness and thermal stability, making them less useful for structural applications. Several existing or potential niche applications have been reported for aliphatic polycarbonates, such as soft blocks for polyurethanes (*1*), ceramic binders (*2,3*), and high resolution photoresists (*4*). The later two applications rely on the clean thermal degradation of aliphatic polycarbonates. Moreover, thermal degradation is accelerated in the presence of photoacids. The photoacid-catalyzed cleavage of pendent *tertiary*-butylcarbonates is the basis for the innovative chemical amplification approach to microlithographic resists reported by Ito and coworkers (*5*).

0097–6156/96/0624–0350$12.00/0

The synthetic approaches to aliphatic polycarbonates are numerous and varied and include both step-growth and chain-growth polymerization mechanisms. The step-growth synthetic routes include polymerization of aliphatic diols with phosgene (*6,7*) or dialkyl carbonates (*8*), and the copolymerization of diols and bischloroformates (*7*). The polymerizations with phosgene are carried out as solution rather than interfacial polymerization since alkoxides of cycloaliphatic alcohols cannot be formed in aqueous solution, in contrast to phenolates. Transesterification between cycloaliphatic diols and diethyl carbonate in the presence of a small amount of sodium is a classic method of aliphatic carbonate formation discovered by Carothers (*8*). This procedure was modified using vacuum to give higher molecular weight aliphatic polycarbonates suitable for fiber and film-forming applications (*9*).

The use of solid-liquid phase transfer catalysis in the conjunction with bis(carbonylimidazolides) (*3*) bis(p-nitrophenylcarbonates) (*10*) as developed by Frechet for the synthesis of novel tertiary copolycarbonates. The instability of tertiary chloroformates renders tertiary polycarbonates inaccessible through conventional chloroformate monomers or intermediates. The bis(carbonylimidazolide) monomer was shown to polymerize with both tertiary (*3*) and secondary alcohols (*11*), demonstrating the utility of the method in forming polycarbonates from less reactive sterically hindered monomers. Most of the examples reported involved benzene dimethanol derivatives, or 1,4-butynediol, indicating that an adjacent site of unsaturation may activate the alcohol in the solid-liquid phase transfer reaction scheme.

Ring-opening polymerization of cyclic carbonates and spiroorthocarbonates has proven to be a flexible route to aliphatic polycarbonates. An early example is the polymerization of neopentyl carbonate under transesterification conditions (*12*). Anionic (*13*), cationic (*14*), and group-transfer (*15*) polymerization have been successfully applied to the synthesis of aliphatic polycarbonates from cyclic carbonate monomers. An alternative catalyst system has been developed based on trialkyltin alkoxides (*16*) and oxides (*17*). Ring-opening polymerization of spiroorthocarbonates (*18,19*) offers another route to aliphatic polycarbonates, and is noteworthy as it occurs with zero shrinkage or expansion in volume.

The discovery by Inoue (*20*) that epoxides and carbon dioxide could be copolymerized in the presence of a diethylzinc:water catalysts offers an additional route to aliphatic polycarbonates. The catalyst types which are effective for this polymerization were expanded upon in later work, demonstrating that zinc:pyrogallol (*21*) and aluminum-porphyrin (*22*) catalyst systems afforded the desired linear polycarbonates. The synthesis has been applied to functional epoxides with pendent hydroxy (*23*) and o-nitrobenzyl ether groups (*24*), the later being investigated as a new photoresist scheme.

One of the most interesting properties of aliphatic polycarbonates is their clean thermal degradation to gaseous products (*3,11*). In this paper we describe the synthesis and properties of polycarbonates derived from *trans*-1,4-cyclohexane units, and related polycarbonate-poly(propylene oxide) block copolymers as part of an investigation of thermally labile polymer systems.

EXPERIMENTAL

Materials.
1,4-Cyclohexanediol (Aldrich) and 20% phosgene in toluene (Fluka) (**CAUTION:** Phosgene is a highly toxic gas) were used as received. The α,ω-diols (Aldrich) were distilled from calcium hydride. Pyridine was distilled from butyl lithium. Methylene chloride and chloroform were distilled from phosporous pentoxide. Tetrahydrofuran (THF) was distilled from sodium-benzophenone ketyl.

***trans*-1,4-bis(chlorooxycarbonyl)cyclohexane (1).**
In a magnetically stirred three neck round-bottom flask fitted with a dry-ice condenser was slurried 41.8 g (0.4 mole) of 1,4-cycloxanediol (cis:trans mixture) in 200 mL of tetrahydrofuran. To this was added 500 mL of 1.93 M phosgene in toluene (0.97 mole) in one portion and the mixture was heated to 60 °C for 2 h, then stirred overnight at room temperature. A strong stream of nitrogen was passed over the solution and exhausted through two sodium hydroxide scrubbers in series to remove the excess phosgene. The remaining toluene was evaporated on a rotary evaporator to leave a mixture of white crystals and purple oil. This was slurried in a minimum of petroleum ether, filtered and rinsed once with 20 mL of petroleum ether. The crude white crystals were recrystallized 2X from 450 mL of hexane, using decolorizing carbon in the second recrsytallization. The isolated yield of white crystalline **1** was 40.0 g (83 %): mp = 113 °C, IR (C=O) 1763 cm $^{-1}$.

***trans*-1,4-Cyclohexanediol.**
The separation of the pure *trans*-1,4-cyclohexanediol from the cis:trans mixture was carried out by analogy to the literature procedure (*25*). A slurry of 46.4g (0.40 mole) of 1,4-cycloxanediol (cis:trans mixture), 125 mL (0.9 mole) triethylamine, and 300 mL of methylene chloride was cooled to 0 °C. To the slurry, 85 mL (0.9 mole) of acetic anhydride was added dropwise. The resulting amber solution was allowed to warm to room temperature overnight, transferred to a separatory funnel, and washed 2X with 10 % aqueous hydrochloric acid, water, and 10% aqueous sodium hydroxide. The organic layer was dried over magnesium sulfate and concentrated on a rotary evaporator to give a mixture of yellow oil and white crystals. The mixture was allowed to stand one day, and the solid was filtered and recrystallized from ethanol to give 11.8g of the *trans*-1,4-diacetoxycyclohexane as clear platelets (Rf = 0.55, 20% ethylacetate/hexane:silica gel)

The diacetate (8g, 0.04 mole) was hydrolyzed by addition to a solution 50g of barium hydroxide octahydride in 100 mL of water heated at the reflux temperature. After 5 h the solution was cooled, filtered, and the resulting solid was dried with a stream of nitrogen. The solid was extracted 3X with hot acetone, concentrated, and recrystallized from 2:1 ethyl acetate:hexane. Filtration afforded 3.2 g (69%) of *trans*-1,4-cyclohexanediol clear platelets: mp = 142.5-143 °C, Lit. (*25*) mp = 142 °C. The *trans*-1,4-cyclohexanediol was subsequently converted to the dichloroformate by analogy to the procedure used for the cis:trans mixture, producing a compound identical to **1**.

Polycarbonate Homopolymers (2-4).
The polymerization procedure is described for the polymerization of **1** with 1,4-butanediol and is representative of the general method. A nitrogen flushed three-neck round-bottom flask fitted with a magnetic stirrer was charged with 3.6048 g (40.000 mmol) of butanediol and 9.6428g (40.000 mmol) of **1** and 50 mL of chloroform. The mixture was heated 40-50 °C in a water bath and 15 mL of pyridine was added dropwise via syringe over a period of 20 min. After complete addition, the reaction was a clear pale yellow color. A white precipitate formed and the reaction thickened significantly after an additional 15 min. The polymerization was stirred 2 h at 50 °C, cooled to room temperature and allowed to proceed overnight. The polymerization solution was precipitated in methanol and the resulting white fibrous polymer was washed 3X in methanol and dried *in vacuo*. The polymer was further purified by dissolving in chloroform and reprecipitation in methanol. The isolated yield of polymer was generally 95 %.

Polycarbonate-PO Copolymers (5).
The procedure is described for the copolymerization of **1,** 1,4-butanediol and a 50 wt. % loading of 4,000 molecular weight PO oligomer with hyroxyl end groups. A nitrogen flushed three-neck flask fitted with a dry-ice condensor was charged with 8.00g (2.00 mmole) of PO followed by 21 mL of 1.93 M phosgene in toluene (40 mmole) in one portion and the mixture was heated to 50 °C for 2 h. A strong stream of nitrogen was passed over the solution and exhausted through two sodium hydroxide scrubbers in series to remove the excess phosgene, followed by removal of the remaining toluene *in vacuo*. The reaction vessel was flushed with nitrogen then charged with 2.884 g (32.00 mmole) of 1,4-butanediol dissolved in 15 mL of THF. Pyridine (0.66 mL, 0.646 g, 8.2 mmole), was added dropwise at room temperature, then the reaction was warmed to 50 °C in a water bath. After one hour at 50 °C, chlorofomate **1** (7.3232 g , 30.00 mmole) was dissolved in 50 mL of dry chloroform and added to the reaction mixture using a dropping funnel. Pyridine (15 mL) was added to the mixture dropwise, allowing the precipitate which was formed to substantially dissolve between drops. The polymerization was observed to increase in viscosity as the addition neared completion. After addition of the pyridine, the polymerization was heated to the reflux temperature for 16 h. The polymer was precipitated in 3:1 methanol/water, filtered, and rinsed with 3:1 methanol/water. The white polymer was washed 3X with 1:1 methanol-water, followed by a brief acetone rinse in the Buchner funnel. The polymer was dried *in vacuo* in the presence of potassium hydroxide as a desiccant. The yield of white fibery polymer was 14.8g (93%). The 75 wt.% PO copolymers was not rinsed with acetone after isolation, and was reprecipitated into 2:1 methanol:water from THF.

Measurements.
Glass transition temperatures were measured on a DuPont dynamic scanning calorimeter with a heating rate of 10 °C/min. Thermal gravimetric analysis was carried out a heating rate of 10 °C/min. Gel Permeation chromatography was carried out with a THF mobile phase and polystyrene standards. Films of homopolymers and copolymers were prepared by doctor-blading chloroform solutions followed by solvent removal on a hot plate.

RESULTS AND DISCUSSION

Our evaluation of new thermally degradable polymer structures focused primarily on aliphatic polycarbonates which could be produced by the chloroformate route. While examining the synthesis of chloroformates from various diols, we found the reaction of the mixed isomers of 1,4-cyclohexanediol resulted in the formation of the *trans*-1,4-bis(chlororoxycarbonyl)cyclohexane (**1**) as a crystalline material, which could be easily separated and handled. This was not an unexpected result, as derivitization of 1,4-cyclohexanediol as an acetate results in a crystalline trans product, and is the basis of a method for separating *cis*- and *trans*-1,4-cyclohexanediol. The synthetic procedure for the preparation of **1** was analogous to literature procedures, and involved the reaction of a the isomeric mixture of 1,4-cyclohexanediol with phosgene in toluene (Scheme I). The slurry of diol reacted readily at 60 °C, as indicated by the disappearance of the diol solid to give a clear solution. After several hours the excess phosgene and solvent were removed to give a mixture of a purple oil and white crystals. The crystals were easily separated and purified by recrystallization to give a single isomer of the dichloroformate. This was shown to be the expected trans isomer by decomposition to the diol, and conversion to the diacetate, which was

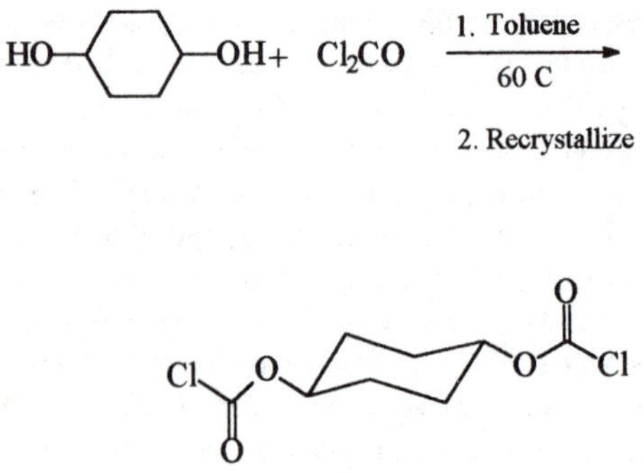

HO–C6H10–OH + Cl_2CO → (1. Toluene, 60 C; 2. Recrystallize)

Cl–C(=O)–O–C6H10–O–C(=O)–Cl

1

SCHEME I

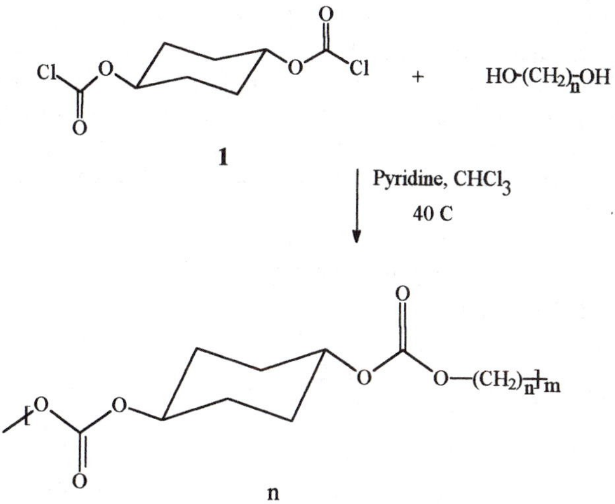

1 + $HO{-}(CH_2)_n{-}OH$

Pyridine, $CHCl_3$
40 C

	n
2a :	4
b:	5
c:	6
d:	7

SCHEME II

compared to an authentic sample. In addition, *trans*-1,4-cyclohexanediol, prepared from *trans*-1,4-diacetoxycyclohexane, was converted to the chloroformate and was found to be identical **1**. Monomer **1** was easily handled in ambient conditions and has been found to be stable for > 6 years in storage.

Polymerization of **1** with primary $\alpha,\omega-$diols in either chloroform or THF produced high molecular weight polymers **2a-d** (Scheme II). The polymerizations were carried out by addition of pyridine to a mixture of the two monomers . The pyridine was added dropwise to the stirred mixture at such a rate that the insoluble chloroformate-pyridine complex (*26*) dissolved prior to the addition of the next drop. The polymerization temperature was maintained at 40-50 °C to assist in the dissolution of the complex. Either THF or chloroform could be used as the solvent, however, chloroform was preferred since the pyridine hydrochloride salt remained soluble for the entire pyridine addition process, making it easier to monitor the dissolution of the chloroformate-pyridine complex. A problem occurred when pyridine contacted the walls of the flask during addition causing adhesion of the insoluble chloroformate-pyridine complex to the reactor walls. The adhered complex would not dissolve in the polymerization regardless of reaction time. This was avoided by adding the pyridine to the vortex of the stirred solution. The polymers were isolated by precipitation to give a fibery white solid in high yield. The polymerization of **1** with primary diols was shown to consistently give high molecular weight polymers (**2a-d**) (Table 1). The only exception observed was a single attempt with ethylene glycol which initially gave the expected viscous solution after pyridine addition, but on continued stirring the viscosity dropped and precipitation yielded no polymer, possibly due to a depolymerization mechanism which affords ethylene carbonate.

TABLE 1
Characterization of Poly(carbonates)

Polymer	Diol[a]	Pzn Solvent	IV[b] (dL/g) ($CHCl_3$)	GPC M_w	GPC M_n	M_w/M_n
2a	n = 4	THF	0.82	52,400	20,600	2.55
2a	4	$CHCl_3$	0.84	53,400	27,200	1.97
2b	5	$CHCl_3$	0.70	47,400	22,600	2.09
2c	6	$CHCl_3$	0.65	49,100	24,500	2.00
2d	7	$CHCl_3$	0.62	50,500	22,600	2.24
3	2,5-hexane-diol	$CHCl_3$	0.15	-	-	-
4	1,3-cyclo-hexanediol	$CHCl_3$	0.12	16,000	8,800	1.81

a) Scheme II, b) intrinsic viscosity, chloroform, 25 °C.

Polymerizations of **1** with 2,5-hexanediol and 1,3-cyclohexanediol afforded polymers **3** and **4**, respectively. These polymers had molecular weights which were substantially lower than **2a-d.** This is attributable to an elimination process involving the choroformate-pyridine complex as a side reaction, which becomes more important with the less reactive secondary diols (*26*). Daly reported polymerizations of cyclohexanediol with tetramethyl cyclobutanediol dichloroformate afforded high molecular weight polymer, this being a case where elimination cannot occur due to the methyl substitution at the β– position (*27*).

With a route to aliphatic polycarbonate homopolymers established, we turned our attention towards the synthesis of thermally degradable block copolymers, with the goal of preparing thermally labile elastomers. We chose to use polycarbonate **2a** as a hard segment and poly(propylene oxide), PO, as the soft segment. PO readily degrades thermally and can be obtained as hydroxyl functional oligomers, where the terminal functional groups are a primary and a secondary alcohol. The copolymer synthetic strategy called for conversion of the hydroxyl end groups to chloroformates, since the lower reactivity associated with the secondary diol may lead to poor PO incorporation. The hydroxyl end groups of a 4,000 molecular weight PO oligomer were converted to the chloroformate with phosgene. The oligomer was allowed to

Table 2
Polycarbonate-PO Block Copolymer

Cop.	%PO	IV (dl/g) (CHCL$_3$)	PO T_g (°C)	PC T_g(°C)	PC T_m (°C)	T_d (°C)[ab]
2a	0	0.82	-	40	205	290
5a	25	0.64	-	30	200	300, 350
5b	50	0.43	-70	10	190	300, 350
5c	75	0.50	-70	-	160	300,350
PO	100	-	-70	-	-	350

(a) decomposition temperature, 10 wt.% loss at 5 °C/min heating rate, nitrogen atmosphere. (b) The two T_d values reported for the block copolymers corre spond to the polycarbonate and PO blocks.

react with an excess of 1,4-butanediol in THF and 2 eq. of pyridine (relative to PO) at 50 °C. A chloroform solution of **1** was then delivered to the reaction vessel, followed by dropwise addition of the remaining pyridine (Scheme III). The copolymers (**5**) were isolated by precipitation in methanol:water and filtration. Copolymers with 25, 50 and 75 wt.% PO (**5a-c**) were prepared by this method (Table 2). High molecular weight materials were obtained, as indicated by the intrinsic viscosities, in close to quantitative yield. The incorporation of PO was close to the amount charged as determined by TGA, and as indicated by the high isolated yields. Copolymerizations targeted for the 50 wt.% composition produced low molecular weight materials in some cases, the reasons for which are not understood.

The thermolysis of polycarbonates derived from tertiary polycarbonates has been studied in detailed by GC-MS (*3*) and TGA-MS (*11*) techniques. Thermal degradation occurs in the 190-210 °C range to afford a mixture of carbon dioxide, dienes, diols, and eneols. The elimination products formed are derived from the tertiary carbonate, consistent with the material design. Thermolysis of polymers **2a-d** occurred with a decomposition temperature of 290 - 300 °C, and therefore, are significantly less labile than the systems derived from tertiary carbonates (Table 3). The higher decomposition temperature relative to the tertiary polycarbonates is consistent with the presence of secondary and primary carbonate linkages. The presence of purely secondary carbonate linkages in polymers **3** and **4** led to only a slight decrease in decomposition temperature to 280 °C. The thermal decomposition of all the polycarbonates was clean with no residue present above 350 °C in the dynamic TGA run.

Polymers **2a-d** were crystalline materials and displayed strong melting transitions. Polyesters derived from *trans* cyclohexane units are known to display crystalline and liquid crystalline characteristics (*28*). These polymers displayed Tg's ranging from -5

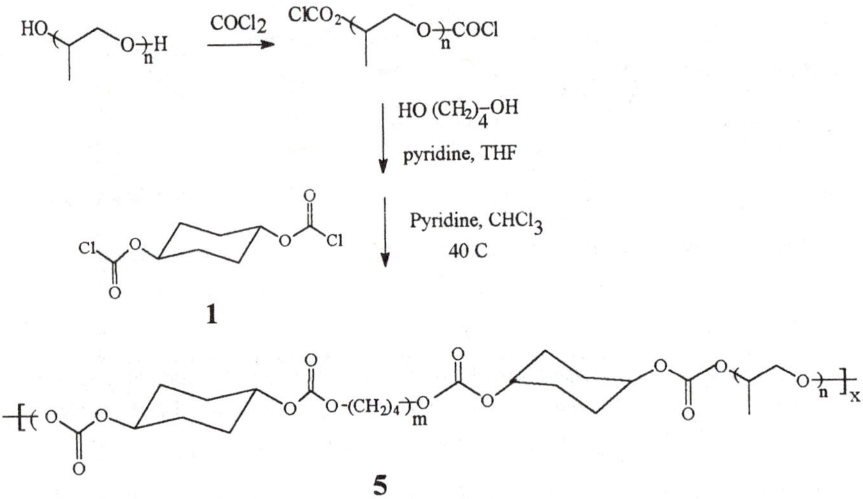
HO
O
H
n
COCl2
ClCO2
O
COCl
n
HO (CH2)4-OH
pyridine, THF
Pyridine, CHCl3
40 C
1
5

Scheme III

- 30 °C and Tm's ranging from 165-205 °C as determined by dynamic scanning calorimetry (Table 3). The polymers readily crystallized from the melt, and crystallization transitions, Tc, were readily observed at a temperature 30-40 °C below Tm as the samples were cooled (Table 3). An example of this thermal behavior is shown for **2a** in Figure 1.

TABLE 3
Characterization of Thermal Properties

Polymer	$-(CH_2)_n-$	T_g (°C)	T_m (°C)	T_c (°C)	T_d (°C)[a]
2a	4	30	205	160	290
2b	5	-5	170	130	300
2c	6	10	175, 185	160	300
2d	7	-5	165	-	300
3	-	80	-	-	280
4	-	60	-	-	275

(a) decomposition temperature, 10 wt.% loss at 5 °C/min heating rate, nitrogen atmosphere.

The Tm's for **2** showed the expected odd-even effect with respect to variation of the diol chain length (*29*), with the highest value for the polycarbonate derived from 1,4-butanediol. When the butanediol based polycarbonate was quench-cooled with liquid nitrogen from the melt, subsequent heating led to a new, broad melting transition at 0 °C, followed by crystallization, then the Tm at 205 °C (Figure 2). The quench-cooling appears to afford a kinetically favored crystal form with a lower Tm.

Films of **2a** were cast from chloroform and were translucent to opaque. The films were very tough and displayed a high modulus, in the 6-8 GPa range as measured by dynamic mechanical thermal analysis (DMTA). Samples could also be prepared by compression molding, however, some degradation occurred when the polymer powder was heated in the press above the Tm and in an air atmosphere. The resulting thermoformed sample displayed a drop in the intrinsic viscosity from 0.70 to 0.48 dL/g, relative to the polymer powder.
Block copolymers **5a-c** displayed a heterogeneous morphology consistent with the formation of discrete PO domains (Table 2). DSC scans of copolymers **5b** and **5c** displayed transitions at -70 °C corresponding to the Tg of a discrete PO phase. The presence of a discrete PO phase was observed for **5a-c** by DMTA, with a damping peak at -50 °C in all cases. Copolymer **5a**, with 25 wt.% PO, showed only a slight decrease in Tg and Tm, and maintained a sharp melting transition, indicative of little phase-mixing. A similar observation was made in the 50 wt.% PO copolymer, **5b**, however, some depression in Tg and Tm were observed. In the case of the 75 wt.% PO copolymer, **5c**, the melting transition was very broad and much lower in magnitude. Due to the relatively low PO molecular weight of 4,000, the theoretical molecular weight of the polycarbonate block is 1,000, which is probably too low and dilute to crystallize. The presence of crystallinity was important for good mechanical properties, with tough films obtainable from **5a** and **5b**, and very weak films from **5c**. The poor mechanical properties observed in the case of **5c** appears to preclude the generation of elastomers with these materials.

The thermal degradation of **5a-c** was followed by dynamic TGA and showed a two-stage decomposition. The polycarbonate block decomposed at 280 °C, while the PO block decomposed at 350 °C (Tables 2 and 3). Since the decomposition points of the blocks were resolvable, this provided a measure of copolymer composition

SUMMARY

The polycondensation of trans-1,4-bis(chlorooxycarbonyl)cyclohexane with diols was found to be a useful method of synthesizing high molecular weight aliphatic polycarbonates (**2**). The polymers derived from primary diols were highly crystalline with melting points up to 205 °C. The method was extended to block copolymers with poly(propylene oxide), which displayed a heterogeneous morphology with discrete PO domains. The expected clean thermal degradation was observed for all homopolymers and copolymers.

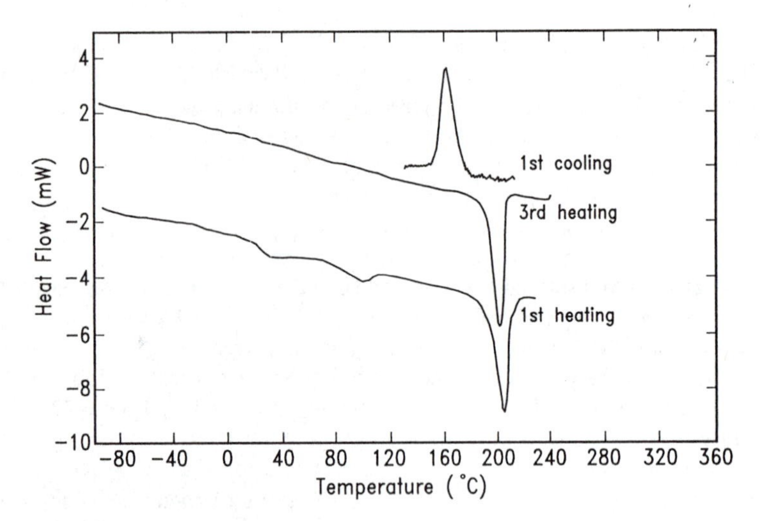

Figure 1. DSC trace for polymer **2a**.

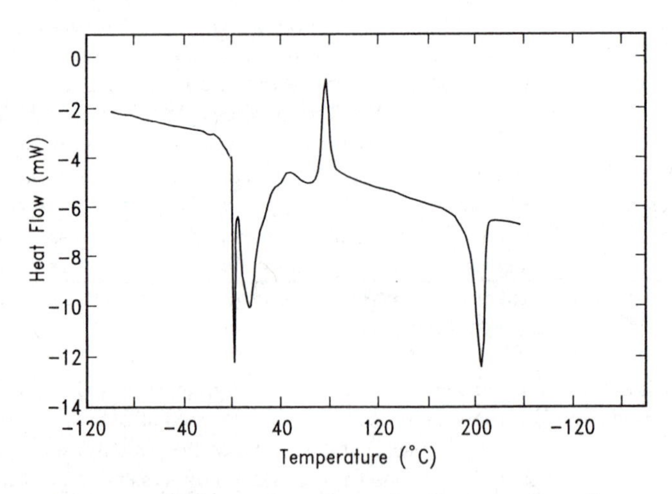

Figure 2. DSC trace for polymer **2a** after quench-cooling.

REFERENCES

1. *Encyclopedia of Polymer Science and Engineering*, Mark, H.F; Bikales, N.M.; Overberger, C.G.; Menges, G., Ed; Wiley-Interscience, Vol. 11, p.648.

2. Nufer; R.W. *Proc. of the 42nd Elec. Comp. Tech. Conf.*, **1992**, 673.

3. Houlihan, F.M.; Bouchard, F.; Frechet, J.M.J.; Willson, C.G . *Macromolecules*, **1986**, *19*, 13.

4. Frechet, J.M.J; Eichler, E.; Stanciulescu, M.; Iizawa, T.; Bouchard, F. ; Houlihan, F.M.; C.G. Willson in *Polymers for High Technology*; ACS Symposium Series, No. 346; Bowden, M.J., Turner, S., Eds; American Chemical Society: Washington D.C., 1987, pp 138-148.

5.(a) Frechet, J.M.J; Ito, H.; Willson, C.G. *Proc. Microcircuit Eng.*, **1982**, 260.(b) Ito, H. in *Radiation Curing in Polymer Science and Technology*, Fouassier, J.P.; Rabek, J.F., eds, Elsevier, London, **1993**, vol.4, ch 11.(c) MacDaonald, S.A.; Willson, C.G.; Frechet, J.M.J. *Acc. Chem Res.*, **1994**, *27*, 151.

6. Schnell, H.; *Ind. Eng. Chem.*, **1959**, *51*, 157 .

7. Wegner, G.; Nakabayashi, N.; Duncan, S.; Cassidy, H.G. *J. Polym. Sci., Part A-1*, **1968**, *6*, 3395.

8. Carothers, W. H.; Van Natta, F.J. *J. Amer. Chem. Soc.*, **1930**, *52*, 314.

9. Gawlak, H.; Palmer, R.P.; Rose, J.B.; Sandiford, D.J.; Turner-Jones, A. *Chem. Ind (London)*, **1962**, *25*, 1148.

10. Frechet, J.M.J.; Bouchard, F.; Houlihan, F.M.; Eichler, E.; Kryczka, B., Willson, C.G. *Macromol. Chem., Rapid Commun.*, **1986**, *7*, 121.

11. Shild, H.G.; Horner, M.G. *J.M.S.-Pure Appl. Chem.*, **1994**, *A31*, 1955.

12. Sarel, S.; Poholoryles; L.A. *J. Amer. Chem. Soc.*, **1958**, *80*, 4596.

13. Kuhling, S.; Keul, H.; Hocker, H.; Buysch, H.J.; Schon, N.; Leitz E. *Macromolecules*, **1991**, *24*, 4229.

14. Kricheldorf, H.R.; Dunsing, R.; Albet, A.S. *Makromol. Chem.*, **1987**, *188*, 2453.

15. Hovestadt, W.; Muller, A.J.; Keul, H.; Hocker, H. *Makromol. Chem.Rapid Commun.*, **1990**, *11*, 271.

16. Kricheldorf, H.R.; Jenssen, J.; Kreiser-Saunders, I. *Makromol. Chem*, **1991**, *192*, 2391.

17. Albertsson, A.-C.; Sjoling, M. *J.M.S.-Pure Appl. Chem.*, **1992**, *A29*, 43.

18. Sakai, S.; Fujinami, T.; Sukarai, S. *J. Polym. Sci., Polm. Lett.*, **1973**, *11*, 631.

19. Endo, T.; Bailey, W.J. *J. Polym. Sci.: Polym.Chem Ed.*, **1975**, *13*, 2525.

20.(a) Inoue, S.; Koinuma, H.; Tsuruta, T. *Makromol. Chem.* **1969**, *130*, 210. (b) US Patents 3,585,168; 3,900,424; 3,953,383.

21. Kuran, W.; Pasynkiewicz, S.; Skupinska, J.; Rokicki, A. *Makromol. Chem.*, **1976**, *177*, 11.

22. Aida, T.; Ishikawa, M.; Inoue, S. *Macromolecules*, **1986**, *19*, 8.

23. Inoue, S.; *J. Macromol. Sci.-Chem.*, **1979**, *A13*, 651.

24. Frechet, J.M.J.; Houlihan, F.M.; Willson, C.G. *Proc. ACS Div. Polym. Matls.: Sci. and Eng.*, **1985**, *53*, 268.

25. Olberg, R.C.; Pines, H.; Ipatieff, V.N. *J. Amer. Chem. Soc.*, **1944**, *66*, 1096.

26. Matzner, M.; Kurkjy, R.P.; Cotter, R.J. *Chem Rev*, **1965**, 645

27. Daly, W.H.; Hahn, B.R. *Polymer Preprints*, **1989**, *30(1)*, 337

28.(a)Polk, M.B.; Bota, K.B.; Akubuiro, E.C.; Phingbodhipakkiya, M. *Macromolecules*, **1981**, *14*, 1626. (b) Kwolek, S.L.; Luise, R.R. *Macromolecules*, **1986**, *19*, 1789.

29. An analogous series of polyesters appears in the *Encyclopedia of Polymer Sciencean Engineering*, Mark, H.F; Bikales, N.M.; Overberger, C.G.; Menges, G., Ed; Wiley-Interscience, Vol. 12., p.1.

RECEIVED January 25, 1996

Chapter 23

Polyurethanes Based on Fluorinated Diols

Tai Ho[1,7], Aslam A. Malik[2], Kenneth J. Wynne[3], Thomas J. McCarthy[4], Kent H.-Z. Zhuang[5], Kurt Baum[6], and Robert V. Honeychuck[1]

[1]Department of Chemistry, George Mason University, Fairfax, VA 22030
[2]Chemical Products, GenCorp Aerojet, Sacramento, CA 95813
[3]Physical Sciences S&T Division, Office of Naval Research, Arlington, VA 22217–5660 and Materials Chemistry Branch, Naval Research Laboratory, Washington, DC 20375–5320
[4]Department of Polymer Science and Engineering, University of Massachusetts, Amherst, MA 01003
[5]Department of Chemistry, State University of New York, Buffalo, NY 14260
[6]Fluorochem, Inc., Azusa, CA 91702

Two types of fluorinated polyurethanes were prepared, and their thermal and surface properties were evaluated. Polyurethanes containing fluorinated segments in the backbone were synthesized based on ethylene-fluoroalkyl-ethylene diols and 1,6-hexamethylene diisocyanate (HDI). Polyurethanes containing fluorinated side chains were prepared from a polyfluorooxetane prepolymer, methylene bis(phenylene isocyanate) (MDI) or isophorone diisocyanate (IPDI), and using benzene dimethanol (BDM) as the chain extender. Both types of polyurethanes show low glass transition temperatures and low surface energies (i.e. high advancing contact angles of water.)

This research is part of an effort directed toward understanding properties required for a "minimally adhesive polymer surface", i.e., one which resists settlement and adhesion of marine organisms. Initially, we have directed our efforts toward polymers which have low surface energies and low glass transition temperatures (T_g).(*1*) To achieve the goal of low T_g, we focussed on polyurethanes based on polyethers. To achieve the goal of low surface energy, we explored polyethers with either fluorinated backbone segments or fluorinated side chains.

[7]Current address: Code 6120, Naval Research Laboratory, 4555 Overlook Avenue, SW, Washington, DC 20375–5320

0097–6156/96/0624–0362$12.00/0

A family of fluorinated diols of the formula $HOC_2H_4\text{-}(CF_2CF_3CF)_x(CF_2)_4\text{-}(CF_2CFCF_3)_y\text{-}C_2H_4OH$ synthesized by Baum and Malik(*2*) presented an opportunity to synthesize polyurethanes containing fluorinated segments in the backbone. Previously, fluorinated diisocyanates,(*3*) diols and diamines(*4 - 6*) have been incorporated into polyurethanes to enhance environmental stability and antithrombogenicity of the materials. Polymerization reactions between members of this family of diols and hexamethylene diisocyanate (HDI) were carried out (Scheme I). HDI was chosen for study because polyurethanes based on aliphatic diisocyanates have been known to exhibit good resistance to weathering, chemicals and abrasion. (*7*)

$HOCH_2CH_2(CF_2CF(CF_3))_x(CF_2)_4(CF_2CF(CF_3))_yCH_2CH2\ OH\ +\ OCN\text{-}(CH_2)_6\text{-}NCO$ --------->

1a, C_7 diol, x + y = 1 **2**
1b, C_9 diol, x + y = 1.67
1c, C_{11} diol, x + y = 2.47

$\text{-}[OCH_2CH_2(CF_2CF(CF_3))_x(CF_2)_4(CF_2CF(CF_3))_yCH_2CH_2O\text{-}OCNH\text{-}(CH_2)_6\text{-}NHCO]_n\text{-}$

3, x + y = 1, polymerized in bulk
4, x + y = 1.67, polymerized in THF
5, x + y = 1.67, polymerized in bulk
6, x + y = 2.47, polymerized in THF
7, x + y = 2.47, polymerized in bulk

Scheme I

Oligomers of fluorooxetanes were synthesized by Malik, et al. (*8*) As shown in Scheme II, thermoplastic polyurethanes elastomers were prepared by reacting these prepolymers with methylene bis(phenylene isocyanate) (MDI) or isophorone diisocyanate (IPDI) and using benzene dimethanol (BDM) as the chain extender .

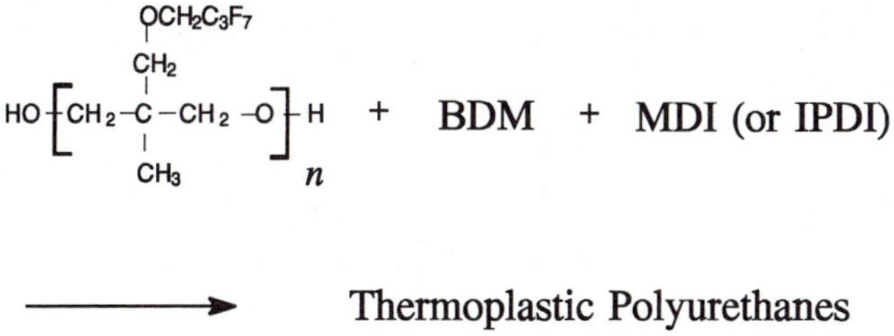

$HO\text{-}[CH_2\text{-}C(CH_3)(CH_2OCH_2C_3F_7)\text{-}CH_2\text{-}O]_n\text{-}H$ + BDM + MDI (or IPDI)

⟶ Thermoplastic Polyurethanes

Scheme II

Thermoset polyurethanes elastomers were prepared by reacting the prepolymers with a diisocyanate and a crosslinking agent, such as trimethylolpropane (Scheme III).

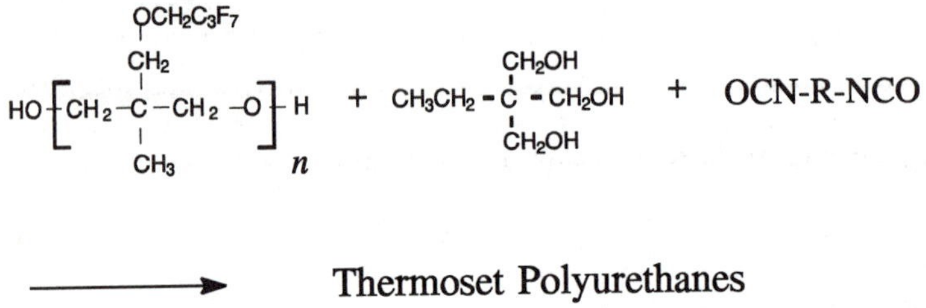

$$HO\left[CH_2-\underset{CH_3}{\overset{OCH_2C_3F_7 \atop |\atop CH_2}{C}}-CH_2-O\right]_n H \; + \; CH_3CH_2-\underset{CH_2OH}{\overset{CH_2OH}{C}}-CH_2OH \; + \; OCN\text{-}R\text{-}NCO$$

$$\longrightarrow \quad \text{Thermoset Polyurethanes}$$

Scheme III

Thermal and surface properties of the resulting polyurethanes were characterized. The polymers all meet the requirement of low T_g and high advancing contact angles of water.

EXPERIMENTAL

Materials. Fluorinated diols, $HOC_2H_4(CF_2CF_3CF)_x(CF_2)_4(CF_2CFCF_3)_yC_2H_4OH$, were derived from the successive addition of hexafluoropropene to the perfluorobutane diiodide starting material, followed by addition of ethylene to the terminal carbon-iodine bonds and conversion of C-I functionality to alcohol.(2)

Diols **1a**, **1b** and **1c**, molecular weight 440, 540 and 660, respectively, were prepared at Fluorochem, Inc., Azusa, CA. We refer to the diols as C_7-, C_9-, and C_{11}- because these designations represent the average number of carbons bearing fluorines. HDI was MONDUR® HX from Mobay Chemical. Diols was used as received; HDI was distilled under vacuum before use.

The synthesis of oxetanes substituted at the 3-position with fluoroalkyloxy groups such as $CF_3(CF_2)_xCH_2OCH_2$- , where x=0,1,2 and 6, has been reported elsewhere.(8) These monomers were polymerized in the presence of a Lewis acid catalyst and an alcohol initiator to give the corresponding hydroxy-terminated polyether prepolymer with fluoroalkyloxy side chains in 90-95% yields.

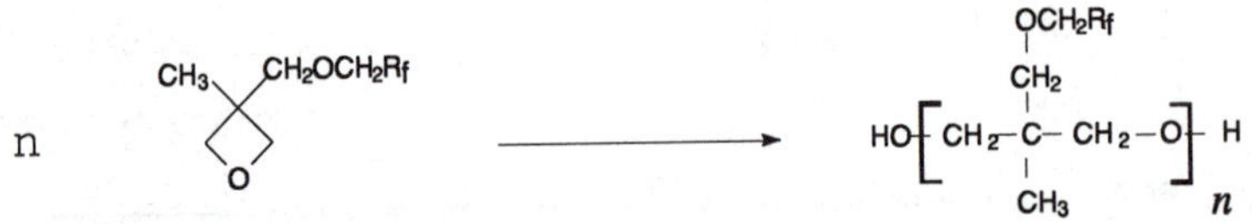

$$n \; \text{(3-}CH_3\text{-3-}CH_2OCH_2R_f\text{ oxetane)} \longrightarrow HO\left[CH_2-\underset{CH_3}{\overset{OCH_2R_f \atop |\atop CH_2}{C}}-CH_2-O\right]_n H$$

These prepolymers were colorless oils that exibited number average molecular weights in the range of 2000 to 20,000, and polydispersities 1.2 to 1.6. The molecular weight of the prepolymer was controlled by factors such as monomer/initiator ratio, reaction temperature, concentration, and the reactivity of the monomer. The equivalent weight of the prepolymer was determined by ^{1}H NMR employing trifluoroacetic acid anhydride (TFAA) and trichloroacetyl isocyanate end group analyses. DSC analysis of the prepolymer revealed the glass transition temperature of the polyether backbone at ca. -45°C. Only a minor dependence of glass transition temperature on the length of the side chain was observed.

The prepolymer used in later syntheses was based on monomers with x=2. The number average molecular weight by gel permeation chromatography (GPC) using polystyrene reference was 20,000 and by NMR was 18,000. (*9*) This prepolymer is designated 7-Fox20K. MDI (Kodak) was distilled under reduced pressure and stored in refrigerator before use. IPDI, BDM, tetrahydrofuran (THF), N,N-dimethyl-acetamide (DMAc) and methanol were purchased from Aldrich and used as received.

Bulk Polymerization. Absolute viscosities of the C_7, C_9, and C_{11} diols at temperatures from 24 to 66 °C were measured to facilitate the selection of the method and conditions of polymerization. The viscosities of the diols exhibit no dependence on the spin rate (the effective shear rate was from 0.046 to 4.6 sec^{-1}) and show an Arrhenius-type dependence on temperature. Correlations between viscosity and the absolute temperature were found using linear regression techniques. They are:

for C_7 diol μ (poise) = $13.9\ e^{7449\,(1/T\ -\ 1/297)}$

for C_9 diol μ (poise) = $41.7\ e^{8588\,(1/T\ -\ 1/298)}$

and for C_{11} diol μ (poise) = $214\ e^{9523\,(1/T\ -\ 1/297.5)}$

The mixing of reactants and catalyst in this investigation was achieved using a magnetic stirrer (Corning Hot Plate Stirrer PC-351). Empirically, it was found that three poise was the upper limit of viscosity for the stirrer to function well. This limit set a threshold temperature for each diol for magnetically stirred bulk polymerizations. Using the above correlations, the threshold temperatures are 43 °C for C_7 diol, 55 °C for C_9 diol, and 70 °C for C_{11} diol.

Mixtures of C_7 diol, **1a**, and HDI exhibit good fluidity and mild reactivity at temperatures between 50 and 70 °C, and the reaction carried out in bulk for 18 hours produced polymers soluble in THF. On the other hand, mixtures of C_9 diol, **1b**, and HDI, and C_{11} diol, **1c**, and HDI are more viscous at those temperatures. To achieve the same level of fluidity in the initial reaction mixture, the reaction was carried out at higher temperatures, 70°C for C_9 diol and 90°C for C_{11} diol. In these systems, bulk polymerization resulted in a mixture of linear polymers and gel after 18 hours. THF was then added into the mixture to extract the soluble linear polymer. The separation of the soluble and insoluble portions was achieved by decantation, and the soluble polymers were recovered as cast films. These latter two polymerizations were also carried out in THF solution.

Solution Polymerization. Fluorinated diols **1b** and **1c** react with hexamethylene diisocyanate in THF in the presence of dibutyltin dilaurate catalyst to give polyurethanes having structures **4** and **6** (Scheme 1). In a typical synthesis, 1.7×10^{-6} mol of dibutyltin dilaurate was added to a solution of 1.91×10^{-3} mol of hexamethylene diisocyanate (HDI, **2**) in 0.5 mL of THF. A solution of 1.250 g of **1c** in 1.5 mL of THF was then added. The solution was stirred for 2 hours at 25°C, after which time the solution was very viscous. THF (22 mL) was added and the mixture was heated at 68°C. The soluble fraction was decanted from the insoluble fraction. The soluble fraction was poured into 500 mL of methanol, yielding a white precipitate. After filtration and removal of solvent *in vacuo*, the ratio of the weight of the soluble fraction to that of the insoluble was 6.6. The combined yield of these fractions of **6** was 59%. Less than complete turnover, which is a general phenomenon in these solution polymerizations (the range is 49 - 89%), could conceivably be attributed to biuret and allophanate formation, crosslinking, and subsequent non-precipitation of unreacted diol or low molecular weight polymer. (*10*) The catalyst is necessary for reasonable reaction time in THF solution. An experiment without it resulted in no methanol-precipitated material after seven days using **1c** and **2**.

Copolymerization of 7-Fox20K Prepolymer. The polymerization reaction involving prepolymer 7-Fox20K , MDI (or IPDI), and BDM is described in Scheme II. In a typical run, 4.09g ($2.1x10^{-4}$ mol) of 7-Fox20K prepolymer was introduced into a one-neck flask equipped with a magnetic stirrer bar, and the material was degassed in a vacuum oven at 50°C over night. At that point, $5.795x10^{-2}$ g ($4.2x10^{-4}$ mol) of BDM in 2 ml THF was added, and the flask was sealed with a rubber septum. The contents were stirred and heated to 60-65°C under nitrogen for one hour, then 10 μl (~0.2%)of dibutyltin dilaurate (DBTDL) was injected into the vessel with a syringe. Solid MDI, 0.161 g ($6.4x10^{-4}$ mol) was added immediately by removing the septum momentarily. The reaction proceeded at 60-67°C for 24 hours. The products, being white elastomers, were removed from the flask by dispersing with THF. The products form white suspension in THF, but dissolve completely in THF/DMAc (4:1) when the solid concentration was less than 1%. The polymer was reprecipitated by pouring the solution into excess methanol. The yield was 91%. This copolymer is designated 7Fox20K-M-B2, where 7Fox20K indicates the composition and molecular weight of the soft segment, M stands for MDI, B for BDM, and 2 represents the molar ratio between BDM and 7Fox20K prepolymer.

When IPDI was used instead of MDI, the products were also white elastomers, but were soluble in THF. The resulting copolymer is designated as 7Fox20K-IP-B2 in the same manner.

Thermoset polyurethane elastomers were prepared by reacting 7Fox20K prepolymers with a diisocyanate such as IPDI or hydrogenated MDI in the presence of a crosslinking agent such as trimethylolpropane or Jeffamine. The reactions were conducted at 65°C in the presence of a catalyst such as dibutyltin dilaurate or a 1:1 mixture of dibutyltin dilaurate and DABCO.

Characterization. Absolute viscosities of the diols were measured with a Brookfield viscometer (model RVTD, spindle no. 4). The viscometer was calibrated with silicone oil viscosity standards supplied by Brookfield Engineering Laboratories, Inc. The diol was contained in a test tube, I.D.= 7.25 mm, and temperature control was achieved through a water bath.

Molecular weight was determined by GPC. The equipment consisted of a Hewlett-Packard Series 1050 pump and two Altex μ-spherogel® columns (size 10^3 and 10^4 Å, respectively) connected in series. Polymer in the effluent was detected with a Wyatt/Optilab 903 interferometric refractometer, and the average molecular weights were determined using polystyrene standards as references. The mobile phase was THF or THF/DMAc 4/1 mixture, and the concentration of the polymer was usually 1 % by weight.

A Perkin-Elmer DSC 7 instrument was used to measure thermal properties. The scanning rate was 10°C/min. Samples were inserted at room temperature, taken immediately to -50°C, scanned to 200°C, quenched (200°C/min) to -50°C, and re-scanned.

Infrared spectra were collected on a Perkin-Elmer 1800 FTIR instrument on KBr plates, using thin films obtained via evaporation of THF solutions.

NMR spectra were obtained on a Bruker AMX 400 spectrometer operating at 400.13 MHz for ^1H acquisition. A 5 second receiver delay was employed and a 30° pulse width was used. All samples were prepared as solutions in chloroform-d, and the spectra were referenced to residual $CHCl_3$.

Advancing contact angle were measured either by goniometry or Wilhelmy plate method. The former was carried out with an NRL A-100 Contact Angle Goniometer (commercially available through Ramé-Hart, Inc). Sample films were prepared from 10 wt% solution in THF, cast into aluminum pans lined with Teflon, and dried in a hood for one week. The films were kept under inverted beakers in the hood to reduce contact with contaminants. Only the surface formed at the air-solution interface was examined. The Wilhelmy plate method was performed with a Cahn dynamic contact angle analyzer (DCA 312).

RESULTS AND DISCUSSION

Polyurethanes Containing Fluorinated Segments in Backbone.

Polymerization Reaction Kinetics. Infrared spectra of the reaction mixture during polymerization reactions were taken, and the isocyanate absorption band centered at 2275 cm^{-1} was monitored to determine the extent of the reaction. The concentration of the isocyanate group is plotted as a function of time, the data suggest a first order reaction at the initial stage, but the apparent rate of reaction decreases significantly during the second hour (Figure 1). Since there is no large disparity between the reactivities of two isocyanate groups separated by an aliphatic chain, (*11*) the decrease in reaction rate is likely due to the swift build-up of viscosity concurrent with polymerization. The rate constant, k, during the first hour of reaction at 58°C was 6.51 x 10^{-4} sec^{-1}.

In order to determine the activation energy, the rate constant at room temperature was determined. A small quantity of reaction mixture (58°C) was withdrawn thirty minutes after mixing and coated onto a KBr window. The specimen was then kept at 22°C and the extent of reaction monitored by FTIR spectroscopy. The sample remained optically clear throughout the measurement. The rate constant at room temperature (22°C) was 0.41 x 10^{-4} sec^{-1}, and the activation energy 14.1 kcal $mole^{-1}$. These data may be compared with that obtained by Wong and Frisch for three aliphatic diisocyanates: trans-cyclohexane-1,4-diisocyanate (CHDI), isophorone diisocyanate (IPDI), and 4,4'-methylene bis(cyclohexyl) diisocyanate (HMDI). (*12*) In a model reaction with n-butanol (in toluene, 50° C, dibutyltin dilaurate catalyst) a second order reaction was found up to 50% conversion with rate constants of 1.40, 0.73, 0.64 l $mole^{-1}min^{-1}$ for CHDI, IPDI, and HMDI, respectively. The activation energy for CHDI was determined to be 17.78 kcal $mole^{-1}$. The initial rate constant obtained in the present study in the absence of solvent was 6.51 x $10^{-4}sec^{-1}$. This implies a half time at 17.7 minutes, which is comparable to Wong and Frisch's data (11.9 to 26.0 minutes); The activation energy for the reaction of **1** with HDI is similar to that reported by Wong and Frisch for CHDI with butanol.

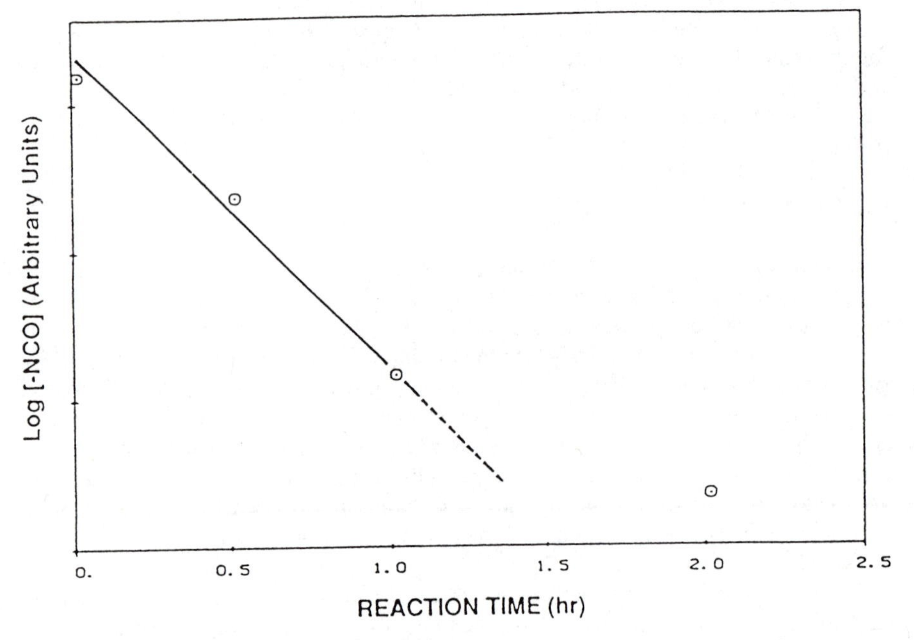

Figure 1. Concentration of the isocyanate group as a function of time during the first two hours of a polymerization reaction between C--diol and HDI at 58°C. The straight line is drawn to guide the eye. Reproduced with permission from reference 1(b).

Proton NMR spectra for the monomers **1,2** and polymers **3-7** indicated with the formation of polymers, the OH peak in the diol monomer at 1.76 ppm disappears, and is replaced by a new peak at 4.91 ppm assigned to NH, (Figure 2). Peaks due to the methylene groups adjacent to oxygen in the diol moved down field by about 0.3 ppm, while methylene protons adjacent to nitrogen in the diisocyanate shifted upfield by ca. 0.15 ppm.

The number average (M_n) and weight average (M_w) molecular weights of the current series of polymers were estimated utilizing gel permeation chromatography (GPC). Since the specific refractive increments (dn/dc) of these polymers in THF solution are very low (<0.01 ml/g), (*13*) high concentration solutions of the polymers (about 2-3 wt%) were used to enhance the signals, and the resulting chromatograms were smoothed by the method of Fourier transform convolution. (*14*) Molecular weights were determined using polystyrene standards as references. Typical values for the weight and number average molecular weights of these polymers, (Table I and Figure 3), showed the following trends. The weight fraction distributions for polymers based on C_7 diol and HDI, exemplified by curve A, Figure 3, indicate that both the molecular weight and polydispersity of the polymers are in satisfactory ranges. The weight fraction distribution curves for polymers based on C_9 diol and HDI, curves B and C, however, represent only the lower ends of the ranges of the molecular weights. This is because gelation occurred during polymerization, and a portion of the product, being crosslinked, was not amenable to GPC measurements. Curves D and E in Figure 3 are weight fraction distributions for polymers based on C_{11} diol and HDI. The curves indicate these polymers have polydispersities at the expected level ($M_w/M_n \approx 2$) and molecular weights somewhat less than polymers based on C_7 diol and HDI. The lower molecular weight is a bias caused by gelation as in the case of polymers based on C_9 diol and HDI. Comparing curves B and C with D and E, one can notice that in polymers based on C_9 diol and HDI, crosslinking involves almost all molecules of molecular weight more than 40,000, while in polymers based on C_{11} diol and HDI, molecules of comparable molecular weight do not always crosslink. The reasons for this lower threshold for crosslinking in the former are not known. On the other hand, molecular weights of these two types of polymers both appear to be of multi-mode distribution. This type of distribution could occur if during the polymerization the catalyst was not uniformly distributed and the reaction proceeded locally at different rates. The localization of reaction could be caused by high viscosity of the diols exacerbated by the rapid progress of the polymerization reaction. Alternatively, the multi-mode distribution could be due to the existence of branched products. Under the reaction conditions, partial trimerization of HDI could occur(7), which would lead to the formation of branched products. We did not investigate these two possibilities further.

Thermal transitions were revealed in DSC traces, (Figure 4). Polymers based on C_7 diol and HDI, **3**, show interesting thermal behavior. The glass transition region, between 10 and 25 °C, appears diminished compared with the two melting peaks. The melting peaks, located at about 46 and 67 °C, respectively, exhibit time-dependent variations in position and height. The distance between these two peaks decreases with time, while the total area under the peaks increases with time. A detailed analysis of the thermal properties of these polymers has been reported separately.(1(b)) DSC traces for polymers based on C_9 diol and HDI, **4** and **5**, and C_{11} diol and HDI, **6** and **7**, indicate

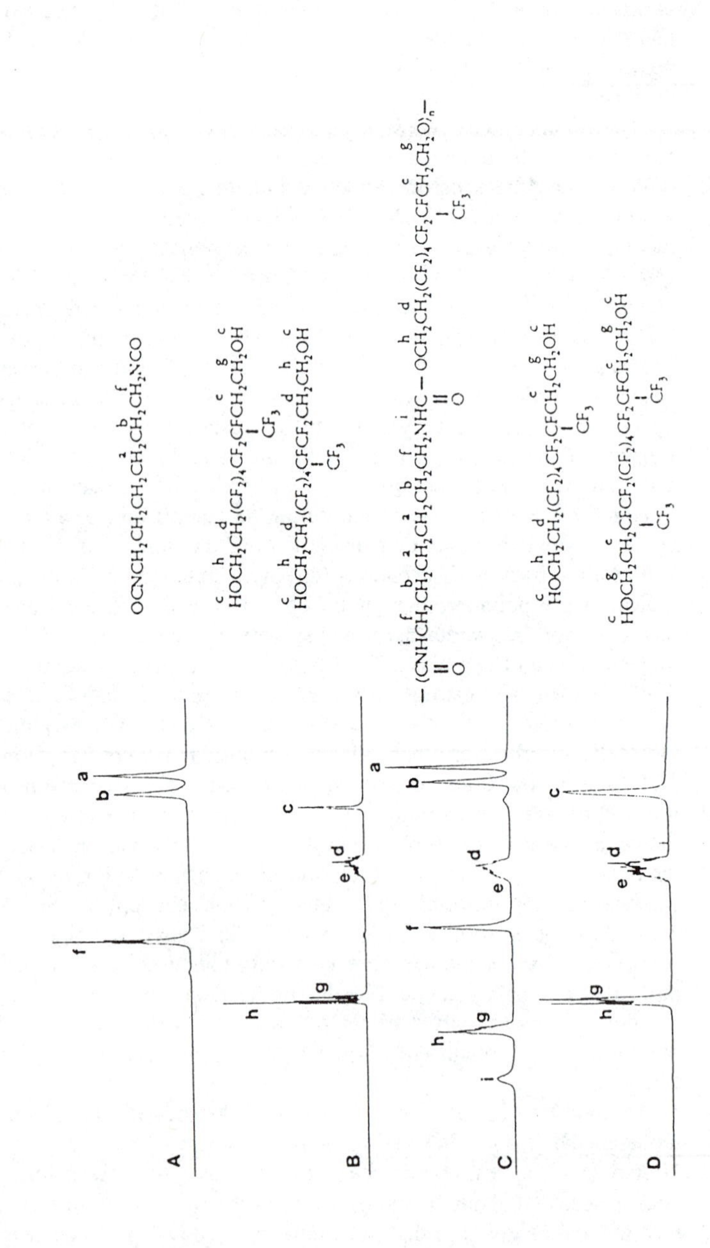
A
B
C
D
OCNCH2CH2CH2CH2CH2CH2NCO
HOCH2CH2(CF2)4CF2CFCH2CH2OH
CF3
HOCH2CH2(CF2)4CFCF2CH2CH2OH
CF3
—(CNHCH2CH2CH2CH2CH2CH2NHC — OCH2CH2(CF2)4CF2CFCH2CH2O)n—
O
O
CF3
HOCH2CH2(CF2)4CF2CFCH2CH2OH
CF3
HOCH2CH2CFCF2(CF2)4CF2CFCH2CH2OH
CF3
CF3

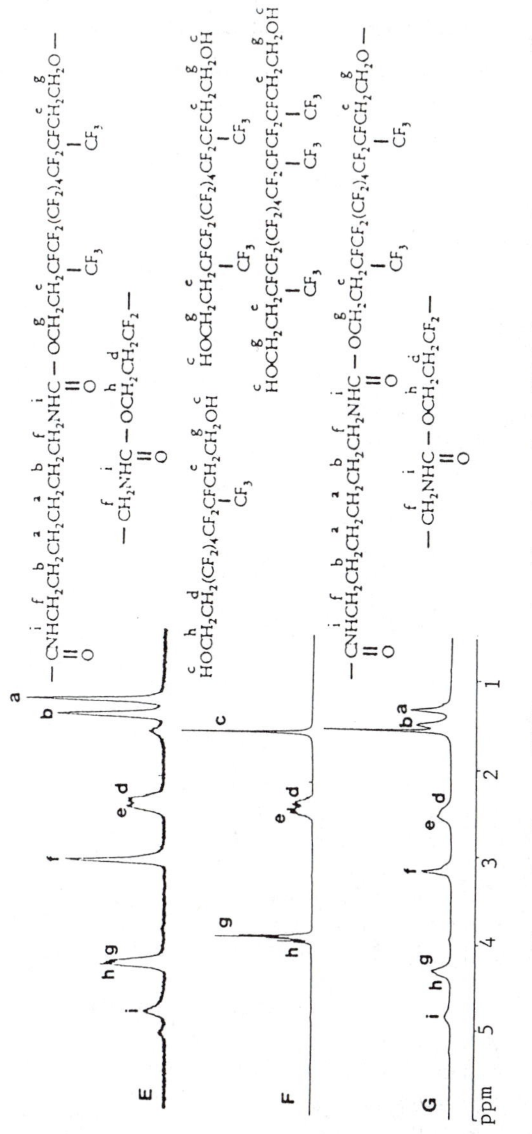

Figure 2. Proton NMR spectra in chloroform-d: A, HDI (2); B, C- diol (1a); C, bulk C-polyurethane (3); D, C_9 diol (1b); E, bulk C_9 polyurethane (5); F, C_{11} diol (1c); G, bulk C_{11} polyurethane (7). Reproduced with permission from reference 1(c).

Table I Typical values of glass transition temperatures and average molecular weights for polymers **3-7**. Reproduced with permission from reference 1(c).

Polymer	T_g, °C	M_w	M_n	M_w/M_n
3	18.2	54,300	34,300	1.6
4	22.0	11,900	9800	1.2
5	18.7	20,000	16,600	1.2
6	26.5	43,000	22,300	1.9
7	20.4	39,400	21,800	1.8

T_g measured with DSC, molecular weights with GPC.

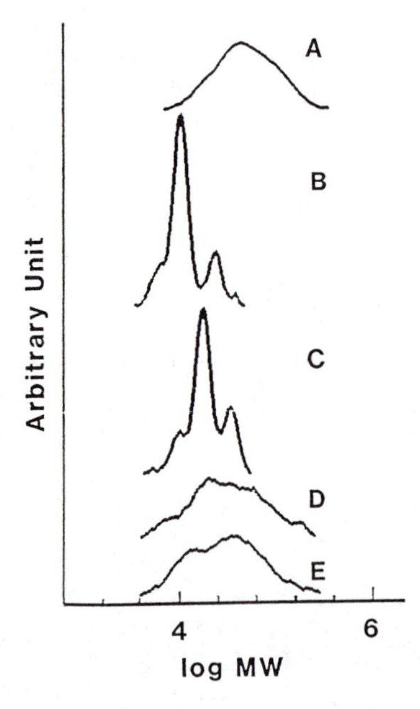

Figure 3. Weight fraction distribution curves, derived from GPC chromatograms, for polymers: A, polyurethane based on C- diol and HDI, polymerized in bulk; B, polyurethane based on C_9 diol and HDI, polymerized in THF solution; C, polyurethane based on C_9 diol and HDI, polymerized in bulk; D, polyurethane based on C_{11} diol and HDI, polymerized in THF solution; and E, polyurethane based on C_{11} diol and HDI, polymerized in bulk. Reproduced with permission from reference 1(c).

no melting phenomena, and the glass transitions for these polymers occur at temperatures between 15 and 30 °C, (Table I).

Advancing contact angles of water and methylene iodide on surfaces of these polymers were measured by goniometry, and surface free energies were calculated using the method suggested by Owens and Wendt. (*15*) By this method, the surface free energy for polyethylene is 33.2 erg/cm^2, polytetrafluoroethylene, 19.1 erg/cm^2, and polyhexafluoropropylene, 12.8 erg/cm^2. (*16*) Polar segment containing polymers, such as poly(methyl methacrylate) and poly(hexamethylene adipamide), have surface free energies higher than 40 erg/cm^2.(15) Surface free energies of the fluorinated polyurethanes reported herein range between 25.5 and 31.5 erg/cm^2. These data are consistent with surfaces consisting of both hydrocarbon and perfluorohydrocarbon segments. Ratner and coworkers have studied the surface compositions of polyurea-urethanes containing fluorinated chain extenders with angle resolved X-ray photoelectron spectroscopy (XPS, also known as ESCA).(4,*17*) They did not observe surface enrichment of fluorine in their systems except when the chain extender was a perfluoroether oligomer.

The higher surface free energy value of 31.5 erg/cm^2, is associated with polymers based on C_7 diol with a fluorocarbon content at 58 wt%. When the fluorocarbon content increases to 64 wt% as in polymers based on C_9 diol the surface free energy decreases to 25.5 erg/cm^2. Further increase in fluorocarbon content (C_{11} diol polymer **7**) does not result in decrease in surface free energy.

Polyurethanes Containing Fluorinated Side Chains.

Average molecular weights determined by GPC for a typical batch of the copolymer 7Fox20K-M-B2 were Mw=196,000 and Mn= 122,000 based on polystyrene standards, while values for 7Fox20K-IP-B2 were Mw=131,000 and Mn=101,000.

DSC traces for 7Fox20K-M-B2 showed a complicated pattern for the endothermic peaks in the first heating-cooling cycle. There is a single peak at -42°C which reflects the glass transition of the 7-Fox soft segment. A series of peaks located at temperatures ranging from 90 to 170°C presumably due to the various types of aggregations of the hard segments. This multi-peak phenomenon occurring in the glass transition region for the hard segment of MDI based polyurethanes has been discussed extensively in the literature.(*18*) In the second cycle, the low temperature peak remains at the same location and intensity, but the high temperature peaks consolidated to a shoulder at 90°C.

DSC traces for 7Fox20K-IP-B2 showed the peak at -42°C that reflects the glass transition of the 7-Fox soft segments. But there was no peak detected that corresponds to the glass transition of the hard segments. We had studied polyurethanes of similar structures that contains polydimethylsiloxane soft segments and IPDI-BDM hard segments. In that materials, we did not detect T_g peak for the hard segments by DSC, either. However, clear evidence for separate domains were shown by dynamic mechanical analysis(*19*) and solid state NMR.(*20*)

The most convenient solvent for copolymer 7Fox20K-M-B2 is the 4:1 mixture of THF and DMAc. THF is a good solvent for the soft segment, while DMAc is a solvent for the hard segment. Films cast from solutions containing about 1% polymers tend to have segregated domains resulting in visible nodular structures on the surface. Reducing the molar ratio of the chain extender to 7-Fox oligomer from 2 to 1 led to films of improved quality, however the resulting copolymer was tackier than 7Fox20K-M-B2.

The IPDI-based polyurethanes have the advantage of being soluble in THF and can be cast into good quality films easily.

Copolymers 7Fox20K-M-B2 and 7Fox20K-IP-B2 were dip-coated on cover glasses for optical microscopes. Contact angles of water on such surface were measured with Cahn DCA. The advancing contact angle of water on freshly prepared copolymer films was about 110° , which is comparable to that for Teflon® (ca. 109°). The same contact angle reduced to 95° after the films were immersed in water for two months.

The thermoset polyurethanes were tack-free elastomers that exhibited good resistance towards common organic solvents and inorganic bases but were not stable towards strong acids and exhibited swelling in tetrahydrofuran and Freon. The glass transition temperatures were in the neighborhood of -45°C, whereas the static contact angle of the polymer surface with distilled water was 110°. Despite the low surface energy, these fluorinated polyurethanes exhibited excellent adhesion to aluminum, steel, and EPDM rubber. An explanation for the unusual behavior is that in fluorinated polyurethanes, non-polar fluorinated side chains phase separate and orient towards the polymer/air interface to provide a low energy surface, whereas the polar urethane linkages orient towards the relative high energy substrate, thus providing a strong bond between the substrate and the polymer.

ESCA Analysis. Angle resolved ESCA data for polyurethanes **3** (based on C_7-diol and HDI) and 7Fox20K-IP-B2, presented in (Table II), clearly show that the distribution of the fluorinated segments is not uniform throughout the film. Compared with the *bulk* composition, the fluorinated segments of each polyurethane contain a much higher proportion of fluorine, lower concentrations of carbon and oxygen, and no nitrogen. At high take-off angles (75° and 90°), ESCA results essentially reflect the bulk composition. At lower take-off angle (15°), the data indicate a disproportionally high presence of the fluorinated segments. Since the probing depth of ESCA increases with increasing take-off angle. We infer that the fluorinated segments distribute preferentially at the polymer/air interface. The data also indicate the surface enrichment of fluorine is more extensive when the fluorinated segments are on the side chains instead of the backbone.

CONCLUSIONS

Polyurethanes based on fluorinated diols of different structures have been prepared in connection with interests in "minimally adhesive polymer surfaces". All of the resulting polymers exhibit low T_g reflecting the flexibility of the polyether backbone. Low energy surfaces are also achieved. ESCA data indicate a preferential distribution of fluorinated segments at the polymer/air interface. Incorporating fluorinated side chains in the polymer is more effective in achieving surface enrichment of fluorine than incorporating fluorinated backbone segments.

Acknowledgements. This research was supported in part by the Office of Naval Research.

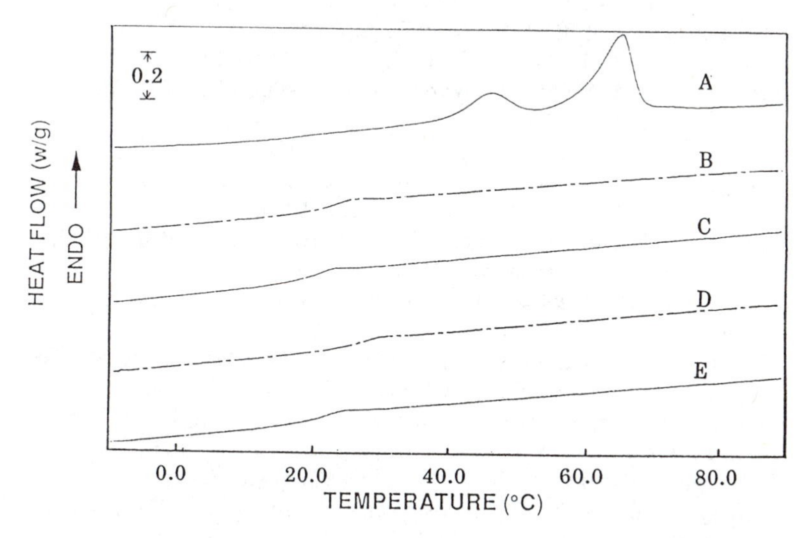

Figure 4. DSC traces for polymers: A, polyurethane based on C- diol and HDI, polymerized in bulk; B, polyurethane based on C_9 diol and HDI, polymerized in THF solution; C, polyurethane based on C_9 diol and HDI, polymerized in bulk; D, polyurethane based on C_{11} diol and HDI, polymerized in THF solution; and E, polyurethane based on C_{11} diol and HDI, polymerized in bulk. Reproduced with permission from reference 1(c).

Table II ESCA data at different take-off angles for polyurethanes C7-HDI(3) and 7Fox20K-IP-B2.

	C7-HDI(3)				7Fox20K-IP-B2			
	Bulk	FBS[%]	15°	75°	Bulk	FSC[•]	15°	90°
C_{1s}	48.7	40.7	50.2	51.1	48.5	33.3	45.1	48.8
F_{1s}	35.9	51.9	35.1	32.6	39.2	58.3	43.6	39.0
O_{1s}	10.3	7.4	10.6	11.3	11.9	8.3	11.3	12.0
N_{1s}	5.1	0.0	4.2	5.1	0.4	0.0	0.0	0.1

[%]Fluorinated backbone segments.
[•]Fluorinated side chains.

REFERENCES

1. Previous work in this field from Naval Research Laboratory includes (a) Brady, R. F.; Griffith, J. R.; Love, K. S.; Field, D. E. *J. Coatings Tech.* **1987**, *59,* 113; (b) Ho, T.; Wynne, K. J. *Macromolecules* **1992**, *25*, 3521; (c) Honeychuck, R. V.; Ho, T.; Wynne, K. J.; Nissan, R. A. *Chemistry of Materials* **1993,** *5*, 1299.
2. Baum, K.; Malik, A.A. *J. Org. Chem.* **1994**, *59*, 6804.
3. Gosnell, R.; Hollander, J. *J. Macromol. Sci., Phys.* **1967,** *B1*, 831.
4. Yoon, S. C.; Ratner, B. D. *Macromolecules* **1986,** *19*, 1068.
5. Takakura, T.; Kato, M.; Yamabe, M. *Makromol. Chem.* **1990**, *191,* 625.
6. (a) Chapman, T.M.; Benrashid, R.; Marra, K.G.; Keener, J.P. *Macromolecules* **1995**, *28*, 331; (b) Chapman, T.M.; Marra, K.G. *Macromolecules* **1995**, *28*, 2081.
7. Frisch, K.C.; Kordomenos, P. in *Applied Polymer Science*, 2nd ed.; Tess, R.W.; Poehlein, G.W., Eds.; ACS Symposium Ser. 285 1985, p 988.
8. Malik, A.A.; Harvey, W.L.; Carlson, R.P.; Wynne, K.J.; Ho, T. paper presented at 12th Winter Fluorine Conference, St. Petersburg, Florida, 1995.
9. GPC was performed at NRL, NMR at GenCorp Aerojet.
10. Munk, P. *Introduction to Macromolecular Science;* John Wiley & Sons: New York, 1989; p 111.
11. Saunders, J. H.; Frisch, K. C. *Polyurethanes: Chemistry and Technology I. Chemistry*; Interscience Publishers: New York, 1962; p 150.
12. Wong, S. W.; Frisch, K. C. In *Advances in Urethane Science and Technology*, vol. 8; Frisch, K. C., Klempner, D., Eds.; Technomic: Westport, CT, 1981; p 75.
13. This value was obtained by comparing spectra for polystyrene and the current polyurethanes at the same concentration.
14. We used programs included in *Mathematica* to perform this operation; Wolfram, S. *Mathematica: a system for doing mathematics by computer*, 2nd ed.; Addison-Wesley Publishing Company, Inc.: Redwood City, CA, 1991; p 679.
15. Owens, D.K.; Wendt, R.C. *J. Appl. Polym. Sci.* **1969**, *13*, 174.
16. Wu, S. *Polymer Interface and Adhesion;* Marcel Dekker: New York, 1982; p 170.
17. Edelman, P. G.; Castner, D. G.; Ratner, B. D. *Polym. Prepr. (Am. Chem. Soc., Div. Polym. Chem.)* **1990**, *31*, 314.
18. Some examples are: (a) Hesketh, T. R.; van Bogart, J. W. C.; Cooper, S. L. *Polym. Eng. Sci.* **1980**, 20(3), 190;(b) Blackwell, J.; Lee, C. D. *J. Polym. Sci., Polym. Phys. Ed.* **1984**, 22, 759.; (c) Leung, L. M.; Koberstein, J. T. *Macromolecules* **1986**, 19, 706; (d) Yoon, S. C.; Sung, Y. K.; Ratner, B. D. *Macromolecules* **1990**, 23, 4351.
19. Ho, T.; Wynne, K. J.; Nissan, R. *Macromolecules* **1993**, *26*, 7029.
20. Cho, G.; Natansohn, A.; Ho, T.; Wynne, K.J. to be published.

RECEIVED December 4, 1995

Chapter 24

Poly(sulfonium cation) for the Synthesis of High-Molecular-Weight Poly(phenylene sulfide)

Kimihisa Yamamoto and Eishun Tsuchida

Department of Polymer Chemistry, Waseda University, Tokyo 169, Japan

High molecular weight poly(phenylene sulfide)(PPS), Mw > 2 x 10^5, was synthesized via a poly(sulfonium cation) through oxidative polymerization of methyl 4-(phenylthio)phenyl sulfoxide. Polymerization proceeds in protic acid homogeneously. The poly(sulfonium cation) was soluble in common solvents such as acetonitrile, acetone, formic acid and dimethyl sulfoxide. The poly(sulfonium cation) was available as a soluble intermediate for the synthesis of high molecular weight poly(phenylene sulfide). It was feasible to convert high molecular weight poly(arylene sulfide) through the Sn_2 reaction with pyridine as a nucleophile. This polymerization is applicable to the synthesis of new block copolymers such as, poly(phenylene sulfide)-*block*-[poly(tetrafluoroethylene-*co*-perfluoro (propylvinyl ether)]}-*block*-poly(phenylene sulfide). The polysulfonium cation is isolatable as a stable salt which acts as an unique polymeric reagent for the methylation of aromatic compounds having mobile hydrogen.

High molecular weight poly(*p*-phenylene sulfide) (PPS) has received much attention as a high-performance engineering plastic due to its excellent chemical, thermal, and mechanical properties which are a result of its high crystallinity(*1-6*) These properties make it difficult to synthesize high molecular weight PPS. Normally, the preparation of PPS employs high-temperature and high-pressure processes in order to overcome poor solvent solubility, *e.g.* commercial preparations of PPS are often carried out at temperature in excess of the boiling point of NMP, and require the use of pressure vessels. To overcome these difficulties a large amount of effort has been put into the formation of high molecular weight poly(arylene sulfide) (PAS)(Eq. 1)(*7-15*). PPS can also be obtained by other reaction mode, *e.g.* electrophilic reaction and

0097–6156/96/0624–0377$12.00/0

oxidative polymerization(Eq. 2)(*16-21*) These polymerizations proceed at room temperature and provides pure PPS, but results in low molecular weights due to poor solubility. These polymers subsequently had to be cured in order to develop useful properties.

$$Cl-C_6H_4-Cl + Na_2S \longrightarrow -[C_6H_4-S]_n- \quad \cdots (1)$$

$$C_6H_5-S-S-C_6H_5 \longrightarrow -[C_6H_4-S]_n- \quad \cdots (2)$$

These synthetic difficulties can be overcome by the use of a soluble precursor for the target polymer. Thus, conductive polymers such as poly(*p*-phenylenevinylene)(*22-25*) and poly(*p*-phenylene)(*26-28*) which are insoluble in common solvents, are synthesized via soluble precursors. Thermostable plastics such as poly(ether ether ketone) (PEEK)(*29*) and poly(arylene sulfide ketone)(*30*) are also obtained via soluble precursors.

On the basis of the electronic structure of the sulfur atom, the poly(phenylsulfonium cation) should be employed as a precursor. We have communicated a preliminary successful attempt to synthesize PPS via a soluble poly[methyl[4-(phenylthio)phenyl]sulfonium cation] intermediate (eq 3)(*31,32*). The process involves the self-condensation of methyl 4-(phenylthio)phenyl sulfoxide in acid by an electrophilic substitution reaction at room temperature.

$$C_6H_5-S-C_6H_4-S(=O)-CH_3 \longrightarrow -[C_6H_4-S-C_6H_4-S^+(CH_3)]_n- \longrightarrow -[C_6H_4-S]_{2n}- \quad \cdots (3)$$

The mild conditions of the precursor method enables synthesis of high performance PPS copolymers. Recently it was reported that addition of perfluoroalkane polymer to PPS increase the toughness and high sliding(*33, 34*). The copolymer of PPS with perfluoroalkane shows more good properties than the PPS alloy because it is difficult to make the PPS alloy with high compatibility due to the high melting point and high crystallinity(*35*), which is restricting the application of PPS alloy. However copolymerization with perfluoroalkanes is not possible due to the high temperature requirements for making the PPS segment. For example polycondensation of dichlorobenzene with sodium sulfide, or halothiophenolate salts results in the decomposition of the perfluoroalkane polymers due to the strong nucleophilicity of the sulfide compounds , which results in elimination HF or HI. This paper describes the synthesis of thiophenylene polymer containing a perfluoroalkane segment by the oxidative polymerization at room temperature.

The properties of the polysulfonium as a new polymer which has alternatively thioether and sulfonium group. The polysulfonium is isolatable as a stable salt which shows interesting properties such as transalkylation.

Experimental Details

Materials. Dichloromethane and chloroform were purchased from Tokyo Kasei Co., and distilled twice in the usual manner. Methyl phenyl sulfide was purchased from Tokyo Kasei Co. and methyl phenyl sulfoxide was obtained by the oxidation of the methyl phenyl sulfide. Trifluoromethanesulfonic acid, and pyridine were purchased from Tokyo Kasei Co. Ltd. and used without further purification.

Polymerization of Methyl 4-(phenylthio)phenyl Sulfoxide. *Poly[methyl[4-(phenylthio)phenyl] sulfonium trifluoromethane sulfonate](PPST)*. A 100 mL, round-bottom flask with a Teflon-covered magnetic stirring bar was charged with methyl 4-(phenylthio)phenyl sulfoxide (1 g, 4 mmol). The flask was cooled to 0 °C. Trifluoro-methane sulfonic acid (5 mL) was added at 0 °C with stirring. The temperature was increased slowly to room temperature over a period of 0.5 - 1 h. The reaction solution turned from colorless to pale blue. The reaction are continued for another 20 h at room temperature. The reaction was then quenched by pouring it into ice water. The precipitated polymer was then chopped in a blender, washed with water, and dried in vacuum at room temperature for 20 h. Yield: 1.53 g (100%). IR (KBr, cm^{-1}): 3086, 3023, 2932(ν_{C-H}); 1570, 1478, 1422($\nu_{C=C}$); 1258, 638(ν_{C-F}); 1161, 1067 ($\nu_{S=O}$); 816 (δ_{C-H}). Anal. Calcd for $C_{14}H_{11}S_3F_3O_3$: C, 44.20, H, 2.91; S, 25.28. Found: C, 44.10; H, 2.73; S, 25.11.

Poly(phenylene sulfide). A 200 mL, three-neck, round-bottom flask equipped with a magnetic stirring bar, reflux condenser, thermometer, and N_2 gas in inlet charged with PPST (1 g, 2.6 mmol) and pyridine (10 mL). The reaction mixture was stirred at room temperature and after a few minutes, the reaction mixture turned into a white suspension. The reaction was continued for 1 h at room temperature and then the temperature was slowly raised to reflux. The reaction was continued for 20 h at reflux temperature. The reaction was quenched by cooling it to room temperature and pouring into methanol (200 mL, 10% HCl). The precipitate was washed with methanol and chloroform. The polymer was purified by continuous extraction in a Sorhlet apparatus with methanol for 5 h and was dried in vacuo at 60 °C for 20 h. The resulting polymer was isolated as a white powder. Yield: 0.56g, 99%. IR(KBr,cm^{-1}): 3065(ν_{C-H}); 1572, 1472, 1387 ($\nu_{C=C}$); 810 (δ_{C-H}) , 1091, 1074, 1009, 554, 481.

CP/MAS ^{13}C-NMR (100 MHz, ppm): 132.1, 134.3 (phenyl C). Anal Calcd for C_6H_4S: C, 66.63; H, 3.73; S, 29.64. Found C, 66.63; H, 3.63; S, 29.61.

Synthesis of Polysulfonium-*block*-Poly(TFE-*co*-PVE)-*block*-Poly sulfonium. A 500 mL three-necked round bottom flask equipped with a mechanical stirring stick (vacuum sealing type), thermometer and dropping funnel was charged with α,ω-bis(phenyleneoxy-1,4-phenylenethio) poly(TFE-*co*-PVE) (30.0 g, 5 mmol as *Mw* : 6,000) and freon-113(1,1,2-trichloro-1,2,2-trifluoroethane) (100 mL). After the poly(TFE-*co*-PVE) had dissolved in freon-113, the flask was cooled to 0 °C and trifluoro-methanesulfonic acid (200 mL) was added. Methyl-4-phenylthio-phenyl sulfoxide (50 g, 200 mmol) was gradually added for 1 h. The reaction mixture became dark blue and was stirred for 1 h at 10 °C. The reaction temperature was increased to room temperature and stirred over 18 h. After the reaction, the reaction mixture was poured into ice water and washed with water to remove excess trifluoromethane sulfonic acid. After stirring in 3 L of water for 20 h, the resulting polymer was suspended in acetonitrile (600 mL), stirred for 2 h at reflux temperature and filtered to remove the homopolymer of polysulfonium. The filtrate was washed with acetonitrile and dried in vacuo for 24 hr at room temperature to give 99.1 g, yield 93 %. IR (KBr, cm^{-1}) 3086, 3023, 2932, 1570, 1478, 1422, 1258, 1161, 1067, 816, 638; ^{1}H-NMR(CD_3CN, 400MHz, ppm) 7.63, 7.66, 7.85, 7.88 (AB quartet, phenyl 8H), 7.3-7.7(m, end phenyl H), 3.62 (s, methyl 3H). ^{19}F-NMR(CD_3CN, 100MHz, ppm) -78 (s, $CF_3SO_3^-$), -79, -81, -117, -119, -120, -127, -140 (s, TFE/PVE).

Synthesis of Poly(phenylene sulfide)-*block*-Poly(TFE-*co*-PVE)-*block*-poly(phenylene sulfide). The block copolymer can be demethylated by refluxing pyridine. A 1 L three-necked round bottom flask equipped with a mechanical stirring stick (vacuum sealing type), thermometer and reflux condenser was charged with obtained polymer (101.4 g) and acetonitrile (600 mL). The solution was stirred at room temperature for 2 h and pyridine (80 g, 1.01 mmol) was added to demethylate the polysulfonium. The reaction mixture was stirred at room temperature for 10 h. The pale yellow suspension changed to a white suspension. The reaction temperature was increased and the mixture was refluxed for 20 h and then precipitated in methanol (10 % HCl, 2 L), washed with water and methanol. The obtained polymer was refluxed in ethanol (1 L) for 5 h to wash the polymer and freon 113 (1 L) to remove the Poly(TFE-*co*-PPVE) homopolymer for 15 h and dried in vacuo for 15 h at 60 °C (66.8 g, yield: 96.5 %). IR (KBr, cm^{-1}) 3065, 1572, 1472, 1387, 1234, 1091, 1074, 1009, 812, 554, 481; CP/MAS ^{13}C-NMR(

100MHz, ppm) 133.4, 136.5(phenyl C); ^{19}F-NMR (100MHz, ppm) -79, -81, -117, -119, -120, -127, -140.

Methylation of Nucleophiles. A 100 ml round-bottomed flask with a Teflon-covered magnetic stirrer bar was charged with poly(sulfonium cation) (1 g, 2.6 mmol as unit of poly(sulfonium cation)). Acetonitrile (20 ml) was added at room temperature and stirred. Nucleophiles such as phenol and aniline were added to the reaction mixture, which was filtrated by silica gel column (2 x 20 cm, benzene eluent) and the filtrate was evaporated. Conversion of obtained compounds were evaluated by gas chromatography.

Measurement of Viscosity of Polysulfonium. CF_3SO_3Na (905 mg, 5.26 mmol) was dissolved in acetonitile / waster = 50/50 (vol %) 100 mL as solution A. Poly(sulfonium cation) 200 mg was dissolved in the solution 10 mL and was added into an Ubbelohde viscometer. For all measurement of the viscosity at 25 °C, the solution was diluted to half concentration with solution A. On the basis of the correlation plot of c(g dl^{-1}, concentration of polycation) and ηsp (specific viscosity)/c, limiting viscosity number [η] can be obtained from the intersection of vertical axis.

Computational Calculation. Reactivity (stability) of the sulfonium cations is evaluated as LUMO energies calculated by a semiempirical molecular orbital calculation PM3(*36*) (MOPAC Version 5.0). Nucleophilicity of the sulfonium cations are compared. LUMO energies and stability of bonds in the sulfonium cation are evaluated as a two center energy using the keyword ENPAR. An optimized structure was obtained using the keyword PULAY. PM3 calculations were performed on a Fujitsu VP2200 super computer.

Measurements. ^{1}H-, ^{13}C-, ^{19}F-NMR were recorded using a FT-NMR (GXS 400, JEOL Co.). IR spectra were obtained with a JASCO Model IR-810 spectrometer using potassium bromide pellets. DSC measurements were done in a nitrogen atmosphere using SSC/220 (Seiko Co.) thermal analyzer: sample size, 7-10 mg; heating rate, 20 °C/min. Thermogravity (TG) measurements were done in a nitrogen atmosphere on a TG/DTA 220 (Seiko Co.) thermal analyzer: sample size, 7-10 mg; heating rate 20 °C / min. The determination of the molecular weight of the resulting polymer was measured using a GPC (Shimadzu Co.: LC-9A, SPD-6AV(265 nm), Column: Asahi Chemical Industry Co. Ltd., GS-510H + GS-310H) with THF (1 mL / min) as the eluent.

Results and Discussion

Computational Evaluation of Sulfonium Cations. The sulfonium cation and poly(sulfonium cation) are isolated as a stable salt. In addition, demethylation of MSP easily occurs through a $S_{n}2$ reaction with a base such as pyridine. Furthermore, PPST converts to PPS by demethylation. The reactivity and stability of the sulfonium cation were evaluated in the LUMO energy and two center energies from the semiempirical molecular orbital calculation PM3. LUMO energies of the methyldiphenylsulfonium cation is compared with sulfonium cations derivatives, *e.g.*, methylbisphenylthio-, methylphenyl(phenylthio)-, and (methylthio)diphenyl- sulfonium cation (Figure 1). LUMO energies of the diarylalkylsulfonium cation were the highest in these cations. Trityl cation is well known to be stably isolated and possesses 4.3 eV as LUMO energy. The calculation supports the idea that MSP also exists as a stable salt because MSP possesses a higher LUMO energy than trityl cation. On the basis of two-center energies of diphenylmethyl sulfonium, the phenyl $C\text{-}S^{+}$ bond(-15.9 eV) is stronger than methyl $C\text{-}S^{+}$ (-14.8 eV). These results suggest that the methyl $C\text{-}S^{+}$ bond is morecleavable than Phenyl $C\text{-}S^{+}$ in $S_{n}2$ reactions. These calculation results support the idea that the poly[methyl[4-(phenylthio)phenyl]sulfonium cation] is a soluble precursor with stability and reaction selectivity for the synthesis of PPS.

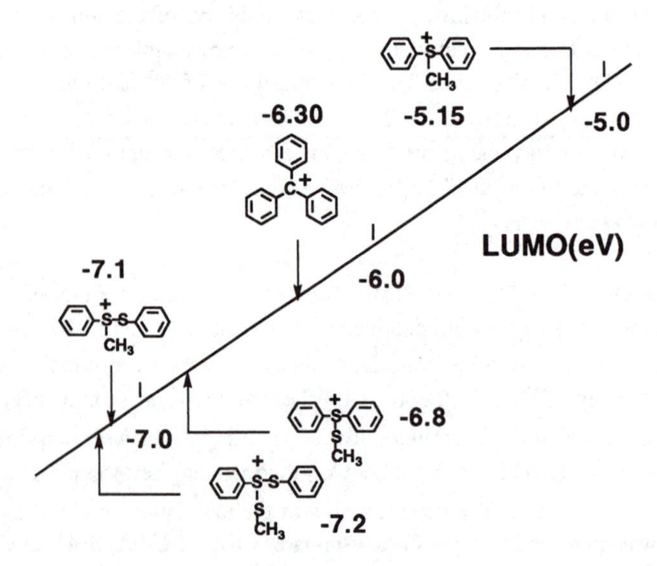

Figure 1 LUMO Energies of Sulfonium Cations Calculated by PM3 Method

Synthesis of High Mw PPS Methyl 4-(phenylthio)phenyl sulfoxide was polymerized in trifluoromethanesulfonic acid at room temperature for 24 h to produce the polymer in quantitative yield. The mixture was then poured into water to give the polymer as a trifluoromethanesulfonate salt. The resulting polymer was isolated as a white powder having the empirical formula $C_{13}H_{11}S_2CF_3SO_3$ and was soluble in sulfolane, nitrobenzene, dimethyl sulfoxide, pyridine, and formic acid. The methyl group in the resulting polymer was confirmed by the IR spectrum. The IR spectrum of the polymer also shows a strong absorption at 1258, 638 ($\nu_{C\text{-}F}$) and 1161, 1067 cm($\delta_{S=O}$), which means that the polymer contains trifluoromethanesulfonate as a counter anion. In the 500 MHz COSY ^1H-^{13}C NMR spectrum (Figure 2), methyl groups are observed at 3.78 and 28.93 ppm, respectively, which are at lower field than that of a neutral thiomethyl group. These results support the formation of poly[methyl[4-(phenylthio)phenyl] sulfonium cation].

Activated sulfoxides are well-known for use as electrophiles (Swern method). The polymerization of the sulfoxide compound is initiated via the sulfonium cation as an active species by the protonation on the oxygen of the sulfoxide bond. The active species electrophilically substitutes on the benzene ring to eliminate water.

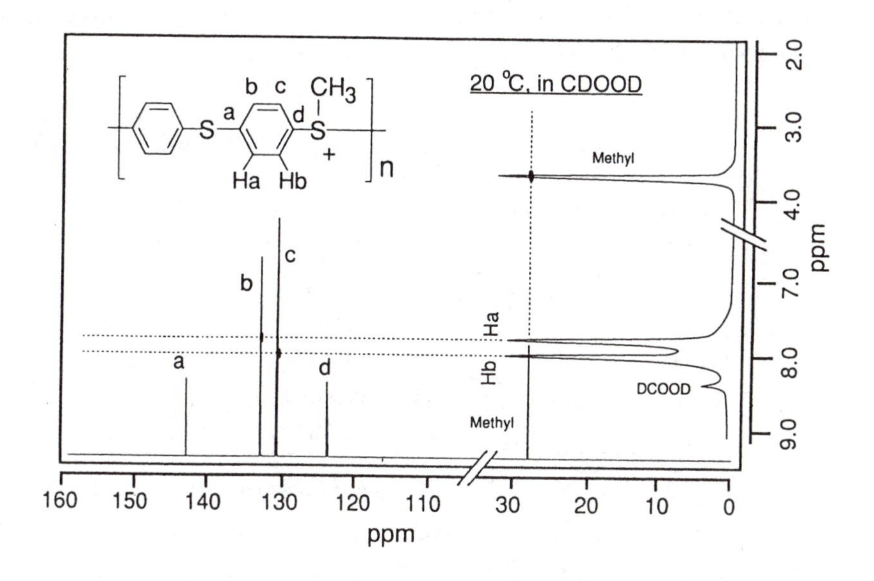

Figure 2 COSY ^1H-^{13}C NMR spectra of Poly[methyl(4-phenylthio)phenyl sulfonium trifluoromethane sulfonate] in DCOOD.

The demethylation of poly[methyl[4-(phenylthio)phenyl] sulfonium cation] was carried out in refluxing pyridine. The precursor polymer is completely soluble in the mixture at room temperature, however on reaction a initially white insoluble powder precipitated which was identified as *N*-Methylpyridinium trifluoromethane sulfonate salt. This was isolated quantitatively from the reaction mixture. The resulting polymer has the empirical formula of C_6H_4S and shows the same IR spectrum and CP/MAS ^{13}C NMR spectra of commercially available PPS. No absorption bands of methyl, sulfone, and sulfoxide bonds are detected in the IR spectrum.

Molecular weight of the polymer was determined to be 1.2×10^5 (20 h, 25 °C) and 6.7×10^4 (5 h, 25 °C, Table I) by high-temperature GPC. Polymerization of methyl 4-(phenylthio)phenyl sulfoxide was came out in various acids. The addition of P_2O_5 as a dehydrate agent in the media promotes the polymerization, e.g., the polymerization in trifluoromethanesulfonic acid P205 results in a high molecular weight PPS with *Mw* 2.4×10^5. The high molecular weight PPS (*MW* 2.4×10^5) has a larger molecular weight distribution (*Mw/Mn* = 5.6), because of the high viscosity of the reaction mixture DSC and TG measurements for *MW* 2.4×10^5 shows Tm 260 °C, Tg 96 °C, Tc 156 °C and $Td_{10\%}$ 535 °C.

The polymerization not only enables PPS formation but also allows the synthesis of poly(arylene sulfide) with high crystallinity. The precursor method also efficiently provides poly(oxyphenylene sulfide) with *Mw* = 4×10^4.

Table I Oxidative Polymerization of Sulfide Compounds

Method	Monomer	Temp. (°C)	Solvent	Mw (x 10^4)	$Td_{10\%}$ (°C)
Ox. Polymn	Ph-S-C6H4-S(=O)-CH3	25	CF_3COOH	—	—
		25	CF_3SO_3H	12.0	525
	ref.38	25	CF_3SO_3H/P_2O_5	24.9	535
	Ph-S-C6H4-S-CH3 / $SbCl_5$	20	CH_2Cl_2	2.0	515
	ref.37 Ph-S-C6H4-S-CH3 / O_2	20	CH_3SO_3H/P_2O_5	20.4	526
		70	CH_3SO_3H/P_2O_5	10.4	525
Phillips	Cl-C6H4-Cl / Na_2S	250	N-methylpyrrolidone	1.9	512
R.Lenz	Br-C6H4-SCu	300	—	1.2	502
A.Hay	I-C6H4-SS-C6H4-I	280	diphenylether	2.2	520

Synthesis of Block Copolymer of PPS with Poly(TFE-*co*-PVE). The copolymerization of methyl 4-(phenylthio)phenyl sulfoxide with poly(TFE-*co*-PVE) was carried out in a mixture of freon-113/ trifluoromethanesulfonic acid as the polymerization solvent for 10 hr (Scheme 1). The solvent viscosity increased at the end stage of the polymerization. The resulting polymer was washed with freon-113 and acetonitrile repeatedly to remove the unreacted poly(TFE-*co*-PVE) and homopolymer of polysulfonium(*39*). The block copolymer was isolated with 93 % yield. It is believed that the degree of the PPS segment is estimated to be ca. 20 as an average through the completely reaction, which depends on the feed concentration ratio of the monomers(*40*).

The resulting polymer has a peak at 638 cm^{-1} in the IR spectrum, which is attributed to trifluoromethanesulfonate and at 816 cm^{-1} that corresponds to the phenylene-1,4-linkage in the polythiophenylene segment. Peaks at 7.63, 7.66, 7.85 and 7.88 ppm in the ^1H-NMR correspond to the aromatic 8H. The methyl resonance was also observed at 3.62 ppm and is shifted up field due to the positively charged sulfur atom. The presence of poly(TFE-*co*-PVE) was confirmed by IR peaks at 1258 and 1067 cm^{-1}. ^{19}F-NMR also shows peaks at -77.7 ($CF_3SO_3^-$), -78.7, -81.3 and -127.4 (perfluoroalkane chain).

These spectroscopic data indicate that diblock polymer formation may be occuring. The poly (TFE-*co*-PVE) has two oxyphenylene groups at the end of the chain, which shows the chemical shift of the phenyl protons at 6.65-7.85 ppm similar to those of low molecular weight model compounds such as α,ω-bis(phenyl-eneoxy-1,4-phenylenethio)perfulorobutane. ^1H-NMR spectrum of the resulting block polymer of polysulfonium and poly(TFE-*co*-PVE) did not show the peaks around 6.65 - 7.3 ppm but around 7.3-7.9 ppm whose chemical shifts are assigned to phenyl protons at the end thiophenyl- ene groups of the polysulfo-nium chain (main peaks at 7.6 and 7.8 are attributed to the polysulfonium chain).

After demethylation with pyridine, the resulting polymer was isolated as a pale yellow powder. In the IR spectrum, the absorption peak at 812 cm^{-1} was attributed to a C-H out-of-plane vibration of the benzene

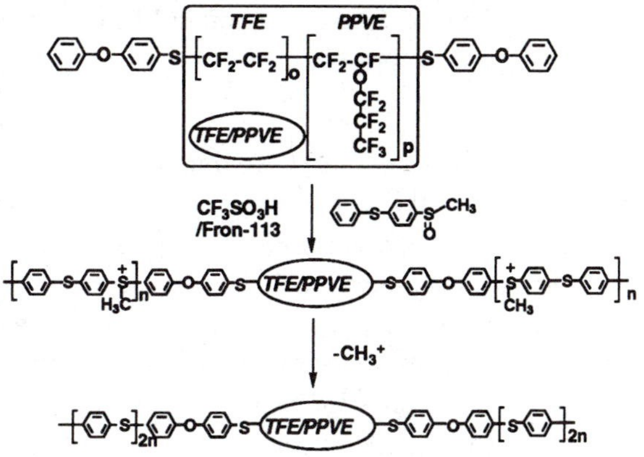

Scheme 1

ring, while an absorption peak at 1234 cm^{-1} attributed to a C-F out-of-plane vibration of the perfluoroalkane chain. The ν_{C-H} absorption peaks of the methyl group and peaks of $CF_3SO_3^-$ disappear in the IR spectrum. The demethylation proceeds through a trans-methylation mechanism from the sulfonium cation to pyridine. Quantitative formation of the methylpyridinium salt was confirmed as a by-product. ^{19}F-NMR shows peaks at -78.7, -81.3, and -127.4 ppm and CP-MAS ^{13}C-NMR shows peaks at 133.4 and 136.5 ppm. These spectroscopic data indicate the presence of the poly(TFE-*co*-PVE) and PPS structure.

After a O_2 plasma etching treatment of the resulting copolymer, a three dimensional mesh is observed (diameter of one mesh is about 2 µm) by means of the SEM analysis(Figure 3), which supports a homogeneous microphase separation between PPS and poly(TFE-*co*-PVE) segment. The DSC measurement shows peaks at Tg = 90 °C, Tc = 138 °C, and Tm = 276 °C which come from the PPS moieties.

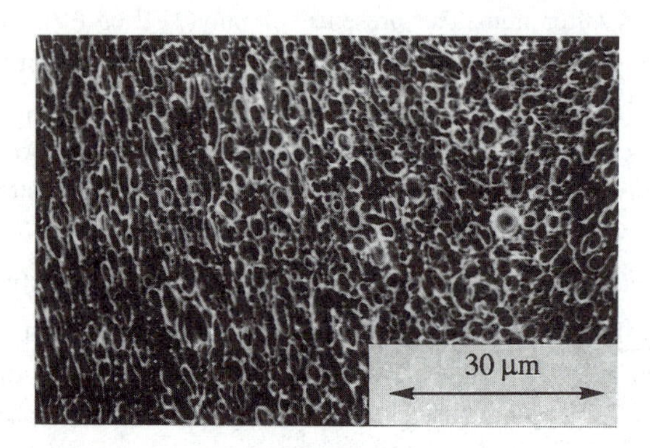

Figure 3 Scanning Electron Micrograph of the Resulting PPS-poly(TFE-*co*-PVEF) Block Copolymer

Solvent Properties of Polysulfonium. The poly(sulfonium cation) is soluble in common solvents such as acetonitrile, DMSO, formic acid and sulfuric acid. However, PPS has poor solubility in various solvents, which not only make it difficult to synthesize under mild conditions but also to characterize, *e.g.*, determination of molecular weight. The viscosity of poly(sulfonium cation) as a novel polyelectrolyte which is a useful precursor of PPS synthesis was determined in acetonitrile solution (50 vol% aqueous) and in formic acid. PPSs with Mw=1.14 x 10^4, 2.01 x 10^4, 2.83 x 10^4, 3.91 x 10^4, 7.01 x 10^4, 9.52 x 10^4 which were determined by high temperature GPC were prepared by polymerization of the sulfoxide.

The poly(sulfonium cation) with trifluoromethanesulfonate anion is a new type polyelectrolyte which has alternative structure of positively charged phenylsulfonium and phenylsulfide. The solution viscosity and solubility of a polyelectrolyte are influenced by a co-existence of supporting electrolyte and solvent species. The poly(sulfonium cation) is insoluble in aqueous solution but soluble in polar organic solvents, *e.g.* acetone, acetonitrile and dimethyl sulfoxide (DMSO). In particular, the polycation is very soluble in acetonitrile and formic acid (solubility: more than 100 mg mL^{-1}).

The viscosity measurements were performed using an Ubbelohde viscometer in the presence of sodium trifluoromethanesulfonate as supporting electrolyte (see experimental section). In the absence of the supporting electrolyte, the viscosity curve in acetonitrile shows a typical polycation characteristic. In the region of the dilute concentration, the viscosity was increased drastically. The viscosity in dilute concentration also increased by additional water since the ionic dissociation is promoted by the addition of water. The solubility in acetonitrile is based on the strong hydrophobic properties of the phenylthio group. The ionic dissociation in the presence of water comes from a strong hydrophilicity of the positively charged phenylsulfonium group. Since the hydrophobicity of the polycation is believed to be stronger than hydrophilicity, the polycation in acetonitrile/water =50/50(vol%/vol%) shows the highest viscosity in the absence of the supporting electrolyte (the polycation is insoluble in acetonitrile with >70% water).

In the presence of the supporting electrolyte (50 mM), the viscosity curve becomes linear like a non-polyelectrolyte. Limiting viscosity of poly(sulfonium cations with various molecular weight were correlated to the molecular weight of the demethylated PPS (Figure 4). The curve is fitted with good agreement as a function: $y= ax^b$. In the relation between the viscosity of the polycation and the molecular weight of PPS, K and α values of Mark - Houwink equation, $[\eta] = K M^{\alpha}$, were determined to be 1.55×10^{-6} and 1.30 respectively. From relation between viscosity of polycation and Mcation which was estimated by an equation $M_{Cation} = M(C_{14}H_{11}S_3F_3O_3)/ M(C_{12}H_8S_2) \times M_{PPS}$ (Molecular weight), K and a were determined to be 7.46×10^{-7} and 1.30 respectively.

Demethylation of Polysulfonium. The demethylation to PPS moiety was also carried out in acetonitrile to examine the reactivity using high-molecular weight poly(methyl-4-phenylthiophenl trifluoromethanesulfonate) (poly(sulfonium cation)) ($Mw > 2 \times 10^5$). The poly(sulfonium cation) was demethylated through nucleophilic reaction with pyridine at reflux. An equimolar amount of pyridine when added to a solution at the poly(sulfonium cation) resulted in about 50% demethylation. In the presence of over five equivalents of pyridine per unit of poly(sulfonium cation), the demethylation of poly(sulfonium cation) proceeds completely, which was confirmed by IR spectra.

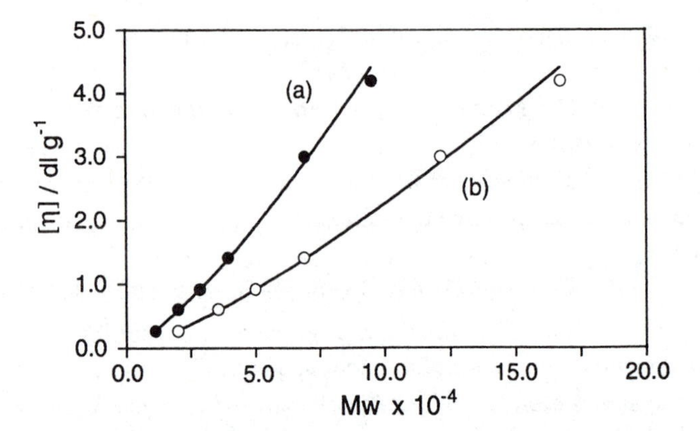

Figure 4 Relationship between Viscosity of Poly(sulfonium cation) and Molecular Weight of PPS(after demethylation)(a, ●), and between the viscosity of Poly(sulfonium cation) (b, ○) and the Molecular Weight of Poly(sulfonium cation) which is calculated on the basis of the molecular weight of PPS. in formic acid containing sodium trifluoromethanesulfonate (50 mM, 23 °C)

Demethylation of poly(sulfonium cation) occurs by nucleophile substitution reaction via $S_{n}2$ mechanism The demethylation is also classified as a transmethylation to nucleophiles. Using nucleophiles that have a mobile proton such as phenol, aniline and benzoic acid derivatives in the presence of potassium carbonate, demethylation of poly(sulfonium cation) proceeds efficiently to yield the methylated products of phenol and benzoic acid with 100 % conversion (Scheme 2). In the case of aniline, both N-methyl and *N,N*-dimethyl aniline are obtained and methylation of *N*-methyl aniline occurs more easily than aniline because of the higher Lewis basicity.

These results reveal that poly(methyl-4-phenyl thiophenylsulfonium trifluromethane sulfonate) acts as an efficient alkylation agent to transfer methyl substituent of the polycation to the nucleophile. After the reaction, PPS as a by-product is removed simply by filtration from reaction mixture because the PPS is insoluble in the solvent. The poly(sulfonium cation) is classified as a new polymeric agent for

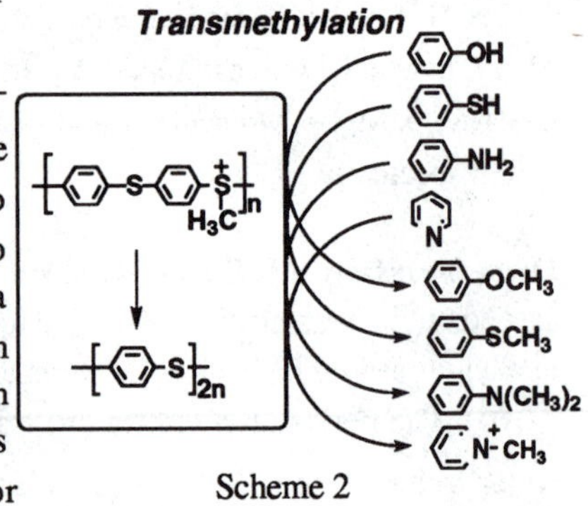

Scheme 2

alkylation to nucleophile and provides an efficient and convenient process to yield methylated aromatic compounds.

Acknowledgment
This work was partially supported by a Grant-in-Aid for Scientific Research.

Literature Cited

1. Koch, W.; Heitz, W. *Macromol. Chem.* **1983**,*184*, 779.
2. Jones, T. P. H.; Mitchell, G. R.; Windle, A. H. *Colloid Polym. Sci.* **1983**, *261*, 110.
3. Hawkins, R. T. *Macromolecules* **1976**, *9*,189.
4. Edmonds, J. T., Jr.; Hill, H. W., Jr. U.S. Patent 3354129, **1967**; *Chem. Abstr.* **1968**, *68*, 13598. Campbell, R. W.; Edmonds, J. T. U.S. Patent 4038259, *Chem. Abstr.* **1977**, *87*, 854v. Hill, H. W., Jr. *Ind. Eng. Chem., Prod. Res. Rev.* **1979**, *18*, 252.
5. Elsenbaumer, R. L.; Shaclkette, L. W.; Sowa, J. W.; Baughman, R. H. *Mol. Cryst. Liq. Cryst.* **1982**, *83*, 229.
6. Black, R. M.; List, C. F.; Wels, R. J. *J. Appl. Chem.* **1967**, *17*, 269
7. Idel, K.; Merten, J. U.S. Patent 4303781, **1979**; *Chem. Abstr.* **1981**, *94*, 175864.
8. Idel, K.; Ostlinning, E.; Freitag, D. DE. Patent 3428985, **1986**; *Chem. Abstr.* **1986**, *104*, 225451.
9. Idel, K.; Ostlinning, E.; Freitag, D.; Alewelt, W. DE. Patent 3428984. **1986**: *Chem. Abstr* **1986**, *104*, 225452
10. Idel, K.; Ostlinning, E.; Freitag, D. EP. Patent 171021, **1986**; *Chem. Abstr.* **1986**,*104*, 225450.
11. Ostlinning, E.; Idel, K. DE. Patent 3243189,**1984**; *Chem. Abstr.* **1984,** *101*, 111595.
12. Schmidt, M.; Tresper, E.; Alewelt, W.; Dorf, E. U.; Ruesseler, W. U.S. Patent 5037952, **1990**; *Chem. Abstr.* **1990**, *113*, 172969.
13. Campbell, R. W. Ger. Offern.2453749,**1975**; *Chem. Abstr.* **1975**, *83*, 115380.
14. Asakura, T.; Noguchi, Y.; Kobayashi, H. US. Patent 4286018, **1978**; *Chem. Abstr.* **1979**, *92*, 23888.
15. Inoue, H.; Kato, T.; Ogawara, K. U.S. Patent 5037913, **1989**; *Chem. Abstr.* **1990**, *112*, 57021.
16. Yamamoto, K.; Jikei, M.; Katoh, J.; Nishide, H.; Tsuchida, E. *Macromolecules* **1992**, *25*, 2698
17. Yamamoto, K.; Jikei, M.; Murakami, Y.; Nishide, H.; Tsuchida, E. *J. Chem. Soc., Chem. Commun.* **1991**, 596.

18. Tsuchida, E.; Yamamoto, K.; Nishide, H.; Yoshida, S.; Jikei, M. *Macromolecules* **1990**, *23*, 2101.
19. Tsuchida, E.; Yamamoto, K.; Jikei, M.; Nishide, H. *Macromolecules* **1990**, *23*, 930.
20. Shouji, E.; Yamamoto, K.; Katoh, J.; Nishide, H.; Tsuchida, E. *Polym. Adv. Technol.* **1991**, *2*, 149.
21. Yamamoto, K.; Tsuchida, E.; Nishide, H.; Yoshida, S.; Park, Y. S. *J. Electrochem. Soc.* **1992**, *139*, 2401.
22. Gagnon, D. G.; Capistran, J. D.; Karasz, F. E.; Lenz, R. W. *Polym. Bull.* **1984**, *12*, 293.
23. Lahti, R. M.; Modarelli, D. A.; Denton, F. R., III; Lenz, R. W.; Karasz. F. E. *J. Am Chem .Soc.*, **1988**, *110*, 7258.
24. Murase, I.; Ohnishi, T.; Noguchi, T.; Hirooka, M. *Polym. Commun.* **1984**, *25*, 327.
25. Wessling, R. A. *J. Polym. Sci., Polym. Symp.* **1985**, *72*, 55.
26. Ballard, D. G. H.; Courtis, A.; Shirley, I. M.; Taylor, S. C. *J. Chem. Soc., Chem. Commun.* **1983**, 954.
27. Ballard, D. G. H.; Courtis, A.; Shirley, I. M.; Taylor, S. C. *Macromolecules* **1988**, *21*, 294
28. McKean, D. R.; Stille, J. K. *Macromolecules* **1987**, *20*, 1787.
29. Kelsey, D. R.; Robeson, L. M.; Clendinning, R. A.; Blackwell, C. S. *Macromolecules* **1987**, *20*, 1204.
30. Lyon, K. R.; McGrath, J. E.; Geibel, *J. F. Polym. Mater. Sci. Eng.* **1993**, *65*, 249
31. Yamamoto, K., Shouji, E.; Nishide, H.; Tsuchida, E. *J. Am. Chem. Soc.* **1993**, *115*, 5819.
32. Tsuchida, E.; Shouji, E.; Yamamoto, K. *Macromolecules*, **1993**, *26*, 7144
33. JP Pat. 1005081, 1429107, 1146392.
34. JP Pat 62-232457, 57-202344; USP 3487454, 4115283.
35. JP Pat 59-155462
36. Stewart, J. J. P. *J. Comput. Chem.* 1989, *10*, 221.
37. Tsuchida, E.; Shouji, E.; Yamamoto, K. *Macromolecules*, **1993**, *27*, 1057
38. Tsuchida, E.; Shouji, E.; Yamamoto, K. *Chem. Lett.* **1993**, 1927
39. The solubility of the polysulfonium salt and poly(TFE-*co*-PVE) homopolymers are quite different After polymerization the homopolymer are removed from the resulting block copolymer by repeated extract ions.
40. The polymerization of methyl 4-(phenylthio) phenyl sulfoxide proceeds quantitatively.

RECEIVED December 14, 1995

POLYIMIDES AND HIGH-TEMPERATURE POLYMERS

Chapter 25

Formation of Transparent Silica–Polymer Hybrids Based on Siloxane-Containing Polyimides

Norbert A. Johnen, Laura L. Beecroft, and Christopher K. Ober[1]

Department of Materials Science and Engineering, Cornell University, Ithaca, NY 14853–1501

The study of hybrid organic-inorganic materials becomes increasingly important as demands on conventional materials exceed their capabilities. Tailoring of organic polymers, using copolymers or polymer blends, will certainly not yield properties which surpass those of inorganic glasses in areas where they dominate, i.e. thermal stability, thermal expansion, or mechanical properties. However, hybrids which combine the advantages of polymers with those of inorganic compounds may offer solutions that overcome the problems associated with spin-on-glass or other sol-gel materials. Advantages and disadvantages of each hybrid component are listed in Tables I and II.

Within such new hybrid materials, the organic polymer or oligomer may be physically or chemically connected to the inorganic network. The homogeneity and the physical properties of hybrid materials depend on the relative size and degree of connectivity of the organic/inorganic domains. Increased size of the minority domain diminishes the physical properties of the material, since as the domain size grows to the micron scale, the material is no longer as intimately connected. Hybrids have been recently reviewed in a paper by Novak (*1*).

A standard technique for hybrid formation is to use sol-gel chemistry to produce nanometer to micron sized inorganic particles in the presence of a dissolved polymer. Confining our comments to the specific example of silica, this process may result in either physical blending of silica with polymer or chemical grafting reactions that connect the silica to polymer. The nature of the dissolved polymer and the reaction conditions for the condensation polymerization of silica are important in determining the character of the hybrid (*2, 3*). Silica produced under basic conditions usually results in a very coarse grained material of large particle size. On the other hand, acidic conditions form much finer grained silica. Therefore the slightly buffered acidic environment provided by the polyamic acid is desirable for carrying out our hybrid syntheses.

Several methods for producing hybrid materials are shown schematically in Figure 1. A variety of inorganic phases have been investigated. Most commonly, alkoxy silicates are used to form a silica network via step-growth polymerization. Besides alkoxy silicates, several alkoxides, (e.g. alkoxy titanates (*4, 5*), alkoxy aluminates (*6*), alkoxy germanates and alkoxy vanadates (*7*)) have been used to prepare inorganic networks with a range of properties. Several different organic

[1]Corresponding author

0097–6156/96/0624–0392$12.00/0

Hybrid Synthesis and Processing

In-Situ Solution Processing

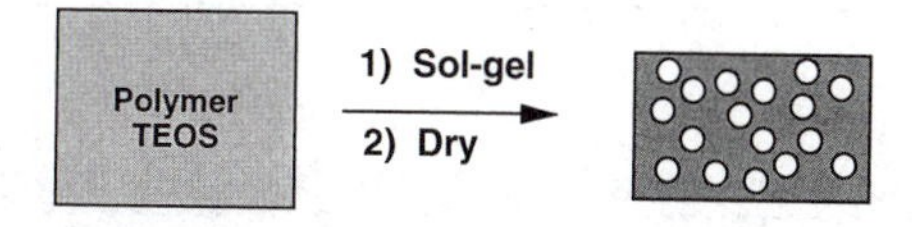

Blending

(e.g. PVAc, PMMA, Poly(phosphazene), etc.)

Grafting via Reactive Groups

(e.g. polymers modified with TEOS groups
Ormosils/Ceramers)

True In-Situ Processing

Ex-Situ Processing

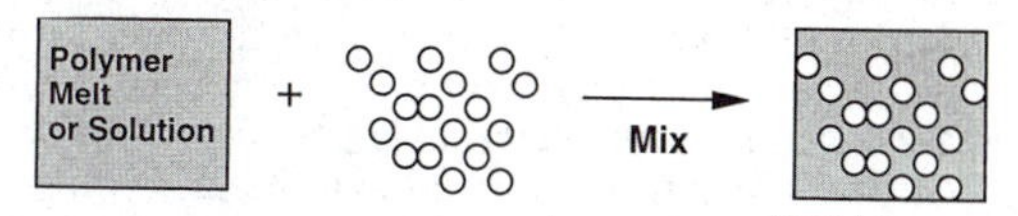

Figure 1. Schematic of methods used for hybrid production.

polymers have been incorporated into sol-gel composites including alkyl methacrylate (*8, 9*), poly(vinyl acetate) (*10*), poly(ethyl oxazoline) (*11*) and poly(ethylene oxide) (*12*).

Table I. Advantages and Disadvantages of Organic Polymer

Advantages :	Disadvantages :
• Low dielectric constant	• High CTE (coefficient of thermal expansion)
• Low refractive index	• Lower hardness, modulus, & strength
• High flexibility	
• Good processability	
• Ease of patterning	

Table II. Advantages and Disadvantages of Inorganic Polymers (Glass)

Advantages :	Disadvantages :
• Low CTE (coefficient of thermal expansion)	• High dielectric constant
• High hardness, modulus, & strength	• Brittleness
• High thermal stability	

A second route involves swelling the host polymer with a sol-gel reagent to produce the hybrid *in situ* (e.g. poly(dimethylsiloxane) (*6, 13*)). Both methods are effective, but the latter requires no additional solvent. Finally, some researchers have blended nanometer size particles which have been produced ex-situ with a host polymer. This last technique has the disadvantage that the mixing process may lead to poor dispersion (*14*). For our studies, we have chosen to investigate hybrid formation using polyimide and silica. These materials are noted for their excellent thermal stability and mechanical properties, and are often chosen in applications involving microelectronics. The high thermal stability of these hybrids will allow them to withstand treatment needed to complete transformation of the inorganic component.

Poly(imide) based hybrids have been investigated previously. Nandi et al. (*15*) prepared poly(imide)/chromium and poly(imide)/iron hybrids using $Cr(C_6H_6)$ or $Fe_3(CO)_{12}$, respectively, as inorganic precursors. Although the poly(imide) was not chemically connected with the inorganic material, cluster formation over 1.5 nm could not be detected using SEM. The authors claimed that the lack of agglomeration was due to the metals prebinding with the carboxylic acid units of the poly(amic acids), PAAs, prepared from oxydianiline (ODA) and pyromellitic dianhydride (PMDA). This prebinding process hindered the formation of inorganic clusters, because the water required for the hydrolysis/condensation reactions of the inorganic was obtained only during the final high temperature imidization step.

Poly(imide)/titania and poly(imide)/silica hybrids prepared using titanium triethoxide or TEOS (tetraethyl orthosilicate), respectively, were investigated by Nandi et al. (*16*). These systems were limited to an inorganic concentration of less than 42 wt-% and the transparency limit was well below that (~10 wt-%). Above this limit, grossly phase separated particles larger than 1 micron were observed. The hybrids of poly(imide)/titania and poly(imide)/silica showed particle sizes up to 0.1 microns when the inorganic content was below 42 wt-%. The second phase size distribution was not uniform throughout films of these hybrids. The concentration of inorganic material was greatest at the hybrid-air interface whereas the substrate-hybrid interface consisted almost exclusively of organic polymer.

Kakimoto et al. (*17*) and Morikawa et al. (*18*) prepared poly(imide)/silica hybrids with various organic/inorganic ratios, using TEOS and a PAA prepared from ODA and PMDA. They reported that below 8 volume-% silica, transparent hybrids with particle sizes in the submicron range were obtained, whereas above 8 volume-% silica the composites showed particle sizes around 5 microns. These latter composites appeared macroscopically opaque.

Morikawa et al. (*19*) later prepared poly(imide)-silica hybrids using polyamic acids synthesized from PMDA, ODA and other diamines. These diamines contained hydrolizable functions, enabling the formation of a chemically interconnected organic/inorganic network. In a sol-gel reaction with TEOS these PAA's formed transparent materials over a wide range of silica compositions. SEM micrographs indicated a silica domain size of 1 to 1.5 microns.

Using a new technique, we have successfully prepared poly(imide)/silica hybrid materials from TEOS and a siloxane-containing poly(amic acid) (PAA) with a range of organic to inorganic ratios from 0 to 100% (*20*). These hybrids showed no gross phase separation over the entire silica range. The chemical nature of the polymer allowed covalent connections between the organic and inorganic segments thereby forming a strongly interconnected organic / inorganic network. Using a second approach we also synthesized a hexafunctional PAA in a condensation reaction using ODA, PMDA and 3-amino-propyltriethoxyl silane (APTES). This hexafunctional PAA was also incorporated into an inorganic network using sol-gel chemistry. Under appropriate conditions hybrids were prepared with a wide range of organic to inorganic ratios without showing phase separation down to the nanoscale regime. Since the quality of the hybrid is coupled to its homogeneity, we investigated all materials obtained from sol-gel reactions initially by their optical appearance. The effect of reaction conditions and in particular pH will be discussed. TGA measurements were used to evaluate thermal stability of the hybrids as well as to determine silica content. Results of solid state ^{29}Si NMR studies of the hybrids will be reported. More detailed investigations will be described in future which use optical microscopy and scanning electron microscopy to determine the nature of the dispersed inorganic phase. The remarkable mechanical properties of these hybrids will be discussed in a future publication.

Experimental

Materials. TEOS (tetraethoxy silane, Aldrich), Ultradel 1414 (PAA from Amoco), ODA (oxydianiline, Aldrich) and APTES (3-aminopropyltriethoxyl silane, Aldrich) were used as received without further purification. DMF (Aldrich), THF (Aldrich), NMP (Aldrich) and acetonitrile (Aldrich) were dried over magnesium sulfate and stored over molecular sieve 4 Å. PMDA (pyromellitic dianhydride, Aldrich) was recrystallized three times from dry acetone and sublimed at least three times under vacuum.

Synthesis of Hybrids from End-functional Poly(amic acids). PMDA and NMP were mixed in a round bottom flask with a stirring magnet. When all the PMDA was dissolved, a solution of ODA in NMP was added. After 10 minutes, APTES was added. A predetermined amount of this PMDA-ODA mixture was pipetted into a vial and the desired amount of TEOS added to this mixture. Finally, the required amount of water was added and the mixture allowed to react for 48 to 96 hours at room temperature. For reactant quantities, see Table III.

Table III. Hybrid Mixtures from End-functional PAA

Entry	PMDA:ODA:APTES (mmol ratio)	TEOS (mmol)	Wt-% SiO_2	f SiO_2	H_2O:TEOS (mol/mol)
1	1.95: 1.28: 1.36	10.0	40	0.25	1.21
2	1.97:1.69:0.59	10.0	42	0.27	2.20
3	1.97:1.68:0.59	10.0	42	0.27	2.40
4	0.53:0.43:0.23	10.0	72	0.56	1.97
5	0.04:0.04:0.02	10.0	97	0.94	2.46

Preparation Of The Sol-Gel Hybrid from Siloxane-containing PAA. The precursor solution was prepared in solvent with a PAA as organic material, TEOS as the silica precursor, as well as either water or dilute aqueous hydrochloric acid (one to nine weight percent) for hydrolysis. The solvent was used to create a single phase mixture. For reactant quantities, see Table IV.

The polymeric PAA was diluted (e.g. in DMF) and stirred for at least one hour before the desired amount was weighed into a screw capped vial containing a stirring bar. Typical quantities are listed in Tables III and IV. After the addition of TEOS, either water or an aqueous hydrochloric acid solution (1-9 wt%) was added at a molar ratio of about 2-4 times water compared to TEOS. With the addition of TEOS and water the mixture became translucent. If 15 minutes of stirring did not recover transparency further solvent was added.

Table IV. Hybrid Mixtures using Siloxane-containing PAA (Ultradel 1414)

Entry	Wt. Fraction TEOS *	Wt. Fraction SiO_2	f SiO_2	H_2O:TEOS (mol/mol)
6	0.70	0.40	0.25	2.00
7	0.76	0.48	0.32	2.07
8	0.90	0.72	0.57	1.60
9	0.95	0.86	0.75	2.10

* Weight fraction TEOS to PAA in reaction mixture. Solvent and H_2O are additional components.

Processing Thick Film and Bulk Hybrids. After stirring the transparent hybrid solution for at least 4 days, a sample was removed and volatile materials were allowed to evaporate to near dryness at room temperature. The sample was then annealed at 60 °C under vacuum. Materials with different structures were obtained depending on the composition and treatment of the precursor solution. Under the right conditions, free-standing transparent films with thicknesses of 1-5 mm were obtained which were hard but not brittle. High concentrations of TEOS combined with fast evaporation of volatile materials and high temperatures during thermal treatment resulted in opaque, translucent, or strongly phase separated composites.

The curing of the dried samples under dry nitrogen was either performed in one step using a temperature of 215-235°C for 1-3 h or by slow thermal ramping and long isothermal periods. The thermal treatment imidized the amic acid units and formed the silica network. In both cases the yellow film containing PAA was converted into a dark, transparent, glassy material containing the poly(imide) as evidenced by IR-spectroscopy. The cured composite possessed a characteristic sound when shaken in a glass vial depending on the content of the inorganic material.

Instrumental. Mass loss measurement for determining silica content was carried out with a DuPont 990 thermogravimetric analyzer. ^{29}Si-solid state NMR spectra of the cured polymeric PAA sample and the cured hybrid material were measured with a Bruker CXP 200 spectrometer.

Results and Discussion

Two approaches to hybrid formation were used and involved reaction of TEOS with either a commercial poly(amic acid) (PAA, Ultradel 1414) which contained short poly(dimethyl siloxane) segments or a PAA synthesized with triethoxy silane end groups. The structures of the poly(amic acid) with siloxy end functions and the siloxane-containing PAA (Ultradel 1414) are shown in Figure 2. Typical conditions for hybrid formation from either of these polymers are given in Table III and IV and are discussed below. In this publication we discuss the nature of the silicon species in the hybrid and the effect of acid catalysis on the quality of the hybrid.

Preparation Of The Hybrid. Table III lists conditions used to produce hybrids from end-functional PAA's with the relative composition of each PAA component given. In this case, NMP was used as solvent and the water/TEOS molar ratio was roughly 2. The weight fraction of silica in each hybrid was determined by TGA measurements and corrected for pre-existing Si in the polyimides. Volume fraction silica content was estimated from the weight data. All mixtures listed in Tables III and IV produced transparent hybrids. Table IV presents data for reaction conditions used in the preparation of hybrids from the siloxane containing PAA. The table shows the weight fraction of TEOS used with respect to the PAA. Added water was used in amounts equal to a mole/mole ratio roughly 2 with respect to TEOS. Ultradel 1414 was found by ^{1}H-NMR to have ~11 mole-% siloxane content. DMF was added as need to keep the mixture in a single phase.

Addition of TEOS and water was of major importance in controlling the miscibility of the reaction mixture. When added in several small portions, less solvent was required to retain a homogeneous solution compared to a simultaneous single addition of the same amount of TEOS and water. Nevertheless a critical amount of high boiling point solvent was required for the formation of transparent films. Transparent solutions which had a critically low solvent concentration formed exclusively opaque films.

We concluded from this behavior that two experimental guidelines must be followed during silica formation. The formation of transparent films (hybrid materials) required that the precursor solution remained uniform at all times, and this solution also was of an appropriate consistency to avoid phase separation during film drying. In cases when the inorganic material was no longer stabilized with solvent, the precursor solution became inhomogeneous resulting in the formation of light scattering second phase droplets. These droplets contained TEOS, water and HCl (if used), and could be dissolved with additional solvent rendering a transparent solution. However, when no additional solvent was added the hydrolysis/condensation reaction of TEOS

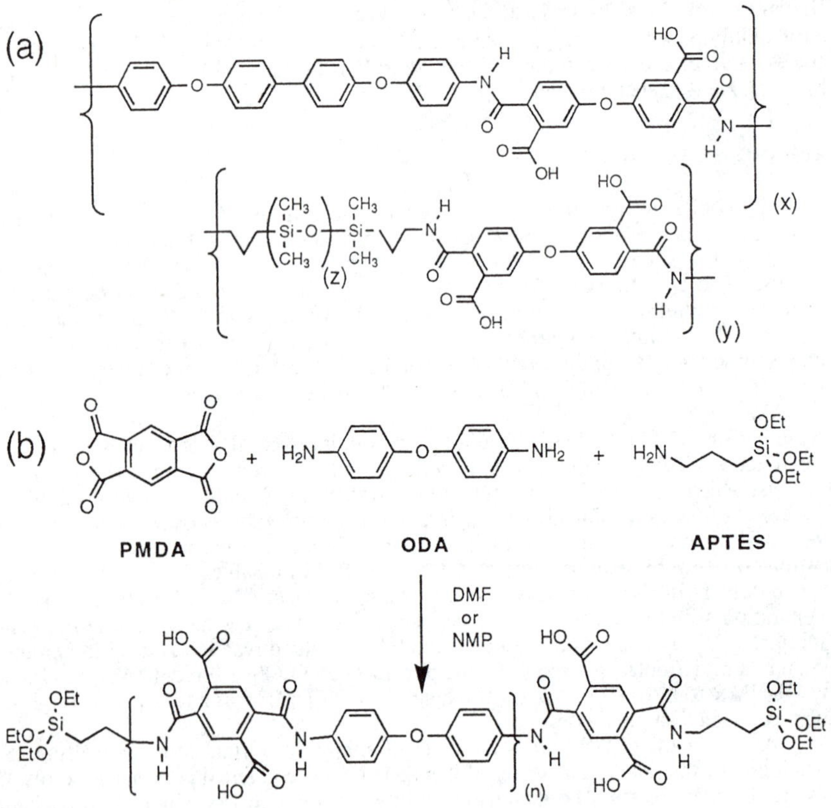

(a)

H N HO O O O O N H O O OH (x)

CH_3 Si-O CH_3 (z) CH_3 Si CH_3 H N HO O O O O N H O O OH (y)

(b)

O O O O O O + H_2N O NH_2 + H_2N Si OEt OEt OEt

PMDA ODA APTES

DMF or NMP

OEt EtO Si EtO N H HO O O HO O O N H O (n) N H HO O O HO O O N H OEt Si OEt OEt

Figure 2. Chemical structure of poly(amic acid)s used in this study: (a) Ultradel 1414; and (b) poly(amic acid) produced from PMDA, ODA and APTES.

within these domains resulted in insoluble silica particles and opaque hybrids. Transparent solutions with a low solvent concentration also formed opaque hybrids, because of the uncontrolled evaporation of the solvent.

Effect of Acidity. We compared the effect of aqueous hydrochloric acid on the efficiency and speed of silica network formation to solutions containing only distilled water. In agreement with published papers (*21, 22*) which described the positive effect of added catalysts (i.e. acids, bases, or salts), we observed an acceleration in silica formation due to hydrolysis of TEOS when hydrochloric acid was added. Concentration of hydrochloric acid above 9 wt% in the added water accelerated silica formation and resulted in rapidly phase separated solutions. Unfortunately this acceleration resulted in opaque solutions and films, whereas comparable mixtures containing only distilled water formed transparent solutions and films.

We concluded from this behavior that acidic conditions favor the hydrolysis/self-condensation reaction of TEOS more than that with either the polymeric PAA or the end-functional PAA. This resulted in the rapid crosslinking of TEOS while incorporating little or no organic polymer, and was followed by phase separation due to immiscibility of the resulting silica particles. Very acidic conditions resulted in a largely inorganic network which scattered light due to large silica particles. Use of NMP or DMF buffered the solution resulting in a decrease of acidity due to the basic character of these solvents resulting in transparent hybrids. THF solution produced opaque composites only, possibly due to its inability to buffer acidic conditions.

Comparing our results with publications in the poly(imide)/silica area showed some similarities. Kakimoto et al. (*23*) recently reported that transparent poly(imide)/silica hybrids were prepared using a triethylamine salt of the PMDA-ODA PAA (PAA-NEt_3). The improvement over the composites prepared earlier by this group (*17*) and Morikawa et al. (*18*) was obtained through the use of this PAA-NEt_3 salt which was methanol soluble requiring less solvent because TEOS is more soluble in methanol than in DMAc. These hybrids appeared macroscopically transparent (up to 42% silica), and SEM microscopy of these organic / inorganic materials indicated phase separation must be in the submicron regime (0.2 to 0.07 microns, depending on the temperature during evaporation). The use of triethylamine appeared to buffer the solution, reducing the acidity in the sol-gel mixture and resulting in a more controlled hydrolysis/condensation reaction of TEOS.

Molecular Uniformity of the Hybrids. The hybrids described here involved the chemical grafting of silica to the polyimide organic phase. In one case, we end-functionalized our PAA with APTES enabling chemical incorporation into the inorganic network. The result of this approach with hexafunctional PAA's was transparent poly(imide)/silica hybrids which showed no large scale phase separation over the silica contents studied. In this case, chemical attachment of the silica phase occurs at the APTES end groups. Similarly, the use of the siloxane-containing polyimide produced remarkably uniform hybrids. We believe that the grafting reaction occurs in this latter case by a scission-insertion reaction involving the siloxane block of the Ultradel 1414. Such high contents of incorporated silica have not been achieved previously for poly(imide)/silica materials to the author's knowledge. Kakimoto et al. (*17*) and Morikawa et al. (*18*) used a PAA prepared from PMDA-ODA which allowed only physical interactions, and thus did not achieve the same loading of silica in their transparent hybrids.

The preparation of the hexafunctional PAA's was a very simple one step reaction in contrast to the synthesis of PAA's used by Morikawa (*19*) which required 7-9 steps. Therefore this specific synthesis has great potential for sol-gel hybrid reactions. This reaction enables adjustment of the organic material to its desired properties by two simple parameters. One is the length of the organic component between the connection

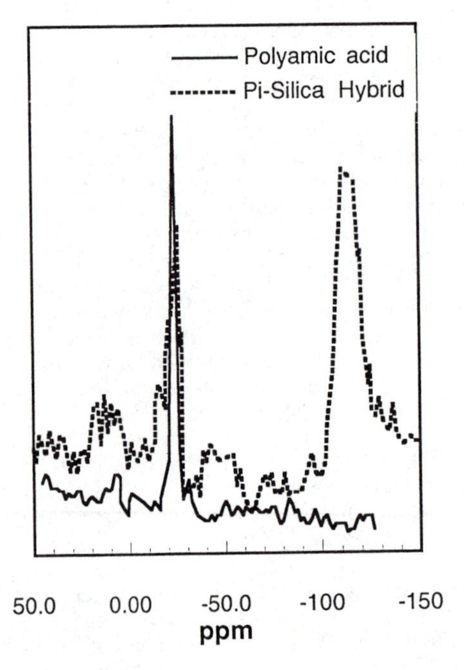

Figure 3. ^{29}Si solid state NMR of Ultradel 1414™ and its sol-gel hybrid containing 50 wt-% silica after vacuum drying and annealing at 250 °C for 10 hours.

points to the inorganic material which can be adjusted by the ratio of dianhydride and diamine. The second is a modification of the properties of the hybrid using selected diamines or dianhydrides.

Characterization of Silicon in the Hybrid. The Ultradel 1414 and its sol-gel hybrid containing 50 wt-% silica were investigated using ^{29}Si-NMR after vacuum drying and annealing at 250 °C for 10 hours. Both the PAA and the composite were preheated up to 300°C in a nitrogen atmosphere prior to annealing for several minutes to minimize crack formation.

In both spectra shown in Figure 3, a peak at -22 ppm was detected which is due to the dimethylsiloxane group. For the hybrid material, an additional broad peak at -111 ppm was detected. Lippmaa et al. (*24*) and Mägi et al. (*25*) published data for different states of silicon identified using ^{29}Si-NMR. The regions for the chemical shifts for different silicon states given by these authors were helpful in identifying the silicon state within our composite. The lack of signal in the region of -60 ppm to -100 ppm indicates that we do not have any Q0 (δ= -60 to -80 ppm), Q1 (δ= -67 to -85 ppm) , or Q2 (δ = -73 to -94 ppm). Silicon in its Q4-state has a chemical shift in the region of (-104 to -120 ppm), thus the peak at δ= -111 ppm for the composite material suggests the formation of a silica network with high Q4 content. The breadth of the peak (δ =-100 to -125 ppm) suggests that some Q3 component may be present in the cured hybrid.

Conclusions

Novel transparent poly(imide)/silica hybrid materials have been synthesized over a full range of organic/inorganic compositions. These transparent hybrids of unusually high inorganic content may be useful in many applications in particular the formation of hard protecting films and barrier layers. Two synthetic strategies were used to create chemical interactions between the organic and inorganic components. In one synthetic scheme, hybrids were prepared from a siloxane containing PAA resulting in transparent materials due to grafting reactions between the siloxane containing blocks and the silica phase. In a second approach, hybrids were prepared from end-functionalized PAAs which also were transparent due to the strong chemical attachment of the PAA end groups to the silica network. During hybrid formation it was necessary that the precursor solution remain transparent which required a critical amount of high boiling solvent to avoid phase separation during drying. Addition of acid catalyst favored self-condensation of the silica network resulting in reduced organic/inorganic interactions and opaque films. DMF and NMP buffering of the acid catalyzed solutions is believed to have aided transparent film formation. ^{29}Si-NMR of the siloxane containing hybrids showed a high Q4 silicon content after thermal treatment.

Acknowledgment. We thank the Industry-Cornell Alliance for Electronic Packaging and the Air Force Office of Sponsored Research for financial support. We thank the Amoco Corporation for the donation of the Ultradel 1414. Some of the sol-gel reactions were carried out by Whitney Hapangama. The ^{29}Si NMR were measured by Dr. J. DuChamps and Prof. T. M. Duncan.

Literature Cited

1. Novak, B.M.; *Adv. Mater.* **1993**, *5*,422.
2. Iler, R. K. ; "The Chemistry of Silica", Wiley & Sons, New York (1979).
3. Brinker, C.J.; Scherer, G.W.; "Sol-Gel Science", Academic Press, New York (1990).

4. Wang, S. B.; Mark, *J. E.; Polym. Bull.* **1987**, *17*, 277.
5. Gunji, T.; Nagao, Y.; Misono, T.; Abe, Y.; *J. Non-Crystalline Solids* **1989**, *107*, 149.
6. Mark, J. E.; Wang, S.-B.; *Polym. Bull.* **1988**, *20*, 443.
7. Hou, L.; Sakka, S.; *J. Non-Crystalline Solids* **1989**, *112*, 424.
8. Wei, Y.; Bakthavatchalam, R.; Whitecar, C. K.; *Chem. Mat.* **1990**, *2*, 337.
9. Landry, C. J. T.; Coltrain, B. K.; *Polymer Preprint (ACS)* **1991**, *32*, 514.
10. Fitzgerald, J. F.; Landry, C. J. T.; Schillace, R. V.; Pochan, J. M.; *Polymer Preprint (ACS)* **1991**, *32,* 532.
11. David, I. A.; Scherer, G. W.; *Polymer Preprint (ACS)* **1991**, *32*, 530.
12. Fujita, M; Honda, K.; *Polym. Commun.* **1989**, *30*, 200.
13. Sun, C.-C.; Mark, J. E.; *Polymer* **1989**, *30*, 104.
14. L. L. Beecroft, C. K. Ober, D. B. Barber, C. R. Pollock, J. L. Mass, and J. M. Burlitch, *Proc. ACS Div.: Polym. Mat.: Sci. & Eng.*, **1995,** *73,* 162 .
15. Nandi, M.; Conklin, J. A.; L. Salvati, L. Jr. and R. Sen.; *Chem Mater.* **1990**, *2*, 772.
16. Nandi, M.; Conklin, J. A.; L. Salvati, L. Jr. and R. Sen.; *Chem Mater.* **1991**, *3*, 201.
17. Kakimoto, M.; Morikawa, A.; Yoshitake, I.; Imai; Y.; *Mat. Res. Soc. Symp. Proc.* **1991**, *22,* 769.
18. Morikawa, A.; Yoshitake, I.; Kakimoto, M.; Imai; Y.; *Polym. J.* **1992**, *24,* 107.
19. Morikawa, A.; Yoshitake, I.; Kakimoto, M.; Imai; Y.; *J. Mater. Chem.* **1992**, *2*, 679.
20. C. K. Ober and N. A. Johnen, *Polymer Preprints: Proc. ACS Div. Polym. Chem.***1995**, *36(1),* 715 .
21. Pope, E. J. A.; Mackenzie, J. D.; *J. Non-Cryst. Sol.* **1986**, *87,* 185.
22. Brinker, C. J.; *J. Non-Cryst. Solids* **1988**, *100,* 31.
23. Kakimoto, M.; Yoshitake, I.; Morikawa, A.; Yamaguchi, H.; Imai; Y.; *Polymer Preprints(ACS)* **1994**, *35,* 393.
24. Lippmaa, E.; Mägi, M.; Sampson, A.; Engelhardt, G.; Grimmer, A.-R.; *J. Am. Chem. Soc.* **1980**, *102,* 4889.
25. Mägi, M.; Lippmaa, E.; Samoson, A.; Engelhardt, G.; Grimmer, A.-R.; *J. Phy. Chem.* **1984**, *88,* 1518.

RECEIVED December 4, 1995

Chapter 26

Thianthrene-Containing Polymers: Polyimides, Aramids, and Polybenzoxazoles

Randy A. Johnson and Lon J. Mathias[1]

Department of Polymer Science, University of Southern Mississippi, Hattiesburg, MS 39406–0076

Thianthrene and thioether-containing aromatic dicarboxylic acids were readily synthesized by nucleophilic aromatic substitution. Polyamides obtained from the diacids displayed good thermal stabilities and solubilities enhanced over typical aramids. Films displayed good toughness and flexibility, consistent with high molecular weight and low crystallinity.

Thianthrene-2,3,7,8-tetracarboxylic dianhydride was prepared starting from dichlorophthalic acid, making the *N*-phenylimide and forming the thianthrene ring with either thioamides or sodium sulfide. Polyimides were synthesized by the low temperature formation of poly(amic acid) solutions which were cast into films and then thermally cyclized. Flexible diamines gave creasible films while rigid diamines gave brittle films, with excellent thermal stability. Polymer solubility was not enhanced by incorporating the bent thianthrene structure.

New thianthrene-containing PBOs were synthesized by the polycondensation reaction of thianthrene dicarbonyl chlorides with bis-o-aminophenols in PPA. The PBOs exhibit high glass transition temperatures and good thermal stabilities. No solubility improvements were observed but films were transparent and tough.

Properties of Thianthrene

Thianthrene is a semi-flexible ring system comprised of two benzene rings bonded through vicinal sulfides (structure and numbering system shown in Figure 1). Initial research suggested thianthrene was folded about the S-S axis, in line with a dipole moment of 1.5 D, although flexibility of the molecule permits the folding to be undergo facile inversion,[1,2] with an energy barrier estimated to be 6-7 kcal/mole.[3,4]

X-Ray measurements on crystalline thianthrene indicated the solid state structure consists of puckered and interleaved layers of molecules (Figure 2);[2,5] the

[1]Corresponding author

0097–6156/96/0624–0403$12.00/0

atoms of each molecule lie in two planes which intersect at the S-S axis to form a dihedral angle of 128° (Figure 1).[6,7,8] This conformation enables the sulfur atoms to retain their "natural" valency angle of ca 100°, similar to that found in 1,4-dithiadiene.[9]

Analogs in which the S atoms of thianthrene are replaced by C(H), N or O to give anthracene, phenazine and dibenzo-p-dioxin, respectively, are all planar molecules (Figure 3 and Table I). Compounds are bent if either or both of the atoms are S, Se or Te (Figure 3 and Table II). This difference depends both on the valency angle of the atoms and the fact that *d* orbitals are used in the bonding of S or S-like atoms.

Table I. Atoms that form planar tricyclic molecules

	X	Y	Reference
Anthracene	C(H)	C(H)	10,11
Acridine	C(H)	N	12
Phenazine	N	N	13
Phenoxazine	N(H)	O	14
Benzo-p-dioxin	O	O	14

* atoms in ( ) accompany atoms not in ( ).

Table II. Atoms that form bent tricyclic molecules

	X	Y	Reference
Phenothiazine	N(H)	S	14
Phenoxthionine	O	S	15
Phenoxselenine	O	Se	15
Phenoxtellurine	O	Te	15
Thianthrene	S	S	5,6
Selenanthrene	Se	Se	2,16

* atoms in ( ) accompany atoms not in ( ).

The use of *d* as well as *p* orbitals gives the C-S bond 28% double bond character without the molecule being planar.[17] The double bond characteristic is estimated from the measured C-S bond length of 1.76 A, reduced from the single-bond value of about 1.81 A.[7] The bond length is in good agreement with C-S values in aromatic substances such as phenyl sulfide.

Initial methods for the synthesis of thianthrene were based on the reaction of benzene with sulfur and its dichloride in the presence of aluminum chloride.[18,19,20]

More recently, thianthrene was synthesized from 1,2-dichlorobenzene with hydrogen sulfide at 550 °C.[21] This method is of particular interest since it demonstrates thianthrene formation in high yields from chloro-displacement reactions, and more importantly indicates thermal stability of the thianthrene nucleus well into the upper temperature range for high-performance polymers.

Derivatized thianthrene molecules (**3**) are prepared by functionalizing preformed thianthrene or starting with the functionality present followed by heterocycle formation (Figure 4). The former involves electrophilic substitution of the thianthrene nucleus (**1**) followed by reaction to give the desired substituents. The most common electrophilic reactions are bromination[22,23] and Friedel-Crafts acylation[24,25,26] which have shown good selectivity in giving thianthrenes substituted primarily in the 2,7-positions with minor amounts of the 2,8-isomers.

The latter method involves a precursor (**2**) which has the desired substituent(s) or protected groups present during formation of the six-member heterocycle.[27,28] Several examples include reactions of substituted bis(o-iodophenyl) disulfide with copper,[29] thermal cyclization of substituted 2-mercaptophenyl phenyl sulfide,[30] aprotic diazotization of substituted 2-(phenylthio)phenyl-2-aminophenyl sulfide,[31] and thermal elimination of nitrogen from substituted 1,2,3-benzothiadiazole.[32] The advantage of using these reactions is the possibility of controlling the position of substitution. However, they suffer from limited availability of starting materials, low reaction yields and/or high material costs.

Two of the target thianthrene compounds, thianthrene-2,7-dicarboxylic acid and thianthrene-2,3,7,8-tetracarboxylic acid, have previously been reported.[24,28,33] The dicarboxylic acid was prepared via direct functionalization of thianthrene, and investigated as a precursor to dyes and aramids as discussed later. The tetracarboxylic acid derivative was prepared by two methods in which the functionality was present before the heterocycle was formed. The first consists of reacting 5-bromo-4-mercaptophthalic acid with Cu_2O. The other involves reacting *N*-phenyl-4,5-dichlorophthalimide (**4**) with thiobenzamide (**5**) to yield *N,N'*-diphenyl-thianthrene-2,3,7,8-tetracarboxylic bisimide (**6**) (Figure 5).[34] Hydrolysis of bisimide (**6**) followed by acidification gave thianthrene-2,3,7,8-tetracarboxylic acid (**7**).

Thianthrene-Containing Macromolecules

Rigid-rod polyimides and aramids are often insoluble and infusible, properties which limit their use to fiber applications.[35,36] Many investigations have succeeded in improving solubility of high-performance polymers through incorporation of flexible groups in the back-bone, pendent substituents or meta-linked backbone moieties.[37,38,39,40,41] However, thermal stability often decreases as solubility increases. Hexafluoroisopropylidene-containing polymers are an exception.[42] The two trifluoromethyl groups enhance solubility by preventing crystallization and chain-packing, and a high T_g results from steric inhibition to rotation by the bulky groups. In addition, the strong C-F bonds help maintain excellent thermal stability. However, hexafluoroisopropylidene-containing polymers have poor UV stability.[43]

Incorporation of aromatic and heterocyclic rings into the polymer backbone is known to enhance thermal stability. It is also thought that thermal stability may be improved by increasing the number of double-strand heterocycles in place of

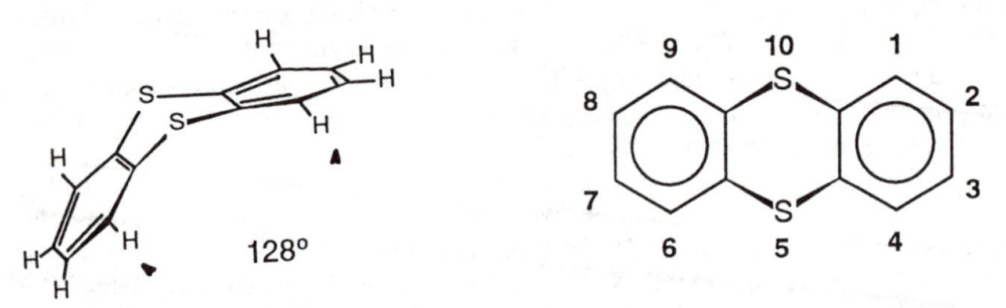

Figure 1. Thianthrene dihedral angles (left) and numbering system (right).

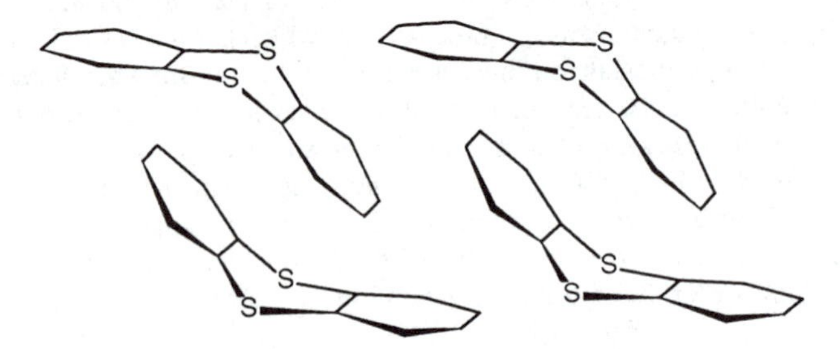

Figure 2. Thianthrene crystal structure.

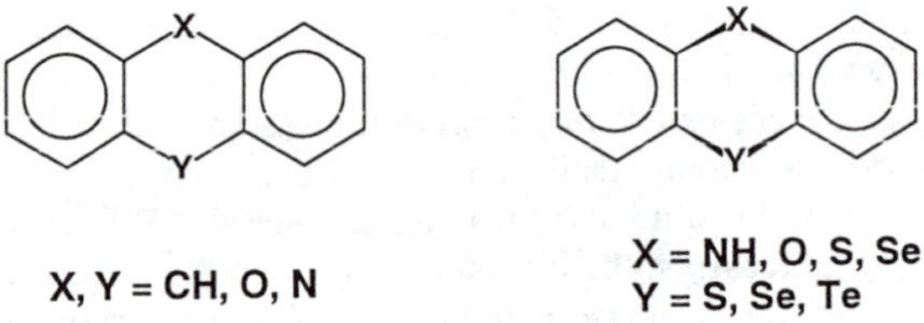

X, Y = CH, O, N

X = NH, O, S, Se
Y = S, Se, Te

Figure 3. Structure of planer and bent tricyclic ring systems.

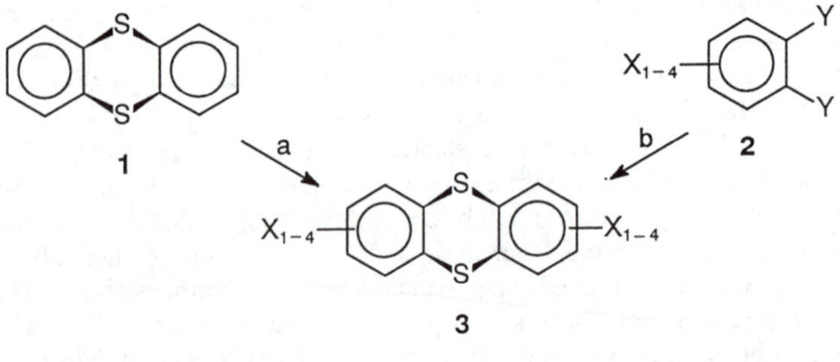

1

2

3

Figure 4. Techniques for obtaining derivatized thianthrenes.

single-strand connecting groups in the polymer backbone.[44] Increased crystallinity, strong intermolecular interaction and tight polymer chain packing also influence thermal behavior. Non-planar, double-strand heterocycles are excellent candidates for improving polymer solubility while retaining good thermal stability. Several articles which focussed on thianthrene-containing polymers are reviewed here.

Thianthrene-containing polyimides have been prepared from thianthrene-2,7- and -2,8-diamine with aromatic tetracarboxylic dianhydrides by the conventional two-step process.[45,46] Their films exhibited amorphous X-ray diffraction patterns, moderately better solubility but reduced thermal stability compared to those derived from 2,7-diaminodibenzo-p-dioxin and the open-chain 4,4'-diaminodiphenyl sulfide. This result was attributed to the dihedral angle of the thianthrene moiety interfering with intermolecular interactions. The more rigid thianthrene-containing polyimides did not exhibit a T_g and those with greater backbone flexibility degraded at lower temperatures, approximately 375 °C and 320 °C in nitrogen and air, respectively.

Thianthrene-containing polyamides were prepared from thianthrene-2,7- and -2,8-diamines with various aromatic dicarboxylic acid chlorides, or by converting thianthrene-containing polyketones to polyamides. Reacting thianthrene diamines with terephthalic and isophthalic acid chlorides at low-temperature in *N*-methylpyrrolidinone (NMP) gave novel aramids.[27,33] X-Ray diffraction indicated that the polyterephthalamides had significant crystallinity, whereas the polyisophthalamides were amorphous. The amide functionality increased solubility compared to polyimides containing thianthrene, benzo-p-dioxin and diphenyl sulfide units. The thianthrene-containing polyisophthalamides had T_g's of 207 °C and 193 °C for the 2,7- and 2,8-isomers, respectively. The polyterephthalamides degraded at approximately 380 °C and 365 °C in nitrogen and air atmospheres, respectively.

Aliphatic polyamides were also prepared from the thianthrene diamines with aliphatic diacid chlorides.[47] These polymers had increased solubility but decreased thermal stability relative to the all-aromatic polymers.

Thianthrene-containing polyamides have also been prepared by reacting thianthrene-2,7-dicarboxylic acid chloride with aromatic and aliphatic diamines at low-temperature.[33] The polymerization of thianthrene-2,7-dicarboxylic acid chloride and 1,4-phenylenediamine yielded 81% polyamide with an inherent viscosity of 0.75 dL/g, although no other polymers or characterization results were discussed.

Thianthrene-containing polyamides were also prepared by converting thianthrene-containing polyketones to polyamides by Schmidt and Beckmann rearrangements.[48] These polymers derived had lower inherent viscosities relative to those prepared by the low-temperature reaction of diamines and diacid chlorides.

Thianthrene-containing polyketones **8** and **9** (Figure 6) were prepared by reacting thianthrene (**1**) with terephthalic and isophthalic diacid chlorides in polyphosphoric acid in the presence of $AlCl_3$.[48] The polyketones were insoluble in almost all organic solvents and had relatively low inherent viscosities, ranging from 0.39 to 0.50 dL/g. Glass transition temperatures were 210 °C and 175 °C, respectively, and thermal stabilities showed degradation onsets of 475 °C in nitrogen.

Overall goals of our research in this area include synthesis of thianthrene di-, tri- and tetracarboxylic acid derivatives (**10, 11, 12, 13**; Figure 7) starting with 3,4-dichlorobenzoic acid and 4,5-dichlorophthalic acid; evaluation of their use in formation of thianthrene-containing polyimides, aramids, and polybenzoxazoles; and

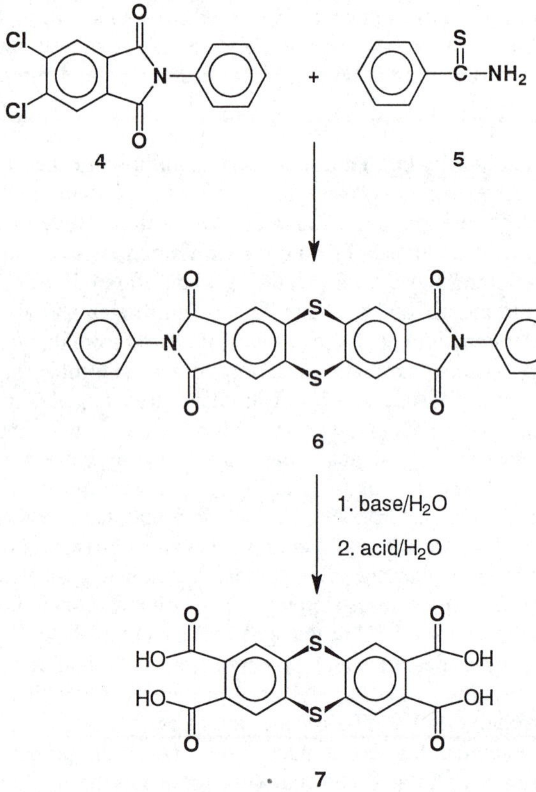

4 + 5

1. base/H_2O

2. acid/H_2O

6

7

Figure 5. Synthesis of thianthrene-2,3,7,8-tetracarboxylic acid (7).

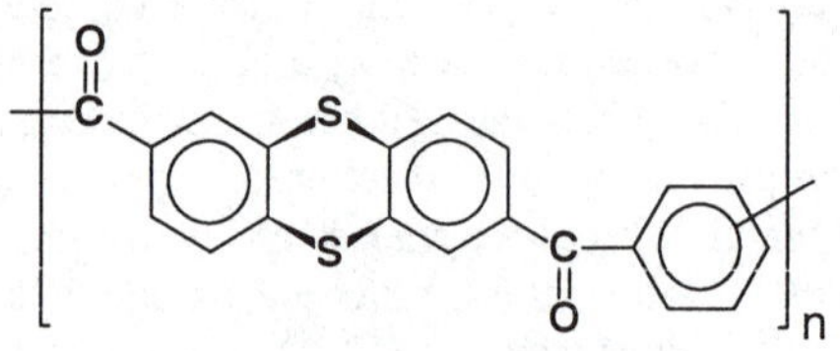

8 = 1,4-phenylene

9 = 1,3-phenylene

Figure 6. Thianthrene-containing polyketones.

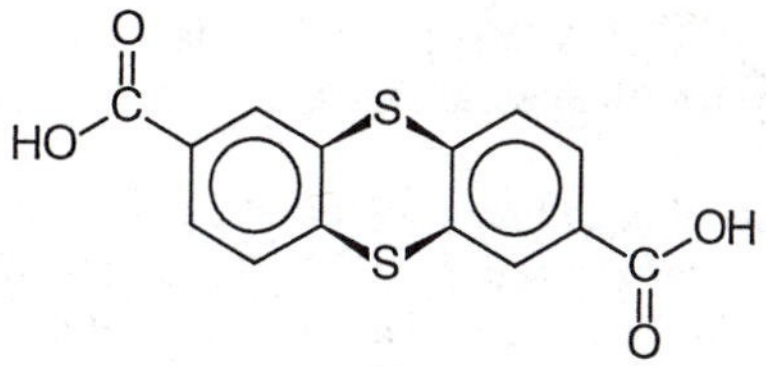

10

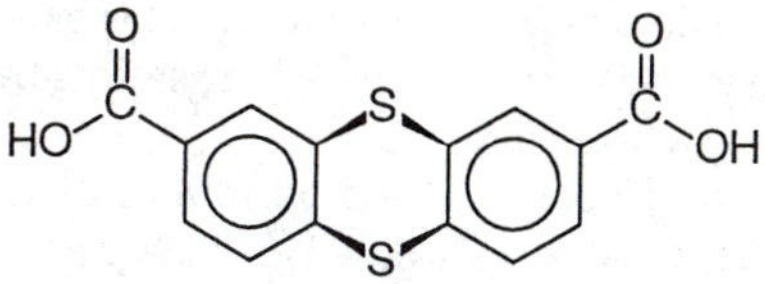

11

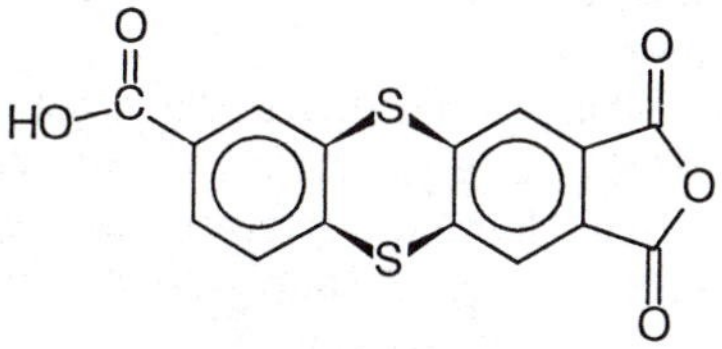

12

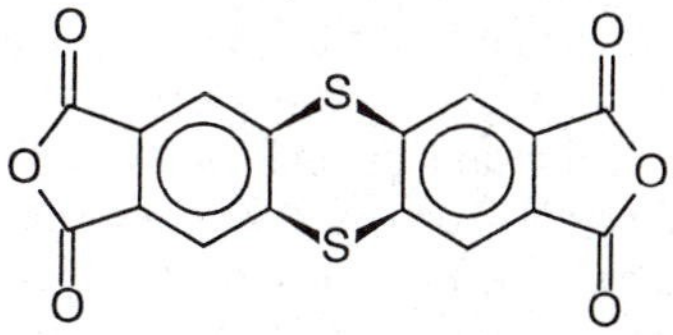

13

Figure 7. Target thianthrene monomers.

characterization of the thianthrene-containing polymers by spectroscopic, physical and mechanical methods. The following sections summarize our overall results.

Thianthrene-based Aramids

Aromatic polyamides (aramids) exhibit high thermal stability and excellent tensile properties because of polymer chain stiffness and intermolecular hydrogen bonding of amide groups. Nomex® and Kevlar® are two of the earliest developed aramids and best examples of these materials. Rigid-rod aramids have excellent properties for fiber applications, although processing into other forms such as films and coatings is difficult since decomposition simultaneously occurs with melting and poor solubility is inherent in these polymers.

Introducing N-substituents improved solubility by removing H-bonding, but at the expense of thermal stability and the desired physical properties. Better results were obtained by introducing flexible groups into the polymer backbone[49,50,51] or including substituents pendent to the backbone.[52,53,54,55] Although thermal stability was moderately reduced, the increase in solubility has enabled aramids to be processed readily from solution.

Thioether-based polyamides have been obtained by incorporating thioether moieties into both polymer backbones and pendent groups. [52,56,57] These aramids displayed moderate increases in solubility but decreases in thermal stability. Incorporation of aromatic and heterocyclic rings into polymer backbones is known to enhance thermal stability. Polymers containing non-planar, double-strand heterocycles, such as thianthrene, should be excellent candidates for improving aramid solubility while retaining good thermal stability.

Thianthrene-2,7-dicarboxylic acid has previously been prepared via direct functionalization of thianthrene, and was investigated as a precursor to dyes and aramids.[58,59] The latter were prepared from the diacid chloride derivative and were obtained with relatively poor yields and viscosities. The solubility and thermal properties were not reported.

The most common method for obtaining aramids is the reaction of diamines with dicarboxylic acid chlorides in a dipolar aprotic solvent at low temperature. *N,N*-Dimethylacetamide (DMAc), *N*-methylpyrrolidinone (NMP) and tetramethylurea (TMU) are the preferred solvents for such polymerizations, giving good polymer solubility and acting as acid acceptors. However, as the molecular weight builds up, organic solvents alone are not sufficient to keep aramids in solution. However, addition of lithium chloride and/or calcium chloride to DMAc has been shown to greatly increase polymer solubility.[60]

More recently, aramids have been obtained by direct polycondensation of aromatic dicarboxylic acids and aromatic diamines in amide solvents using triphenyl phosphite and pyridine as condensing agents.[61,62] This method uses dicarboxylic acids which are easier to handle than acid chlorides and eliminates the additional step of converting the dicarboxylic acids into acid chlorides. Lithium chloride and calcium chloride have also been added to these polymerization mixtures to increase polymer solubility. This method has been used to synthesize both flexible and rod-like polyamides with high inherent viscosity values (ie, up to 6.2 dL/g).[63] It has also been shown that additional triphenyl phosphite ensures complete polymerization with minimal effect on molecular weight.[64] Thus, trace water, which degrades acid chlorides and reduces molecular weight, can easily be consumed by the excess triphenyl phosphite with no molecular weight decrease.

Monomer Synthesis

Syntheses of the functionalized thianthrenes and the thioether intermediate are shown in Figure 8. The thioether bonds of thianthrene and the mono-thioether precursor were formed by nucleophilic aromatic substitution of activated aromatic dichlorides with Na_2S in DMAc. The overall process involves incorporation of the desired aryl functionality first (the carboxylic acid groups), followed by formation of the heterocycle or thioether linkage. This route uses less expensive starting materials than alternatives, and the reaction yields are higher than previously described techniques used for synthesizing thianthrene monomers.

The reaction giving *N,N,N',N'*-tetramethylthianthrene-2,7- and -2,8-dicarboxamides) generates both isomers in essentially equivalent amounts. The bisamide mixture was hydrolyzed to the thianthrene-2,7- and -2,8-dicarboxylic acids (**10** and **11**) which could be separated and purified by fractional recrystallization from DMAc/AcOH. The first step of the reaction has been found to give predominantly the 4,4'-thiobis[*N,N*-dimethyl-3-chlorobenzamide] (**14**), formed first due to the amide group selectively activating the para position for nucleophilic aromatic substitution; no evidence was seen for the 3,4' and 3,3'-isomers.

Formation of the thianthrene moiety evidently involves two routes since both possible derivatives are formed in equal amounts. Displacement of the 3,3'-chlorine substituents of **14** with Na_2S directly yields thianthrene substituted in the 2,8-position. Thioether scrambling of **14** with Na_2S followed by subsequent formation of vicinal sulfide bonds is required to explain formation of thianthrene substituted in the 2,7-positions. Thioether exchange in the intermediate is possible due to the activating influence of amide functionality which strongly influences initial thioether formation. Ether exchange with sulfur nucleophiles has been quite effective for other systems which are activated for nucleophilic aromatic substitution.[65]

By using only half the amount of Na_2S and lower temperatures, the thioether intermediate **14** was obtained as the major product. Unreacted starting material was removed by column chromatography, and since the thianthrene bisamides were present in only small amounts, recrystallization from ethyl acetate allowed isolation of bisamide **14**. No evidence of the 3,4'- or 3,3'-thioether isomers being formed was found. Base hydrolysis of **14** followed by recrystallization gave 4,4'-thiobis[3-chlorobenzoic acid] **15**. For comparison, 4,4'-thiobis[benzoic acid] **16** was synthesized in a similar fashion and its NMR and FTIR spectra obtained.

Polyamide Synthesis and Properties

The preparation of aramids by direct polycondensation of aromatic diamines with aromatic dicarboxylic acids using triphenyl phosphite and pyridine as condensing agents has been well documented.[61,64] This method was used here to polymerize the commercially available diamines, 4,4'-oxydianiline (ODA) and 1,4-phenylenediamine (PDA), with **10**, **11**, **15**, and **16**. Polymers obtained with this method formed fibrous precipitates on pouring the reaction mixtures into stirring methanol. Essentially quantitative yields were obtained for all systems evaluated. High molecular weight

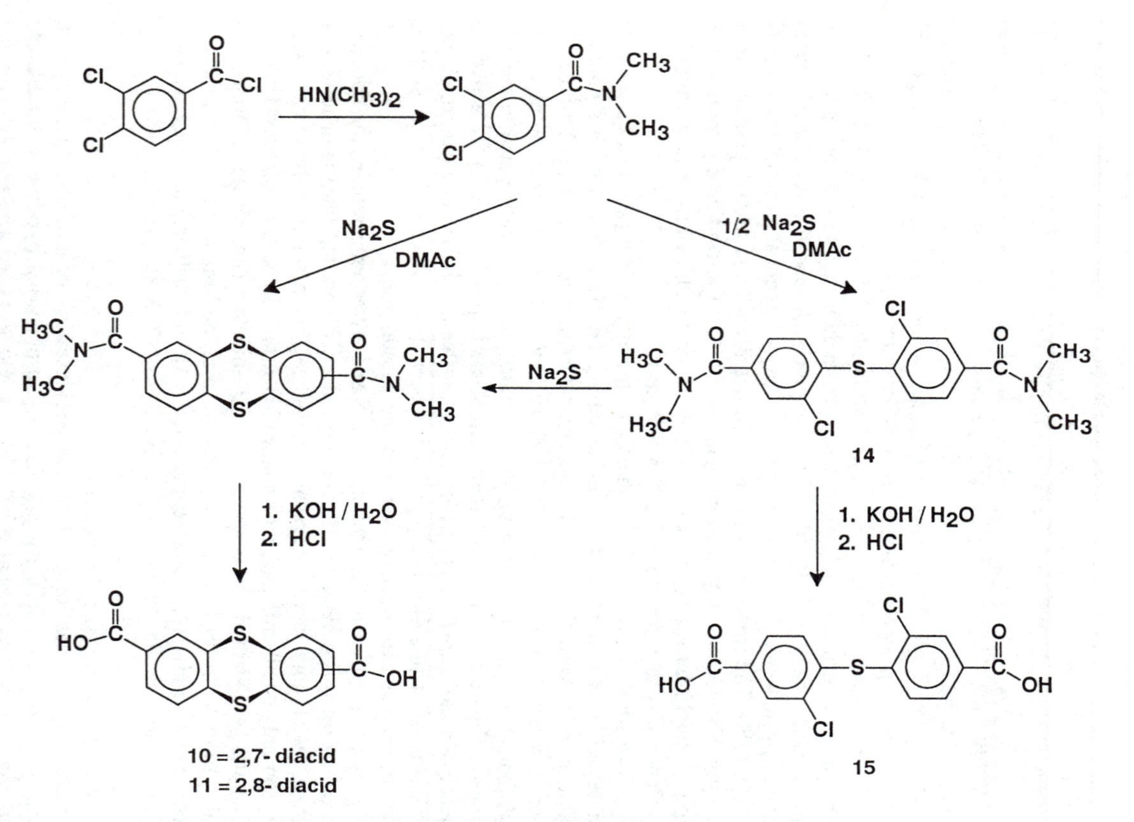

Figure 8. Synthesis of dicarboxylic acid monomers.

polymers were obtained with inherent viscosities of 1.29-2.46 dL/g (in 5 wt-% LiCl-DMAc at 30 °C). Lower molecular weight polymers (n_{inh} < 0.5 dL/g) were obtained when no excess triphenyl phosphite was used. Polyamide molecular weights are strongly affected by the ratio of diacid monomer to triphenyl phosphite.[64]

The polymers gave ^{13}C NMR spectra consistent with the expected structures. Peak positions for the aramids were consistent with those of model bisamides (not shown). Polyamide FTIR spectra also confirmed product structures. All IR spectra showed absorption bands at 3280 cm^{-1} and 1660 cm^{-1} corresponding to the N-H and C=O amide stretching, respectively.

The thianthrene and thioether-containing polymers were moderately more soluble than typical rigid aramids; although solubility depends on the structure of both monomers, the diamines appear to be the limiting factor. The ODA-based polymers have more flexibility than the rigid PDA-based polymers, and were more soluble. Solubility with respect to the aromatic diacids generally decreased in the order **15** > **11** > **10** > **16**. This order was expected based on the molecular symmetry and flexibility of the diacids. The ortho dichloro substituents of monomer **15** appear to create the greatest hindrance to polymer chain packing by causing the benzene rings to twist out-of-plane. Monomers **10** and **11**, which are pseudo-trans and pseudo-cis substituted thianthrenes, respectively, emulate the solubility characteristics for linear and bent polymers synthesized from para and meta functionalized benzene monomers. However, the thianthrene polymers were more soluble than typical aramids presumably due to the dihedral angle of thianthrene somewhat disrupting polymer chain packing. The thioether monomer, with the lowest amount of steric and rotational hindrance of the aryl rings, gave the least soluble polymers. Polymers prepared from thianthrene dicarboxylic acids have very similar solubility characteristics to those prepared from thianthrene diamines. In both cases, the pseudo-cis isomer was more soluble than the pseudo-trans.

Only a few of these polymers displayed an obvious T_g and these values were in the range expected for relatively rigid backbones (265-311 °C). More importantly, thermal stabilities of the polymers were excellent with decomposition beginning above 400 °C in both nitrogen and air. Surprisingly, the thianthrene polyamides had degradation temperatures very close to that of the open chain thioether analog. This differs from the polyamides synthesized from thianthrene diamines with aromatic dicarboxylic acids for which thermal degradation occurred as much as 75 °C below the thioether analogs. This behavior was also observed for thianthrene-based polyimides prepared from thianthrene dianhydride which were thermally more stable than the polyimides prepared from thianthrene diamines.[45,34] The low thermal stability of thianthrene diamine-containing polymers was attributed to the low crystallinity caused by the non-planar thianthrene unit.[47]

The polyamides prepared here also have low crystallinity as determined by wide angle X-ray (no or weak crystalline peaks seen above strong broad amorphous bases). The thermal stability of thianthrene-based polymers may depend on the basic chemical structure (ie, thianthrene dicarboxylic acid derivatives give aramids which are more stable than those derived from thianthrene diamines) as well as on crystallinity, intermolecular interactions and polymer chain packing. However, the possibility of trace amounts of impurity within the thianthrene diamines could have significantly affected the onset of degradation in earlier polymers.

Finally, polyamide films were prepared by spreading filtered solutions of the ODA-based polymers in DMF, and the PDA-based polymers in DMAc/LiCl, onto a teflon coated surface. The films were dried at 60-125 °C in a dust-free chamber equipped with a nitrogen gas inlet. The films prepared with LiCl were immersed in H_2O to remove residual salts and DMAc. The films were light beige in color, and all were tough and flexible. The DMF prepared films were transparent and those derived from DMAc/LiCl were partially cloudy. Absence of residual solvent was demonstrated by TGA.

Thianthrene-containing Polyimides

Polyimides are an important class of commercial polymers possessing outstanding thermal behavior combined with excellent mechanical properties.[66] A large number of derivatives have been synthesized starting from a variety of aromatic or aliphatic diamines or diisocyanates in reaction with tetracarboxylic acids and dianhydrides.[67] Most successful syntheses involve ring-opening of a dianhydride by a diamine to give poly(amic acid)s that are soluble and can be processed by solution methods. Thianthrene-base polyimides were prepared by the conventional two-step process. The first step consisted of reacting dianhydride with aromatic diamines in DMAc at low temperatures to give poly(amic acid)s. The second step involved thermal cyclization of solution-cast films to give the respective polyimides. Polymerization of thianthrene-2,3,7,8-tetracarboxylic dianhydride (**13**) with aromatic diamines gave essentially quantitative yields in all cases and the poly(amic acid) intermediates had intrinsic viscosities in the range of 0.15-0.5 in DMAc at 30 °C (Table III).

The free-standing films were made by casting DMAc solutions of the poly(amic acid)s onto a dust-free glass plate and vacuum drying at ambient temperature until tack-free. Films were cured under a dynamic nitrogen atmosphere at 30 °C for 0.5 h and at 100, 200, 300 °C each for 1 h. In situ FTIR spectra (using a heated IR cell) confirmed the cyclization process. Disappearance of peaks at ca 1220 and 1544 cm^{-1} were diagnostic for reaction of the amide and acid residues while appearance of new peaks at ca 1365 and 1775 cm^{-1} confirmed imide formation.

Table III. Poly(amic acid) and polyimide properties

POLYIMIDE	$[\eta]^a$ dL / g	TGA[b] loss temperature (°C)			
		onset temp. (N_2 / air atm.)		10% wt. loss (N_2 / air atm.)	
7 - ODA	0.82	428	428	562	566
7 - MDA	0.25	435	427	573	563
7 - pPDA	0.46	434	414	564	543
7 - mPDA	0.18	414	421	576	561

[a] Intermediate poly(amic acid)s measured in DMAc at 30 °C.
[b] Measured at heating rate of 10 °C/min.

After cooling to ambient temperature, the amber colored films were removed by soaking in warm water. Films obtained from thianthrene dianhydride and ODA and MDA were creasible as opposed to those derived from 1,3- and 1,4-phenylenediamine which were brittle. Table III includes the thermal analysis data for the prepared polyimides. Degradation onset was above 425 °C in both N_2 and air atmospheres and char yields were > 60 % for all the polymers in N_2 atmosphere at 800 °C. In fact, even in air, significant weight loss is not seen below 500 °C. Although the thianthrene containing polyimides have low crystallinity (as determined by wide angle X-ray analysis performed on the films), the polymers were soluble only in sulfuric acid.

Poly(thianthrene-benzoxazole)s

Aromatic poly(benzoxazole)s (PBOs) are a class of heterocyclic polymers that are known to have excellent thermal stability, high mechanical properties, and good environmental resistance.[68,69,70,71] These outstanding properties have been the driving force to investigate their use in fibers, films, coatings and composites. However, uses of PBOs has been limited since they generally have poor solubility in organic solvents, high glass transition temperatures, and decompose below their melting points.

Processing of PBOs has been improved by the incorporation of the 2,2-hexafluoroisopropylidene (6F) group into the polymer backbone.[71,72,73] These polymers have increased flexibility and solubility plus decreased color, dielectric constant, and crystallinity. However, high-performance polymers containing the 6F group have been shown to be less resistant to UV degradation.[74] Other bridging groups such as ethers and ketones have slightly increased the solubility of PBOs.[75] The thianthrene moiety has not been incorporated into the backbone of poly(benzoxazole)s, although it offers potential for improving solubility without sacrificing other properties.

Aromatic poly(benzoxazole)s are commonly prepared by one of two methods. The first is very similar to the low temperature, two-step synthesis of polyimides. Bis(o-aminophenol)s are reacted with aromatic diacid chlorides to form soluble poly(o-hydroxy amide)s which are processed. Thermal cyclodehydration then converts the backbone groups into benzoxazoles.[73]

The second method is a direct polycondensation reaction of bis(o-aminophenol)s with aromatic dicarboxylic acids or diacid chlorides in polyphosphoric acid (PPA).[76] Dicarboxylic acids have lower costs and are easier to handle than acid chlorides, but are less reactive. The PPA acts both as reaction medium and condensing agent. It has been shown that a high P_2O_5 content provides good monomer and polymer solubility. Specifically, at the end of the polymerization, the PPA should contain ~83.0 % P_2O_5 to ensure high molecular weight PBOs that are still soluble and not too viscous to process.

Thianthrene-2,7- and -2,8-dicarbonyl chlorides were prepared from the dicarboxylic acids with thionyl chloride in approximately 65-80% yield after purification. Poly(phosphoric acid) was used as the polycondensation medium for these polymers. The initial 77.3% P_2O_5 content kept the viscosity low enough to thoroughly disperse the monomers and was adjusted after dehydrochlorination to give a value of 83.0 % at the end of polymerization. This value has been shown to solubilize monomers and polymers, drive the condensation reaction to completion, and keep the viscosity low enough to process directly from the polymerization mixture.[76]

Initial model compound syntheses and polymerizations were done directly with the thianthrene dicarboxylic acids. These monomers had poor solubility in PPA even after P_2O_5 adjustment and with temperatures above 175 °C. Particle size was reduced as much as possible to help solubilization, but apparently the carboxylic acid-to-monomer molecular weight ratio limits solubility of rigid, high melting dicarboxylic acids in the acidic medium. High molecular weight PBOs were obtained from dicarboxylic acids which had good solubility in PPA. The diacid chlorides were more reactive and dissolved readily in the acidic medium at lower temperatures. Polymerizations were usually completed within 24 h at 195 °C due to the viscous reaction mixtures stopping the mechanical stirrer. Monomer solubility and polymer viscosity restricted polymer concentration to < 5%, and these mixtures did not exhibit birefringence or other signs of lyotropic liquid crystallinity. Solid-state ^{13}C NMR of model compound and PBOs correlate well with each other and expected structures. FTIR spectra for both the model compound and polymer showed no O-H and N-H (3000-3500 cm-1) or C=O (1690 cm-1) stretching bands of the monomers.

The PBOs derived from HAB were insoluble in common organic solvents. This required processing of the PBOs directly from the polymerization mixture into films. The films were transparent and tough, but flawed after leaching out the PPA. Solubility could be achieved in strong acids and $AlCl_3/NO_2R$ systems, which form complexes with the benzoxazole that enhance solubility. The 6F-containing polymers had good solubility in m-cresol and films were obtained by casting from 5% m-cresol solutions. Intrinsic viscosities of the 6F-containing polymers were measured in m-cresol at 30 °C and were between 0.88-1.51 dL/g (Table IV).

Table IV. Polymer Characterization

POLYMERS	[η]	T_g (°C)	TGA[a] (°C)	
			onset temp. (N_2 / air atm.)	10% wt. loss (N_2 / air atm.)
10 - HAB	--	450[c]	451 / 450	643 / 551
10 - 6F	0.92[b]	372[d]	478 / 455	533 / 528
11 - HAB	--	--	462 / 391	655 / 590
11 - 6F	1.51[b]	357[d]	435 / 442	545 / 526
15 - HAB	--	--	473 / 459	533 / 511
15 - 6F	0.88[b]	298[d]	453 / 449	500 / 488

[a] Measured at heating rate of 10 °C/min.
[b] Measured in m-cresol at 30 °C.
[c] DMA, measured at a heating rate of 4 °C/min.
[d] DSC, measured at a heating rate of 10 °C/min.

Glass transition temperatures ranged from 298-450 °C (Table IV). No melt transitions were observed for the polymers. The bent thianthrene structure limited crystallinity and chain packing as determined by wide angle X-ray (no crystalline peaks seen above strong broad amorphous bases). Thermogravimetric analysis showed good thermal stabilities in both air and nitrogen atmospheres. PBOs derived from 4,4'-thiobis[3-chlorobenzoic acid] showed an initial weight loss corresponding to elimination of both chlorine substituents followed by catastrophic degradation of the polymer.

Conclusions

Thianthrene-di-, tri-, and tetracarboxylic acids, and a variety of their derivatives, were prepared and polymerized with co-monomers to obtain thianthrene-containing polyimides, aramids and polybenzoxazoles. The multiply substituted thianthrene derivatives were prepared starting with dichloro-substituted benzamide or phthalimide via chlorine displacement by sulfur nucleophiles. The protected carboxyl groups enhanced the displacement reaction to give thianthrene bisamides and imides in good yields. Deprotection with base gave carboxylic acid derivatives.

Thianthrene-2,3,7,8-tetracarboxylic dianhydride polymerized with aromatic diamines by the conventional low temperature technique to yield soluble poly(amic acid)s. Polyimides were obtained by thermal cyclization of poly(amic acid) films. Rigid diamines gave brittle films whereas 4,4'-oxydianiline and 4,4'-methylene dianiline gave films that were creasable. The polyimides had excellent thermal stability in nitrogen and air and a moderate increase in solubility over the more linear and rigid analogs obtained from pyromellitic dianhydride.

Thianthrene-2,7- and -2,8-dicarboxylic acids plus a synthetic intermediate, 4,4'-thiobis[3-chlorobenzoic acid], were converted to new aromatic polyamides having inherent viscosities of 1.29 to 2.39 dL/g by direct polycondensation with 4,4'-oxydianiline and 1,4-phenylenediamine in *N*-methyl-2-pyrrolidinone using triphenyl phosphite and pyridine. Thianthrene-based polyamides were more soluble than analogous poly(thioether amide)s. Polymer films were cast from either DMF or DMAc/LiCl solutions and analyzed by FTIR and NMR. All prepared aramids displayed good thermal stability by DSC and TGA.

Thianthrene and thioether dicarboxylic acids were converted to acid chlorides and polymerized with bis-o-aminophenols to yield new poly(benzoxazole)s (PBOs). Polymers were prepared via solution polycondensation in poly(phosphoric acid) at 90-200 °C. Transparent PBO films were obtained directly from polymerization mixtures or cast from m-cresol solutions. The films were flexible and tough. Non-fluorinated PBOs were soluble only in strong acids and $AlCl_3/NO_2R$ systems which form complexes with the benzoxazole heterocycle to enhance solubility. Glass transition temperatures of these PBOs ranged from 298-450 °C. Thermogravimetric analysis showed good thermal stabilities in both air and nitrogen atmospheres.

Acknowledgment This research was supported in part by a grant from the Office of Naval Research.

References

1. Bergmann, E.; Tschudnowsky, M. *Berichte* **1932**, *65*, 457.
2. Wood, R. G.; Crackston, J. E. *Phil. Mag.* **1941**, *31*, 62.
3. Chandra, A. K. *Tetrahedron* **1963**, *19*, 471.
4. Castrillon, J. P. A.; Szmant, H. H. *J. Org. Chem.* **1967**, *32(4)*, 976.
5. Rowe, I; Post, B. *Acta Cryst.* **1958**, *11*, 372.
6. Lynton, H.; Cox, E. G. *J. Chem. Soc.* **1956**, *954*, 4886.
7. Rowe, I.; Post, B. *Acta Cryst.* **1956**, *9*, 827.
8. Hosoya, B. *Acta Cryst.* **1963**, *16*, 310.
9. Howell, P. A.; Curtis, R. M.; Lipscomb, W. M. *Acta Cryst.* **1954**, *7*, 498.
10. Robertson, J. M. *Proc. Roy. Soc. A* **1933**, *140*, 79.
11. Mathieson, A. M.; Robertson, J. M.; Sinclair, V. C. *Acta Cryst.* **1950**, *3*, 245.
12. Phillips, D. C. *Acta Cryst.* **1956**, *9*, 237.
13. Herstein, F. H.; Schmidt, G. M. J. *Acta Cryst.* **1955**, *8*, 399, 406.
14. Cullinane, N. M.; Rees, W. T. *Trans. Faraday Soc.* **1940**, *36*, 507.
15. Wood, R. G.; McCale, C. H.; Williams, G. *Phil. Mag.* **1941**, *31*, 71.
16. Wood, R. G.; Williams, G. *Nature* **1942**, *150*, 321.
17. Sutton, In *Determination of Organic Structures by Physical Methods*; Braude; Nachod, Ed.; Academic: New York, 1955; p 402.
18. Voronkov, M. G.; Faitel'son, D. D. *Khim. Geterotsikl. Soedin.* **1967**, 245.
19. Schmight, E. V. *Berichte* **1878**, *11*, 1175.
20. Kraft, F.; Lyons, R. E. *Berichte* **1896**, *29*, 437.
21. Deryagina, E. N.; Shagun, L. G.; Ivanova, G. M.; Vakul'skaya, T. I.; Modonov, V. B.; Vitkovskii, V. Y.; Voronkov, M. G. *Zhurnal Organicheskoi Khimii* **1978**, *14(12)*, 2611.
22. Gilman, H.; Swayampati, D. R. *J. Am. Chem. Soc.* **1955**, *77*, 5944.
23. Gilman, H.; Swayampati, D. R. *J. Org. Chem.* **1958**, *23*, 313.
24. Ciba Ltd. Swiss Patent 238 628, 1945; *Chem. Abstr.* **1949**, *43*, 4484c.
25. Vasiliu, G.; Cohn, E. *Rev. Chim.* **1964**, *15(3)*, 139.
26. Prema, S.; Srinivasan, M. *Eur. Polym. J.* **1987**, *23(11)*, 897.
27. Niume, K; Nakamichi, K.; Toda, F.; Uno, K.; Hasegawa, M.; Iwakura, Y. *J. Polym. Sci.: Part A: Polym. Chem.* **1980**, *18*, 2163.
28. Tilika, V.; Polmane, G.; Meirovics, I.; Neilands, O. *Zinat. Akad. Vestis, Kim. Ser.* **1982**, *2*, 201.
29. Berber, H. J.; Smiles, S. *J. Chem. Soc.* **1928**, 1141.
30. Collinane, N. M.; Davies, C. G. *Rev. Cl. Trav. Chim. Pays-Bas* **1936**, *55*, 881.
31. Benati, L.; Montevecchi, P. C.; Tundo, A.; Zanardi, G. *J. Chem. Soc., Perkin Trans.* **1974**, *1*, 1272.
32. Montevecchi, P. C.; Tundo, A. *J. Org. Chem.* **1981**, *46*, 4998.
33. Niume, K.; Nakamichi, K. Japan. Kokai 96 893, 1976; *Chem. Abstr.* **1976**, *85*, 178197n.
34. Yoneyama, M.; Cei, G.; Mathias, L. *Polym. Prepr. (Am. Chem. Soc., Div. Polym. Chem.)* **1991**, *32(2)*, 195.

35. Bessonov, M. I.; Koton, M. M.; Kudryavtsev, V. V.; Laius, L. A. *Polyimides: Thermally Stable Polymers*; Consultants Bureau: New York, 1987; Chapter 2.
36. Penn, L.; Larson, F. *J. Appl. Polym. Sci.* **1979**, *23*, 59.
37. Yang, H. H. *Aromatic High Strength Fibers*, Wiley Interscience: New York, 1989; pp 47-50.
38. Scroog, C. E. *J. Polym. Sci.: Macromol. Rev.* **1976**, *11*, 161.
39. Lubowitz, H. R.; Farrissey, W. J.; Rose, J. S. U.S. Patent 3 708 458, 1973.
40. Heath, D. R.; Takekoshi, T. U.S. Patent 3 879 428, 1975.
41. Adam, A.; Spiess, H. W. *Makromol. Chem., Rapid Commun.* **1990**, *11*, 249.
42. Cassidy, P. E. *Polym. Prepr. (Am. Chem. Soc., Div. Polym. Chem.)* **1990**, *31(1)*, 338.
43. Anzures, E. T., Ph.D. Thesis, University of Southern Mississippi, Aug. 1991.
44. Overberger, C. G.; Moore, J. A. *Adv. Polym. Sci.* **1970**, *7*, 113.
45. Niume, K.; Nakamichi, K.; Takatuka, R.; Toda, F.; Uno, K.; Iwakura, Y. *J. Polym. Sci.: Part A: Polym. Chem.* **1979**, *17*, 2371.
46. Niume, K.; Hirohashi, R.; Toda, F.; Hasegawa, M.; Iwakura, Y. *Polymer* **1981**, *22*, 649.
47. Niume, K.; Toda, F.; Uno, K.; Hasegawa, M.; Iwakura, Y. *Makromol. Chem.* **1981**, *182*, 2399.
48. Niume, K.; Toda, F.; Uno, K.; Hasegawa, M.; Iwakura, Y. *J. Polym. Sci.: Part A: Polym. Chem.* **1982**, *20*, 1965.
49. Cassidy, P. E. *Polym. Prepr. (Am. Chem. Soc., Div. Polym. Chem.)* **1990**, *31(1)*, 338.
50. Oishi, Y.; Takado, H.; Yoneyama, M.; Kakimoto, M.; Imai, Y. *J. Polym. Sci.: Part A: Polym. Chem.* **1990**, *28*, 1763.
51. Srinivasan, R.; Moy, T.; Saikumar, J.; McGrath, J. E. *Polym. Prepr. (Am. Chem. Soc., Div. Polym. Chem.* **1992**, *33(2)*, 225.
52. Lozano, A. E.; De Abajo, J.; De La Campa, G.; Preston, J. *J. Polym. Sci.: Part A: Polym. Chem.* **1992**, *30*, 1327.
53. Hatke, W.; Schmidt, H. W. *Polym. Prepr. (Am. Chem. Soc., Div. Polym. Chem.)* **1991**, *32(1)*, 214.
54. Rogers, H. G.; Gaudiana, R. A.; Hollinsed, W. C.; Kalyanaraman, P. S.; Manello, J. S.; McGowan, C.; Minns, R. A.; Sahatjian, R. *Macromolecules* **1985**, *18*, 1058.
55. Lozano, A.; Preston, J. *Polym. Prepr. (Am. Chem. Soc., Div. Polym. Chem.)* **1993**, *34(1)*, 517.
56. Nykolyszak, T.; Fradet, A.; Marechal, E. *Makromol. Chem.* **1992**, *193*, 2221.
57. Hogo, T.; Kato, R.; Inoe, H.; Ogawara, K. Japan Patent 01 172 427, 1989; *Chem. Abstr.* **1990**, *112*, 57085s.
58. Ciba Ltd. Swiss patent 243 008, 1946; *Chem. Abstr.* **1949**, *43*, 5966h.
59. Niume, K.; Nakamichi, K. Japan. Kokai 96 893, 1976; *Chem. Abstr.* **1976**, *85*, 178197.

60. Preston, J. in *Encyclopedia of Polymer Science and Engineering, 2nd Edition*; Klingsberg, A; Piccininni, R. M.; Salvatore, A.; Baldwin, T., Eds.; John Wiley and Sons: New York, 1988; Vol. 11, pp 397-401.
61. Yamazaki, N.; Matsumoto, M.; Higashi, F. *J. Polym. Sci.: Part A: Polym. Chem.* **1975**, *13*, 1373.
62. Higashi, F.; Goto, M.; Kakinoki, H. *J. Polym. Sci.: Part A: Polym. Chem.* **1980**, *18*, 1711.
63. Krigbaum, W. R.; Kotek, R.; Mihara, Y.; Preston, J. *J. Polym. Sci.: Part A: Polym. Chem.* **1985**, *23*, 1907.
64. Yang, C. P.; Hsiao, S. H.; Huang, C. J. *J. Polym. Sci.: Part A: Polym. Chem.* **1992**, *30*, 597.
65. Cella, J. A.; Fukuyama, J.; Guggenheim, T. L. *Polym. Prepr. (Am. Chem. Soc., Div. Polym. Chem.)* **1989**, *30(2)*, 142.
66. *Polyimides: Synthesis, Characterization and Applications*; Mittal, K. L., Ed.; Plenum: New York, 1984; Vol. 1 and 2.
67. Sroog, C. E. *J. Polym. Sci.: Macromol. Rev.* **1976,** *11*, 161.
68. Arnold, C. *J. Polym. Sci., Macromol. Rev.* **1979**, *14*, 265.
69. Cassidy, P. E. *Thermally Stable Polymers: Syntheses and Properties*; Marcel Dekker: New York, 1980; pp 154-156.
70. Ueda, M.; Sugita, H.; Sato, M. *J. Polym. Sci.: Part A: Polym. Chem.* **1986**, *24*, 1019.
71. Maruyama, Y.; Oishi, Y; Kakimoto, M.; Imai, Y. *Macromolecules* **1988**, *21(8)*, 2305.
72. Dotrong, M.; Dotrong, M. H.; Evers, R. C.; Moore, G. J. *Polym. Prepr. (Am. Chem. Soc., Div. Polym. Chem.)* **1990**, *31(2)*, 675.
73. Joseph, W. S.; Abed, J. C.; Mercier, R.; McGrath, J. E. *Polym. Prepr. (Am. Chem. Soc., Div. Polym. Chem.)* **1993**, *34(1)*, 397.
74. Hoyle, C. E.; Creed, D.; Subramanian, P.; Nagarajan, R.; Pandey, C.; Anzures, E. T. *Polym. Prepr. (Am. Chem. Soc., Div. Polym. Chem.)* **1993**, *34(1)*, 369.
75. Hilborn, J. G.; Labadie, J. W.; Hedrick, J. L. *Macromolecules* **1990**, *23*, 2854.
76. Wolfe, J. F. *Encyclopedia of Polymer Science and Engineering* 1988; Vol. 11, pp 601-635.

RECEIVED December 4, 1995

Chapter 27

A New Facile and Rapid Synthesis of Polyamides and Polyimides by Microwave-Assisted Polycondensation

Yoshio Imai

Department of Organic and Polymeric Materials, Tokyo Institute of Technology, Meguro-ku, Tokyo 152, Japan

A new facile method for the rapid synthesis of aliphatic polyamides and polyimides was developed by using a domestic microwave oven to facilitate the polycondensation of both ω-amino acids and nylon salts as well as of the salt monomers composed of aliphatic diamines and pyromellitic acid or its diethyl ester in the presence of a small amount of a polar organic medium. Suitable organic media for the polyamide synthesis were tetramethylene sulfone, amide-type solvents such as *N*-cyclohexyl-2-pyrrolidone (CHP) and 1,3-dimethyl-2-imidazolidone (DMI), and phenolic solvents like *m*-cresol and *o*-chlorophenol, and for the polyimide synthesis amide-type solvents such as *N*-methyl-2-pyrrolidone, CHP, and DMI. In the case of the polyamide synthesis, the polycondensation was almost complete within 5 min, producing a series of polyamides with inherent viscosities around 0.5 dL/g, whereas the polyimides having the viscosity values above 0.5 dL/g were obtained quite rapidly by the microwave-assisted polycondensation for only 2 min.

Recently there has been growing interest in applying microwave energy to synthetic organic chemistry and synthetic polymer chemistry as well. In the latter field, microwave energy has been utilized for the radical polymerization of vinyl monomers such as 2-hydroxyethyl methacrylate (*1*), methyl methacrylate (*2*), and styrene (*3*), and for the curing of polymers such as epoxy resins (*4-10*) and polyurethanes (*11,12*), as well as for the imidization of polyamic acids (*13,14*). Among them, much efforts have been directed toward the curing of epoxy resins from the practical viewpoint. In most cases, the high heat efficiency gave rise to remarkable rate enhancements and dramatic reduction of reaction times. However, there is no report so far except for our studies (*15,16*) on the synthesis of condensation polymers by using microwave energy. We have already reported the rapid synthesis of aromatic polyamides by microwave-assisted direct polycondensation of aromatic diamines and aromatic dicarboxylic acids with condensing agents(*15*), and a preliminary study on the microwave-assisted rapid synthesis of polyamides from nylon salts (16). This article reviews the first successful rapid synthesis of aliphatic polyamides and polyimides by using a microwave oven to facilitate the polycondensation of both ω-amino acids and nylon salts, as well as of the salt monomers composed of aliphatic diamines and pyromellitic acid or its diethyl ester (*16-19*).

0097–6156/96/0624–0421$12.00/0

Experimental Section

Apparatus. The apparatus used for the polycondensation was a Mitsubishi RR-32 domestic microwave oven (500 W, 2.45 GHz). We made a minor modification to drill a small hole on the top of the microwave oven, and adopted an open reaction system using a 30 mL wide-mouth vial as a reaction vessel.

Monomers and Reagents. Commercial 6-aminohexanoic acid (ε-aminocaproic acid) and 11-aminoundecanoic acid were purified by recrystallization with water. Hexamethylenediamine and dodecamethylenediamine were obtained commercially and purified by distillation under reduced pressure. 12-Aminododecanoic acid, hexanedioic acid (adipic acid), octanedioic acid (suberic acid), decanedioic acid (sebacic acid), and dodecanedioic acid, as well as the solvents employed for the polymerization, were obtained commercially and used without further purification.

A series of nylon salts were prepared by a previously reported procedure (*20*) which involved mixing an ethanol solution of an aliphatic diamine with an ethanol solution of an aliphatic dicarboxylic acid, followed by recrystallization of the resultant precipitates with water.

Pyromellitic acid (PMA) was prepared by a usual hydrolysis of pyromellitic dianhydride, purified by sublimation, with water. 1,4-Diethyl ester of pyromelitic acid, i.e. 2,5-di(ethoxycarbonyl)terephthalic acid (PME), was synthsized according to the reported procedure (*21*) by the reaction of PMDA with absolute ethanol. A series of nylon-salt-type monomers XPMA and XPME, where X stands for the number of methylene unit in the aliphatic diamine, were prepared by a usual procedure by mixing an ethanol solution of an aliphatic diamine with an ethanol solution of PMA and PME, respectively, followed by recrystallization of the resultant precipitates with water (*22,23*).

Polymerization Procedure. The reaction vessel containing a mixture of reactants and a solvent was placed on the center of the turn table in the microwave oven, then nitrogen gas was introduced from the top of the reaction vessel through a thin Teflon tube to minimize danger of fire in the reaction system. The microwave was irradiated for a prescribed time. The temperature of the reaction mixture was determined just immediately after removal of the reaction vessel from the microwave oven by using a thermocouple and a temperature recorder. The polymer formed was isolated by washing the reaction mixture with methanol, followed by drying under vacuum.

Results and Discussion

Microwave Heating of Solvents. It is well known that the heat generation by microwave irradiation is proportional to the product of dielectric constant ε and dielectric loss tangent tan δ of the material (*4*). It is also known that a material of larger dielcetric constant ε generally has a larger dielectric loss tangent tan δ. Hence, a polar material generates much heat quickly by microwave irradiation.

In fact, highly polar solvents such as water, 1,3-dimethyl-2-imidazolidone (DMI), and tetramethylene sulfone (sulfolane) generated heat quickly by the microwave irradiation. For example, water (ε = 78) was heated within 1 min to reach the boiling point near 100°C. Highly polar and high-boiling-point solvents such as DMI (ε = 37 and bp 225°C) and sulfolane (ε = 43 and bp 287°C) were heated to 180°C after 1 min of the microwave irradiation. The heat generation of less polar *m*-cresol (ε = 12 and bp 202°C) was intermediate between that of water and DMI or sulfolane.

Polyamide Synthesis from ω-Amino Acids. The polycondensation of ω-amino acids was carried out with a domestic microwave oven under nitrogen atmosphere (Eq 1).

$$H_2N(CH_2)_xCOOH \longrightarrow [\text{–}NH(CH_2)_xCO\text{–}]_n + H_2O \qquad (1)$$

$x = 5, 10, 11$

In order to induce effective homogeneous heating of the monomers and hence cause efficient polycondensation, we employed a small amount of an organic medium that acts as solvent for both the starting monomers and the resultant polyamides. Figure 1 exhibits the solvent effect on the polymerization temperature and the inherent viscosity of the polyamide formed by the polycondensation of 12-aminododecanoic acid as a function of microwave irradiation time. Here we used *m*-cresol, DMI, and sulfolane as solvents. The temperature of the reaction mixture rose quickly and reached tenperatures in excess of 270°C after 4 min of microwave irradiation. At that time, most of the solvent evaporated leaving a melt (melting temperature of the polymer is 179°C). During this initial period, the polycondensation proceeded rapidly, producing the polyamide having inherent viscosities of around 0.4 dL/g. Between 4 min and 10 min, the reaction temperature remained almost constant, while the inherent viscosity of the polymer gradually increased, finally reaching a value of about 0.7 dL/g.

Table 1 summarizes the solvent effect on the inherent viscosity of the polyamide formed by the polycondensation of 12-aminododecanoic acid for the microwave irradiation time of 4 or 5 min. From this Table, large temperature differences were observed clearly between the final temperature of the solvent alone and the final temperature of the polymerization mixture. This is a clear indication of the participation of the polar amino acid monomer dissolved in the solvent to generate heat during the microwave irradiation polymerization. When water, which has the highest dielectric constant, was used as the solvent, the polyamide having an inherent viscosity of 0.35 dL/g was obtained. In general, the use of the solvents having both high dielectric constant and high boiling point, reaching high final polymerization temperature, led to the formation of the polyamides having inherent visocisties higher than 0.5 dL/g. Such suitable solvents were sulfolane, and amide-type solvents such as *N*-cyclohexyl-2-pyrrolidone and DMI. Hydroxyl-containing solvents having high dielectric constant like ethanediol, and those with less polarity such as benzyl alcohol, *m*-cresol, and *o*-chlorophenol, were also effective for producing the polyamide with the viscosity values around 0.5 dL/g or higher. The fact that *m*-cresol and *o*-chlorophenol were very good solvents for the aliphatic polyamides is probably related to the ready formation of the polyamide with high viscosity values. The solvent with a high boiling point but low dielectric constant, like diphenyl ether, was not heated too much, probably due to limited solubility of the monomer, thereby giving no polymer.

It is apparent again from Table 1 that the viscosity values were greatly affected by the final reaction temperature, and higher final temperature afforded the polyamide having higher viscosity value. In general, the final reaction temperatures in excess of 280°C were required to obtain this type of polyamide with inherent viscosities exceeding 0.5 dL/g. Thus, the solvents played a very important role for the microwave-assisted polycondensation.

Figure 2 shows a comparison of the time dependence curve of the inherent viscosity of the polyamide formed by the microwave-assisted polycondensation of 12-aminododecanoic acid with the same polymer as obtained by conventional melt polycondensation. It is evident that the microwave-assisted polycondensation, curve A, proceeded much faster than the melt polycondensation, curve B. Thus, it is concluded that the internal heat generation of both solvent and the monomer under the microwave irradiation was much more effective for the progress of the polycondensation, producing the polyamide having a high viscosity value in a shorter polymerization time, compared with conventional external heating.

The microwave-assisted polycondensation was extended to the polycondensation of ω-amino acids such as 6-hexanoic acid and 11-aminoundecanoic acid in addition to 12-aminododecanoic acid. Table 2 summarizes the results of the microwave-assisted synthesis of various polyamides. All three amino acids readily afforded the polyamides having reasonable inherent viscosities around 0.5 dL/g in such solvents as *m*-cresol, DMI, and sulfolane for only 4 min of the microwave irradiation.

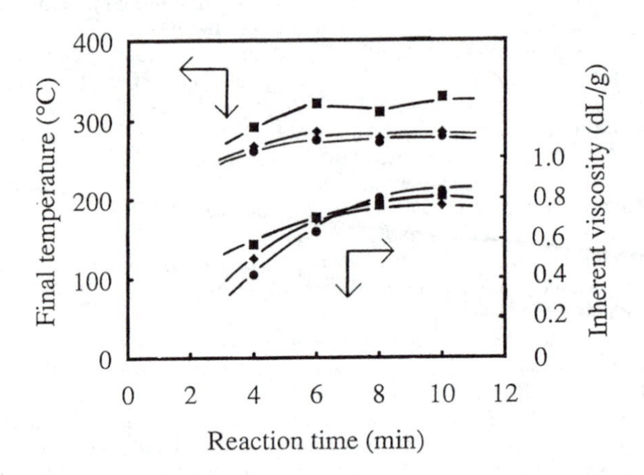

Figure 1. Time dependence of final temperature and inherent viscosity of the polyamide formed by the microwave-assisted polycondensation of 12-aminododecanoic acid (2 g) in different solvents (2 mL): (●) m-cresol, (◆) DMI, and (■) sulfolane.

Table 1. Solvent Effect on Microwave-Assisted Polycondensation of 12-Aminododecanoic Acid [a]

Solvent				Reaction		Polymer
Type	ε	Bp (°C)	Ft [b] (°C)	Time (min)	Ft [c] (°C)	η_{inh} [d] (dL/g)
Water	78	100	97	5	220	0.35
Dimethyl sulfoxide	47	189	172	4	259	0.24
Tetramethylene sulfone	43	287	224	5	282	0.53
Dimethylacetamide	38	166	163	5	281	0.23
1,3-Dimethylimidazolidone	37	225	202	5	300	0.60
N-Methylpyrrolidone	32	202	179	4	267	0.39
N-Cyclohexylpyrrolidone	---	---- [e]	224	5	266	0.46
Nitrobenzene	35	211	198	5	264	0.42
Ethanediol	38	197	193	5	317	0.59
1,4-Butanediol	31	229	189	5	242	0.24
Benzyl alcohol	13	205	128	5	248	0.50
m-Cresol	12	202	153	5	308	0.63
o-Chlorophenol	6	176	110	5	340	0.63
Diphenyl ether	4	258	66	5	109	-----

[a] The polymerization was carried out with 2 g of the monomer and 2 mL of the solvent under microwave irradiation. [b] Final temperature of the solvent alone after 2 min of microwave irradiation. [c] Final temperature of the reaction mixture. [d] Measured at a concentration of 0.5 g/dL in *m*-cresol at 30°C. [e] Bp = 154°C/ 7 Torr.

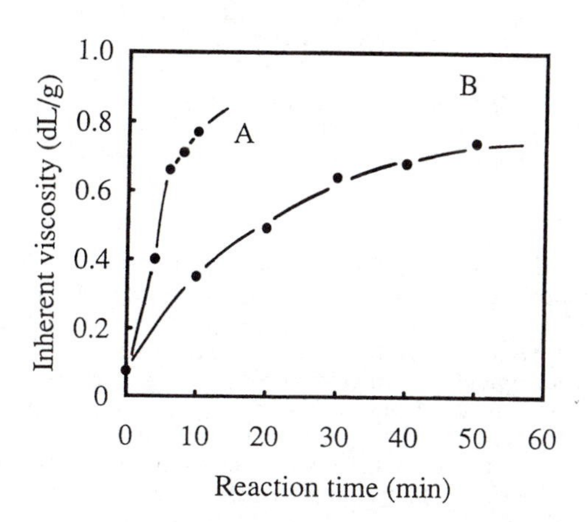

Figure 2. Time dependence of inherent viscosity of the polyamide formed by the polycondensation of 12-aminododecanoic acid (2 g): (A) the microwave-assisted polycondensation in m-cresol, and (B) the melt polycondensation at 210°C.

Table 2. Microwave-Assisted Synthesis of Various Polyamides from Both ω-Amino Acids and Nylon Salts [a]

Polymer type	Polymer η_{inh} [b] (dL/g)		
	m-Cresol	DMI [c]	Sulfolane [d]
Nylon 6	0.50	0.52	0.41
11	0.58	0.59	0.54
12	0.35	0.62	0.48
66	0.27	0.29	0.17
68	0.22	0.24	0.26
610	0.20	0.21	0.21
612	0.24	0.25	0.21
126	0.58	0.49	0.57
128	0.33	0.36	0.32
1210	0.52	0.36	0.51
1212	0.41	0.39	0.33

[a] The polymerization was carried out with 2 g of the ω-amino acid or nylon salt and 1 mL of the solvent under microwave irradiation for 4 min. [b] Measured at a concentration of 0.5 g/dL in *m*-cresol at 30°C. [c] DMI: 1,3-dimethylimidazolidone, and the polymerization was conducted for 5 min. [d] Sulfolane: tetramethylene sulfone.

Polyamide Synthesis from Nylon Salts. The polycondensation of nylon salts was also accelerated by the microwave irradiation (Eq 2).

$$H_2N(CH_2)_xNH_2 + HOOC(CH_2)_yCOOH$$

$$\longrightarrow {}^+H_3N(CH_2)_xNH_3{}^{+-}OOC(CH_2)_yCOO^-$$

$$\longrightarrow [-NH(CH_2)_xNHCO(CH_2)_yCO-]_n + 2\,H_2O \qquad (2)$$

$$x = 6, 12 \qquad y = 4, 6, 8, 10$$

Table 2 also lists the results of the microwave-assisted syntheses of a series of polyamides starting from the nylon salts. The polycondensation of the nylon salts was again accelerated by the microwave irradiation in the presence of polar organic solvents, yielding the polyamides having viscosity values between 0.2 and 0.5 dL/g. In more detail, the hexamethylenediamine-based nylon salts such as 66, 68, 610, and 612 tended to discolor partly during the microwave heating, thereby giving the polyamides with rather lower viscosity values around 0.2 dL/g. On the contrary, such a discoloration was not observed for the polycondensations of the dodecamethylenediamine-bearing nylon salts such as 126, 128, 1210, and 1212, resulting in the polyamides with higher inherent viscosities in the range of 0.3–0.6 dL/g.

Polyimide Synthesis from Nylon-Salt-Type Monomers. The microwave-assisted polycondensation was further applicable to the effective, rapid synthesis of aliphatic polypyromellitimides from the salt monomers XPMA and XPME composed of aliphatic diamines and both pyromellitic acid PMA and its diethyl ester PME (Eqs 3 and 4).

$${}^+H_3N(CH_2)_xNH_3{}^+ \quad {}^-OOC,\ COOH,\ HOOC,\ COO^- \text{ (on benzene ring)} \qquad \text{XPMA}$$

$$\longrightarrow \left[-(CH_2)_x-N\langle^{CO}_{CO}\rangle C_6H_2 \langle^{CO}_{CO}\rangle N-\right]_n + 4\,H_2O \qquad (3)$$

$${}^+H_3N(CH_2)_xNH_3{}^+ \quad {}^-OOC,\ COOEt,\ EtOOC,\ COO^- \text{ (on benzene ring)} \qquad \text{XPME}$$

$$\longrightarrow \left[-(CH_2)_x-N\langle^{CO}_{CO}\rangle C_6H_2 \langle^{CO}_{CO}\rangle N-\right]_n + 2\,H_2O + 2\,EtOH \qquad (4)$$

$$x = 6 \sim 12$$

For the polycondensation of the nylon-salt-type monomers leading to aliphatic polypyromelitimides, we selected here high-boiling-point and polar organic media such as N-methyl-2-pyrrolidone (NMP, ε = 32 and bp 202°C), N-cyclohexyl-2-pyrrolidone (CHP, bp 154°C/7 Torr), 1,3-dimethyl-2-imidazolidone (DMI), and tetramethylene sulfone (sulfolane). Figure 3 shows the time dependence of the inherent viscosity of the polyimide formed by the microwave-assisted polycondensation of the dodecamethylenediamine-based salt monomers, 12 PMA and 12 PME, in the presence of DMI. The polycondensation proceeded very rapidly, and only 2 min of the microwave irradiation afforded readily the polyimide having an inherent viscosity of 0.7 dL/g. After that time, the increase in the inhernt viscosity was rather gradual with prolonged microwave irradiation. No appreciable differences in the reactivity were observed between these two types of the salt monomers, 12 PMA and 12 PME, with respect to the inherent viscosities of the resultant polyimides.

The solvent effect on the microwave-assisted polycondensation of two salt monomers 12PMA and 12PME is summarized in Table 3. The best polymerization medium for this polycondensation was DMI, which gave the polyimides with inherent viscosities about 0.7 dL/g or higher. Other amide-type solvents such as NMP and CHP were also effective for producing the polyimides with high viscosity values. Sulfolane was the next suitable solvent. However, *m*-cresol, which was found to be one of the best solvent for the microwave-assisted polyamide synthesis, was inadequite for the polyimide synthesis because of lower solubility of these pyromellitic acid-based salt monomers.

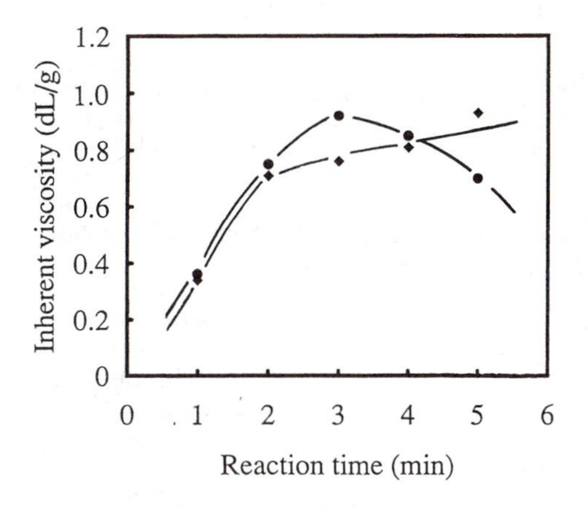

Figure 3. Time dependence of inherent viscosity of the polyimide formed by the microwave-assisted polycondensation of 12PMA (-●-) and 12PME (-◆-) in DMI solvent (1 mL), where the monomer was 1 g.

Table 3. Solvent Effect on Microwave-Assisted Polycondensation of Salt Monomers of 12PMA and 12PME [a]

Solvent			Polymer η_{inh} [b] (dL/g)	
Type	ε	Bp (°C)	From 12PMA	From 12PME
N-Methylpyrrolidone	32	202	0.51	0.59 [c]
N-Cyclohexylpyrrolidone	---	---- [d]	0.74 [c]	0.74 [c]
1,3-Dimethylimidazolidone	37	225	0.68	0.86
Tetramethylene sulfone	43	287	0.31 [c]	0.12 [c]

[a] The polymerization was carried out with 1 g of the monomer and 2 mL of the solvent under microwave irradiation for 2 min. [b] Measured at a concentration of 0.5 g/dL in concentrated sulfuric acid at 30°C. [c] A partial gelation occurred and the viscosity measurement was conducted after filtration of the solution. [d] Bp = 154°C/7 Torr.

The microwave-assisted polycondensation was extended to the polycondensation of a series of PMA- and PME-based salt monomers having 6 to 12 methylene units. The results of the polycondensation are tabulated in Table 4. All the polycondensations were carried out in the presence of DMI, and proceeded very rapidly under microwave irradiation for only 2 min, yielding quantitatively the polyimides with high inherent viscosity values. When the pyromellitic acid based XPMA salt monomers were used, a series of polyimides having inherent viscosities in the range of 0.5 and 0.8 dL/g were readily obtained. In the case of the XPME salt monomers composed of aliphatic diamines and pyromellitic acid diethyl ester, these salt monomers dissolved more readily in DMI compared with the parent mPMA salt monomers, and hence the microwave-assisted polycondensation rapidly produced a series of polyimides with higher viscosity values between 0.8 and 1.6 dL/g. In more detail, when the salt monomers having 11 and 12 methylene units were used, the polymerization mixture remained in a melt state even after the solvent was evaporated, because the resultant polyimides had melting temperatures around 300°C. In the case of the polycondensation of the salt monomer having 6 methylene unit, the powdery polymer was obtained due to having high polymer melt temperature about 450°C, while all the other salt monomers having 7 to 10 methylene units afforded the polyimides with partly melted appearance. Hence the polymerization temperature must have reached to 300°C but not exceeded 400°C.

Table 4. Microwave-Assisted Synthesis of A Series of Polyimides from Nylon-Salt-Type Monomers [a]

Monomer		Polymer	
Code	Mp [b] (°C)	η_{inh} [c] (dL/g)	Tm [b] (°C)
6PMA	242	0.51	447
7PMA	232	0.82	345
8PMA	272	0.71	379
9PMA	258	0.70	310
10PMA	224	0.73	337
11PMA	226	0.67	297
12PMA	234	0.68	300
6PME	204	1.44	—
7PME	210	1.49 [d]	—
8PME	180	1.61 [d]	—
9PME	183	1.32	—
10PME	179	0.98	—
11PME	178	1.14	—
12PME	173	0.86	—

[a] The polymerization was carried out with 1 g of the monomer and 2 mL of 1,3-dimethylimidazolidone under microwave irradiation for 2 min. [b] An endothemic peak temperature determined by DTA at a heating rate of 10°C/min in nitrogen. [c] Measured at a concentration of 0.5 g/dL in concentrated sulfuric acid at 30°C. [d] A partial gelation occurred and the viscosity measurement was conducted after filtration of the solution.

Figure 4 shows a comparison of the time dependence curve of the inherent viscosity of the polyimide formed by the microwave-assisted polycondensation of 12PMA with that by a conventional solid-state polycondensation at 250°C. It is obvious that the microwave-assisted polycondensation for curve A proceeded much faster than the solid-state polycondensation for curve B. Thus it is concluded that the internal heating by the microwave irradiation was highly effective compared with a conventional external heating, yielding the polyimide with a high inherent visocisty in a very short polymerization time.

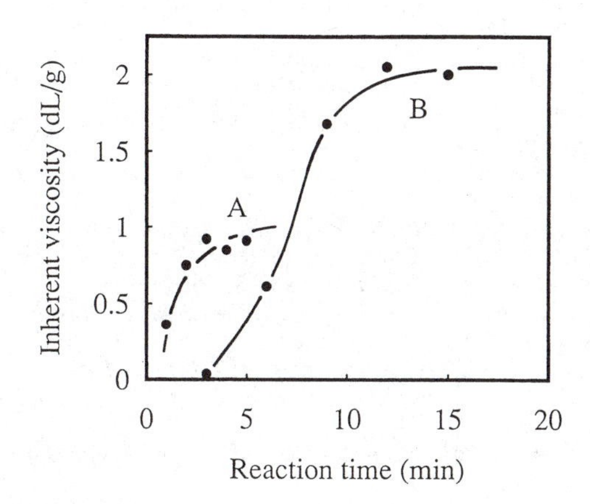

Figure 4. Time dependence of inherent viscosity of the polyimide formed by the polycondensation of 12PMA (1 g): (A) the microwave-assisted polycondensation in DMI (1 mL), and (B) the thermal polycondensation at 250°C.

Conclusions

We have developed a new facile method for the rapid synthesis of aliphatic polyamides and aliphatic polypyromellitimides by using a domestic microwave oven by the polycondensation of both ε-amino acids and nylon salts as well as nylon-salt-type monomers in the presence of an appropriate polar organic medium. The use of such an organic medium was necessary to induce effective homogeneous heating of the monomers and hence cause efficient polycondensation leading to the formation of the polyamides and polyimides having high inherent viscosities around 0.5 dLg or above. The microwave-assisted polycondensation proceeded rapidly and was almost complete within 5 min for the polyamide synthesis and within 2 min for the polyimide. This method should become more and more important from the viewpoint of high efficiency utilization of heat energy. Further studies on this project for the rapid synthesis of other types of condensation polymers are now in progress.

References and Notes

1. Teffal, M.; Gourdenne, A. *Eur. Polym. J.* **1983**, *19*, 543.
2. Al Doori, D.; Huggett, R.; Bates, J. F.; Brooks, S. C. *Dent. Mater.* **1988**, *4*, 25.
3. Stoffer, J. O.; Sitaram, S. P. *Am. Chem. Soc., Proc. Polym. Mater. Sci. Eng.* **1994**, *71*, 55.
4. Mijovic, J.; Wijaya, J. *Polym. Compos.* **1990**, *11*, 184.
5. Thuillier, F. M.; Jullien, H. Grenier-Loustalot, M. F. *Polym. Commun.* **1986**, *27*, 206.

6. Lewis, D. A.; Hedrick, J. C.;McGrath, J. E.; Ward, T. C. *Am Chem. Soc., Polym. Prepr.* **1987**, *28* [2], 330.
7. Le Van, Q.; Gourdenne, A. *Eur. Polym. J.* **1987**, *23*, 777.
8. Beldjoudi, N.; Bouazizi, A.; Douibi, D.; Gourdenne, A. *Eur. Polym. J.* **1988**, *24*, 49.
9. Singer, S. M.; Jow, J.; Delong, J. D.; Hawley, M. C. *SAMPE Quat.* **1989**, *20*, 14.
10. Mijovic, J.; Wijaya, J. *Macromolecules,* **1990**, *23*, 3671.
11. Silinski, B.; Kuzmycz, C.; Gourdenne, A. *Eur. Polym. J.* **1987**, *23*, 273.
12. Jullien, H.; Valot, H. *Polymer,* **1985**, *26*, 506.
13. Lewis, D. A.; Summers, J. D.; Ward, T. C.; McGrath, J. E. *J. Polym. Sci., Part A, Polym. Chem.* **1992**, *30*, 1647.
14. Kishanprasad, V. S.; Gedam, P. H. *J. Appl. Polym. Sci.* **1993**, *50*, 419.
15. Park, K. H.; Watanabe, S.; Kakimoto, M.; Imai, Y. *Polym. J.* **1993**, *25*, 209.
16. Watanabe, S.; Hayama, K.; Park, K. H.; Kakimoto, M.; Imai, Y. *Makromol. Chem., Rapid Commun.* **1993**, *14*, 481.
17. Imai, Y.; Nemoto, H.; Watanabe, S.; Kakimoto, M. *Polym. Prepr. Jpn.* **1994**, *43*, 377; *Polym. J.*, in press.
18. Imai, Y.; Nemoto, H.; Kakimoto, M. *Polym. Prepr. Jpn.* **1994**, *43*, 1994; *J. Polym. Sci., Part A, Polym. Chem.*, submitted.
19. Imai, Y. *Am. Chem. Soc., Polym. Prepr.* **1995**, *36* [1], 711.
20. Sorenson, W. R.; Campbell, T. W. *Preparative Methods of Polymer Chemistry, 2nd Ed.*; Interscience, New York, 1968, p.74.
21. Kumagai, Y.; Itoya, K.; Kakimoto, M.; Imai, Y. *Polymer,* **1995**, *36*, 2827.
22. Itoya, K.; Kumagai, Y.; Kakimoto, M.; Imai, Y. *Polym. Prepr. Jpn.* **1992**, *41*, 2131.
23. Imai, Y.; Inoue, T.; Watanabe, S.; Kakimoto, M. *Polym. Prepr. Jpn.* **1994**, *43*, 371.

RECEIVED December 4, 1995

Chapter 28

Novel Step-Growth Polymers from the Thermal [2π + 2π] Cyclopolymerization of Aryl Trifluorovinyl Ether Monomers

David A. Babb[1], R. Vernon Snelgrove[1], Dennis W. Smith, Jr.[1], and Scott F. Mudrich[2]

[1]Central Research and Development, Organic Product Research, and [2]Analytical and Engineering Sciences, Dow Chemical Company, Freeport, TX 77541

Thermoplastic and thermosetting polymers comprising alternating arylene ether and 1,2-hexafluorocyclobutane subunits are prepared from aromatic bis- and tris-trifluorovinyl ether monomers. Step growth, addition type polymerization proceeds by the thermally induced [$2\pi + 2\pi$] cyclodimerization of the trifluorovinyl functionality. Monomers are conveniently prepared in two steps from phenolic starting materials and 1,2-dibromotetrafluoroethane or from Grignard reagents containing the flourinated vinyl group. Preparation and performance of these novel polymers are described with references to sources of further information.

Polyarylene ethers containing the hexafluorocyclobutane ring constitute a new and important family of high performance polymers. In general, the hexafluorocyclobutane linkage provides good thermal stability, improved dielectric insulation, enhanced ignition resistance, and increased solubility (for thermoplastics and pre-gel thermosets). The radical mediated [$2\pi + 2\pi$] cyclopolymerization reaction by which the polymers are formed provides a zero-volatile polymerization process with classical step-growth behavior. The trifluorovinyl ether monomer precursors are conveniently prepared in two steps from phenolic starting materials (*1*) (Figure 1).

The combination of processability and performance provided by these polymers makes them natural candidates for applications such as microelectronics laminates as well as structural and dielectric composites (*2*). The versatility provided by the use of phenolic feedstocks has allowed the extension of this chemistry into polyesters (*3*), classical photoresist chemistry (*4*), and thermally crosslinkable thermoplastics (*5*). It has also provided a method for thermally processing low molecular weight imide-containing oligomers into high molecular weight polyetherimides (*6*). In addition, oligomers which are useful as fluorinated lubricant fluids have been prepared (*7*).

0097–6156/96/0624–0431$12.00/0

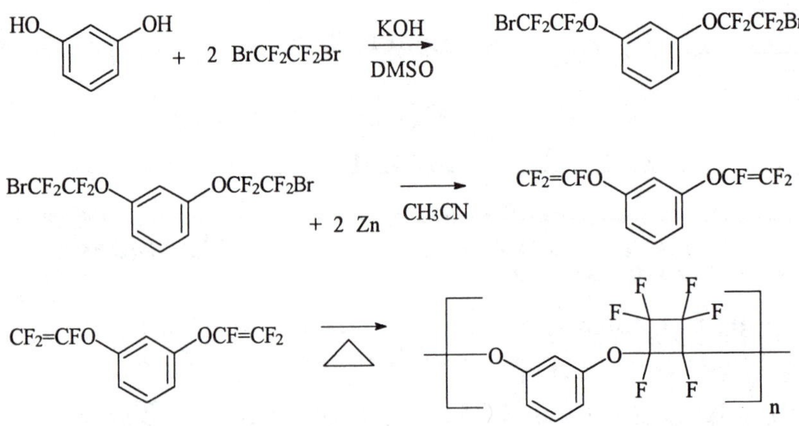

Figure 1. Preparation of hexafluorocyclobutane arylene ether polymers.

Recently a new synthetic method for preparing monomers has been developed using the Grignard reagent prepared from 4-bromotrifluorovinyloxybenzene, which is prepared in high yield from 4-bromophenol by the classical preparative method outlined in Figure 1. The ability to prepare a Grignard reagent from a trifluorovinyl-containing monomer was somewhat unexpected, given the sensitivity of fluorinated olefins to attack by nucleophilic reagents. The stability of this Grignard reagent has allowed the preparation of monomers that are somewhat more difficult to prepare directly from phenolics. In one example, reaction with chlorodimethylsilane has produced 4-trifluorovinyloxyphenyl dimethyl silane, from which a variety of siloxane monomers are accessible (*8*) (Figure 2).

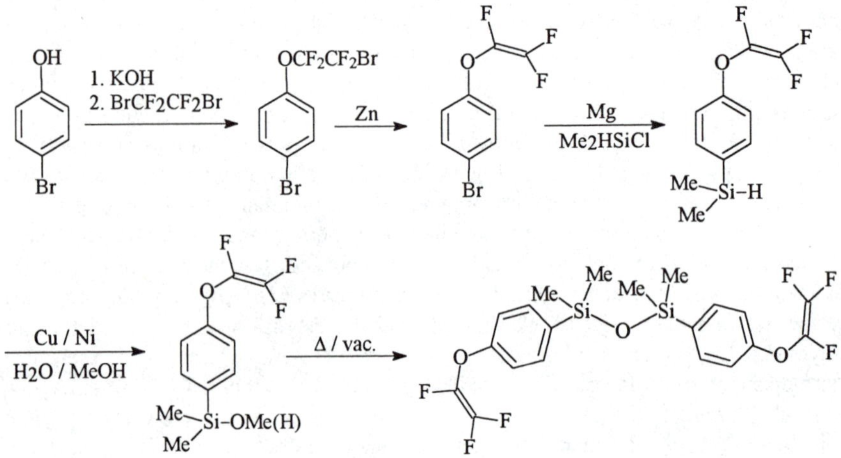

Figure 2. Siloxane monomer synthesis via Grignard chemistry.

Thermal polymerization of these monomers produces new types of fluorosilicone polymers which are just now being explored. The Grignard reagent has also been reacted with phosphorous trichloride to give the tris(4-trifluorovinyloxyphenyl)phosphine monomer which, after oxidation with hydrogen peroxide in ethanol gives a quantitative yield of the tris(4-trifluorovinyloxyphenyl)-phosphine oxide monomer (*9*). The polymer resulting from this monomer is under investigation for it's thermal and thermal / oxidative properties.

The Dimerization Reaction

Although the thermal [2π + 2π] cyclodimerization of the trifluorovinyl functionality is typically quite sensitive to the nature of the substituent on the trifluorovinyl group (*10*), our work with aromatic trifluorovinyl ethers has provided results consistent with a single well-behaved cycloaddition mechanism. Cyclodimerization of the fluorinated olefin takes place in a predominantly head-to-head fashion, creating a 1,2-substituted hexafluorocyclobutane ring. The corresponding 1,3-substituted ring is usually present in less than 2% of the total (*1*). The selectivity of this reaction is consistent with a stepwise mechanism (Figure 3), beginning with head-to-head bond formation between the terminal vinyl carbons, proceeding through the most stable diradical intermediate which then collapses to form the 1,2-substituted hexafluorocyclobutane ring (*11*).

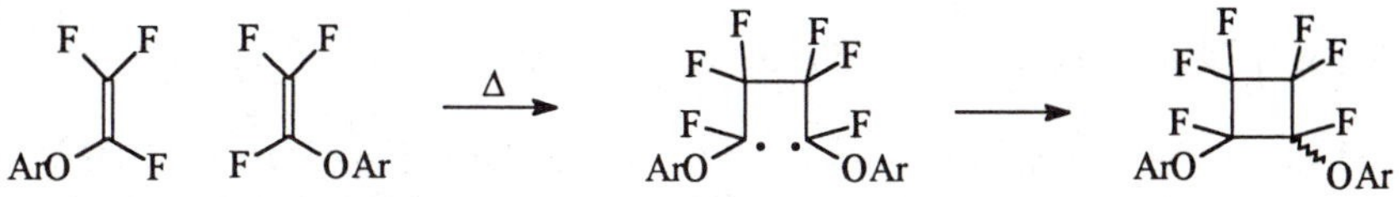

Figure 3. Cyclodimerization of Aromatic Trifluorovinyl Ethers.

The mechanism of the thermal [2π + 2π] cyclodimerization is therefore not a violation of the Woodward-Hoffmann rules for pericyclic reactions wherein concerted thermal [2π + 2π] cyclodimerizations are disallowed suprafacially, but is instead a radical-mediated step-wise ring formation. The resulting hexafluorocyclobutane ring has either a *cis*-1,2 or a *trans*-1,2 conformation. The geometry of these two linkages and the effect that they have on the polymer behavior are suspected to be quite dissimilar (*vide infra*).

The substituent sensitivity of this reaction is such that the trifluorostyrenes (proceeding through a benzylic diradical) have been shown to dimerize very efficiently at room temperature (*12*). Insertion of the oxygen atom to form an aryl trifluorovinyl ether pushes the kinetically-significant dimerization temperature to over 100°C and provides monomers with sufficient shelf stability for most applications. Figure 4 illustrates a typical neat polymerization by differential scanning calorimetry (DSC) where the onset of exothermic polymerization is detected near 140 °C as is observed similarly for most monomers studied in our labs.

The high efficiency and selectivity of the [2π +2π] cyclopolymerization reaction can be readily demonstrated by the preparation of high molecular weight thermoplastics via this reaction. Given the proper reaction conditions, thermoplastics with Mn $\geq$ 100,000 and polydispersities approaching 2.0 are common (*13*). Monomer purity is, of course, of paramount importance in gaining high molecular weights in these step-growth polymerizations. The primary impurity which affects the

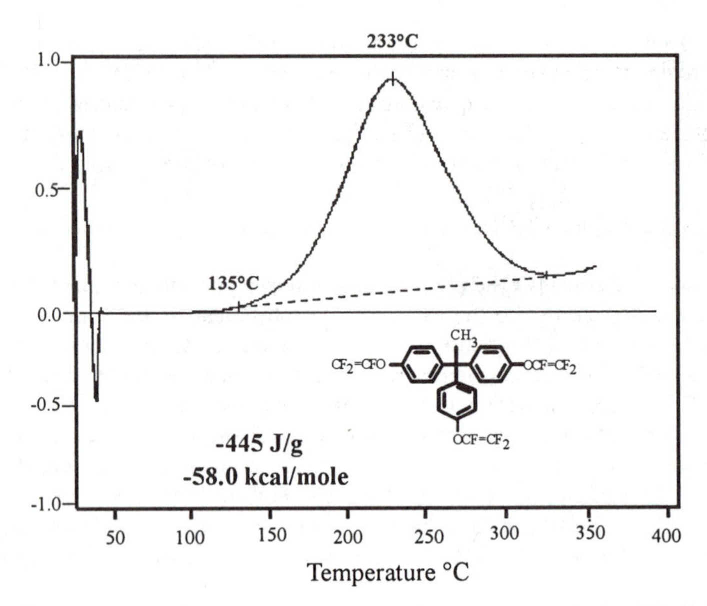

Figure 4. DSC of neat Tris(triflurovinyloxyphenyl)ethane monomer (10 °C/min).

polymerization is a monomer with one functional site occupied by a 1,1,2,2-tetra-fluoroethyl ether functional group ($-OCF_2CF_2H$) in place of the desired trifluorovinyl-ether group ($-OCF=CF_2$). This impurity acts as a chain terminator in thermoplastics. Taking into consideration the presence of this functionality, polymers have been prepared which appear to approach the upper limits of degree of polymerization allowed by their purities.

One rather curious polymer can be prepared from resorcinol, as illustrated in Figure 1 (*3*). For the resulting poly(1,3-phenylene-1,2-hexafluorocyclobutyl ether) a Tg = 27°-32°C was measured. Although glassy and brittle at room temperature (it is difficult to obtain high molecular weight homopolymer from this monomer), when held in a warm hand for a few moments (approx. 35°C) the polymer becomes soft and pliable, even elastomeric. If stretched and allowed to cool (elongations >200% are possible), it freezes into the distorted shape. If then passed under warm tap water it returns to it's original shape with a remarkable degree of elastic memory. While this is not particularly surprising behavior, the presence of a glass transition just above room temperature makes this a convenient material for the demonstration of basic polymer thermal properties.

The hexafluorocyclobutane linking structure which is the product of the cyclodimerization is an interesting study in itself. The two possible isomers (*cis* and *trans*) which can be formed from the reaction are quite different in nature. Space-filling models and *ab-initio* calculations suggest that the lowest energy conformation of the *trans* isomer provides essentially linear chain propagation, while that of the *cis* isomer imposes a roughly 90° dihedral angle in the polymer backbone. *Ab-initio* calculations indicate the lowest energy conformations of the two isomers are less than 0.5 kcal/ mole different in energy (the trans being slightly lower) (*14*), suggesting that they may be expected to form in roughly equal amounts in the solution polymerized polymer. This is in accord with a low molecular weight oligomer model study performed in our lab (*1*) and ^{13}C NMR analysis of a solution-polymerized thermoplastic (*8,15*).

It should be noted here that it is difficult to obtain high molecular weights with some monomers despite their high purities. This may be especially true of monomers that impart a dihedral angle to the direction of chain propagation such as those derived from bisphenol A (isopropylidene-2,2-diphenol) and resorcinol. It is easier to form high polymer from monomers with a more linear functional group orientation such as hydroquinone, 4,4'-biphenol and 2,6-dihydroxynaphthalene. The reason for this is not completely understood, but may involve decreased availability of the polymer chain end due to compounded chain entanglement under the effects of a dihedral angle in the bisphenol and a roughly 50% population of *cis* isomer in the cyclobutane linkage. This postulate is consistent with the high degree of elastic memory mentioned above for the thermoplastic homopolymer prepared from resorcinol. It is also supported by the observation that many of these "bent" monomers will form highly swellable gels at relatively low molecular weights, and will essentially cease further polymerization before ductile macromolecular properties are attained. The extractable soluble polymer from these gels invariably has a low molecular weight, typically with weight averaged molecular weights (Mw) in the range of 20,000 to 40,000 as measured by gel permeation chromatography (GPC) standardized against polystyrene (PS) standards. Range-finding experiments conducted in our labs on the polymer prepared from 4,4'-biphenol have indicated that a minimum of 70,000 Mw (GPC-PS) is required to move this polymer from the glassy prepolymer state into the ductile mechanical behavior of high molecular weight polymers.

Thermoplastics

The ability to prepare perfluorocyclobutane ring-containing polymers from phenolic starting materials (*1,3*) has led to the preparation of a variety of thermoplastic polymers from commercially available bisphenols. In general, those thermoplastics which contain no strongly polarizing groups elsewhere in the monomer exhibit low dielectric constants, low dissipation factors, good thermal stability, and enhanced solvent solubility. These polymers tend to have glass transition temperatures between 80° - 200 °C, thereby limiting their utility in high temperature high performance applications. This property of low Tg is in accordance with previously prepared fluorocarbon arylene ethers which contain linear fluorocarbon segments (*16*).

A table of the properties of a variety of perfluorocyclobutane thermoplastic polymers can be found in the literature (*13,17*). As an example of the types of properties that can be achieved with this polymer chemistry, Table I lists the properties of the thermoplastic homopolymer prepared from 4,4'-bis(trifluorovinyloxy)biphenyl (monomer CAS #134130-19-1, polymer CAS #134174-05-3) (see Figure 5). This particular thermoplastic has been studied more than any other thermoplastic because of its combination of properties, including a relatively higher Tg and good ductile mechanical properties.

Thermoplastic perfluorocyclobutane arylene ether polymers offer a solvent soluble, melt processable, low dielectric alternative to conventional fluoropolymers. Relatively little time has been spent on their development, however. The bifunctional monomers have proven quite useful as comonomers in modifying the thermal and mechanical properties of high Tg thermoset copolymers. In particular, the siloxane PFCB polymer discussed earlier (Figure 3) exhibits a Tg = 16 °C (DSC) thereby providing a reactive toughening or flexibilizing additive either as block type copolymers or comonomer for random incorporation (*8*).

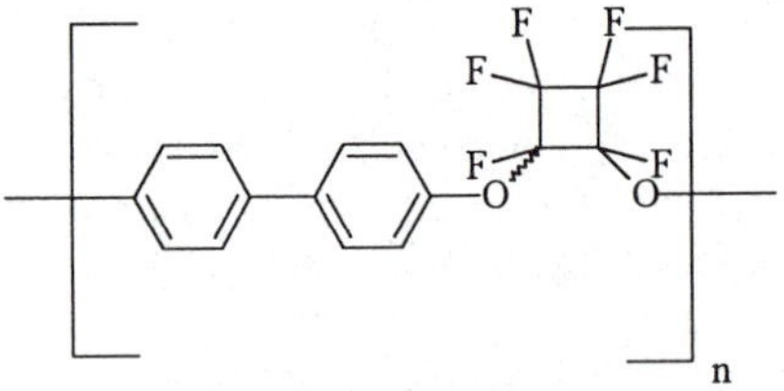

Figure 5. Poly(4,4'-biphenyl-1,2-hexafluorocyclobutyl ether).

Table I. Properties of Poly(4,4'-biphenyl-1,2-hexafluorocyclobutyl ether)

Tensile Strength (MPa)	50.3 ± 1.4	Tg (DMA)	160°-170°C
Tensile Modulus (MPa)	1,770 ± 79	Dielectric Constant (10 kHz)	2.35
Flexural Strength (MPa)	92.4 ± 2.1	Dissipation Factor (10 kHz)	0.0003
Flexural Modulus (MPa)	1,779 ± 85	Percent Water Absorption (24 hr)	0.040
Percent Elongation (break)	18.5	Limiting Oxygen Index	0.419

Thermosets

At this writing only one thermoset system based on a trifunctional monomer has been extensively studied. Until very recently it has not been necessary to investigate other thermoset polymers on a performance basis. This polymer is based on the trifluorovinyl ether monomer prepared from 1,1,1-tris(4-hydroxyphenyl)ethane as a starting phenolic (*1,3*). The tris(4-triflurovinyloxyphenyl)ethane monomer (CAS #134130-24-8) is cured in solution at 140°C-150°C for five to eight hours. The thermally induced addition reaction combined with step growth kinetics allows solution viscosity to be precisely controlled. The resulting prepolymer (CAS #154854-78-0) exhibits classical broad polydispersity behavior (*14*). As an example, the prepolymer with a weight average molecular weight (Mw) of 2,100 (against polystyrene standards) has a polydispersity (Mw/Mn) of 1.8. As the Mw grows to 7,030 the polydispersity diverges to 3.8, and finally at a Mw of 22,400 the polydispersity has grown to 9.25.

The prepolymer solution can be coated by classical application techniques such as by spin-coating. The prepolymer coating is then given a post-bake cycle to as low as 235°C for one hour or as high as 325°C for two hours, depending on the thermal requirements of the final application. The final coating thus deposited does an exceptional job of planarizing topographical features (*18*), and has been shown to function well in smoothing substrates such as color pixels for flat panel displays (*19*). In addition the prepolymer has been shown to planarize Canasite glass ceramic substrates, used as an alternative to aluminum substrates in magnetic data storage devices, to an average roughness (Ra) of less than 2 nm (*20*). After post-bake the resulting thermoset polymer (CAS #136615-09-3) (Figure 6) has good thermal and thermal / oxidative stability to elevated temperatures for extended processing times (*21*). Table II lists a set of selected properties for the thermoset polymer.

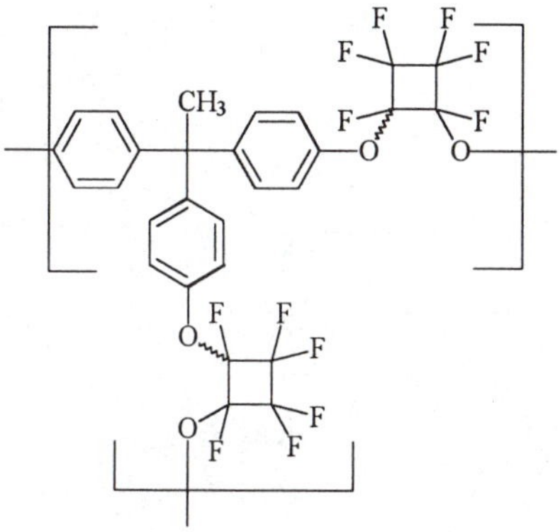

Figure 6. Homopolymer of 1,1,1-Tris(4-trifluorovinyloxy)phenyl Ethane

Table II. Properties of the Homopolymer of 1,1,1-Tris(4-trifluorovinyloxy)phenyl Ethane

Tensile Strength (MPa)	66.0 ± 1.4	Tg (DMA)	380°C
Tensile Modulus (MPa)	2,270 ± 79	Dielectric Constant (10 kHz)	2.45
Flexural Strength (MPa)	74 ± 12	Dissipation Factor (10 kHz)	0.0004
Flexural Modulus (MPa)	2,320 ± 13	Percent Water Absorption (24 hr)	0.021
Percent Elongation (break)	4.1	Limiting Oxygen Index	0.47

When the triphenylethane monomer is copolymerized with the bifunctional 4,4'-bis(trifluorovinyloxy)biphenyl monomer, significant improvements are obtained in the flexural strength with a moderate compromise in modulus values. When the triphenylethane monomer is copolymerized with the monomer 9,9-bis(4-trifluoro-vinyloxyphenyl)fluorene (CAS #134130-20-4), improvements are obtained in both the flexural strength and modulus. Figure 7 illustrates the compositions of the random copolymers prepared for the mechanical properties study. Table III lists the properties of three different molar compositions of the random copolymers of 1,1,1-tris(4-trifluorovinyloxyphenyl)ethane with 4,4'-bis(trifluorovinyloxy)biphenyl and 9,9-bis(4-trifluorovinyloxyphenyl) fluorene monomers.

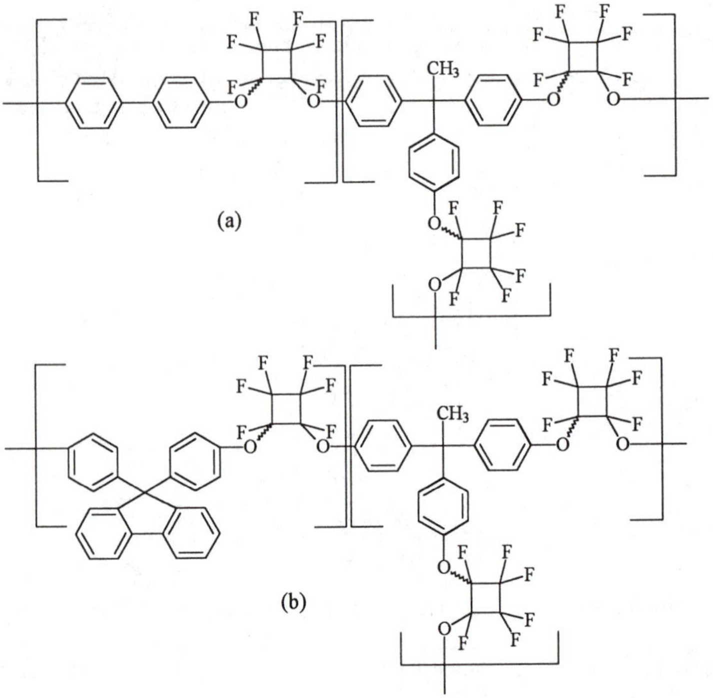

Figure 7. Random copolymers of 1,1,1-tris(4-trifluorovinyloxyphenyl)ethane with (a) 4,4'-bis(trifluorovinyloxy)biphenyl and (b) 9,9-bis(4-trifluorovinyloxyphenyl) fluorene.

As mentioned earlier, the Grignard chemistry which led to the preparation of the siloxane elastomers has also yielded a new thermoset polymer system based on a triphenylphosphine oxide core structure (*9*). The Grignard reagent is prepared from 4-bromotrifluorovinyl ether in THF and then reacted with phosphorous trichloride to provide tris(4-trifluorovinyloxyphenyl)phosphine. This monomer is then converted to tris(4-trifluorovinyloxyphenyl)phosphine oxide by treatment with one equivalent of hydrogen peroxide in ethanol solution. Attempts to prepare this monomer via fluoroalkylation of tris(4-hydroxyphenyl)phosphine oxide were unsuccesful due to problems in purification of the final monomer. Figure 8 illustrates the use of the Grignard chemistry in the preparation of the triphenylphosphine oxide monomer.

Triarylphosphine oxide containing polymers have recently received considerable attention due to their enhanced thermal / oxidative and ignition resistant

Table III. Properties of Random Copolymers with 1,1,1-Tris(4-trifluorovinyloxy)-phenyl Ethane

(a) Biphenyl Monomer Mole %	70%	80%	90%
Tensile Strength (MPa)	68.0 + 0.6	66.0 + 0.4	60.5 + 0.5
Tensile Modulus (MPa)	1,844 + 51	1,787 + 35	1,732 + 99
Flexural Strength (MPa)	112.4 + 5.0	108.8 + 2.3	102.0 + 2.0
Flexural Modulus (MPa)	1,822 + 46	1,757 + 57	1,768 + 36.2
(b) Fluorene Monomer Mole %	40%	50%	90%
Tensile Strength (MPa)	35.2 + 5.5	40.0 + 2.8	48.5 + 3.1
Tensile Modulus (MPa)	2,494 + 9	2,634 + 12	3,243 + 145
Flexural Strength (MPa)	85.9 + 18.8	85.5 + 4.0	*
Flexural Modulus (MPa)	2,699 + 9	2,745 + 25	3,243 + 39

* Samples failed to break at 10% strain and average stress of 92.5 MPa

performance (*22*). Initial results of the thermal and thermal / oxidative performance of the perfluorocyclobutane triphenylphosphine oxide polymer indicate improvements in thermal / oxidative performance as compared to the triphenylethane thermoset homopolymer, with little or no improvements in the anaerobic thermal stability. This observation is consistent with the postulated mechanisms for the thermal and thermal / oxidative decomposition of the perfluorocyclobutane polymers which have been recently published (*21*).

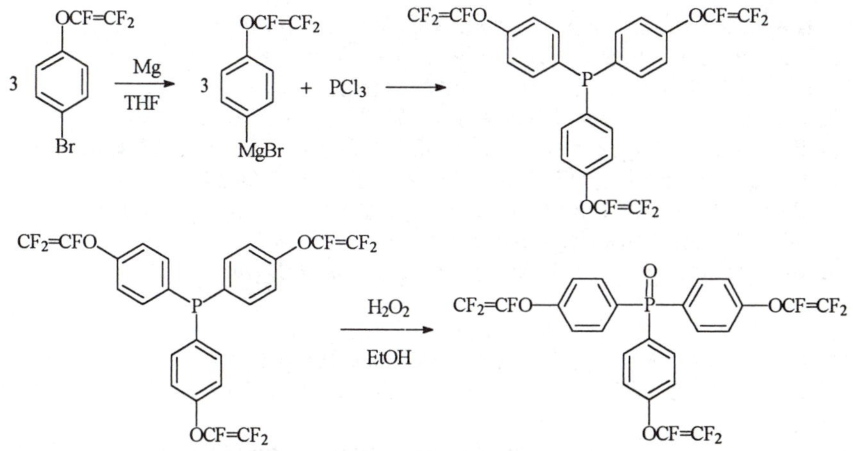

Figure 8. Synthesis of tris(4-trifluorovinyloxyphenyl)phosphine oxide.

Conclusion

The cyclopolymerization of aromatic trifloroviny1 ethers has emerged as a new addition to the chemistry of step-growth polymerization. The synthetic methodology that makes these monomers available from phenolic precursors is a key element to the value of the technology, providing a versatile, efficient and convenient way to incorporate fluorocarbon segments into a wide variety of polymer systems. In addition to the homopolymer and copolymer approaches outlined above, the chemistry has been used to end-cap the oligomers of other polymers with a thermally-activated chain extending group (*23*). In addition, trifluorovinyl ethers with pendant reactive groups (e.g. 4-trifluorovinyloxybenzoyl chloride) have been cyclodimerized to provide bifunctional perfluorocyclobutane ring-containing monomers for use in classical condensation polymerization reactions (*24*). The 1,2-substituted hexafluorocyclobutane ring is a thermally and chemically stable linkage that can enhance polymer performance in numerous applications where the polymer is a critical component of overall system performance.

References

1. Babb, D.A.; Ezzell, B.R.; Clement, K.S.; Richey, W.F.; Kennedy, A.P. *J. Polym. Sci., Part A: Polym. Chem.* **1993**, *31*, 3465-3477.
2. Kennedy, A.P.; Bratton, L.D.; Jezic, Z.; Lane, E.R.; Perettie, D.J.; Richey, W.F.; Babb, D.A.; Clement; K.S. *US Patent #5,460,782* **Sept. 21, 1993**.
3. Babb, D.A.; Clement, K.S.; Richey, W.F.; Ezzell; B.R. *US Patent #5,037,917* **Aug. 6, 1991**.
4. Babb, D.A.; Richey, W.F.; Clement, K.S.; Moyer, E.S.; Sorenson, M.W. *US Patent #5,426,164* **June 20, 1995**.
5. Babb; D.A. *US Patent #5,159,036*; **Oct. 27, 1992**.
6. Babb, D.A.; Richey, W.F.; Clement, K.S.; Ezzell, B.R. *US Patent #5,364,917*; **Nov. 15, 1994**.
7. Babb, D.A.; Morgan, T.A. *US Patent #5,364,547*; **Nov. 15, 1994**.
8. Smith, Jr., D.W.; Babb, D.A. *Macromolecules*, in press.
9. Babb, D.A.; Smith, Jr., D.W. unpublished results.
10. Sharkey, W.H. *Fluorine Chem. Rev.*, **1968**, *2*, 1, and references therein.
11. Bartlett, P.D.; Montgomery, L.K.; Seidel, B. *J. Am. Chem. Soc.*, **1964**, *86*, 616.
12. Heinze, P.L.; Burton, D.J. *J. Org. Chem.*, **1988**, *53*, 2714-2720.
13. Babb, D.A.; Ezzell, B.R.; Clement, K.S.; Richey, W.F. *Polymeric Materials Encyclopedia: Synthesis, Properties and Applications,* CRC Press, in press.
14. Babb, D.A.; Rondan, N.G.; Smith, Jr., D.W. *Polym. Preprints*, **1995**, *36*, (1), 721.
15. Smith, Jr., D.W. unpublished results.
16. Labadie, J.W.; Hedrick, J.L.; *Macromolecules*, **1990**, *23*, 5371.
17. Babb, D.A.; Clement, K.S.; Richey, W. F.; Ezzell, B.R. *US Patent #5,364,917*; **Nov. 15, 1994**.
18. Babb, D.A. in *Fluoropolymers: Synthesis and Properties,* Plenum Publishing Co., in press.

19. Perettie, D.J.; Bratton, L.D.; Bremmer, J.N.; Babb, D.A. *Proceedings Int'l Symp. on Electronic Imaging: Sci. and Tech., IS&T/SPIE*, **1993**, 1911-03.
20. Perettie, D.J.; Babb, D.A.; Chen, Q.; Judy, J.H. *IEEE Magnetics Proceedings, MMM-Intermag Conference*; **1994**, DQ-08.
21. Kennedy, A.P. ; Babb, D.A.; Bremmer, J.N.; Pasztor, Jr., A.J. *J. Polym. Sci., Part A: Polym. Chem.*, **1995**, *33*, 1859.
22. Yang, H.; Rogers, M.E.; McGrath, J.E. *Polym. Preprints*, **1995**, *36*, (1), 205, and references therein.
23. Clement, K.S.; Babb, D.A.; Ezzell, B.R.; Richey; W.F. *US Patent #5,198,513*; **March 30, 1993**.
24. Clement, K.S.; Babb, D.A.; Ezzell, B.R. *US Patent # 5,210,265*; **May 11, 1993**.

RECEIVED December 6, 1995

Chapter 29

Synthesis of Ordered Polymer by Direct Polycondensation

6. Ordered Poly(acylhydrazide–amide–thioether)

Mitsuru Ueda and Takashi Okada

Department of Materials Science and Engineering, Faculty of Engineering, Yamagata University, Yonezawa, Yamagata 992, Japan

Ordered poly(acylhydrazide-amide-thioether) was prepared by the poly(addition-direct condensation) of two nonsymmetric monomers, 4-nitrophenyl acrylate (XabX) and 4-aminobenzhydrazide (ZefZ) with a pair of symmetric monomers (YcdY), 4,4'-thiobisbenzenethiol (YccY) and isophthalic acid (YddY), using the condensing agent diphenyl(2,3-dihydro-2-thioxo-3-benzoxazolyl) phosphonate (**1**). The polymerization was carried out by mixing all monomers in the presence of **1** and triethylamine in dipolar aprotic solvents at room temperature, yielding the ordered poly(acylhydrazide-amide-thioether) with inherent viscosity of 0.21 dL/g. The authentic ordered polymer was prepared to verify the structure of the ordered polymer. The microstructures of polymers obtained were investigated by ^{13}C-NMR spectroscopy, and it has been found that the polymer had the expected ordered structure. Furthermore, the model reactions were studied in detail to demonstrate the feasibility of polymer formation.

Most condensation polymers are prepared by reactions between two different bifunctional symmetric monomers. However, the synthesis of condensation polymers from a symmetric (YccY) and a nonsymmetric (XabX) monomer has been only slightly explored.

Our group has initiated a synthesis of ordered polymer from nonsymmetric monomers by direct polycondensation. In the previous papers, we reported a successful synthesis of ordered polyamides (head-to-head or tail-to-tail) from a symmetric monomer and a nonsymmetric monomer, (*1,2*), and ordered polyamides (head-to-tail) from a symmetric monomer and a nonsymmetric monomer or a pair of two symmetric monomers (*3*).

Recently, we succeeded on the synthesis of ordered(-abcd-) poly(amide-thioether) by poly(addition-direct condensation) of a pair of two nonsymmetric monomers, 4-(acryloxy)benzoic acid (XabX) and 4-aminobenzenethiol (YcdY), where a polyaddtion was combined with a polycondensation to prepare the ordered polymer (*4*). Because, the selective amidation of carboxylic acids would be difficult using condensing agents due to the small difference of pKa values between carboxylic acids. An active ester is also useful as a substitute for a carboxylic acid component,

0097–6156/96/0624–0442$12.00/0

and a combination of a carboxylic acid and an active ester were successfully applied to the synthesis of ordered (-abcd-) polyamide by a pair of symmetric monomers with a nonsymmetric monomer (*5*).

On the basis of the above findings, we initiated the synthesis of ordered (-abcdef-) polymer from three nonsymmetric monomers, XabX, YcdY and ZefZ by direct polycondensation.

This article describes the successful synthesis of ordered poly(acylhydrazide-amide-thioether) from two nonsymmetric monomers, 4-nitrophenyl acrylate (XabX) and 4-aminobenzhydrazide (ZefZ) with a pair of symmetric monomers (YcdY), 4,4'-thiobisbenzenethiol (YccY) and isophthalic acid (YddY) in the presence of the condensing agent diphenyl(2,3-dihydro-2-thioxo-3-benzoxazolyl)-phosphonate (**1**).

Experimental

Materials. N,N-Dimethylformamide (DMF) and N-methyl-2-pyrrolidone (NMP) were stirred over powdered calcium hydride overnight, distilled under reduced pressure, and then stored over 4-Å molecular sieves. 4-Aminobenzhydrazide (**5**) was prepared from methyl 4-aminobenzoate and hydrazine monohydrate according to the reported procedure (*6*). 4,4'-Thiobisbenzenthiol (**6**) and isophthalic acid (**7**) were purified by recrystallization. Phenyl acrylate and 4-nitrophenyl acetate were prepared by the reactions of acryloyl chloride or acetyl chloride with phenol or 4-nitrophenol in the presence of triethylamine and tetahydrofuran, respectively. Triethylamine (TEA) and tetahydrofuran (THF) were purified by the usual method. Other reagents and solvents were obtained commercially and used as received.

The condensing agent diphenyl(2,3-dihydro-2-thioxo-3-benzoxazolyl) phosphonate (**1**) was prepared according to the reported procedure (*7*).

Model Compounds.

Phenyl 3-phenylthiopropionate. Benzenethiol (0.11 mL, 1.0 mmol) was added to a solution of phenyl acrylate (0.148 g, 1.0 mmol) in THF (2 mL) at room temperature. The solution was stirred for 5 min and poured into 10% aqueous sodium hydrogen carbonate. The solution was extracted with ether. The extract was dried ($MgSO_4$) and evaporated. The residue was subjected to silica gel column chromatography using benzene as eluent to give a colorless liquid. The yield was 0.219 g (85%). IR (NaCl): ν 1760 cm^{-1} (C=O). ^{1}H-NMR ($CDCl_3$): δ 2.84 (t,2H,-CH_2-), 3.25 ppm (t,2H,-SCH_2-). ^{13}C-NMR ($CDCl_3$): δ 28.3 (-CH_2-), 34.0 (-SCH_2-), 169.6 ppm (C=O). Anal. Calcd for $C_{15}H_{14}O_2S$: C,69.74; H,5.46. Found: C,69.56; H,5.55.

Phenyl 3-phenylaminopropionate. The mixture of phenyl acrylate (0.148 g, 1.0 mmol) and aniline (0.093 g, 1.0 mmol) was heated at 60°C for 4h and poured into water. The product that precipitated was filtered out, washed with water, and dried. The product was purified by silica gel column chromatography using ethyl acetate-n-hexane (2:3) as eluent to afford white plates. The yield was 0.214 g (89%). mp 82-84°C. IR (KBr): ν 3400 (N-H), 1750 cm^{-1} (C=O). ^{1}H-NMR ($CDCl_3$): δ 2.83 (t,2H,-CH_2-), 3.54 (t,2H,-$NHC\underline{H_2}$-), 4.06 ppm (s,1H,N-H). ^{13}C-NMR ($CDCl_3$): δ 34.0 (-CH_2-), 39.3 (-$NHCH_2$-), 170.9 ppm (C=O). Anal. Calcd for $C_{15}H_{15}NO_2$: C,74.66; H,6.27; N,5.81. Found: C,74.40; H,6.14; N,5.75.

Phenyl 3-(N'-benzoylhydrazino)propionate. A solution of phenyl acrylate (1.48 g, 10 mmol) and benzhydrazide (1.36 g, 10 mmol) in THF (30 mL) was stirred at room temperature for 7days and the solvent was evaporated. The product was purified by silica gel column chromatography using ethyl acetate-n-hexane (3:1) as eluent to afford white plates. The yield was 1.33 g (47%). mp 89-90 °C. IR (KBr): ν 1745 (C=O,ester), 1640 cm^{-1} (C=O,amide). ^{1}H-NMR ($CDCl_3$): δ 2.76 (t,2H,-CH_2-), 3.32 (t,2H,-NH$\underline{H_2}$-), 5.08 (s,1H,N-H), 8.51 ppm (s,1H,CONH). ^{13}C-NMR (CDCl3): δ 33.5 (-CH_2-), 47.4 (-NHCH_2-), 167.6 (C=O,amide), 171.1 ppm (C=O,ester). Anal. Calcd for $C_{16}H_{16}N_2O_3$: C,67.59; H,5.67; N,9.85. Found: C,67.65; H,5.59; N,9.70.

The following model compounds were prepared from the corresponding nucleophiles and acyl chlorides.

S-Phenyl isophthalate: mp 119-121 °C (from benzene). IR (KBr): ν 1675 cm^{-1} (C=O). ^{13}C-NMR (DMSO-*d6*): δ 188.4 ppm (C=O). Anal. Calcd for $C_{20}H_{14}O_2S_2$: C,68.55; H,4.03. Found: C,68.77; H,4.10.

Isophthalanilide: mp 291-293 °C (Lit.(*8*) 295-296°C) (from acetic acid). IR (KBr): ν 3400 (N-H), 1640 cm^{-1} (C=O). ^{1}H-NMR (DMSO-*d6*): 10.50 ppm (s,2H,N-H). ^{13}C-NMR (DMSO-*d6*): δ 165.2 ppm (C=O).

N',N'-Dibenzoylisophthalodihydrazide: mp 293-294 °C (from acetic acid). IR (KBr): ν 3200 (N-H), 1640 cm^{-1} (C=O). ^{1}H-NMR (DMSO-*d6*): δ 10.60 and 10.70 ppm (s,2H,N-H). ^{13}C-NMR (DMSO-*d6*): δ 165.5 and 166.0 ppm (C=O). Anal. Calcd for $C_{22}H_{18}N_4O_4$: C,65.66; H,4.50; N,13.92. Found: C,65.57; H,4.67; N,13.72.

S-Phenyl acetate: bp 95 °C/4 mmHg (Lit. (*9)*100 °C/6 mmHg). IR (KBr): ν 1710 cm^{-1} (C=O). ^{1}H-NMR (DMSO-*d6*): δ 2.45 ppm (s,3H,CH_3). ^{13}C-NMR (DMSO-*d6*): δ 30.0 (CH_3), 193.2 ppm (C=O).

N'-Acetylbenzohydrazide: mp 180-182 °C (from ethyl acetate). IR (KBr): ν 3200 (N-H), 1650 cm^{-1} (C=O). ^{1}H-NMR (DMSO-*d6*): δ 1.92 (s,3H,CH_3), 9.89 and 10.29 ppm (s,2H,N-H). ^{13}C-NMR (DMSO-*d6*): δ 20.6 (CH_3), 165.6 and 168.6 ppm (C=O). Anal. Calcd for $C_9H_{10}N_2O_2$: C,60.67; H,5.66; N,15.72. Found: C,60.40; H,5.71; N,15.48.

4-Nitrophenyl acrylate (4). To an ice-cooled solution of 4-nitrophenol (6.96 g, 50 mmol) and triethylamine (7.0 mL, 50 mmol) in THF (20 mL) was added dropwise acryloyl chloride (4.1mL , 50 mmol). After being stirred for 30 min, triethylamine hydrogenchloride was filtered and the solution was evaporated. The product recrystallized from n-hexane to afford white plates. The yield was 9.66 g (50 %). mp 60 °C (Lit. (*10*)59-60 °C). IR (KBr): ν 1715 cm^{-1} (C=O). ^{1}H-NMR ($CDCl_3$): δ 6.08-6.69 (dd,2H,H_2C=, t,1H,H_2C=C$\underline{H}$-). ^{13}C-NMR ($CDCl_3$): δ 163.5 ppm (C=O).

Competitive reaction of benzenethiol, benzhydrazide and aniline with phenyl acrylate. Pheny acrylate (0.148 g, 1.0 mmol) was added to a solution of benzenethiol (0.11 mL, 1.0 mmol), benzhydrazide (0.136 g, 1.0 mmol) and aniline (0.093 g, 1.0 mmol) in THF at room temperature. The solution was stirred for 5 min and was evaporated. The crude product was purified by silica gel column chromatography using benzene as eluent to give pure phenyl 3-phenylthiopropionate (**2**). The yield was 0.237 g (92%).

Competitive reaction between benzoic acid and 4-nitrophenyl acetate with benzhydrazide. Condensing agent **1** (0.422 g, 1.1 mmol) was added to a solution of benzoic acid (0.122 mL, 1.0 mmol), 4-nitrophenyl acetate (0.181 g, 1.0 mmol) and benzhydrazide (0.136 g, 1.0 mmol) in NMP at room temperature. The solution was stirred for 10 min and poured into 10 % aqueous sodium hydrogen carbonate. The precipitate was filtered, washed with water, and dried. The crude product was stirred with n-hexane to remove 4-nitrophenyl acetate and the insoluble compound was filtered and dried to give pure dibenzoylhydrazine (**3**). The yield was 0.225 g (95%).

Authentic ordered polyamide (10a).
N',N'-Bis(4-aminobenzoyl)isophthalohydrazide (8). A solution of isophthalic acid (0.498 g, 3 mmol), triethylamine (0.84 mL, 6 mmol) and condensing agent **1** (2.53 g, 6.6 mmol) in NMP (4 mL) was added dropwise a solution of 4-aminobenzhydrazide (0.907 g, 6 mmol) in NMP (4 mL). The solution was stirred at room temperature for 1h and poured into 10% aqueous sodium hydrgen carbonate. The product was filtered out and dried. Then this product refluxed in chloroform. The residue was 1.19 g (92%). mp 295 °C (by DTA). IR (KBr): ν 3225 (N-H), 1640 cm^{-1} (C=O). ^{1}H-NMR (DMSO-d_6): δ 5.57 ppm (s,4H,N-H). ^{13}C-NMR (DMSO-d_6): δ 165.7 and 166.1 ppm (C=O). Anal. Calcd for $C_{22}H_{20}N_6O_4$: C,61.11; H,4.66; N,19.43. Found: C,59.53; H,4.69; N,18.61.

Di(4-Nitrophenyl) 3,3'-(thiodi-p-phenylenethio)dipropanoate (9). A solution of 4-nitrophenyl acrylate (3.68 g, 20 mmol), 4,4'-thiobisbenzenethiol (2.50 g, 10 mmol) and a small amount of triethylamine in THF (25 mL) was stirred at room temperature for 2h and was evaporated. The product was purified by silica gel column chromatography using dichloromethane-n-hexane (9:1) as eluent to afford yellow plates. The yield was 6.37 g (100%). mp 77-78 °C. IR (KBr): ν 1760 cm^{-1} (C=O). ^{1}H-NMR ($CDCl_3$): δ 2.94 (t,2H,-CH_2-), 3.28 ppm (t,2H,-SCH_2-). ^{13}C-NMR ($CDCl_3$): δ 28.9 (-CH_2-), 34.4 (-SCH_2-), 169.2 ppm (C=O). Anal. Calcd for $C_{30}H_{24}N_2O_8S_3$: C,56.59; H,3.80; N,4.40. Found: C,56.76; H,3.87; N,4.22.

A solution of **8** (0.216 g, 0.5 mmol), **9** (0.318 g, 0.5 mmol), 1-hydroxybenzotriazole (HOBt) (0.03 g, 0.2 mmol) and lithium chloride (0.127 g, 3 mmol) in DMF (2 mL) was stirred at 80 °C for 7d and poured into 10 % aqueous sodium hydrogen carbonate. After through washing with water and drying, the polymer weighed 0.395 g (100%). The inherent viscosity of polymer in NMP was 0.30 dL/g, measured at a concentration of 0.5 g/dL at 30°C. IR (KBr): ν 3300 (N-H), 1650 cm^{-1} (C=O). ^{13}C-NMR (DMSO-d_6): δ 27.7 (-CH_2-), 36.1 (-SCH_2-),165.1, 165.4 and 169.7 ppm (C=O). Anal. Calcd for $(C_{40}H_{36}N_6O_7S_3)_n$: C,59.39; H,4.49; N,10.39. Found: C,60.08; H,4.60; N,9.97.

Polyamide prepared by direct polycondensation (10b). Compound **5** (0.151 g, 1 mmol) and **6** (125 g, 0.5 mmol) were dissolved in DMF (2 mL). To this solution was added **7** (0.083 g, 0.5 mmol), **4** (0.193 g, 1 mmol), triethylamine (0.14 mL, 1 mmol) and condensing agent **1** (0.412 g, 1.1 mmol). The solution was stirred at room temperature for 1h. To the resulting solution was added 1-hydroxybenzotriazole (HOBt) (0.03 g, 0.2 mmol) and lithium chloride (0.127 g, 3 mmol). Then the solution was stirred at 80°C for 7d and poured into 10% aqueous sodium hydrogen carbonate. After thorough washing with water and drying, the polymer weighed 0.395 g (100%). The inherent viscosity of polymer in NMP was 0.21 dL/g, measured at a concentration of 0.5 g/dL at 30°C. IR (KBr): ν 3300 (N-H), 1650 cm^{-1} (C=O). ^{13}C-NMR (DMSO-d_6): δ 27.7 ($-CH_2-$), 36.1 ($-SCH_2-$), 165.1, 165.4 and 169.7 ppm (C=O). Anal. Calcd for $(C_{40}H_{36}N_6O_7S_3)_n$: C,59.39; H,4.49; N,10.39. Found: C,58.92; H,4.53; N,9.73.

Measurements. The infrared spectra were recorded on a Hitachi I-5020-FT-IR spectrophotometer, and the NMR spectra on a JEOL EX270 (270MHz) and JNM-A-500 (500MHz) spectrometers. Viscosity measurements were carried out by using an Ostwald viscometer at 30°C. Thermal analyses were performed on a Seiko SSS 5000 TG/DTA 200 thermal analyzer at a heating rate of 10°C/min for thermogravimtric analysis.

Results and Discussion

We considered three kinds of nucleophiles and electrophiles for the preparation of ordered polymer from three nonsymmetric monomers by poly(addition-direct condensation). Then, we thought to combine the two concenpts in our papers (ref. *4* and *5*) for the selection of these components, and decided to use acylhydrazide, aromatic amine, and thiol as nucleophiles, and carboxylic acid, active ester and vinyl ester as electrophiles, respectively.

Prior to the synthesis of the ordered polymer, the following model compound work was performed to determine if the model compounds were formed in quantitative yields to constitute an ordered polymer forming reaction.

Selection of Nucleophiles. In the previous papers (*4*), we reported the selective addition reaction of benzenethiol occurred in the competitive reaction of phenyl acrylate between benzenethiol and aniline. The competitive addition reaction of phenyl acrylate with benzenethiol, benzhydrazide and aniline was carried out in THF at room temperature to clarify the difference in their reactivities (eq 1).

$$H_2C{=}CH{-}C({=}O){-}O{-}C_6H_5 + \left\{\begin{array}{l} C_6H_5{-}SH \\ C_6H_5{-}NH_2 \\ C_6H_5{-}C({=}O)NHNH_2 \end{array}\right. \longrightarrow \underset{\mathbf{2}}{C_6H_5{-}S{-}CH_2CH_2C({=}O){-}O{-}C_6H_5} \quad (1)$$

The selective addition reaction of benzenethiol was observed, and the desired product, phenyl 3-phenylthiopropanoate (**2**) was obtained in quantitative yield.

Selection of Carboxylic Acid Derivatives. The competitive reaction between benzoic acid and 4-nitrophenyl acetate with amines, such as benzhydrazide and aniline, was carried out at room temperature for 10 min in the presence of condensing agent **1** (eq 2).

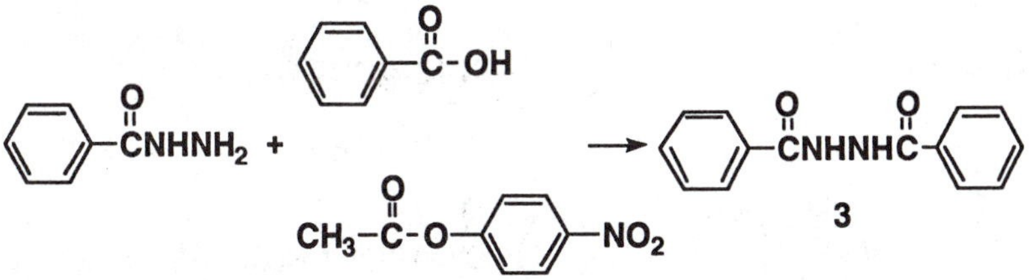

(2)

Selective amidations were observed in both cases, and dibenzoylhydrazine (**3**) and benzanilide were obtained in quantitative yields.

Selection of Amines. The competitive reaction between benzhydrazide (about pK_a=7) and aniline (about pK_a=4) with benzoic acid in the presence of condensing agent **1** has been reported to yield compound **3** in quantitative yield (eq 3).

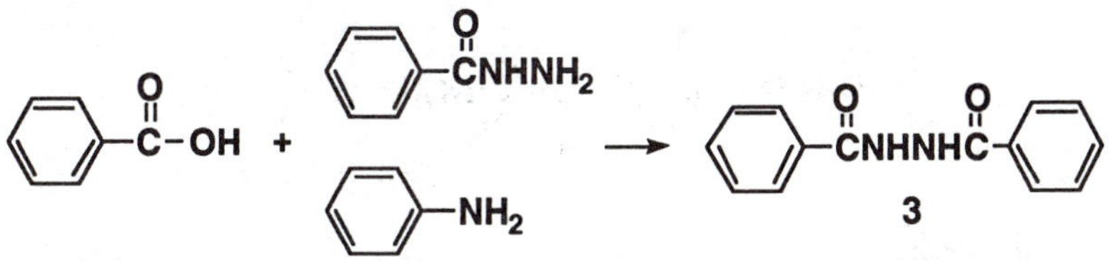

(3)

On the basis of model reactions and the availability of reactants, we decided to use two nonsymmetric monomers, 4-nitrophenyl acrylate **4** (XabX) and 4-aminobenzhydrazide **5** (ZefZ) with a pair of symmetric monomers (YcdY), 4,4'-thiobisbenzenethiol **6** (YccY) and isophthalic acid **7** (YddY).

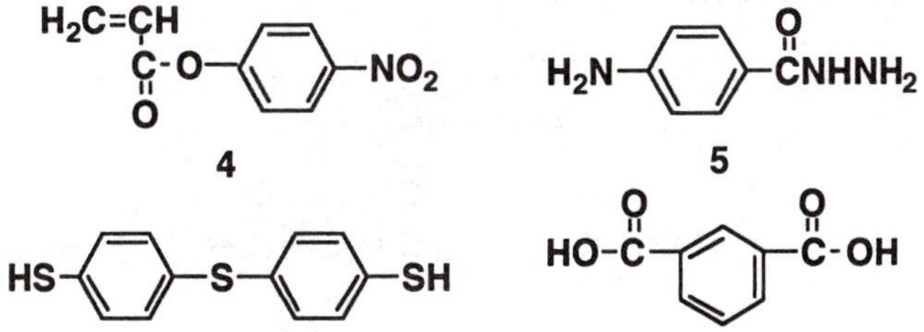

The model compounds as follows were prepared from the corresponding acyl chloride or phenyl acrylate with nucleophiles, respectively, in order to clarify the structure of polymer obtained.

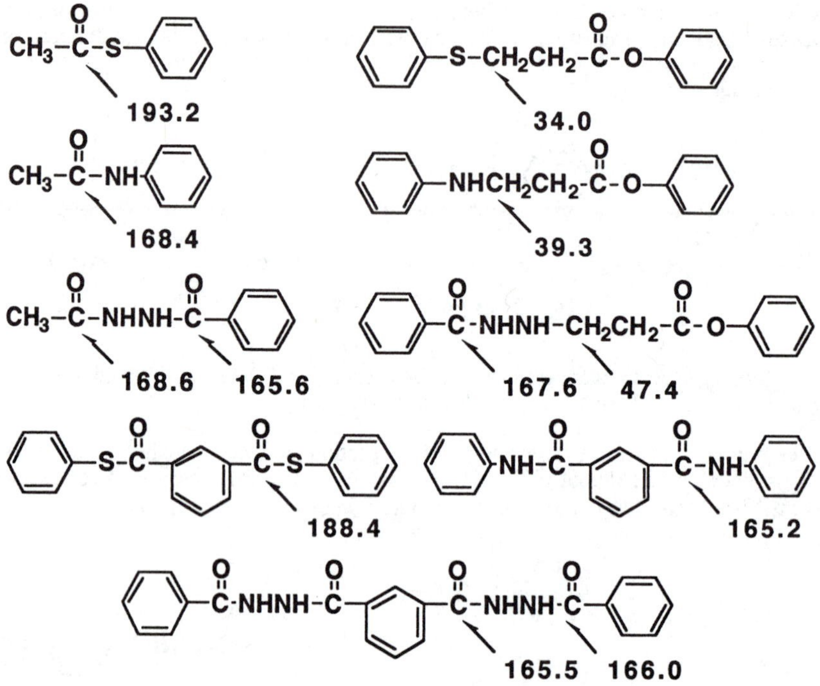

CH_3-C(=O)-S-Ph 193.2

Ph-S-CH_2CH_2-C(=O)-O-Ph 34.0

CH_3-C(=O)-NH-Ph 168.4

Ph-NHCH_2CH_2-C(=O)-O-Ph 39.3

CH_3-C(=O)-NHNH-C(=O)-Ph 168.6 165.6

Ph-C(=O)-NHNH-CH_2CH_2-C(=O)-O-Ph 167.6 47.4

Ph-S-C(=O)-C_6H_4-C(=O)-S-Ph 188.4

Ph-NH-C(=O)-C_6H_4-C(=O)-NH-Ph 165.2

Ph-C(=O)-NHNH-C(=O)-C_6H_4-C(=O)-NHNH-C(=O)-Ph 165.5 166.0

Polymer Synthesis.

Synthesis of Authentic Ordered Polyamide (10a). In order to clarify the structure of ordered polymer, N',N'-bis(4-aminobenzoyl)isophthalohydrazide (**8**) and di(4-nitrophenyl) 3,3'-(thiodi-p-phenylenethio)dipropanoate (**9**) were prepared by the condensation of isophthalic acid with 4-aminophenylhydrazide in the presence of **1**, and the addition reaction of 4-nitrophenyl acrylate with 4,4'-thiobisbenzenethiol, respectively (eq. 4, 5).

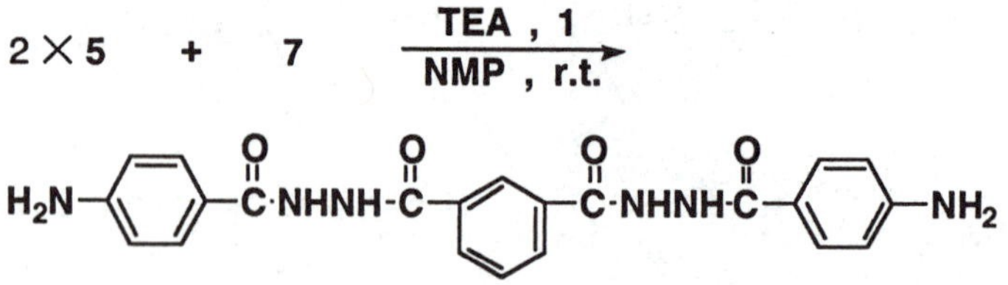

$$2\times 5 + 7 \xrightarrow[\text{NMP , r.t.}]{\text{TEA , 1}} \quad (4)$$

H_2N-C_6H_4-C(=O)-NHNH-C(=O)-C_6H_4-C(=O)-NHNH-C(=O)-C_6H_4-NH_2

8

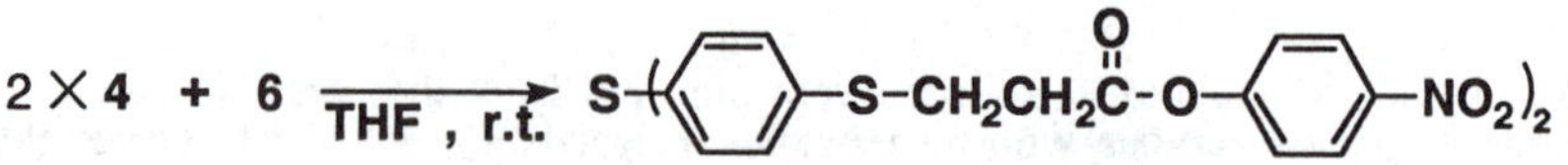

$$2\times 4 + 6 \xrightarrow[\text{THF , r.t.}]{} S\!\left(C_6H_4\text{-S-}CH_2CH_2\text{C(=O)-O-}C_6H_4\text{-}NO_2\right)_2 \quad (5)$$

9

The authentic ordered polymer was prepared by the solution polycondensation of **8** and **9** in DMF at 80 °C in the presence of HOBt. The polycondensation proceeded smoothly, giving polymer **10a** with inherent viscosities of 0.30 dL/g (eq. 6)

$$8 + 9 \xrightarrow[\text{DMF , 80°C}]{\text{HOBt}} \left(\overset{O}{\overset{\|}{C}}CH_2CH_2S\text{-}C_6H_4\text{-}S\text{-}CH_2CH_2\overset{O}{\overset{\|}{C}}\text{-}NH\text{-}C_6H_4\text{-}\overset{O}{\overset{\|}{C}}\text{-}NHNH\text{-}\overset{O}{\overset{\|}{C}}\text{-}C_6H_4\text{-}\overset{O}{\overset{\|}{C}}\text{-}NHNH\text{-}\overset{O}{\overset{\|}{C}}\text{-}C_6H_4\text{-}NH \right) \quad (6)$$

10a

Synthesis of Ordered Polyamide by Direct Polycondensation (10b). The synthesis of the ordered polymer was carried out by mixing four monomers all at once in the presence of **1** in DMF at room temperature and then warming to 80 °C. The polycondensation proceeded slowly and gave polymer **10b** with inherent viscosities of 0.21 dL/g (eq.7)

$$2 \times 4 + 2 \times 5 + 6 + 7 \xrightarrow[\text{DMF , r.t.~80°C}]{\text{TEA , 1 , HOBt}} \mathbf{10b} \quad (7)$$

Polymer Characterization. The IR spectra of the poly(acylhydrazide-amide-thioether)s (**10**) were consistent with those of model compounds and known analogues. Both polymers **10a** and **10b** prepared showed characteristic NH, amide I and amide II bands in the ranges 3220-3320, 1630-1640 and 1520-1540 cm^{-1}, respectively. Elemental analyses also supported the formation of the expected polymers.

The microstructure of polymers **10** was determined by means of ^{13}C-NMR spectroscopy. The ^{13}C-NMR spectra of authentic ordered polymer **10a** and polymer **10b** prepared by poly(addition-direct condensation) are presented in Figure 1 and 2. The signals of carbon nuclei in amide carbonyl groups for polymers **10a** and **10b** appeared at 165.1, 165.4 and 169.7 ppm. These peaks are assigned, as shown in the inset Figure 1, on the basis of chemical shifts for model compounds. On the other hand, the eight peaks of carbon nuclei in amide carbonyl groups for random polymer would be expected from its random structure. The signals of carbon nuclei in methylene groups for polymers **10a** and **10b** appeared at 27.7 and 36.1 ppm. A few small signals are observed at the spectra of both polymers between at 27.7 and 36.1 ppm. If these peaks are attributed to the random structure, extra peaks of amide carbonyl groups should be appeared at around 165-170 ppm. However, no extra peaks were observed in this region. So, these extra small peaks would be not derived from its random structure, but some impurities. Furthermore, the spectrum of polymer **10a** is identical to that of polymer **10b** prepared by poly(addition-direct condensation). These findings indicate that polymer **10b** prepared by poly(addition-direct condensation) was the desired ordered poly(acylhydrazide-amide-thioether).

Polymers **10** were light yellow solids, soluble in sulfuric acid and dipolar aprotic solvents, such as NMP, DMF and DMSO, and insoluble in other common organic solvents.

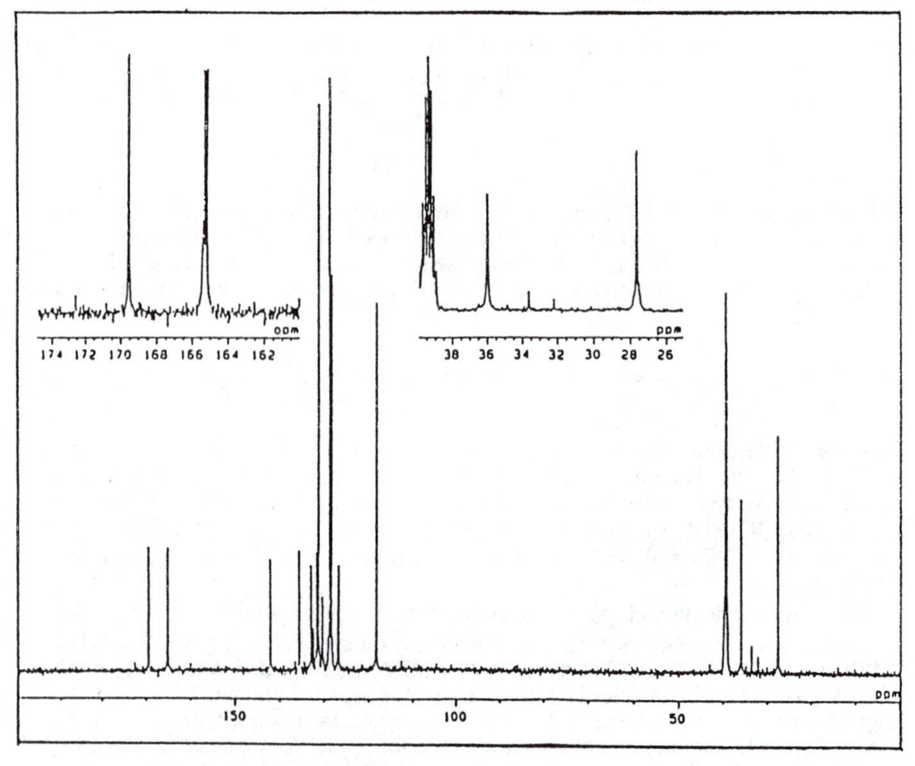

Figure 1. 125 MHz ^{13}C-NMR spectrum of polymer **10a** in [$(CD_3)_2SO$] at 25°C.

The thermal stability of polymer **10** was examined by thermogravimetry (TG) and differential thermal analysis (DTA). The rapid weight loss, observed at 190-270°C in the TG trace, amounted to 6.0% of the weight of polymer. This value is in good agreement with the value of weight loss (6.6%) calculated from the elimination of water due to the 1,3,4-oxadiazole ring formation.

Conclusions

We have demonstrated that the synthesis of ordered poly(acylhydrazide-amide-thioether) can be achieved by the combination of polyaddition and direct polycondensation from two nonsymmetric monomers, **4** ((XabX) and **5** (ZefZ) with a pair of symmetric monomers (YcdY), **6** (YccY) and **7** (YddY) using condensing agent **1**.

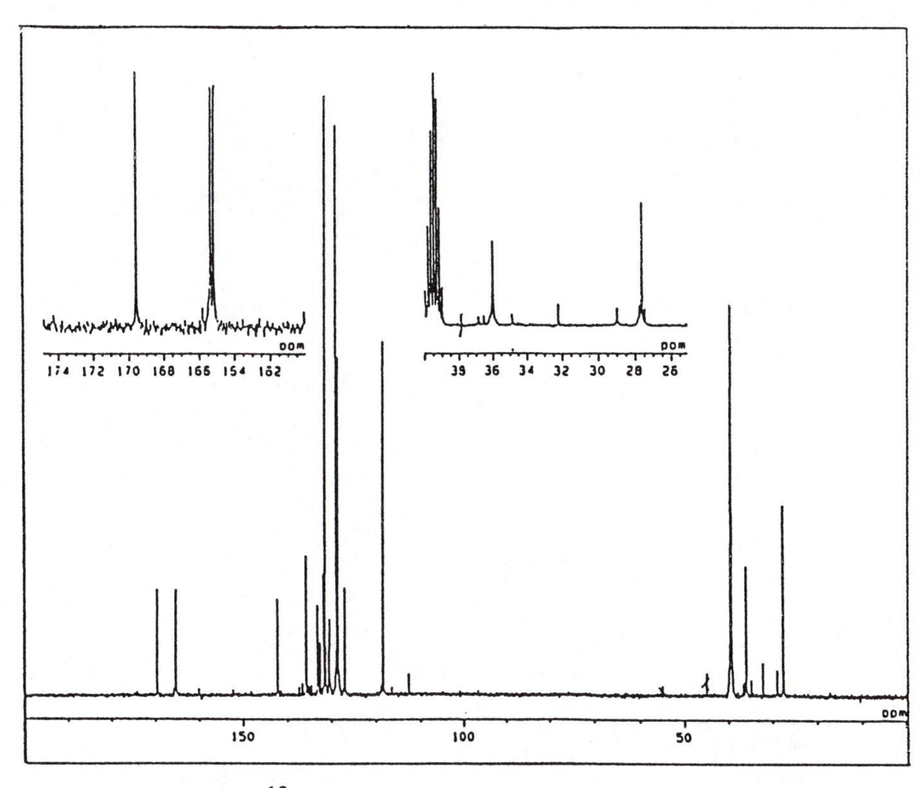

Figure 2. 125 MHz ^{13}C-NMR spectrum of polymer **10b** in [$(CD_3)_2SO$] at 25°C.

Acknowledgment.

We thank Hitoshi Nagasawa and Sadao Kato for their assistance and Takeyoshi Takahashi for performing the elemental analyses.

Literature Cited

1. Ueda, M.; Morosumi, T.; Sato, R. *Polym. J.* ,**1991**, 23, 167.
2. Ueda, M.; Morishima, M.; Kakuta, M. *Polym. J.* **1991**, 23, 1151.
3. Ueda, M.; Morishima, M.; Kakuta, M.; Sugiyama, J. *Macromolecules* , **1992**, 25, 6580.
4. Ueda, M.; Okada, T. *Macromolecules* , **1994**, 27, 3449.
5. Ueda, M.; Sugiyama, H. *Macromolecules* , **1994**, 27, 240.
6. Culbertson, B. M.; Dietz, S. *J. Polym. Sci., Polym. Lett. Ed.* **1968**, 5, 247.
7. Ueda, M.; Kameyama, M.; Hasimoto, K. *Macromolecules* , **1988**, 21, 19.
8. Kinsanov, A. V.; Egorova, N. L. *Zh. Obshch. Khim.* **1953**, 23, 1920.
9. John, C. S.; Cart, W. B. *J. Am. Chem. Soc.* **1955**, 77, 4875.
10. N. N. Lebedev; L. V. Anderianova. *J. Gen. Chem. USSR.* **1955**, 25, 195.

RECEIVED December 4, 1995

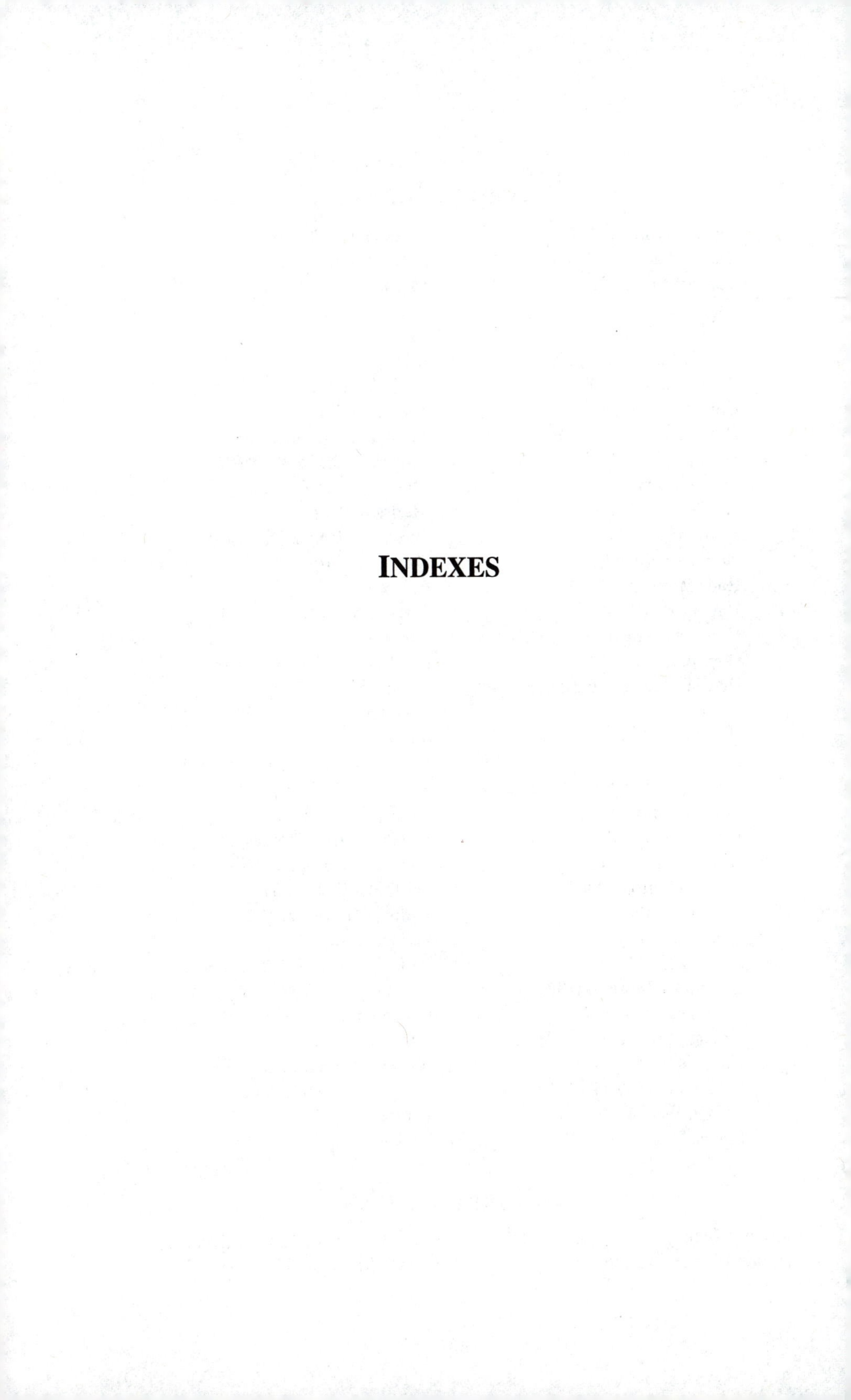

INDEXES

Author Index

Affiliation Index

Subject Index